Lecture Notes in Computer Science　　12375

More information about this series at http://www.springer.com/series/7412

Andrea Vedaldi · Horst Bischof ·
Thomas Brox · Jan-Michael Frahm (Eds.)

Computer Vision – ECCV 2020

16th European Conference
Glasgow, UK, August 23–28, 2020
Proceedings, Part XXX

Springer

Editors
Andrea Vedaldi [iD]
University of Oxford
Oxford, UK

Horst Bischof [iD]
Graz University of Technology
Graz, Austria

Thomas Brox [iD]
University of Freiburg
Freiburg im Breisgau, Germany

Jan-Michael Frahm
University of North Carolina at Chapel Hill
Chapel Hill, NC, USA

ISSN 0302-9743   ISSN 1611-3349   (electronic)
Lecture Notes in Computer Science
ISBN 978-3-030-58576-1   ISBN 978-3-030-58577-8   (eBook)
https://doi.org/10.1007/978-3-030-58577-8

LNCS Sublibrary: SL6 – Image Processing, Computer Vision, Pattern Recognition, and Graphics

This Springer imprint is published by the registered company Springer Nature Switzerland AG
The registered company address is: Gewerbestrasse 11, 6330 Cham, Switzerland

Foreword

Hosting the European Conference on Computer Vision (ECCV 2020) was certainly an exciting journey. From the 2016 plan to hold it at the Edinburgh International Conference Centre (hosting 1,800 delegates) to the 2018 plan to hold it at Glasgow's Scottish Exhibition Centre (up to 6,000 delegates), we finally ended with moving online because of the COVID-19 outbreak. While possibly having fewer delegates than expected because of the online format, ECCV 2020 still had over 3,100 registered participants.

Although online, the conference delivered most of the activities expected at a face-to-face conference: peer-reviewed papers, industrial exhibitors, demonstrations, and messaging between delegates. In addition to the main technical sessions, the conference included a strong program of satellite events with 16 tutorials and 44 workshops.

Furthermore, the online conference format enabled new conference features. Every paper had an associated teaser video and a longer full presentation video. Along with the papers and slides from the videos, all these materials were available the week before the conference. This allowed delegates to become familiar with the paper content and be ready for the live interaction with the authors during the conference week. The live event consisted of brief presentations by the oral and spotlight authors and industrial sponsors. Question and answer sessions for all papers were timed to occur twice so delegates from around the world had convenient access to the authors.

As with ECCV 2018, authors' draft versions of the papers appeared online with open access, now on both the Computer Vision Foundation (CVF) and the European Computer Vision Association (ECVA) websites. An archival publication arrangement was put in place with the cooperation of Springer. SpringerLink hosts the final version of the papers with further improvements, such as activating reference links and supplementary materials. These two approaches benefit all potential readers: a version available freely for all researchers, and an authoritative and citable version with additional benefits for SpringerLink subscribers. We thank Alfred Hofmann and Aliaksandr Birukou from Springer for helping to negotiate this agreement, which we expect will continue for future versions of ECCV.

August 2020

Vittorio Ferrari
Bob Fisher
Cordelia Schmid
Emanuele Trucco

Preface

Welcome to the proceedings of the European Conference on Computer Vision (ECCV 2020). This is a unique edition of ECCV in many ways. Due to the COVID-19 pandemic, this is the first time the conference was held online, in a virtual format. This was also the first time the conference relied exclusively on the Open Review platform to manage the review process. Despite these challenges ECCV is thriving. The conference received 5,150 valid paper submissions, of which 1,360 were accepted for publication (27%) and, of those, 160 were presented as spotlights (3%) and 104 as orals (2%). This amounts to more than twice the number of submissions to ECCV 2018 (2,439). Furthermore, CVPR, the largest conference on computer vision, received 5,850 submissions this year, meaning that ECCV is now 87% the size of CVPR in terms of submissions. By comparison, in 2018 the size of ECCV was only 73% of CVPR.

The review model was similar to previous editions of ECCV; in particular, it was double blind in the sense that the authors did not know the name of the reviewers and vice versa. Furthermore, each conference submission was held confidentially, and was only publicly revealed if and once accepted for publication. Each paper received at least three reviews, totalling more than 15,000 reviews. Handling the review process at this scale was a significant challenge. In order to ensure that each submission received as fair and high-quality reviews as possible, we recruited 2,830 reviewers (a 130% increase with reference to 2018) and 207 area chairs (a 60% increase). The area chairs were selected based on their technical expertise and reputation, largely among people that served as area chair in previous top computer vision and machine learning conferences (ECCV, ICCV, CVPR, NeurIPS, etc.). Reviewers were similarly invited from previous conferences. We also encouraged experienced area chairs to suggest additional chairs and reviewers in the initial phase of recruiting.

Despite doubling the number of submissions, the reviewer load was slightly reduced from 2018, from a maximum of 8 papers down to 7 (with some reviewers offering to handle 6 papers plus an emergency review). The area chair load increased slightly, from 18 papers on average to 22 papers on average.

Conflicts of interest between authors, area chairs, and reviewers were handled largely automatically by the Open Review platform via their curated list of user profiles. Many authors submitting to ECCV already had a profile in Open Review. We set a paper registration deadline one week before the paper submission deadline in order to encourage all missing authors to register and create their Open Review profiles well on time (in practice, we allowed authors to create/change papers arbitrarily until the submission deadline). Except for minor issues with users creating duplicate profiles, this allowed us to easily and quickly identify institutional conflicts, and avoid them, while matching papers to area chairs and reviewers.

Papers were matched to area chairs based on: an affinity score computed by the Open Review platform, which is based on paper titles and abstracts, and an affinity score computed by the Toronto Paper Matching System (TPMS), which is based on the paper's full text, the area chair bids for individual papers, load balancing, and conflict avoidance. Open Review provides the program chairs a convenient web interface to experiment with different configurations of the matching algorithm. The chosen configuration resulted in about 50% of the assigned papers to be highly ranked by the area chair bids, and 50% to be ranked in the middle, with very few low bids assigned.

Assignments to reviewers were similar, with two differences. First, there was a maximum of 7 papers assigned to each reviewer. Second, area chairs recommended up to seven reviewers per paper, providing another highly-weighed term to the affinity scores used for matching.

The assignment of papers to area chairs was smooth. However, it was more difficult to find suitable reviewers for all papers. Having a ratio of 5.6 papers per reviewer with a maximum load of 7 (due to emergency reviewer commitment), which did not allow for much wiggle room in order to also satisfy conflict and expertise constraints. We received some complaints from reviewers who did not feel qualified to review specific papers and we reassigned them wherever possible. However, the large scale of the conference, the many constraints, and the fact that a large fraction of such complaints arrived very late in the review process made this process very difficult and not all complaints could be addressed.

Reviewers had six weeks to complete their assignments. Possibly due to COVID-19 or the fact that the NeurIPS deadline was moved closer to the review deadline, a record 30% of the reviews were still missing after the deadline. By comparison, ECCV 2018 experienced only 10% missing reviews at this stage of the process. In the subsequent week, area chairs chased the missing reviews intensely, found replacement reviewers in their own team, and managed to reach 10% missing reviews. Eventually, we could provide almost all reviews (more than 99.9%) with a delay of only a couple of days on the initial schedule by a significant use of emergency reviews. If this trend is confirmed, it might be a major challenge to run a smooth review process in future editions of ECCV. The community must reconsider prioritization of the time spent on paper writing (the number of submissions increased a lot despite COVID-19) and time spent on paper reviewing (the number of reviews delivered in time decreased a lot presumably due to COVID-19 or NeurIPS deadline). With this imbalance the peer-review system that ensures the quality of our top conferences may break soon.

Reviewers submitted their reviews independently. In the reviews, they had the opportunity to ask questions to the authors to be addressed in the rebuttal. However, reviewers were told not to request any significant new experiment. Using the Open Review interface, authors could provide an answer to each individual review, but were also allowed to cross-reference reviews and responses in their answers. Rather than PDF files, we allowed the use of formatted text for the rebuttal. The rebuttal and initial reviews were then made visible to all reviewers and the primary area chair for a given paper. The area chair encouraged and moderated the reviewer discussion. During the discussions, reviewers were invited to reach a consensus and possibly adjust their ratings as a result of the discussion and of the evidence in the rebuttal.

After the discussion period ended, most reviewers entered a final rating and recommendation, although in many cases this did not differ from their initial recommendation. Based on the updated reviews and discussion, the primary area chair then made a preliminary decision to accept or reject the paper and wrote a justification for it (meta-review). Except for cases where the outcome of this process was absolutely clear (as indicated by the three reviewers and primary area chairs all recommending clear rejection), the decision was then examined and potentially challenged by a secondary area chair. This led to further discussion and overturning a small number of preliminary decisions. Needless to say, there was no in-person area chair meeting, which would have been impossible due to COVID-19.

Area chairs were invited to observe the consensus of the reviewers whenever possible and use extreme caution in overturning a clear consensus to accept or reject a paper. If an area chair still decided to do so, she/he was asked to clearly justify it in the meta-review and to explicitly obtain the agreement of the secondary area chair. In practice, very few papers were rejected after being confidently accepted by the reviewers.

This was the first time Open Review was used as the main platform to run ECCV. In 2018, the program chairs used CMT3 for the user-facing interface and Open Review internally, for matching and conflict resolution. Since it is clearly preferable to only use a single platform, this year we switched to using Open Review in full. The experience was largely positive. The platform is highly-configurable, scalable, and open source. Being written in Python, it is easy to write scripts to extract data programmatically. The paper matching and conflict resolution algorithms and interfaces are top-notch, also due to the excellent author profiles in the platform. Naturally, there were a few kinks along the way due to the fact that the ECCV Open Review configuration was created from scratch for this event and it differs in substantial ways from many other Open Review conferences. However, the Open Review development and support team did a fantastic job in helping us to get the configuration right and to address issues in a timely manner as they unavoidably occurred. We cannot thank them enough for the tremendous effort they put into this project.

Finally, we would like to thank everyone involved in making ECCV 2020 possible in these very strange and difficult times. This starts with our authors, followed by the area chairs and reviewers, who ran the review process at an unprecedented scale. The whole Open Review team (and in particular Melisa Bok, Mohit Unyal, Carlos Mondragon Chapa, and Celeste Martinez Gomez) worked incredibly hard for the entire duration of the process. We would also like to thank René Vidal for contributing to the adoption of Open Review. Our thanks also go to Laurent Charling for TPMS and to the program chairs of ICML, ICLR, and NeurIPS for cross checking double submissions. We thank the website chair, Giovanni Farinella, and the CPI team (in particular Ashley Cook, Miriam Verdon, Nicola McGrane, and Sharon Kerr) for promptly adding material to the website as needed in the various phases of the process. Finally, we thank the publication chairs, Albert Ali Salah, Hamdi Dibeklioglu, Metehan Doyran, Henry Howard-Jenkins, Victor Prisacariu, Siyu Tang, and Gul Varol, who managed to compile these substantial proceedings in an exceedingly compressed schedule. We express our thanks to the ECVA team, in particular Kristina Scherbaum for allowing open access of the proceedings. We thank Alfred Hofmann from Springer who again serve as the publisher. Finally, we thank the other chairs of ECCV 2020, including in particular the general chairs for very useful feedback with the handling of the program.

August 2020

Andrea Vedaldi
Horst Bischof
Thomas Brox
Jan-Michael Frahm

Organization

General Chairs

Vittorio Ferrari   Google Research, Switzerland
Bob Fisher   University of Edinburgh, UK
Cordelia Schmid   Google and Inria, France
Emanuele Trucco   University of Dundee, UK

Program Chairs

Andrea Vedaldi   University of Oxford, UK
Horst Bischof   Graz University of Technology, Austria
Thomas Brox   University of Freiburg, Germany
Jan-Michael Frahm   University of North Carolina, USA

Industrial Liaison Chairs

Jim Ashe   University of Edinburgh, UK
Helmut Grabner   Zurich University of Applied Sciences, Switzerland
Diane Larlus   NAVER LABS Europe, France
Cristian Novotny   University of Edinburgh, UK

Local Arrangement Chairs

Yvan Petillot   Heriot-Watt University, UK
Paul Siebert   University of Glasgow, UK

Academic Demonstration Chair

Thomas Mensink   Google Research and University of Amsterdam,
                                 The Netherlands

Poster Chair

Stephen Mckenna   University of Dundee, UK

Technology Chair

Gerardo Aragon Camarasa   University of Glasgow, UK

Tutorial Chairs

Carlo Colombo   University of Florence, Italy
Sotirios Tsaftaris   University of Edinburgh, UK

Publication Chairs

Albert Ali Salah   Utrecht University, The Netherlands
Hamdi Dibeklioglu   Bilkent University, Turkey
Metehan Doyran   Utrecht University, The Netherlands
Henry Howard-Jenkins   University of Oxford, UK
Victor Adrian Prisacariu University of Oxford, UK
Siyu Tang   ETH Zurich, Switzerland
Gul Varol   University of Oxford, UK

Website Chair

Giovanni Maria Farinella University of Catania, Italy

Workshops Chairs

Adrien Bartoli   University of Clermont Auvergne, France
Andrea Fusiello   University of Udine, Italy

Area Chairs

Lourdes Agapito   University College London, UK
Zeynep Akata   University of Tübingen, Germany
Karteek Alahari   Inria, France
Antonis Argyros   University of Crete, Greece
Hossein Azizpour   KTH Royal Institute of Technology, Sweden
Joao P. Barreto   Universidade de Coimbra, Portugal
Alexander C. Berg   University of North Carolina at Chapel Hill, USA
Matthew B. Blaschko   KU Leuven, Belgium
Lubomir D. Bourdev   WaveOne, Inc., USA
Edmond Boyer   Inria, France
Yuri Boykov   University of Waterloo, Canada
Gabriel Brostow   University College London, UK
Michael S. Brown   National University of Singapore, Singapore
Jianfei Cai   Monash University, Australia
Barbara Caputo   Politecnico di Torino, Italy
Ayan Chakrabarti   Washington University, St. Louis, USA
Tat-Jen Cham   Nanyang Technological University, Singapore
Manmohan Chandraker   University of California, San Diego, USA
Rama Chellappa   Johns Hopkins University, USA
Liang-Chieh Chen   Google, USA

Yung-Yu Chuang   National Taiwan University, Taiwan
Ondrej Chum   Czech Technical University in Prague, Czech Republic
Brian Clipp   Kitware, USA
John Collomosse   University of Surrey and Adobe Research, UK
Jason J. Corso   University of Michigan, USA
David J. Crandall   Indiana University, USA
Daniel Cremers   University of California, Los Angeles, USA
Fabio Cuzzolin   Oxford Brookes University, UK
Jifeng Dai   SenseTime, SAR China
Kostas Daniilidis   University of Pennsylvania, USA
Andrew Davison   Imperial College London, UK
Alessio Del Bue   Fondazione Istituto Italiano di Tecnologia, Italy
Jia Deng   Princeton University, USA
Alexey Dosovitskiy   Google, Germany
Matthijs Douze   Facebook, France
Enrique Dunn   Stevens Institute of Technology, USA
Irfan Essa   Georgia Institute of Technology and Google, USA
Giovanni Maria Farinella University of Catania, Italy
Ryan Farrell   Brigham Young University, USA
Paolo Favaro   University of Bern, Switzerland
Rogerio Feris   International Business Machines, USA
Cornelia Fermuller   University of Maryland, College Park, USA
David J. Fleet   Vector Institute, Canada
Friedrich Fraundorfer   DLR, Austria
Mario Fritz   CISPA Helmholtz Center for Information Security,
                          Germany
Pascal Fua   EPFL (Swiss Federal Institute of Technology
                          Lausanne), Switzerland
Yasutaka Furukawa   Simon Fraser University, Canada
Li Fuxin   Oregon State University, USA
Efstratios Gavves   University of Amsterdam, The Netherlands
Peter Vincent Gehler   Amazon, USA
Theo Gevers   University of Amsterdam, The Netherlands
Ross Girshick   Facebook AI Research, USA
Boqing Gong   Google, USA
Stephen Gould   Australian National University, Australia
Jinwei Gu   SenseTime Research, USA
Abhinav Gupta   Facebook, USA
Bohyung Han   Seoul National University, South Korea
Bharath Hariharan   Cornell University, USA
Tal Hassner   Facebook AI Research, USA
Xuming He   Australian National University, Australia
Joao F. Henriques   University of Oxford, UK
Adrian Hilton   University of Surrey, UK
Minh Hoai   Stony Brooks, State University of New York, USA
Derek Hoiem   University of Illinois Urbana-Champaign, USA

Timothy Hospedales	University of Edinburgh and Samsung, UK
Gang Hua	Wormpex AI Research, USA
Slobodan Ilic	Siemens AG, Germany
Hiroshi Ishikawa	Waseda University, Japan
Jiaya Jia	The Chinese University of Hong Kong, SAR China
Hailin Jin	Adobe Research, USA
Justin Johnson	University of Michigan, USA
Frederic Jurie	University of Caen Normandie, France
Fredrik Kahl	Chalmers University, Sweden
Sing Bing Kang	Zillow, USA
Gunhee Kim	Seoul National University, South Korea
Junmo Kim	Korea Advanced Institute of Science and Technology, South Korea
Tae-Kyun Kim	Imperial College London, UK
Ron Kimmel	Technion-Israel Institute of Technology, Israel
Alexander Kirillov	Facebook AI Research, USA
Kris Kitani	Carnegie Mellon University, USA
Iasonas Kokkinos	Ariel AI, UK
Vladlen Koltun	Intel Labs, USA
Nikos Komodakis	Ecole des Ponts ParisTech, France
Piotr Koniusz	Australian National University, Australia
M. Pawan Kumar	University of Oxford, UK
Kyros Kutulakos	University of Toronto, Canada
Christoph Lampert	IST Austria, Austria
Ivan Laptev	Inria, France
Diane Larlus	NAVER LABS Europe, France
Laura Leal-Taixe	Technical University Munich, Germany
Honglak Lee	Google and University of Michigan, USA
Joon-Young Lee	Adobe Research, USA
Kyoung Mu Lee	Seoul National University, South Korea
Seungyong Lee	POSTECH, South Korea
Yong Jae Lee	University of California, Davis, USA
Bastian Leibe	RWTH Aachen University, Germany
Victor Lempitsky	Samsung, Russia
Ales Leonardis	University of Birmingham, UK
Marius Leordeanu	Institute of Mathematics of the Romanian Academy, Romania
Vincent Lepetit	ENPC ParisTech, France
Hongdong Li	The Australian National University, Australia
Xi Li	Zhejiang University, China
Yin Li	University of Wisconsin-Madison, USA
Zicheng Liao	Zhejiang University, China
Jongwoo Lim	Hanyang University, South Korea
Stephen Lin	Microsoft Research Asia, China
Yen-Yu Lin	National Chiao Tung University, Taiwan, China
Zhe Lin	Adobe Research, USA

Haibin Ling	Stony Brooks, State University of New York, USA
Jiaying Liu	Peking University, China
Ming-Yu Liu	NVIDIA, USA
Si Liu	Beihang University, China
Xiaoming Liu	Michigan State University, USA
Huchuan Lu	Dalian University of Technology, China
Simon Lucey	Carnegie Mellon University, USA
Jiebo Luo	University of Rochester, USA
Julien Mairal	Inria, France
Michael Maire	University of Chicago, USA
Subhransu Maji	University of Massachusetts, Amherst, USA
Yasushi Makihara	Osaka University, Japan
Jiri Matas	Czech Technical University in Prague, Czech Republic
Yasuyuki Matsushita	Osaka University, Japan
Philippos Mordohai	Stevens Institute of Technology, USA
Vittorio Murino	University of Verona, Italy
Naila Murray	NAVER LABS Europe, France
Hajime Nagahara	Osaka University, Japan
P. J. Narayanan	International Institute of Information Technology (IIIT), Hyderabad, India
Nassir Navab	Technical University of Munich, Germany
Natalia Neverova	Facebook AI Research, France
Matthias Niessner	Technical University of Munich, Germany
Jean-Marc Odobez	Idiap Research Institute and Swiss Federal Institute of Technology Lausanne, Switzerland
Francesca Odone	Universita di Genova, Italy
Takeshi Oishi	The University of Tokyo, Tokyo Institute of Technology, Japan
Vicente Ordonez	University of Virginia, USA
Manohar Paluri	Facebook AI Research, USA
Maja Pantic	Imperial College London, UK
In Kyu Park	Inha University, South Korea
Ioannis Patras	Queen Mary University of London, UK
Patrick Perez	Valeo, France
Bryan A. Plummer	Boston University, USA
Thomas Pock	Graz University of Technology, Austria
Marc Pollefeys	ETH Zurich and Microsoft MR & AI Zurich Lab, Switzerland
Jean Ponce	Inria, France
Gerard Pons-Moll	MPII, Saarland Informatics Campus, Germany
Jordi Pont-Tuset	Google, Switzerland
James Matthew Rehg	Georgia Institute of Technology, USA
Ian Reid	University of Adelaide, Australia
Olaf Ronneberger	DeepMind London, UK
Stefan Roth	TU Darmstadt, Germany
Bryan Russell	Adobe Research, USA

Mathieu Salzmann	EPFL, Switzerland
Dimitris Samaras	Stony Brook University, USA
Imari Sato	National Institute of Informatics (NII), Japan
Yoichi Sato	The University of Tokyo, Japan
Torsten Sattler	Czech Technical University in Prague, Czech Republic
Daniel Scharstein	Middlebury College, USA
Bernt Schiele	MPII, Saarland Informatics Campus, Germany
Julia A. Schnabel	King's College London, UK
Nicu Sebe	University of Trento, Italy
Greg Shakhnarovich	Toyota Technological Institute at Chicago, USA
Humphrey Shi	University of Oregon, USA
Jianbo Shi	University of Pennsylvania, USA
Jianping Shi	SenseTime, China
Leonid Sigal	University of British Columbia, Canada
Cees Snoek	University of Amsterdam, The Netherlands
Richard Souvenir	Temple University, USA
Hao Su	University of California, San Diego, USA
Akihiro Sugimoto	National Institute of Informatics (NII), Japan
Jian Sun	Megvii Technology, China
Jian Sun	Xi'an Jiaotong University, China
Chris Sweeney	Facebook Reality Labs, USA
Yu-wing Tai	Kuaishou Technology, China
Chi-Keung Tang	The Hong Kong University of Science and Technology, SAR China
Radu Timofte	ETH Zurich, Switzerland
Sinisa Todorovic	Oregon State University, USA
Giorgos Tolias	Czech Technical University in Prague, Czech Republic
Carlo Tomasi	Duke University, USA
Tatiana Tommasi	Politecnico di Torino, Italy
Lorenzo Torresani	Facebook AI Research and Dartmouth College, USA
Alexander Toshev	Google, USA
Zhuowen Tu	University of California, San Diego, USA
Tinne Tuytelaars	KU Leuven, Belgium
Jasper Uijlings	Google, Switzerland
Nuno Vasconcelos	University of California, San Diego, USA
Olga Veksler	University of Waterloo, Canada
Rene Vidal	Johns Hopkins University, USA
Gang Wang	Alibaba Group, China
Jingdong Wang	Microsoft Research Asia, China
Yizhou Wang	Peking University, China
Lior Wolf	Facebook AI Research and Tel Aviv University, Israel
Jianxin Wu	Nanjing University, China
Tao Xiang	University of Surrey, UK
Saining Xie	Facebook AI Research, USA
Ming-Hsuan Yang	University of California at Merced and Google, USA
Ruigang Yang	University of Kentucky, USA

Kwang Moo Yi	University of Victoria, Canada
Zhaozheng Yin	Stony Brook, State University of New York, USA
Chang D. Yoo	Korea Advanced Institute of Science and Technology, South Korea
Shaodi You	University of Amsterdam, The Netherlands
Jingyi Yu	ShanghaiTech University, China
Stella Yu	University of California, Berkeley, and ICSI, USA
Stefanos Zafeiriou	Imperial College London, UK
Hongbin Zha	Peking University, China
Tianzhu Zhang	University of Science and Technology of China, China
Liang Zheng	Australian National University, Australia
Todd E. Zickler	Harvard University, USA
Andrew Zisserman	University of Oxford, UK

Technical Program Committee

Sathyanarayanan N. Aakur	Samuel Albanie	Pablo Arbelaez
Wael Abd Almgaeed	Shadi Albarqouni	Shervin Ardeshir
Abdelrahman Abdelhamed	Cenek Albl	Sercan O. Arik
Abdullah Abuolaim	Hassan Abu Alhaija	Anil Armagan
Supreeth Achar	Daniel Aliaga	Anurag Arnab
Hanno Ackermann	Mohammad S. Aliakbarian	Chetan Arora
Ehsan Adeli	Rahaf Aljundi	Federica Arrigoni
Triantafyllos Afouras	Thiemo Alldieck	Mathieu Aubry
Sameer Agarwal ᛏ	Jon Almazan	Shai Avidan
Aishwarya Agrawal	Jose M. Alvarez	Angelica I. Aviles-Rivero
Harsh Agrawal	Senjian An	Yannis Avrithis
Pulkit Agrawal	Saket Anand	Ismail Ben Ayed
Antonio Agudo	Codruta Ancuti	Shekoofeh Azizi
Eirikur Agustsson	Cosmin Ancuti	Ioan Andrei Bârsan
Karim Ahmed	Peter Anderson	Artem Babenko
Byeongjoo Ahn	Juan Andrade-Cetto	Deepak Babu Sam
Unaiza Ahsan	Alexander Andreopoulos	Seung-Hwan Baek
Thalaiyasingam Ajanthan	Misha Andriluka	Seungryul Baek
Kenan E. Ak	Dragomir Anguelov	Andrew D. Bagdanov
Emre Akbas	Rushil Anirudh	Shai Bagon
Naveed Akhtar	Michel Antunes	Yuval Bahat
Derya Akkaynak	Oisin Mac Aodha	Junjie Bai
Yagiz Aksoy	Srikar Appalaraju	Song Bai
Ziad Al-Halah	Relja Arandjelovic	Xiang Bai
Xavier Alameda-Pineda	Nikita Araslanov	Yalong Bai
Jean-Baptiste Alayrac	Andre Araujo	Yancheng Bai
	Helder Araujo	Peter Bajcsy
		Slawomir Bak

Mahsa Baktashmotlagh
Kavita Bala
Yogesh Balaji
Guha Balakrishnan
V. N. Balasubramanian
Federico Baldassarre
Vassileios Balntas
Shurjo Banerjee
Aayush Bansal
Ankan Bansal
Jianmin Bao
Linchao Bao
Wenbo Bao
Yingze Bao
Akash Bapat
Md Jawadul Hasan Bappy
Fabien Baradel
Lorenzo Baraldi
Daniel Barath
Adrian Barbu
Kobus Barnard
Nick Barnes
Francisco Barranco
Jonathan T. Barron
Arslan Basharat
Chaim Baskin
Anil S. Baslamisli
Jorge Batista
Kayhan Batmanghelich
Konstantinos Batsos
David Bau
Luis Baumela
Christoph Baur
Eduardo
    Bayro-Corrochano
Paul Beardsley
Jan Bednavr'ik
Oscar Beijbom
Philippe Bekaert
Esube Bekele
Vasileios Belagiannis
Ohad Ben-Shahar
Abhijit Bendale
Róger Bermúdez-Chacón
Maxim Berman
Jesus Bermudez-cameo

Florian Bernard
Stefano Berretti
Marcelo Bertalmio
Gedas Bertasius
Cigdem Beyan
Lucas Beyer
Vijayakumar Bhagavatula
Arjun Nitin Bhagoji
Apratim Bhattacharyya
Binod Bhattarai
Sai Bi
Jia-Wang Bian
Simone Bianco
Adel Bibi
Tolga Birdal
Tom Bishop
Soma Biswas
Mårten Björkman
Volker Blanz
Vishnu Boddeti
Navaneeth Bodla
Simion-Vlad Bogolin
Xavier Boix
Piotr Bojanowski
Timo Bolkart
Guido Borghi
Larbi Boubchir
Guillaume Bourmaud
Adrien Bousseau
Thierry Bouwmans
Richard Bowden
Hakan Boyraz
Mathieu Brédif
Samarth Brahmbhatt
Steve Branson
Nikolas Brasch
Biagio Brattoli
Ernesto Brau
Toby P. Breckon
Francois Bremond
Jesus Briales
Sofia Broomé
Marcus A. Brubaker
Luc Brun
Silvia Bucci
Shyamal Buch

Pradeep Buddharaju
Uta Buechler
Mai Bui
Tu Bui
Adrian Bulat
Giedrius T. Burachas
Elena Burceanu
Xavier P. Burgos-Artizzu
Kaylee Burns
Andrei Bursuc
Benjamin Busam
Wonmin Byeon
Zoya Bylinskii
Sergi Caelles
Jianrui Cai
Minjie Cai
Yujun Cai
Zhaowei Cai
Zhipeng Cai
Juan C. Caicedo
Simone Calderara
Necati Cihan Camgoz
Dylan Campbell
Octavia Camps
Jiale Cao
Kaidi Cao
Liangliang Cao
Xiangyong Cao
Xiaochun Cao
Yang Cao
Yu Cao
Yue Cao
Zhangjie Cao
Luca Carlone
Mathilde Caron
Dan Casas
Thomas J. Cashman
Umberto Castellani
Lluis Castrejon
Jacopo Cavazza
Fabio Cermelli
Hakan Cevikalp
Menglei Chai
Ishani Chakraborty
Rudrasis Chakraborty
Antoni B. Chan

Kwok-Ping Chan
Siddhartha Chandra
Sharat Chandran
Arjun Chandrasekaran
Angel X. Chang
Che-Han Chang
Hong Chang
Hyun Sung Chang
Hyung Jin Chang
Jianlong Chang
Ju Yong Chang
Ming-Ching Chang
Simyung Chang
Xiaojun Chang
Yu-Wei Chao
Devendra S. Chaplot
Arslan Chaudhry
Rizwan A. Chaudhry
Can Chen
Chang Chen
Chao Chen
Chen Chen
Chu-Song Chen
Dapeng Chen
Dong Chen
Dongdong Chen
Guanying Chen
Hongge Chen
Hsin-yi Chen
Huaijin Chen
Hwann-Tzong Chen
Jianbo Chen
Jianhui Chen
Jiansheng Chen
Jiaxin Chen
Jie Chen
Jun-Cheng Chen
Kan Chen
Kevin Chen
Lin Chen
Long Chen
Min-Hung Chen
Qifeng Chen
Shi Chen
Shixing Chen
Tianshui Chen

Weifeng Chen
Weikai Chen
Xi Chen
Xiaohan Chen
Xiaozhi Chen
Xilin Chen
Xingyu Chen
Xinlei Chen
Xinyun Chen
Yi-Ting Chen
Yilun Chen
Ying-Cong Chen
Yinpeng Chen
Yiran Chen
Yu Chen
Yu-Sheng Chen
Yuhua Chen
Yun-Chun Chen
Yunpeng Chen
Yuntao Chen
Zhuoyuan Chen
Zitian Chen
Anchieh Cheng
Bowen Cheng
Erkang Cheng
Gong Cheng
Guangliang Cheng
Jingchun Cheng
Jun Cheng
Li cheng
Ming-Ming Cheng
Yu Cheng
Ziang Cheng
Anoop Cherian
Dmitry Chetverikov
Ngai-man Cheung
William Cheung
Ajad Chhatkuli
Naoki Chiba
Benjamin Chidester
Han-pang Chiu
Mang Tik Chiu
Wei-Chen Chiu
Donghyeon Cho
Hojin Cho
Minsu Cho

Nam Ik Cho
Tim Cho
Tae Eun Choe
Chiho Choi
Edward Choi
Inchang Choi
Jinsoo Choi
Jonghyun Choi
Jongwon Choi
Yukyung Choi
Hisham Cholakkal
Eunji Chong
Jaegul Choo
Christopher Choy
Hang Chu
Peng Chu
Wen-Sheng Chu
Albert Chung
Joon Son Chung
Hai Ci
Safa Cicek
Ramazan G. Cinbis
Arridhana Ciptadi
Javier Civera
James J. Clark
Ronald Clark
Felipe Codevilla
Michael Cogswell
Andrea Cohen
Maxwell D. Collins
Carlo Colombo
Yang Cong
Adria R. Continente
Marcella Cornia
John Richard Corring
Darren Cosker
Dragos Costea
Garrison W. Cottrell
Florent Couzinie-Devy
Marco Cristani
Ioana Croitoru
James L. Crowley
Jiequan Cui
Zhaopeng Cui
Ross Cutler
Antonio D'Innocente

Rozenn Dahyot
Bo Dai
Dengxin Dai
Hang Dai
Longquan Dai
Shuyang Dai
Xiyang Dai
Yuchao Dai
Adrian V. Dalca
Dima Damen
Bharath B. Damodaran
Kristin Dana
Martin Danelljan
Zheng Dang
Zachary Alan Daniels
Donald G. Dansereau
Abhishek Das
Samyak Datta
Achal Dave
Titas De
Rodrigo de Bem
Teo de Campos
Raoul de Charette
Shalini De Mello
Joseph DeGol
Herve Delingette
Haowen Deng
Jiankang Deng
Weijian Deng
Zhiwei Deng
Joachim Denzler
Konstantinos G. Derpanis
Aditya Deshpande
Frederic Devernay
Somdip Dey
Arturo Deza
Abhinav Dhall
Helisa Dhamo
Vikas Dhiman
Fillipe Dias Moreira
  de Souza
Ali Diba
Ferran Diego
Guiguang Ding
Henghui Ding
Jian Ding

Mingyu Ding
Xinghao Ding
Zhengming Ding
Robert DiPietro
Cosimo Distante
Ajay Divakaran
Mandar Dixit
Abdelaziz Djelouah
Thanh-Toan Do
Jose Dolz
Bo Dong
Chao Dong
Jiangxin Dong
Weiming Dong
Weisheng Dong
Xingping Dong
Xuanyi Dong
Yinpeng Dong
Gianfranco Doretto
Hazel Doughty
Hassen Drira
Bertram Drost
Dawei Du
Ye Duan
Yueqi Duan
Abhimanyu Dubey
Anastasia Dubrovina
Stefan Duffner
Chi Nhan Duong
Thibaut Durand
Zoran Duric
Iulia Duta
Debidatta Dwibedi
Benjamin Eckart
Marc Eder
Marzieh Edraki
Alexei A. Efros
Kiana Ehsani
Hazm Kemal Ekenel
James H. Elder
Mohamed Elgharib
Shireen Elhabian
Ehsan Elhamifar
Mohamed Elhoseiny
Ian Endres
N. Benjamin Erichson

Jan Ernst
Sergio Escalera
Francisco Escolano
Victor Escorcia
Carlos Esteves
Francisco J. Estrada
Bin Fan
Chenyou Fan
Deng-Ping Fan
Haoqi Fan
Hehe Fan
Heng Fan
Kai Fan
Lijie Fan
Linxi Fan
Quanfu Fan
Shaojing Fan
Xiaochuan Fan
Xin Fan
Yuchen Fan
Sean Fanello
Hao-Shu Fang
Haoyang Fang
Kuan Fang
Yi Fang
Yuming Fang
Azade Farshad
Alireza Fathi
Raanan Fattal
Joao Fayad
Xiaohan Fei
Christoph Feichtenhofer
Michael Felsberg
Chen Feng
Jiashi Feng
Junyi Feng
Mengyang Feng
Qianli Feng
Zhenhua Feng
Michele Fenzi
Andras Ferencz
Martin Fergie
Basura Fernando
Ethan Fetaya
Michael Firman
John W. Fisher

Matthew Fisher
Boris Flach
Corneliu Florea
Wolfgang Foerstner
David Fofi
Gian Luca Foresti
Per-Erik Forssen
David Fouhey
Katerina Fragkiadaki
Victor Fragoso
Jean-Sébastien Franco
Ohad Fried
Iuri Frosio
Cheng-Yang Fu
Huazhu Fu
Jianlong Fu
Jingjing Fu
Xueyang Fu
Yanwei Fu
Ying Fu
Yun Fu
Olac Fuentes
Kent Fujiwara
Takuya Funatomi
Christopher Funk
Thomas Funkhouser
Antonino Furnari
Ryo Furukawa
Erik Gärtner
Raghudeep Gadde
Matheus Gadelha
Vandit Gajjar
Trevor Gale
Juergen Gall
Mathias Gallardo
Guillermo Gallego
Orazio Gallo
Chuang Gan
Zhe Gan
Madan Ravi Ganesh
Aditya Ganeshan
Siddha Ganju
Bin-Bin Gao
Changxin Gao
Feng Gao
Hongchang Gao

Jin Gao
Jiyang Gao
Junbin Gao
Katelyn Gao
Lin Gao
Mingfei Gao
Ruiqi Gao
Ruohan Gao
Shenghua Gao
Yuan Gao
Yue Gao
Noa Garcia
Alberto Garcia-Garcia
Guillermo
    Garcia-Hernando
Jacob R. Gardner
Animesh Garg
Kshitiz Garg
Rahul Garg
Ravi Garg
Philip N. Garner
Kirill Gavrilyuk
Paul Gay
Shiming Ge
Weifeng Ge
Baris Gecer
Xin Geng
Kyle Genova
Stamatios Georgoulis
Bernard Ghanem
Michael Gharbi
Kamran Ghasedi
Golnaz Ghiasi
Arnab Ghosh
Partha Ghosh
Silvio Giancola
Andrew Gilbert
Rohit Girdhar
Xavier Giro-i-Nieto
Thomas Gittings
Ioannis Gkioulekas
Clement Godard
Vaibhava Goel
Bastian Goldluecke
Lluis Gomez
Nuno Gonçalves

Dong Gong
Ke Gong
Mingming Gong
Abel Gonzalez-Garcia
Ariel Gordon
Daniel Gordon
Paulo Gotardo
Venu Madhav Govindu
Ankit Goyal
Priya Goyal
Raghav Goyal
Benjamin Graham
Douglas Gray
Brent A. Griffin
Etienne Grossmann
David Gu
Jiayuan Gu
Jiuxiang Gu
Lin Gu
Qiao Gu
Shuhang Gu
Jose J. Guerrero
Paul Guerrero
Jie Gui
Jean-Yves Guillemaut
Riza Alp Guler
Erhan Gundogdu
Fatma Guney
Guodong Guo
Kaiwen Guo
Qi Guo
Sheng Guo
Shi Guo
Tiantong Guo
Xiaojie Guo
Yijie Guo
Yiluan Guo
Yuanfang Guo
Yulan Guo
Agrim Gupta
Ankush Gupta
Mohit Gupta
Saurabh Gupta
Tanmay Gupta
Danna Gurari
Abner Guzman-Rivera

JunYoung Gwak
Michael Gygli
Jung-Woo Ha
Simon Hadfield
Isma Hadji
Bjoern Haefner
Taeyoung Hahn
Levente Hajder
Peter Hall
Emanuela Haller
Stefan Haller
Bumsub Ham
Abdullah Hamdi
Dongyoon Han
Hu Han
Jungong Han
Junwei Han
Kai Han
Tian Han
Xiaoguang Han
Xintong Han
Yahong Han
Ankur Handa
Zekun Hao
Albert Haque
Tatsuya Harada
Mehrtash Harandi
Adam W. Harley
Mahmudul Hasan
Atsushi Hashimoto
Ali Hatamizadeh
Munawar Hayat
Dongliang He
Jingrui He
Junfeng He
Kaiming He
Kun He
Lei He
Pan He
Ran He
Shengfeng He
Tong He
Weipeng He
Xuming He
Yang He
Yihui He

Zhihai He
Chinmay Hegde
Janne Heikkila
Mattias P. Heinrich
Stéphane Herbin
Alexander Hermans
Luis Herranz
John R. Hershey
Aaron Hertzmann
Roei Herzig
Anders Heyden
Steven Hickson
Otmar Hilliges
Tomas Hodan
Judy Hoffman
Michael Hofmann
Yannick Hold-Geoffroy
Namdar Homayounfar
Sina Honari
Richang Hong
Seunghoon Hong
Xiaopeng Hong
Yi Hong
Hidekata Hontani
Anthony Hoogs
Yedid Hoshen
Mir Rayat Imtiaz Hossain
Junhui Hou
Le Hou
Lu Hou
Tingbo Hou
Wei-Lin Hsiao
Cheng-Chun Hsu
Gee-Sern Jison Hsu
Kuang-jui Hsu
Changbo Hu
Di Hu
Guosheng Hu
Han Hu
Hao Hu
Hexiang Hu
Hou-Ning Hu
Jie Hu
Junlin Hu
Nan Hu
Ping Hu

Ronghang Hu
Xiaowei Hu
Yinlin Hu
Yuan-Ting Hu
Zhe Hu
Binh-Son Hua
Yang Hua
Bingyao Huang
Di Huang
Dong Huang
Fay Huang
Haibin Huang
Haozhi Huang
Heng Huang
Huaibo Huang
Jia-Bin Huang
Jing Huang
Jingwei Huang
Kaizhu Huang
Lei Huang
Qiangui Huang
Qiaoying Huang
Qingqiu Huang
Qixing Huang
Shaoli Huang
Sheng Huang
Siyuan Huang
Weilin Huang
Wenbing Huang
Xiangru Huang
Xun Huang
Yan Huang
Yifei Huang
Yue Huang
Zhiwu Huang
Zilong Huang
Minyoung Huh
Zhuo Hui
Matthias B. Hullin
Martin Humenberger
Wei-Chih Hung
Zhouyuan Huo
Junhwa Hur
Noureldien Hussein
Jyh-Jing Hwang
Seong Jae Hwang

Sung Ju Hwang
Ichiro Ide
Ivo Ihrke
Daiki Ikami
Satoshi Ikehata
Nazli Ikizler-Cinbis
Sunghoon Im
Yani Ioannou
Radu Tudor Ionescu
Umar Iqbal
Go Irie
Ahmet Iscen
Md Amirul Islam
Vamsi Ithapu
Nathan Jacobs
Arpit Jain
Himalaya Jain
Suyog Jain
Stuart James
Won-Dong Jang
Yunseok Jang
Ronnachai Jaroensri
Dinesh Jayaraman
Sadeep Jayasumana
Suren Jayasuriya
Herve Jegou
Simon Jenni
Hae-Gon Jeon
Yunho Jeon
Koteswar R. Jerripothula
Hueihan Jhuang
I-hong Jhuo
Dinghuang Ji
Hui Ji
Jingwei Ji
Pan Ji
Yanli Ji
Baoxiong Jia
Kui Jia
Xu Jia
Chiyu Max Jiang
Haiyong Jiang
Hao Jiang
Huaizu Jiang
Huajie Jiang
Ke Jiang

Lai Jiang
Li Jiang
Lu Jiang
Ming Jiang
Peng Jiang
Shuqiang Jiang
Wei Jiang
Xudong Jiang
Zhuolin Jiang
Jianbo Jiao
Zequn Jie
Dakai Jin
Kyong Hwan Jin
Lianwen Jin
SouYoung Jin
Xiaojie Jin
Xin Jin
Nebojsa Jojic
Alexis Joly
Michael Jeffrey Jones
Hanbyul Joo
Jungseock Joo
Kyungdon Joo
Ajjen Joshi
Shantanu H. Joshi
Da-Cheng Juan
Marco Körner
Kevin Köser
Asim Kadav
Christine Kaeser-Chen
Kushal Kafle
Dagmar Kainmueller
Ioannis A. Kakadiaris
Zdenek Kalal
Nima Kalantari
Yannis Kalantidis
Mahdi M. Kalayeh
Anmol Kalia
Sinan Kalkan
Vicky Kalogeiton
Ashwin Kalyan
Joni-kristian Kamarainen
Gerda Kamberova
Chandra Kambhamettu
Martin Kampel
Meina Kan

Christopher Kanan
Kenichi Kanatani
Angjoo Kanazawa
Atsushi Kanehira
Takuhiro Kaneko
Asako Kanezaki
Bingyi Kang
Di Kang
Sunghun Kang
Zhao Kang
Vadim Kantorov
Abhishek Kar
Amlan Kar
Theofanis Karaletsos
Leonid Karlinsky
Kevin Karsch
Angelos Katharopoulos
Isinsu Katircioglu
Hiroharu Kato
Zoltan Kato
Dotan Kaufman
Jan Kautz
Rei Kawakami
Qiuhong Ke
Wadim Kehl
Petr Kellnhofer
Aniruddha Kembhavi
Cem Keskin
Margret Keuper
Daniel Keysers
Ashkan Khakzar
Fahad Khan
Naeemullah Khan
Salman Khan
Siddhesh Khandelwal
Rawal Khirodkar
Anna Khoreva
Tejas Khot
Parmeshwar Khurd
Hadi Kiapour
Joe Kileel
Chanho Kim
Dahun Kim
Edward Kim
Eunwoo Kim
Han-ul Kim

Gil Levi
Evgeny Levinkov
Aviad Levis
Jose Lezama
Ang Li
Bin Li
Bing Li
Boyi Li
Changsheng Li
Chao Li
Chen Li
Cheng Li
Chenglong Li
Chi Li
Chun-Guang Li
Chun-Liang Li
Chunyuan Li
Dong Li
Guanbin Li
Hao Li
Haoxiang Li
Hongsheng Li
Hongyang Li
Houqiang Li
Huibin Li
Jia Li
Jianan Li
Jianguo Li
Junnan Li
Junxuan Li
Kai Li
Ke Li
Kejie Li
Kunpeng Li
Lerenhan Li
Li Erran Li
Mengtian Li
Mu Li
Peihua Li
Peiyi Li
Ping Li
Qi Li
Qing Li
Ruiyu Li
Ruoteng Li
Shaozi Li

Sheng Li
Shiwei Li
Shuang Li
Siyang Li
Stan Z. Li
Tianye Li
Wei Li
Weixin Li
Wen Li
Wenbo Li
Xiaomeng Li
Xin Li
Xiu Li
Xuelong Li
Xueting Li
Yan Li
Yandong Li
Yanghao Li
Yehao Li
Yi Li
Yijun Li
Yikang LI
Yining Li
Yongjie Li
Yu Li
Yu-Jhe Li
Yunpeng Li
Yunsheng Li
Yunzhu Li
Zhe Li
Zhen Li
Zhengqi Li
Zhenyang Li
Zhuwen Li
Dongze Lian
Xiaochen Lian
Zhouhui Lian
Chen Liang
Jie Liang
Ming Liang
Paul Pu Liang
Pengpeng Liang
Shu Liang
Wei Liang
Jing Liao
Minghui Liao

Renjie Liao
Shengcai Liao
Shuai Liao
Yiyi Liao
Ser-Nam Lim
Chen-Hsuan Lin
Chung-Ching Lin
Dahua Lin
Ji Lin
Kevin Lin
Tianwei Lin
Tsung-Yi Lin
Tsung-Yu Lin
Wei-An Lin
Weiyao Lin
Yen-Chen Lin
Yuewei Lin
David B. Lindell
Drew Linsley
Krzysztof Lis
Roee Litman
Jim Little
An-An Liu
Bo Liu
Buyu Liu
Chao Liu
Chen Liu
Cheng-lin Liu
Chenxi Liu
Dong Liu
Feng Liu
Guilin Liu
Haomiao Liu
Heshan Liu
Hong Liu
Ji Liu
Jingen Liu
Jun Liu
Lanlan Liu
Li Liu
Liu Liu
Mengyuan Liu
Miaomiao Liu
Nian Liu
Ping Liu
Risheng Liu

Sheng Liu
Shu Liu
Shuaicheng Liu
Sifei Liu
Siqi Liu
Siying Liu
Songtao Liu
Ting Liu
Tongliang Liu
Tyng-Luh Liu
Wanquan Liu
Wei Liu
Weiyang Liu
Weizhe Liu
Wenyu Liu
Wu Liu
Xialei Liu
Xianglong Liu
Xiaodong Liu
Xiaofeng Liu
Xihui Liu
Xingyu Liu
Xinwang Liu
Xuanqing Liu
Xuebo Liu
Yang Liu
Yaojie Liu
Yebin Liu
Yen-Cheng Liu
Yiming Liu
Yu Liu
Yu-Shen Liu
Yufan Liu
Yun Liu
Zheng Liu
Zhijian Liu
Zhuang Liu
Zichuan Liu
Ziwei Liu
Zongyi Liu
Stephan Liwicki
Liliana Lo Presti
Chengjiang Long
Fuchen Long
Mingsheng Long
Xiang Long

Yang Long
Charles T. Loop
Antonio Lopez
Roberto J. Lopez-Sastre
Javier Lorenzo-Navarro
Manolis Lourakis
Boyu Lu
Canyi Lu
Feng Lu
Guoyu Lu
Hongtao Lu
Jiajun Lu
Jiasen Lu
Jiwen Lu
Kaiyue Lu
Le Lu
Shao-Ping Lu
Shijian Lu
Xiankai Lu
Xin Lu
Yao Lu
Yiping Lu
Yongxi Lu
Yongyi Lu
Zhiwu Lu
Fujun Luan
Benjamin E. Lundell
Hao Luo
Jian-Hao Luo
Ruotian Luo
Weixin Luo
Wenhan Luo
Wenjie Luo
Yan Luo
Zelun Luo
Zixin Luo
Khoa Luu
Zhaoyang Lv
Pengyuan Lyu
Thomas Möllenhoff
Matthias Müller
Bingpeng Ma
Chih-Yao Ma
Chongyang Ma
Huimin Ma
Jiayi Ma

K. T. Ma
Ke Ma
Lin Ma
Liqian Ma
Shugao Ma
Wei-Chiu Ma
Xiaojian Ma
Xingjun Ma
Zhanyu Ma
Zheng Ma
Radek Jakob Mackowiak
Ludovic Magerand
Shweta Mahajan
Siddharth Mahendran
Long Mai
Ameesh Makadia
Oscar Mendez Maldonado
Mateusz Malinowski
Yury Malkov
Arun Mallya
Dipu Manandhar
Massimiliano Mancini
Fabian Manhardt
Kevis-kokitsi Maninis
Varun Manjunatha
Junhua Mao
Xudong Mao
Alina Marcu
Edgar Margffoy-Tuay
Dmitrii Marin
Manuel J. Marin-Jimenez
Kenneth Marino
Niki Martinel
Julieta Martinez
Jonathan Masci
Tomohiro Mashita
Iacopo Masi
David Masip
Daniela Massiceti
Stefan Mathe
Yusuke Matsui
Tetsu Matsukawa
Iain A. Matthews
Kevin James Matzen
Bruce Allen Maxwell
Stephen Maybank

Helmut Mayer
Amir Mazaheri
David McAllester
Steven McDonagh
Stephen J. Mckenna
Roey Mechrez
Prakhar Mehrotra
Christopher Mei
Xue Mei
Paulo R. S. Mendonca
Lili Meng
Zibo Meng
Thomas Mensink
Bjoern Menze
Michele Merler
Kourosh Meshgi
Pascal Mettes
Christopher Metzler
Liang Mi
Qiguang Miao
Xin Miao
Tomer Michaeli
Frank Michel
Antoine Miech
Krystian Mikolajczyk
Peyman Milanfar
Ben Mildenhall
Gregor Miller
Fausto Milletari
Dongbo Min
Kyle Min
Pedro Miraldo
Dmytro Mishkin
Anand Mishra
Ashish Mishra
Ishan Misra
Niluthpol C. Mithun
Kaushik Mitra
Niloy Mitra
Anton Mitrokhin
Ikuhisa Mitsugami
Anurag Mittal
Kaichun Mo
Zhipeng Mo
Davide Modolo
Michael Moeller

Pritish Mohapatra
Pavlo Molchanov
Davide Moltisanti
Pascal Monasse
Mathew Monfort
Aron Monszpart
Sean Moran
Vlad I. Morariu
Francesc Moreno-Noguer
Pietro Morerio
Stylianos Moschoglou
Yael Moses
Roozbeh Mottaghi
Pierre Moulon
Arsalan Mousavian
Yadong Mu
Yasuhiro Mukaigawa
Lopamudra Mukherjee
Yusuke Mukuta
Ravi Teja Mullapudi
Mario Enrique Munich
Zachary Murez
Ana C. Murillo
J. Krishna Murthy
Damien Muselet
Armin Mustafa
Siva Karthik Mustikovela
Carlo Dal Mutto
Moin Nabi
Varun K. Nagaraja
Tushar Nagarajan
Arsha Nagrani
Seungjun Nah
Nikhil Naik
Yoshikatsu Nakajima
Yuta Nakashima
Atsushi Nakazawa
Seonghyeon Nam
Vinay P. Namboodiri
Medhini Narasimhan
Srinivasa Narasimhan
Sanath Narayan
Erickson Rangel
  Nascimento
Jacinto Nascimento
Tayyab Naseer

Lakshmanan Nataraj
Neda Nategh
Nelson Isao Nauata
Fernando Navarro
Shah Nawaz
Lukas Neumann
Ram Nevatia
Alejandro Newell
Shawn Newsam
Joe Yue-Hei Ng
Trung Thanh Ngo
Duc Thanh Nguyen
Lam M. Nguyen
Phuc Xuan Nguyen
Thuong Nguyen Canh
Mihalis Nicolaou
Andrei Liviu Nicolicioiu
Xuecheng Nie
Michael Niemeyer
Simon Niklaus
Christophoros Nikou
David Nilsson
Jifeng Ning
Yuval Nirkin
Li Niu
Yuzhen Niu
Zhenxing Niu
Shohei Nobuhara
Nicoletta Noceti
Hyeonwoo Noh
Junhyug Noh
Mehdi Noroozi
Sotiris Nousias
Valsamis Ntouskos
Matthew O'Toole
Peter Ochs
Ferda Ofli
Seong Joon Oh
Seoung Wug Oh
Iason Oikonomidis
Utkarsh Ojha
Takahiro Okabe
Takayuki Okatani
Fumio Okura
Aude Oliva
Kyle Olszewski

Björn Ommer
Mohamed Omran
Elisabeta Oneata
Michael Opitz
Jose Oramas
Tribhuvanesh Orekondy
Shaul Oron
Sergio Orts-Escolano
Ivan Oseledets
Aljosa Osep
Magnus Oskarsson
Anton Osokin
Martin R. Oswald
Wanli Ouyang
Andrew Owens
Mete Ozay
Mustafa Ozuysal
Eduardo Pérez-Pellitero
Gautam Pai
Dipan Kumar Pal
P. H. Pamplona Savarese
Jinshan Pan
Junting Pan
Xingang Pan
Yingwei Pan
Yannis Panagakis
Rameswar Panda
Guan Pang
Jiahao Pang
Jiangmiao Pang
Tianyu Pang
Sharath Pankanti
Nicolas Papadakis
Dim Papadopoulos
George Papandreou
Toufiq Parag
Shaifali Parashar
Sarah Parisot
Eunhyeok Park
Hyun Soo Park
Jaesik Park
Min-Gyu Park
Taesung Park
Alvaro Parra
C. Alejandro Parraga
Despoina Paschalidou

Nikolaos Passalis
Vishal Patel
Viorica Patraucean
Badri Narayana Patro
Danda Pani Paudel
Sujoy Paul
Georgios Pavlakos
Ioannis Pavlidis
Vladimir Pavlovic
Nick Pears
Kim Steenstrup Pedersen
Selen Pehlivan
Shmuel Peleg
Chao Peng
Houwen Peng
Wen-Hsiao Peng
Xi Peng
Xiaojiang Peng
Xingchao Peng
Yuxin Peng
Federico Perazzi
Juan Camilo Perez
Vishwanath Peri
Federico Pernici
Luca Del Pero
Florent Perronnin
Stavros Petridis
Henning Petzka
Patrick Peursum
Michael Pfeiffer
Hanspeter Pfister
Roman Pflugfelder
Minh Tri Pham
Yongri Piao
David Picard
Tomasz Pieciak
A. J. Piergiovanni
Andrea Pilzer
Pedro O. Pinheiro
Silvia Laura Pintea
Lerrel Pinto
Axel Pinz
Robinson Piramuthu
Fiora Pirri
Leonid Pishchulin
Francesco Pittaluga

Daniel Pizarro
Tobias Plötz
Mirco Planamente
Matteo Poggi
Moacir A. Ponti
Parita Pooj
Fatih Porikli
Horst Possegger
Omid Poursaeed
Ameya Prabhu
Viraj Uday Prabhu
Dilip Prasad
Brian L. Price
True Price
Maria Priisalu
Veronique Prinet
Victor Adrian Prisacariu
Jan Prokaj
Sergey Prokudin
Nicolas Pugeault
Xavier Puig
Albert Pumarola
Pulak Purkait
Senthil Purushwalkam
Charles R. Qi
Hang Qi
Haozhi Qi
Lu Qi
Mengshi Qi
Siyuan Qi
Xiaojuan Qi
Yuankai Qi
Shengju Qian
Xuelin Qian
Siyuan Qiao
Yu Qiao
Jie Qin
Qiang Qiu
Weichao Qiu
Zhaofan Qiu
Kha Gia Quach
Yuhui Quan
Yvain Queau
Julian Quiroga
Faisal Qureshi
Mahdi Rad

Filip Radenovic
Petia Radeva
Venkatesh
  B. Radhakrishnan
Ilija Radosavovic
Noha Radwan
Rahul Raguram
Tanzila Rahman
Amit Raj
Ajit Rajwade
Kandan Ramakrishnan
Santhosh
  K. Ramakrishnan
Srikumar Ramalingam
Ravi Ramamoorthi
Vasili Ramanishka
Ramprasaath R. Selvaraju
Francois Rameau
Visvanathan Ramesh
Santu Rana
Rene Ranftl
Anand Rangarajan
Anurag Ranjan
Viresh Ranjan
Yongming Rao
Carolina Raposo
Vivek Rathod
Sathya N. Ravi
Avinash Ravichandran
Tammy Riklin Raviv
Daniel Rebain
Sylvestre-Alvise Rebuffi
N. Dinesh Reddy
Timo Rehfeld
Paolo Remagnino
Konstantinos Rematas
Edoardo Remelli
Dongwei Ren
Haibing Ren
Jian Ren
Jimmy Ren
Mengye Ren
Weihong Ren
Wenqi Ren
Zhile Ren
Zhongzheng Ren

Zhou Ren
Vijay Rengarajan
Md A. Reza
Farzaneh Rezaeianaran
Hamed R. Tavakoli
Nicholas Rhinehart
Helge Rhodin
Elisa Ricci
Alexander Richard
Eitan Richardson
Elad Richardson
Christian Richardt
Stephan Richter
Gernot Riegler
Daniel Ritchie
Tobias Ritschel
Samuel Rivera
Yong Man Ro
Richard Roberts
Joseph Robinson
Ignacio Rocco
Mrigank Rochan
Emanuele Rodolà
Mikel D. Rodriguez
Giorgio Roffo
Grégory Rogez
Gemma Roig
Javier Romero
Xuejian Rong
Yu Rong
Amir Rosenfeld
Bodo Rosenhahn
Guy Rosman
Arun Ross
Paolo Rota
Peter M. Roth
Anastasios Roussos
Anirban Roy
Sebastien Roy
Aruni RoyChowdhury
Artem Rozantsev
Ognjen Rudovic
Daniel Rueckert
Adria Ruiz
Javier Ruiz-del-solar
Christian Rupprecht

Chris Russell
Dan Ruta
Jongbin Ryu
Ömer Sümer
Alexandre Sablayrolles
Faraz Saeedan
Ryusuke Sagawa
Christos Sagonas
Tonmoy Saikia
Hideo Saito
Kuniaki Saito
Shunsuke Saito
Shunta Saito
Ken Sakurada
Joaquin Salas
Fatemeh Sadat Saleh
Mahdi Saleh
Pouya Samangouei
Leo Sampaio
  Ferraz Ribeiro
Artsiom Olegovich
  Sanakoyeu
Enrique Sanchez
Patsorn Sangkloy
Anush Sankaran
Aswin Sankaranarayanan
Swami Sankaranarayanan
Rodrigo Santa Cruz
Amartya Sanyal
Archana Sapkota
Nikolaos Sarafianos
Jun Sato
Shin'ichi Satoh
Hosnieh Sattar
Arman Savran
Manolis Savva
Alexander Sax
Hanno Scharr
Simone Schaub-Meyer
Konrad Schindler
Dmitrij Schlesinger
Uwe Schmidt
Dirk Schnieders
Björn Schuller
Samuel Schulter
Idan Schwartz

William Robson Schwartz
Alex Schwing
Sinisa Segvic
Lorenzo Seidenari
Pradeep Sen
Ozan Sener
Soumyadip Sengupta
Arda Senocak
Mojtaba Seyedhosseini
Shishir Shah
Shital Shah
Sohil Atul Shah
Tamar Rott Shaham
Huasong Shan
Qi Shan
Shiguang Shan
Jing Shao
Roman Shapovalov
Gaurav Sharma
Vivek Sharma
Viktoriia Sharmanska
Dongyu She
Sumit Shekhar
Evan Shelhamer
Chengyao Shen
Chunhua Shen
Falong Shen
Jie Shen
Li Shen
Liyue Shen
Shuhan Shen
Tianwei Shen
Wei Shen
William B. Shen
Yantao Shen
Ying Shen
Yiru Shen
Yujun Shen
Yuming Shen
Zhiqiang Shen
Ziyi Shen
Lu Sheng
Yu Sheng
Rakshith Shetty
Baoguang Shi
Guangming Shi

Hailin Shi
Miaojing Shi
Yemin Shi
Zhenmei Shi
Zhiyuan Shi
Kevin Jonathan Shih
Shiliang Shiliang
Hyunjung Shim
Atsushi Shimada
Nobutaka Shimada
Daeyun Shin
Young Min Shin
Koichi Shinoda
Konstantin Shmelkov
Michael Zheng Shou
Abhinav Shrivastava
Tianmin Shu
Zhixin Shu
Hong-Han Shuai
Pushkar Shukla
Christian Siagian
Mennatullah M. Siam
Kaleem Siddiqi
Karan Sikka
Jae-Young Sim
Christian Simon
Martin Simonovsky
Dheeraj Singaraju
Bharat Singh
Gurkirt Singh
Krishna Kumar Singh
Maneesh Kumar Singh
Richa Singh
Saurabh Singh
Suriya Singh
Vikas Singh
Sudipta N. Sinha
Vincent Sitzmann
Josef Sivic
Gregory Slabaugh
Miroslava Slavcheva
Ron Slossberg
Brandon Smith
Kevin Smith
Vladimir Smutny
Noah Snavely

Roger
   D. Soberanis-Mukul
Kihyuk Sohn
Francesco Solera
Eric Sommerlade
Sanghyun Son
Byung Cheol Song
Chunfeng Song
Dongjin Song
Jiaming Song
Jie Song
Jifei Song
Jingkuan Song
Mingli Song
Shiyu Song
Shuran Song
Xiao Song
Yafei Song
Yale Song
Yang Song
Yi-Zhe Song
Yibing Song
Humberto Sossa
Cesar de Souza
Adrian Spurr
Srinath Sridhar
Suraj Srinivas
Pratul P. Srinivasan
Anuj Srivastava
Tania Stathaki
Christopher Stauffer
Simon Stent
Rainer Stiefelhagen
Pierre Stock
Julian Straub
Jonathan C. Stroud
Joerg Stueckler
Jan Stuehmer
David Stutz
Chi Su
Hang Su
Jong-Chyi Su
Shuochen Su
Yu-Chuan Su
Ramanathan Subramanian
Yusuke Sugano

Masanori Suganuma
Yumin Suh
Mohammed Suhail
Yao Sui
Heung-Il Suk
Josephine Sullivan
Baochen Sun
Chen Sun
Chong Sun
Deqing Sun
Jin Sun
Liang Sun
Lin Sun
Qianru Sun
Shao-Hua Sun
Shuyang Sun
Weiwei Sun
Wenxiu Sun
Xiaoshuai Sun
Xiaoxiao Sun
Xingyuan Sun
Yifan Sun
Zhun Sun
Sabine Susstrunk
David Suter
Supasorn Suwajanakorn
Tomas Svoboda
Eran Swears
Paul Swoboda
Attila Szabo
Richard Szeliski
Duy-Nguyen Ta
Andrea Tagliasacchi
Yuichi Taguchi
Ying Tai
Keita Takahashi
Kouske Takahashi
Jun Takamatsu
Hugues Talbot
Toru Tamaki
Chaowei Tan
Fuwen Tan
Mingkui Tan
Mingxing Tan
Qingyang Tan
Robby T. Tan

Xiaoyang Tan
Kenichiro Tanaka
Masayuki Tanaka
Chang Tang
Chengzhou Tang
Danhang Tang
Ming Tang
Peng Tang
Qingming Tang
Wei Tang
Xu Tang
Yansong Tang
Youbao Tang
Yuxing Tang
Zhiqiang Tang
Tatsunori Taniai
Junli Tao
Xin Tao
Makarand Tapaswi
Jean-Philippe Tarel
Lyne Tchapmi
Zachary Teed
Bugra Tekin
Damien Teney
Ayush Tewari
Christian Theobalt
Christopher Thomas
Diego Thomas
Jim Thomas
Rajat Mani Thomas
Xinmei Tian
Yapeng Tian
Yingli Tian
Yonglong Tian
Zhi Tian
Zhuotao Tian
Kinh Tieu
Joseph Tighe
Massimo Tistarelli
Matthew Toews
Carl Toft
Pavel Tokmakov
Federico Tombari
Chetan Tonde
Yan Tong
Alessio Tonioni

Andrea Torsello
Fabio Tosi
Du Tran
Luan Tran
Ngoc-Trung Tran
Quan Hung Tran
Truyen Tran
Rudolph Triebel
Martin Trimmel
Shashank Tripathi
Subarna Tripathi
Leonardo Trujillo
Eduard Trulls
Tomasz Trzcinski
Sam Tsai
Yi-Hsuan Tsai
Hung-Yu Tseng
Stavros Tsogkas
Aggeliki Tsoli
Devis Tuia
Shubham Tulsiani
Sergey Tulyakov
Frederick Tung
Tony Tung
Daniyar Turmukhambetov
Ambrish Tyagi
Radim Tylecek
Christos Tzelepis
Georgios Tzimiropoulos
Dimitrios Tzionas
Seiichi Uchida
Norimichi Ukita
Dmitry Ulyanov
Martin Urschler
Yoshitaka Ushiku
Ben Usman
Alexander Vakhitov
Julien P. C. Valentin
Jack Valmadre
Ernest Valveny
Joost van de Weijer
Jan van Gemert
Koen Van Leemput
Gul Varol
Sebastiano Vascon
M. Alex O. Vasilescu

Subeesh Vasu
Mayank Vatsa
David Vazquez
Javier Vazquez-Corral
Ashok Veeraraghavan
Erik Velasco-Salido
Raviteja Vemulapalli
Jonathan Ventura
Manisha Verma
Roberto Vezzani
Ruben Villegas
Minh Vo
MinhDuc Vo
Nam Vo
Michele Volpi
Riccardo Volpi
Carl Vondrick
Konstantinos Vougioukas
Tuan-Hung Vu
Sven Wachsmuth
Neal Wadhwa
Catherine Wah
Jacob C. Walker
Thomas S. A. Wallis
Chengde Wan
Jun Wan
Liang Wan
Renjie Wan
Baoyuan Wang
Boyu Wang
Cheng Wang
Chu Wang
Chuan Wang
Chunyu Wang
Dequan Wang
Di Wang
Dilin Wang
Dong Wang
Fang Wang
Guanzhi Wang
Guoyin Wang
Hanzi Wang
Hao Wang
He Wang
Heng Wang
Hongcheng Wang

Hongxing Wang
Hua Wang
Jian Wang
Jingbo Wang
Jinglu Wang
Jingya Wang
Jinjun Wang
Jinqiao Wang
Jue Wang
Ke Wang
Keze Wang
Le Wang
Lei Wang
Lezi Wang
Li Wang
Liang Wang
Lijun Wang
Limin Wang
Linwei Wang
Lizhi Wang
Mengjiao Wang
Mingzhe Wang
Minsi Wang
Naiyan Wang
Nannan Wang
Ning Wang
Oliver Wang
Pei Wang
Peng Wang
Pichao Wang
Qi Wang
Qian Wang
Qiaosong Wang
Qifei Wang
Qilong Wang
Qing Wang
Qingzhong Wang
Quan Wang
Rui Wang
Ruiping Wang
Ruixing Wang
Shangfei Wang
Shenlong Wang
Shiyao Wang
Shuhui Wang
Song Wang

Tao Wang
Tianlu Wang
Tiantian Wang
Ting-chun Wang
Tingwu Wang
Wei Wang
Weiyue Wang
Wenguan Wang
Wenlin Wang
Wenqi Wang
Xiang Wang
Xiaobo Wang
Xiaofang Wang
Xiaoling Wang
Xiaolong Wang
Xiaosong Wang
Xiaoyu Wang
Xin Eric Wang
Xinchao Wang
Xinggang Wang
Xintao Wang
Yali Wang
Yan Wang
Yang Wang
Yangang Wang
Yaxing Wang
Yi Wang
Yida Wang
Yilin Wang
Yiming Wang
Yisen Wang
Yongtao Wang
Yu-Xiong Wang
Yue Wang
Yujiang Wang
Yunbo Wang
Yunhe Wang
Zengmao Wang
Zhangyang Wang
Zhaowen Wang
Zhe Wang
Zhecan Wang
Zheng Wang
Zhixiang Wang
Zilei Wang
Jianqiao Wangni

Anne S. Wannenwetsch
Jan Dirk Wegner
Scott Wehrwein
Donglai Wei
Kaixuan Wei
Longhui Wei
Pengxu Wei
Ping Wei
Qi Wei
Shih-En Wei
Xing Wei
Yunchao Wei
Zijun Wei
Jerod Weinman
Michael Weinmann
Philippe Weinzaepfel
Yair Weiss
Bihan Wen
Longyin Wen
Wei Wen
Junwu Weng
Tsui-Wei Weng
Xinshuo Weng
Eric Wengrowski
Tomas Werner
Gordon Wetzstein
Tobias Weyand
Patrick Wieschollek
Maggie Wigness
Erik Wijmans
Richard Wildes
Olivia Wiles
Chris Williams
Williem Williem
Kyle Wilson
Calden Wloka
Nicolai Wojke
Christian Wolf
Yongkang Wong
Sanghyun Woo
Scott Workman
Baoyuan Wu
Bichen Wu
Chao-Yuan Wu
Huikai Wu
Jiajun Wu

Jialin Wu
Jiaxiang Wu
Jiqing Wu
Jonathan Wu
Lifang Wu
Qi Wu
Qiang Wu
Ruizheng Wu
Shangzhe Wu
Shun-Cheng Wu
Tianfu Wu
Wayne Wu
Wenxuan Wu
Xiao Wu
Xiaohe Wu
Xinxiao Wu
Yang Wu
Yi Wu
Yiming Wu
Ying Nian Wu
Yue Wu
Zheng Wu
Zhenyu Wu
Zhirong Wu
Zuxuan Wu
Stefanie Wuhrer
Jonas Wulff
Changqun Xia
Fangting Xia
Fei Xia
Gui-Song Xia
Lu Xia
Xide Xia
Yin Xia
Yingce Xia
Yongqin Xian
Lei Xiang
Shiming Xiang
Bin Xiao
Fanyi Xiao
Guobao Xiao
Huaxin Xiao
Taihong Xiao
Tete Xiao
Tong Xiao
Wang Xiao

Yang Xiao
Cihang Xie
Guosen Xie
Jianwen Xie
Lingxi Xie
Sirui Xie
Weidi Xie
Wenxuan Xie
Xiaohua Xie
Fuyong Xing
Jun Xing
Junliang Xing
Bo Xiong
Peixi Xiong
Yu Xiong
Yuanjun Xiong
Zhiwei Xiong
Chang Xu
Chenliang Xu
Dan Xu
Danfei Xu
Hang Xu
Hongteng Xu
Huijuan Xu
Jingwei Xu
Jun Xu
Kai Xu
Mengmeng Xu
Mingze Xu
Qianqian Xu
Ran Xu
Weijian Xu
Xiangyu Xu
Xiaogang Xu
Xing Xu
Xun Xu
Yanyu Xu
Yichao Xu
Yong Xu
Yongchao Xu
Yuanlu Xu
Zenglin Xu
Zheng Xu
Chuhui Xue
Jia Xue
Nan Xue

Tianfan Xue
Xiangyang Xue
Abhay Yadav
Yasushi Yagi
I. Zeki Yalniz
Kota Yamaguchi
Toshihiko Yamasaki
Takayoshi Yamashita
Junchi Yan
Ke Yan
Qingan Yan
Sijie Yan
Xinchen Yan
Yan Yan
Yichao Yan
Zhicheng Yan
Keiji Yanai
Bin Yang
Ceyuan Yang
Dawei Yang
Dong Yang
Fan Yang
Guandao Yang
Guorun Yang
Haichuan Yang
Hao Yang
Jianwei Yang
Jiaolong Yang
Jie Yang
Jing Yang
Kaiyu Yang
Linjie Yang
Meng Yang
Michael Ying Yang
Nan Yang
Shuai Yang
Shuo Yang
Tianyu Yang
Tien-Ju Yang
Tsun-Yi Yang
Wei Yang
Wenhan Yang
Xiao Yang
Xiaodong Yang
Xin Yang
Yan Yang

Yanchao Yang
Yee Hong Yang
Yezhou Yang
Zhenheng Yang
Anbang Yao
Angela Yao
Cong Yao
Jian Yao
Li Yao
Ting Yao
Yao Yao
Zhewei Yao
Chengxi Ye
Jianbo Ye
Keren Ye
Linwei Ye
Mang Ye
Mao Ye
Qi Ye
Qixiang Ye
Mei-Chen Yeh
Raymond Yeh
Yu-Ying Yeh
Sai-Kit Yeung
Serena Yeung
Kwang Moo Yi
Li Yi
Renjiao Yi
Alper Yilmaz
Junho Yim
Lijun Yin
Weidong Yin
Xi Yin
Zhichao Yin
Tatsuya Yokota
Ryo Yonetani
Donggeun Yoo
Jae Shin Yoon
Ju Hong Yoon
Sung-eui Yoon
Laurent Younes
Changqian Yu
Fisher Yu
Gang Yu
Jiahui Yu
Kaicheng Yu

Ke Yu
Lequan Yu
Ning Yu
Qian Yu
Ronald Yu
Ruichi Yu
Shoou-I Yu
Tao Yu
Tianshu Yu
Xiang Yu
Xin Yu
Xiyu Yu
Youngjae Yu
Yu Yu
Zhiding Yu
Chunfeng Yuan
Ganzhao Yuan
Jinwei Yuan
Lu Yuan
Quan Yuan
Shanxin Yuan
Tongtong Yuan
Wenjia Yuan
Ye Yuan
Yuan Yuan
Yuhui Yuan
Huanjing Yue
Xiangyu Yue
Ersin Yumer
Sergey Zagoruyko
Egor Zakharov
Amir Zamir
Andrei Zanfir
Mihai Zanfir
Pablo Zegers
Bernhard Zeisl
John S. Zelek
Niclas Zeller
Huayi Zeng
Jiabei Zeng
Wenjun Zeng
Yu Zeng
Xiaohua Zhai
Fangneng Zhan
Huangying Zhan
Kun Zhan

Xiaohang Zhan
Baochang Zhang
Bowen Zhang
Cecilia Zhang
Changqing Zhang
Chao Zhang
Chengquan Zhang
Chi Zhang
Chongyang Zhang
Dingwen Zhang
Dong Zhang
Feihu Zhang
Hang Zhang
Hanwang Zhang
Hao Zhang
He Zhang
Hongguang Zhang
Hua Zhang
Ji Zhang
Jianguo Zhang
Jianming Zhang
Jiawei Zhang
Jie Zhang
Jing Zhang
Juyong Zhang
Kai Zhang
Kaipeng Zhang
Ke Zhang
Le Zhang
Lei Zhang
Li Zhang
Lihe Zhang
Linguang Zhang
Lu Zhang
Mi Zhang
Mingda Zhang
Peng Zhang
Pingping Zhang
Qian Zhang
Qilin Zhang
Quanshi Zhang
Richard Zhang
Rui Zhang
Runze Zhang
Shengping Zhang
Shifeng Zhang

Shuai Zhang
Songyang Zhang
Tao Zhang
Ting Zhang
Tong Zhang
Wayne Zhang
Wei Zhang
Weizhong Zhang
Wenwei Zhang
Xiangyu Zhang
Xiaolin Zhang
Xiaopeng Zhang
Xiaoqin Zhang
Xiuming Zhang
Ya Zhang
Yang Zhang
Yimin Zhang
Yinda Zhang
Ying Zhang
Yongfei Zhang
Yu Zhang
Yulun Zhang
Yunhua Zhang
Yuting Zhang
Zhanpeng Zhang
Zhao Zhang
Zhaoxiang Zhang
Zhen Zhang
Zheng Zhang
Zhifei Zhang
Zhijin Zhang
Zhishuai Zhang
Ziming Zhang
Bo Zhao
Chen Zhao
Fang Zhao
Haiyu Zhao
Han Zhao
Hang Zhao
Hengshuang Zhao
Jian Zhao
Kai Zhao
Liang Zhao
Long Zhao
Qian Zhao
Qibin Zhao

Qijun Zhao
Rui Zhao
Shenglin Zhao
Sicheng Zhao
Tianyi Zhao
Wenda Zhao
Xiangyun Zhao
Xin Zhao
Yang Zhao
Yue Zhao
Zhichen Zhao
Zijing Zhao
Xiantong Zhen
Chuanxia Zheng
Feng Zheng
Haiyong Zheng
Jia Zheng
Kang Zheng
Shuai Kyle Zheng
Wei-Shi Zheng
Yinqiang Zheng
Zerong Zheng
Zhedong Zheng
Zilong Zheng
Bineng Zhong
Fangwei Zhong
Guangyu Zhong
Yiran Zhong
Yujie Zhong
Zhun Zhong
Chunluan Zhou
Huiyu Zhou
Jiahuan Zhou
Jun Zhou
Lei Zhou
Luowei Zhou
Luping Zhou
Mo Zhou
Ning Zhou
Pan Zhou
Peng Zhou
Qianyi Zhou
S. Kevin Zhou
Sanping Zhou
Wengang Zhou
Xingyi Zhou

Additional Reviewers

Hanxiao Liu
Hongyu Liu
Huidong Liu
Miao Liu
Xinxin Liu
Yongfei Liu
Yu-Lun Liu
Amir Livne
Tiange Luo
Wei Ma
Xiaoxuan Ma
Ioannis Marras
Georg Martius
Effrosyni Mavroudi
Tim Meinhardt
Givi Meishvili
Meng Meng
Zihang Meng
Zhongqi Miao
Gyeongsik Moon
Khoi Nguyen
Yung-Kyun Noh
Antonio Norelli
Jaeyoo Park
Alexander Pashevich
Mandela Patrick
Mary Phuong
Bingqiao Qian
Yu Qiao
Zhen Qiao
Sai Saketh Rambhatla
Aniket Roy
Amelie Royer
Parikshit Vishwas
   Sakurikar
Mark Sandler
Mert Bülent Sarıyıldız
Tanner Schmidt
Anshul B. Shah

Ketul Shah
Rajvi Shah
Hengcan Shi
Xiangxi Shi
Yujiao Shi
William A. P. Smith
Guoxian Song
Robin Strudel
Abby Stylianou
Xinwei Sun
Reuben Tan
Qingyi Tao
Kedar S. Tatwawadi
Anh Tuan Tran
Son Dinh Tran
Eleni Triantafillou
Aristeidis Tsitiridis
Md Zasim Uddin
Andrea Vedaldi
Evangelos Ververas
Vidit Vidit
Paul Voigtlaender
Bo Wan
Huanyu Wang
Huiyu Wang
Junqiu Wang
Pengxiao Wang
Tai Wang
Xinyao Wang
Tomoki Watanabe
Mark Weber
Xi Wei
Botong Wu
James Wu
Jiamin Wu
Rujie Wu
Yu Wu
Rongchang Xie
Wei Xiong

Yunyang Xiong
An Xu
Chi Xu
Yinghao Xu
Fei Xue
Tingyun Yan
Zike Yan
Chao Yang
Heran Yang
Ren Yang
Wenfei Yang
Xu Yang
Rajeev Yasarla
Shaokai Ye
Yufei Ye
Kun Yi
Haichao Yu
Hanchao Yu
Ruixuan Yu
Liangzhe Yuan
Chen-Lin Zhang
Fandong Zhang
Tianyi Zhang
Yang Zhang
Yiyi Zhang
Yongshun Zhang
Yu Zhang
Zhiwei Zhang
Jiaojiao Zhao
Yipu Zhao
Xingjian Zhen
Haizhong Zheng
Tiancheng Zhi
Chengju Zhou
Hao Zhou
Hao Zhu
Alexander Zimin

Contents – Part XXX

Representative-Discriminative Learning for Open-Set Land Cover Classification of Satellite Imagery

Razieh Kaviani Baghbaderani[1], Ying Qu[1(✉)], Hairong Qi[1], and Craig Stutts[2]

[1] The University of Tennessee, Knoxville, TN, USA
{rkavian1,yqu3,hqi}@utk.edu
[2] Applied Research Associates, Raleigh, NC, USA
cstutts@ara.com

Abstract. Land cover classification of satellite imagery is an important step toward analyzing the Earth's surface. Existing models assume a closed-set setting where both the training and testing classes belong to the same label set. However, due to the unique characteristics of satellite imagery with extremely vast area of versatile cover materials, the training data are bound to be non-representative. In this paper, we study the problem of open-set land cover classification that identifies the samples belonging to unknown classes during testing, while maintaining performance on known classes. Although inherently a classification problem, both representative and discriminative aspects of data need to be exploited in order to better distinguish unknown classes from known. We propose a representative-discriminative open-set recognition (RDOSR) framework, which 1) projects data from the raw image space to the embedding feature space that facilitates differentiating similar classes, and further 2) enhances both the representative and discriminative capacity through transformation to a so-called abundance space. Experiments on multiple satellite benchmarks demonstrate effectiveness of the proposed method. We also show the generality of the proposed approach by achieving promising results on open-set classification tasks using RGB images.

Keywords: Hyperspectral image classification · Open-set recognition

1 Introduction

Recent advancements in computer vision, especially the advent of Convolutional Neural Networks (CNN), have significantly improved the performance of image classification [1–3], detection [4,5], and segmentation [6,7] tasks, enabling their deployment in many different fields. One of such field of applications is satellite

Electronic supplementary material The online version of this chapter (https:// doi.org/10.1007/978-3-030-58577-8_1) contains supplementary material, which is available to authorized users.

© Springer Nature Switzerland AG 2020
A. Vedaldi et al. (Eds.): ECCV 2020, LNCS 12375, pp. 1–17, 2020.
https://doi.org/10.1007/978-3-030-58577-8_1

image analysis that includes resource management, urban development planning, and climate control. Land cover classification or material classification is one of the building blocks of satellite image analysis, providing essential inputs to a series of subsequent tasks including object segmentation, 3D reconstruction and modeling, as well as texture mapping. Supervised land cover classification involves categorization of multispectral or hyperspectral image pixels into predefined material classes, e.g., asphalt, tree, concrete, water, metal, soil, etc. Note that both multispectral and hyperspectral images (MSI and HSI) try to provide additional spectral information, beyond the visible spectra, to reveal extra details and compensate for the coarse spatial resolution of these images.

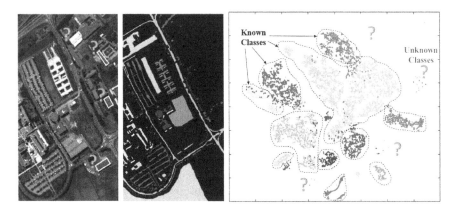

Fig. 1. Open-set land-cover classification: Data samples corresponding to ground truth categories are from the known class set (**K**). It is likely that some categories are not known during training and will be encountered at testing, *i.e.*, samples from unknown class set (**U**). The goal is to identify pixels coming from (**U**), while correctly classify any pixel belonging to (**K**). From left to right, a satellite image from the Pavia University dataset [8] showing unknown materials surfaces with yellow bounding boxes, the ground truth labels, and visualization of the feature space for both known and unknown classes using tSNE [9].

Although inherently a classification problem, material classification in satellite imagery faces a unique challenge: the vast area covered by the satellite imagery makes the task of generating representative training samples almost impossible, as there are a large variety of materials existed on the Earth's surface, especially those not well-exploited regions. Therefore, one of the most essential capabilities of land cover classification is to be able to automatically identify which test image and which area or pixel location of the image, has a higher probability of hosting new classes of materials. This would provide essential guideline to human operators in collecting training samples for the new classes.

The vast majority of existing works for land cover classification have been done under the "static closed world" assumption, meaning that the training and testing sets are drawn from the same label set. As a result, a system observing any unknown class is forced to misclassify it as one of the known classes, thus, weakening the recognition performance. A more realistic scenario is to work in a non-stationary and open environment that not all categories are known *a priori* and testing samples from unseen classes can emerge unexpectedly. Recognition of known and unknown pixels in a given image and correctly classifying known pixels is defined as "open-set land cover classification". Figure 1 explains this process using a real-world satellite image.

In this paper, we present a multi-tasking representative-discriminative open-set recognition (RDOSR) framework to address the challenging land cover classification problem, where both the representative and discriminative aspects of data are exploited in order to best characterize the differences between known and unknown classes. We propose the representative and discriminative learning among three spaces, as shown in Fig. 2, including 1) the transformation from the raw image space to the embedding feature space, and 2) the transformation from the embedding feature space to a so-called abundance space. See Supplement A for an illustration of the effect in different spaces.

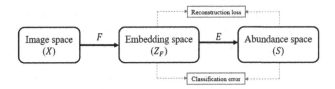

Fig. 2. Representative-discriminative learning through the transformation among 3 spaces: the raw image space, the embedding space, and the abundance space.

The contributions of this paper are thus summarized as follows: First, unlike other open-set recognition methods applied on the raw image space directly, we propose to first learn a classification network that would transform from the raw image space to an embedding feature space such that a more discernible input is fed into the subsequent open-set learning network. Second, we propose to use the so-called Dirichlet-net to transform data from the embedding feature space to the abundance space. Due to the resolution issue, each pixel in a satellite image covers a large area with more than one constituent material, resulting in "mixed pixel". The mixtures are generally assumed to be a linear combination of a few spectral bases, with the corresponding mixing coefficients (or abundances). This way, instead of looking at the mixed pixel, we study the mixing coefficients of each spectral basis in making up the mixture. Thus the abundance space provides a finer-scale representation. Third, to the best of our knowledge, this work is the first attempt to address the critical open-set land cover classification problem essential for analyzing the Earth's surface. Fourth, while the proposed method was motivated by satellite imagery analysis, it is generalizable to RGB images and achieves promising results.

2   Related Work

Conventional Land Cover Classification. These methods mainly employ a traditional classifier on spectral information, with its discriminative power further enhanced through feature engineering algorithms such as minimum noise fraction (MNF) [10], independent component analysis (ICA) [11], morphological profiles [12], and spectral unmixing [13–15]. The advent of deep learning has enabled the extraction of hierarchical features automatically and achieved unprecedented performance. [16,17] applied a 1D-CNN framework in the spectral domain to take into account the correlation between adjacent spectral bands. Several works use a patch surrounding the desired pixels by adopting a 2D-CNN structure [18,19] to incorporate the spatial correlation as well. More recently, integration of both spectral and spatial domains using 3D-CNN structures have been employed to further improve the classification accuracy [20,21].

Although each approach has its own merit, all the existing land cover classification approaches work under the closed-set assumption where the training and testing sets share the same label set.

Open-Set Recognition. Open-set recognition has gained considerable attention due to its handling of unknown class samples based on incomplete knowledge of the data during model training. Early studies are based on traditional classification models including Nearest Neighbor, Support Vector Machine (SVMs), Sparse Representation, etc. The open-set version of Nearest Neighbor was developed based on the distance of the testing samples to the known samples [22]. The SVM-based approaches employed different regularization terms or kernels to detect unknown samples [23,24]. In [25], the residuals from the Sparse Representation-based Classification (SRC) algorithm were used as the score for unknown class detection.

In the context of deep networks, [26] employed a statistical model to calibrate the SoftMax scores and produced an alternative layer, called OpenMax. [27] improved upon the OpenMax layer approach by maximizing the inter-class distance and minimizing the intraclass distance on the penultimate layer. The work of [28] proposed a k-sigmoid activation-based loss function for training a neural network to be able to find an operating threshold on the final activation layer. [29] incorporated the latent representation for reconstruction along with the discriminative features obtained from a classification model to enhance the feature vector used for open-set detection. Unlike previous methods, [30] utilized reconstruction error obtained from a multi-task learning framework as a detection score. Recently, [31] proposed using self-supervision and augmented the input image to learn richer features to improve separation between classes.

More recent works try to simulate open-set classes in order to provide explicit probability estimation over unknown classes. Ge et al. [32] extended OpenMax [26] by synthesizing the unknown samples using a Generative Adversarial Network (GAN) based framework. Along the same line, [33] proposed the counterfactual image generation (OSRCI) framework which employs a GAN to generate samples placing between decision boundaries that can be treated as unknown

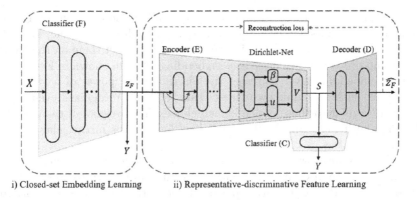

i) Closed-set Embedding Learning   ii) Representative-discriminative Feature Learning

Fig. 3. An overview of the proposed framework: i) Closed-set embedding learning: the classifier F is trained on the spectral domain X to produce latent discriminative embedding $\mathbf{z_F}$. ii) Representative-discriminative feature learning: the encoder E takes the embedding feature $\mathbf{z_F}$ and derives the representative features S using a Dirichlet-Net. The classifier C applied on S enhances the discriminative aspect of S, and the reconstruction error between the decoder output ($\mathbf{\hat{z}_F}$) and input to encoder ($\mathbf{z_F}$) enhances the representative aspect of S.

examples. [34] proposed class-conditioned auto-encoder (C2AE) algorithm where conditional reconstruction helps learning of both known and unknown score distributions.

It should be noted that there are related problems in the literature including outlier detection [35,36] and anomaly detection [37,38] that have some overlap with open-set recognition and can be treated as a relaxed version of open-set recognition. These problems assume the availability of one abnormal class during training. However, general open-set recognition problems usually do not provide information about the type or the number of the unknown classes in advance.

3   Proposed Approach

We propose a representative-discriminative open-set recognition (RDOSR) structure, as shown in Fig. 3. The network mainly consists of two components, 1) a closed-set embedding component to project the data from the original image domain to the embedding domain, such that different classes with similar spectral characteristics are more distinguishable, and 2) a multi-task representative-discriminative learning component to learn a better representation scheme at a finer scale in the abundance space, such that unknown classes can be better differentiated from known classes.

3.1   Network Architecture

One challenging issue of the open-set satellite land cover classification problem is that different classes may possess similar spectral characteristics. Thus, it is likely that an unknown class, whose spectral profile is close to that of a known class, may be misclassified as the known class. To address this issue, instead of detecting unknown classes on the image domain, we detect them on the embedding domain projected by a closed-set embedding layer, as shown in Fig. 3.i. The closed-set embedding layer increases the discriminative power of network to a large extent, such that unknown classes can be better recognized even if their spectra are similar to those of the known classes. The weights of the closed-set embedding layer are trained with a classifier F, which is further elaborated in Sect. 3.2.

To recognize unknown classes in the embedding domain, we propose a multi-task representative-discriminative feature learning framework to boost both the representative and discriminative power of the extracted feature vector, such that it is more informative and effective to recognize unknown samples. This is shown in Fig. 3.ii. The network consists of an encoder-decoder architecture with the representative features S extracted using a sparse Dirichlet encoder E, and a decoder formed by the bases shared among known classes. A classifier C applied on S is also included to further increase its discriminative capability. In this way, the data from unknown classes fed into the network would produce higher reconstruction error, thus can be detected accordingly. The details of network design are further elaborated in Sect. 3.3.

3.2   Closed-Set Embedding Learning

Given the set of sample pixels $X_k = \{\mathbf{x}_1, \mathbf{x}_2, \ldots, \mathbf{x}_{N_k}\}$ from the known classes with each pixel, $\mathbf{x}_i$, being a high-dimensional vector recording the reflectance readings of different spectral bands in the hyperspectral image, the corresponding labels are denoted with $Y_k = \{y_1, y_2, \ldots, y_{N_k}\}$, where N_k is the number of known pixels and $\forall y_i \in \{1, 2, \ldots, L\}$, where L is the number of known classes. To distinguish classes with similar spectral distributions, we project the input data X_k from the image domain to the embedding domain Z_F. The projection is learned through a classifier, F, with parameters Θ_F and the embedded features, $\mathbf{z_F}$ is forced to be discriminative through the cross-entropy loss,

$$\mathcal{L}_f(\Theta_F) = -\frac{1}{N_k} \sum_{i=1}^{N_k} y_i \log[F(\mathbf{x}_i)], \qquad (1)$$

where y_i is a one-hot encoded label and $F(\mathbf{x}_i)$ denotes the vector carrying the predicted probability score of the ith known sample. Such vector is generated by applying a softmax function on the features $\mathbf{z_F}$ in the embedding domain.

This general structure is sufficient for a common classification problem where the encountered classes are known. However, our goal is to increase the discriminative power of the features from classes with similar spectral characteristics. Therefore, we further increase the discriminative capacity of the embedded features with the l_1-norm sparse constraint defined by

$$\mathcal{L}_z(\Theta_F) = \frac{1}{N_k} \sum_{i=1}^{N_k} \|\mathbf{z}_{\mathbf{F}i}\|, \tag{2}$$

where $\mathbf{z}_{\mathbf{F}i}$ is the embedding feature vector learned by the classifier F. With such constraint, the embedded features from samples of different classes are more discriminative, even if their spectra are similar in the image domain.

3.3 Multi-task Representative-Discriminative Feature Learning

With the proposed closed-set embedding layer, the samples are projected from the image domain to the embedding domain possessing more distinguishable features. In order to better identify unknown samples, both discriminative and representative nature of the samples need to be exploited. Previous approaches [30,34] usually train a general auto-encoder to reconstruct samples from known classes. When the samples from unknown classes are fed into the network, the reconstruction error is expected to be larger than that of the known classes in the ideal case, since the network weights are optimized during the training procedure using known samples. However, the challenge lies in the scenario when the unknown classes especially the ones close to the known classes may potentially contribute to small reconstruction error too which would lead to failure in detection.

In this work, instead of adopting a general-purpose auto-encoder, we propose a multi-task representative-discriminative feature learning framework to improve the detection accuracy. The purpose of this network is to decrease the reconstruction error of the known classes while increasing the reconstruction error of unknown classes intensively.

Due to the resolution issue, each pixel in a satellite image usually covers a large geographical area or footprint (e.g., 30×30 m for Landsat-8), resulting in the so-called "mixed pixel" (i.e., each pixel tends to cover more than one constituent materials). These mixtures are generally assumed to be a linear combination of a few spectral bases with the corresponding mixing coefficients (or abundance). The proposed method is designed based on this assumption as shown in Eq. 3. Assume that the feature vector of a sample from known classes $\mathbf{z}_{\mathbf{F}}$ is a linear combination of a few bases B, and such bases are shared among features of the known classes. Thus, each sample of known classes can be decomposed with

$$\mathbf{z}_{\mathbf{F}} = \mathbf{s}B, \tag{3}$$

where $\mathbf{s}$ denotes the proportional coefficients of the shared bases, which serves as a form of "representation" of the embedding feature, that we refer to as the abundance. The abundance vector, or representation, should satisfy two physical constraints, i.e., non-negative and sum-to-one. The samples from unknown classes are also able to be decomposed by Eq. 3 using shared bases of the known classes, B. However, since B does not include the bases of the unknown classes, the distribution of its representations $\mathbf{s}$ should deviate from that of the known classes.

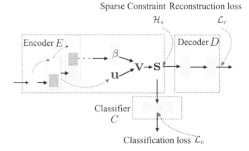

Fig. 4. The flowchart of the multi-task representative-discriminative feature learning framework.

Therefore, we design a network following the model of Eq. 3, which enforces **s** from known classes to follow a certain distribution. And if the network can extract **s** from unknown classes with similar distributions, then we expect they have high reconstruction errors.

The flowchart of the proposed multi-task representative-discriminative feature learning is detailed in Fig. 4. The network performs both the reconstruction task and the classification task. The reconstruction branch consists of a sparse Dirichlet-based encoder E with weights Θ_E and a decoder D with weights Θ_D. The encoder and decoder can be defined by the functions $E : Z_F \rightarrow S$ and $D : S \rightarrow Z_F$, respectively, where Z_F is the embedding space obtained by the closed-set classifier F, and S is the abundance space of latent representations projected by the encoder E. The representations **s** in the latent space S is enforced to follow a Dirichlet distribution. And a sparse constraint is introduced to enhance the representativeness of **s**. More details and justifications are provided below. In addition, S is also enforced to be discriminative by the classifier C, which can be defined by the function $C : S \rightarrow Y$ with weights Θ_C, where Y is the space of known labels.

Representative Feature Learning with Reconstruction The reconstruction branch is constructed according to Eq. 3, where the shared bases are embedded in the decoder D of the network and the corresponding representation **s** is extracted with the encoder E. Since **s** denotes the proportional coefficients of the bases, we enforce it to follow a Dirichlet distribution meeting the non-negative and sum-to-one physical constraints. Following the work of [39–41], we adopt the stick-breaking structure in the encoder to enforce the representations **s** to follow the Dirichlet distribution.

In the stick-breaking structure, a single element s_j in **s** can be expressed by

$$s_j = \begin{cases} v_1 & \text{for} \quad j = 1 \\ v_j \prod_{o<j}(1 - v_o) & \text{for} \quad j > 1, \end{cases} \tag{4}$$

where v_j is drawn from a Kumaraswamy distribution, i.e., $v_j \sim \text{Kuma}(u, 1, \beta)$ as shown in Eq. (5),

$$v_o \sim (1 - (1 - u^{\frac{1}{\beta}})). \tag{5}$$

Then, there are two parameters used to extract representations $\mathbf{s}$, i.e., u and β, both of which are hidden layers in the encoder of the network. A softplus activation function is adopted on the layer β due to its non-negative property, and a sigmoid is used to map u into the $(0, 1)$ range at the layer u. More details of the stick-breaking structure can be found in [40] and [41].

In addition, the entropy function [42] is adopted to reinforce the sparsity of the representation layer. Let $\hat{s}_j = \frac{|s_j|}{\|\mathbf{s}\|}$, for each pixel, the entropy function is defined as,

$$\mathcal{H}_s(\Theta_E) = -\sum_{j=1}^{c} \hat{s}_j \log \hat{s}_j. \tag{6}$$

where c is the dimension of the representation $\mathbf{s}$. The reconstruction loss $\mathcal{L}_r$ is adopted to reduce the reconstruction error of the known classes. It is defined by,

$$\mathcal{L}_r(\{\Theta_E, \Theta_D\}) = \frac{1}{N_k} \sum_{i=1}^{N_k} \|\mathbf{z}_{\mathbf{F}i} - \hat{\mathbf{z}}_{\mathbf{F}}\|_2, \tag{7}$$

where $\mathbf{z}_{\mathbf{F}i}$ is the embedding feature vector fed into the encoder E, and $\hat{\mathbf{z}}_{\mathbf{F}}$ is the reconstructed $\mathbf{z}_{\mathbf{F}i}$ obtained from the decoder D.

Discriminative Feature Learning with Classification Branch. To further increase the discriminative capacity of the representations, a classifier is adopted on the representations $\mathbf{s}$ with the classification loss $\mathcal{L}_c$ defined as,

$$\mathcal{L}_c(\{\Theta_E, \Theta_C\}) = \frac{1}{N_k} \sum_{i=1}^{N_k} y_i \log[E(\mathbf{z}_{\mathbf{F}i})], \tag{8}$$

where y_i is the ground truth label and $E(\mathbf{z}_{\mathbf{F}i})$ denotes the representative feature vector of the ith known sample. Note that the weights of both the reconstruction branch and classifier C are updated together, such that the learned representations can be both representative and discriminative.

3.4   Training Procedure and Network Settings

We first learn the embedding projection by optimizing the weights Θ_F of the classifier F with the loss function,

$$\min_{\Theta_F} \lambda_f \mathcal{L}_f + \lambda_z \mathcal{L}_z, \tag{9}$$

where λ_f and λ_z are two parameters to balance the trade-off between the cross-entropy loss and the sparsity loss.

Then, having the learned embedding layer, the multi-task representative-discriminative feature learning network is trained to minimize both the reconstruction loss and the classification error of the known classes with loss function,

$$\min_{\Theta_E,\Theta_D,\Theta_C} \lambda_r \mathcal{L}_r + \lambda_s \mathcal{H}_s + \lambda_c \mathcal{L}_c, \qquad (10)$$

where λ_r, λ_s, and λ_c are parameters balancing the trade-off between the reconstruction loss, the sparsity loss, and the classification loss.

The structures of the four networks, F, E, D, and C are listed in Table 1.

Table 1. The nodes in the proposed network

Networks	F	E	D	C
Nodes	$[512,\ 1024,\ 512,\ 32,\ L]$	$[3{,}3,\ 3{,}3,\ 10]$	$[10,\ 10,\ L]$	$[L]$

4 Experiments and Results

In this section, the effectiveness of the proposed RDOSR method is evaluated on several widely used benchmark hyperspectral image datasets. In addition, we demonstrate the generalization capacity of the proposed approach on RGB image datasets. Furthermore, the contribution from each component of the proposed framework is analyzed through ablation study.

4.1 Implementation Details

We train the network, described in Sect. 3.1, using an Adam optimizer [43], with a learning rate of 10^{-3}. The classifier F and the joint structure of encoder-decoder-classifier (E-D-C) are trained separately for a total number of 15K epochs. However, the other methods which do not have two separate components were trained for 6K epochs.

For training the classifier F, λ_f and λ_z are set equal to 1 and 0.1, respectively. The weights for the reconstruction λ_r, sparsity λ_s, and classification λ_c losses in training the E-D-C structure are set equal to 0.5, 10^{-3}, and 0.5, respectively. The sparsity weight λ_s is decayed with a weight decay of 0.9977. The classifier F is trained until its accuracy reaches 0.9988. It should be noted that all input data of the datasets are normalized to their mean value and unit variance. In addition, the feature vector obtained from the classifier F is divided by 10 to avoid divergence.

One of the factors that affects the performance of the open-set recognition algorithm is Openness [44] of the problem, defined as,

$$Openness = 1 - \sqrt{\frac{2 \times N_{train}}{N_{test} + N_{target}}}, \qquad (11)$$

where N_{train}, N_{test} and N_{target} are the number of classes known during training, the number of classes given during testing, and the number of classes that need to be recognized correctly during testing phase, respectively. In the experiments, classes of each dataset is partitioned into known and unknown sets according to the Openness.

The code is written in TensorFlow, and all the experiments are performed on a desktop computer having GeForce GPU of 10 GB Memory. The code is available at https://github.com/raziehkaviani/rdosr.

4.2 Metrics

To compare performance of different methods, there are several metrics including overall accuracy or F-score on a combination of known and unknown classes, and Receiver Operating Characteristic (ROC). The first two metrics do not characterize the performance of the model well due to their sensitivity not only to the performance of the model in classifying the known classes, but also an arbitrary operating threshold for detecting unknown samples.

On the other hand, the ROC curve would illustrate the ability of a binary classification system (here, known vs. unknown detection) as a discrimination threshold is varied from the minimum to the maximum value of the given detection measure (here, reconstruction error). Thus, it provides a measure free from calibration. To have a quantitative comparison, the area under the ROC (AUC) is computed in the experiments.

4.3 Open-Set Recognition for Hyperspectral Data

The experiments are conducted on three hyperspectral image datasets:

Pavia University (PU) and Pavia Center (PC). Both PU and PC datasets were gathered over Northern Italy in 2011 by the Reflective Optics Systems Imaging Spectrometer which has a resolution of 1.3 m. The dimension of the PU dataset is 1096×715 pixels with 103 spectral bands, ranging from 430 to 860 nm. The PC dataset has 610×340 pixels with 102 spectral bands. The PU and PC datasets both include nine land cover categories.

Indian Pines (IN). The IN dataset was collected over Northwest Indiana in 1912 by Airborne Visible/Infrared Imaging Spectrometer (AVIRIS). It has a dimension of 145×145 with a resolution of 20 m by pixel and 200 spectral bands. Its ground truth consists of 16 land cover classes.

We compare the performance of the proposed approach with three methods described in the following:

SoftMax: In a neural network classifier, a common confidence-based approach to detect open-set examples is thresholding the SoftMax scores. We use network structure of the classifier F without considering sparsity constraint.

OpenMax [26]: This approach calibrates the SoftMax scores in a classifier and augments them with a $N_k + 1$ class for an unknown category. The replaced

SoftMax layer with an OpenMax layer is used for open-set recognition. We adopt the classifier mentioned earlier in the SoftMax method, and use the Weibull fitting approach with parameter $Weibull\, tail\, size = 10$ to generate OpenMax layer values.

AE+CLS: Fully connected version of MLOSR [30] which utilizes a multi-task learning framework, composed of a classifier and a decoder with a shared feature extraction part, to detect open-set examples. To have a fair comparison with our approach, the encoder, decoder, and classifier are designed as our E (without the Dirichlet-Net), D, C, and trained with $\mathcal{L}_r$ and $\mathcal{L}_c$ loss, with weights of 0.5.

First, each of the L classes is assumed to be unknown which equates to an openness of 2.99%, 2.99%, and 1.63% for PU, PC, IN, respectively. The AUC values corresponding to choosing each of the L labels as unknown are averaged and reported for each method in Table 2. It can be observed that the proposed method outperforms other methods on all three datasets. See Supplement B for detailed comparison on the PU dataset. The minor improvement on the PC dataset can be justified by its differentiated spectrum of different classes which diminishes the effect of the classifier F.

Second, for the Openness equal to 6.46%, the ROC curves of different methods for the PU and IN datasets are illustrated in Fig. 5. As seen from the results on both datasets, the $AE+CLS+Dirichlet$ method which adopts the Dirichlet net to the $AE+CLS$ framework and the proposed method lie above all other methods. It should be noticed that our proposed method is able to detect unknown classes with 60% and above 90% accuracy and almost zero false detection for the PU and PC datasets, respectively.

Table 2. Area under the ROC curve for open-set detection. Results are averaged over L partitioning of the selected dataset to $L - 1$ known and 1 unknown classes.

Method	PU	PC	IN
SoftMax	0.385	0.816	0.555
OpenMax [26]	0.441	0.884	0.415
AE+CLS [30]	0.586	0.757	0.669
AE+CLS+Dirichlet	0.714	0.927	0.681
RDOSR (Ours)	**0.773**	**0.963**	**0.802**

Third, the histograms of reconstruction error for both known and unknown sets with Openness = 2.99% are shown in Fig. 6. It can be observed that the reconstruction errors corresponding to the known set have small values. However, the unknown set produces larger error due to mismatches in terms of representative and discriminative features learned from the known classes examples.

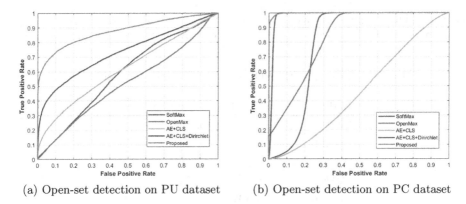

(a) Open-set detection on PU dataset   (b) Open-set detection on PC dataset

Fig. 5. Receiver Operating Curve curves for open-set recognition for PU and PC datasets, for $L = 7$ (openness $= 6.46\%$).

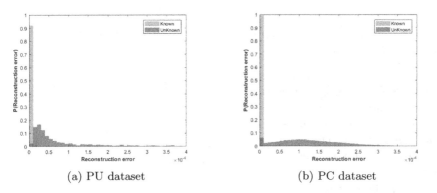

(a) PU dataset   (b) PC dataset

Fig. 6. Reconstruction error distribution of known and unknown classes using the proposed method for PU and PC datasets, for $L = 8$.

4.4   Open-Set Recognition for RGB Images

To show the generalization capacity of the proposed method, we evaluate the performance of the proposed approach on two RGB datasets and compare with several state-of-the-art methods. For this purpose, the classifier F performing pixel-wise classification is substituted with a DenseNet structure which takes a 2D image as input.

Following the protocol in [33], we sample 4 known classes from CIFAR10 [45] to have Openness $= 13.39\%$ and 20 known classes out of 200 categories of TinyImageNet [46] resulting in an Openness of 57.35%. Table 3 summarizes the results where the values other than the proposed RDOSR are taken from [31]. It can be observed that the proposed method has better performance over the compared methods, except for GDOSR [31], on CIFAR10. However, it achieves significant improvement on TinyImageNet. It may be due to the similarity between the classes in TinyImageNet which hinders detecting unknown samples in the image space while RDOSR addresses this issue by operating in the embedding space.

Table 3. Area under the ROC curve for Open-set recognition.

Method	CIFAR10	TinyImageNet
SoftMax	0.677	0.577
OpenMax [26]	0.695	0.576
OSRCI [33]	0.699	0.586
C2AE [34]	0.711	0.581
GDOSR [31]	**0.807**	0.608
RDOSR (Ours)	0.744	**0.752**

4.5   Ablation Study

Starting with a baseline, $AE+CLS$, each component is gradually added to the framework to show its effectiveness. The results corresponding to the ablation study are shown in Fig. 7. It can be seen that employing the baseline structure applied on the spectra domain has the worst performance. However, adding the Dirichlet-based network makes a major improvement due to applying physical con-

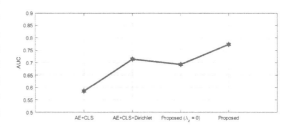

Fig. 7. Ablation study of the proposed method on PU dataset

straints on the latent space learned by the encoder E. Directly performing open-set recognition on an embedding space causes instability problem which is confirmed by a performance drop compared to the $AE+CLS+Dirichlet$ method. Our proposed method addresses the instability issue by adopting a sparsity constraint on the embedding feature vector $\mathbf{z_F}$. As seen from Fig. 7, our proposed method achieves the highest AUC value compared to three other baseline methods.

5   Conclusions

We studied the challenging problem of open-set land cover recognition in satellite images. Although inherently a classification problem, both representative and discriminative features need to be learned in order to best characterize the difference between known and unknown classes. We presented the transformation among three spaces, that is, the original image space, the embedding feature space, and the abundance space, where features with both representative and discriminative capacity can be learned to maximize success rate. The proposed multi-tasking representative-discriminative learning structure was evaluated on three hyperspectral and two RGB image datasets and exhibited significant improvement over state-of-the-art open-set recognition algorithms.

References

1. Krizhevsky, A., Sutskever, I., Hinton, G.E.: ImageNet classification with deep convolutional neural networks. In: Advances in Neural Information Processing Systems, pp. 1097–1105 (2012)
2. He, K., Zhang, X., Ren, S., Sun, J.: Deep residual learning for image recognition. In: Proceedings of the IEEE Conference on Computer Vision and Pattern Recognition, pp. 770–778 (2016)
3. Simonyan, K., Zisserman, A.: Very deep convolutional networks for large-scale image recognition. arXiv preprint arXiv:1409.1556 (2014)
4. Redmon, J., Divvala, S., Girshick, R., Farhadi, A.: You only look once: unified, real-time object detection. In: Proceedings of the IEEE Conference on Computer Vision and Pattern Recognition, pp. 779–788 (2016)
5. Ren, S., He, K., Girshick, R., Sun, J.: Faster R-CNN: towards real-time object detection with region proposal networks. In: Advances in Neural Information Processing Systems, pp. 91–99 (2015)
6. Long, J., Shelhamer, E., Darrell, T.: Fully convolutional networks for semantic segmentation. In: Proceedings of the IEEE Conference on Computer Vision and Pattern Recognition, pp. 3431–3440 (2015)
7. Chen, L.C., Papandreou, G., Kokkinos, I., Murphy, K., Yuille, A.L.: DeepLab: semantic image segmentation with deep convolutional nets, atrous convolution, and fully connected CRFs. IEEE Trans. Pattern Anal. Mach. Intell. **40**(4), 834–848 (2017)
8. http://lesun.weebly.com/hyperspectral-data-set.html
9. van der Maaten, L., Hinton, G.: Visualizing data using t-SNE. J. Mach. Learn. Res. **9**(Nov), 2579–2605 (2008)
10. Green, A.A., Berman, M., Switzer, P., Craig, M.D.: A transformation for ordering multispectral data in terms of image quality with implications for noise removal. IEEE Trans. Geosci. Remote Sens. **26**(1), 65–74 (1988)
11. Comon, P.: Independent component analysis, a new concept? Sig. Process. **36**(3), 287–314 (1994)
12. Plaza, A., Martinez, P., Plaza, J., Perez, R.: Dimensionality reduction and classification of hyperspectral image data using sequences of extended morphological transformations. IEEE Trans. Geosci. Remote Sens. **43**(3), 466–479 (2005)
13. Luo, B., Chanussot, J.: Unsupervised classification of hyperspectral images by using linear unmixing algorithm. In: 2009 16th IEEE International Conference on Image Processing (ICIP), pp. 2877–2880. IEEE (2009)
14. Dópido, I., Zortea, M., Villa, A., Plaza, A., Gamba, P.: Unmixing prior to supervised classification of remotely sensed hyperspectral images. IEEE Geosci. Remote Sens. Lett. **8**(4), 760–764 (2011)

15. Baghbaderani, R.K., Wang, F., Stutts, C., Qu, Y., Qi, H.: Hybrid spectral unmixing in land-cover classification. In: International Geoscience and Remote Sensing Symposium (IGARSS) (2019)
16. Hu, W., Huang, Y., Wei, L., Zhang, F., Li, H.: Deep convolutional neural networks for hyperspectral image classification. J. Sensors **2015**, 1–13 (2015)
17. Song, Y., Zhang, Z., Baghbaderani, R.K., Wang, F., Stutts, C., Qi, H.: Land cover classification for satellite images through 1D CNN. In: Workshop on Hyperspectral Image and Signal Processing: Evolution in Remote Sensing (WHISPERS) (2019)
18. Cao, X., Zhou, F., Xu, L., Meng, D., Xu, Z., Paisley, J.: Hyperspectral image classification with Markov random fields and a convolutional neural network. IEEE Trans. Image Process. **27**(5), 2354–2367 (2018)
19. Sharma, V., Diba, A., Tuytelaars, T., Van Gool, L.: Hyperspectral CNN for image classification & band selection, with application to face recognition. Technical report KUL/ESAT/PSI/1604, KU Leuven, ESAT, Leuven, Belgium (2016)
20. Hamida, A.B., Benoit, A., Lambert, P., Amar, C.B.: 3-D deep learning approach for remote sensing image classification. IEEE Trans. Geosci. Remote Sens. **56**(8), 4420–4434 (2018)
21. Zhong, Z., Li, J., Luo, Z., Chapman, M.: Spectral-spatial residual network for hyperspectral image classification: a 3-D deep learning framework. IEEE Trans. Geosci. Remote Sens. **56**(2), 847–858 (2017)
22. Mendes Júnior, P.R., et al.: Nearest neighbors distance ratio open-set classifier. Mach. Learn. **106**(3), 359–386 (2016). https://doi.org/10.1007/s10994-016-5610-8
23. Scheirer, W.J., Jain, L.P., Boult, T.E.: Probability models for open set recognition. IEEE Trans. Pattern Anal. Mach. Intell. **36**(11), 2317–2324 (2014)
24. Júnior, P.R.M., Boult, T.E., Wainer, J., Rocha, A.: Specialized support vector machines for open-set recognition. arXiv preprint arXiv:1606.03802 (2016)
25. Zhang, H., Patel, V.M.: Sparse representation-based open set recognition. IEEE Trans. Pattern Anal. Mach. Intell. **39**(8), 1690–1696 (2016)
26. Bendale, A., Boult, T.E.: Towards open set deep networks. In: Proceedings of the IEEE Conference on Computer Vision and Pattern Recognition, pp. 1563–1572 (2016)
27. Hassen, M., Chan, P.K.: Learning a neural-network-based representation for open set recognition. arXiv preprint arXiv:1802.04365 (2018)
28. Shu, L., Xu, H., Liu, B.: DOC: deep open classification of text documents. arXiv preprint arXiv:1709.08716 (2017)
29. Yoshihashi, R., Shao, W., Kawakami, R., You, S., Iida, M., Naemura, T.: Classification-reconstruction learning for open-set recognition. In: Proceedings of the IEEE Conference on Computer Vision and Pattern Recognition, pp. 4016–4025 (2019)
30. Oza, P., Patel, V.M.: Deep CNN-based multi-task learning for open-set recognition. arXiv preprint arXiv:1903.03161 (2019)
31. Perera, P., et al.: Generative-discriminative feature representations for open-set recognition. In: Proceedings of the IEEE/CVF Conference on Computer Vision and Pattern Recognition, pp. 11814–11823 (2020)
32. Ge, Z., Demyanov, S., Chen, Z., Garnavi, R.: Generative OpenMax for multi-class open set classification. arXiv preprint arXiv:1707.07418 (2017)
33. Neal, L., Olson, M., Fern, X., Wong, W.-K., Li, F.: Open set learning with counterfactual images. In: Ferrari, V., Hebert, M., Sminchisescu, C., Weiss, Y. (eds.) ECCV 2018. LNCS, vol. 11210, pp. 620–635. Springer, Cham (2018). https://doi.org/10.1007/978-3-030-01231-1_38

34. Oza, P., Patel, V.M.: C2AE: class conditioned auto-encoder for open-set recognition. In: Proceedings of the IEEE Conference on Computer Vision and Pattern Recognition, pp. 2307–2316 (2019)
35. Xia, Y., Cao, X., Wen, F., Hua, G., Sun, J.: Learning discriminative reconstructions for unsupervised outlier removal. In: Proceedings of the IEEE International Conference on Computer Vision, pp. 1511–1519 (2015)
36. You, C., Robinson, D.P., Vidal, R.: Provable self-representation based outlier detection in a union of subspaces. In: Proceedings of the IEEE Conference on Computer Vision and Pattern Recognition, pp. 3395–3404 (2017)
37. Chalapathy, R., Menon, A.K., Chawla, S.: Robust, deep and inductive anomaly detection. In: Ceci, M., Hollmén, J., Todorovski, L., Vens, C., Džeroski, S. (eds.) ECML PKDD 2017. LNCS (LNAI), vol. 10534, pp. 36–51. Springer, Cham (2017). https://doi.org/10.1007/978-3-319-71249-9_3
38. Golan, I., El-Yaniv, R.: Deep anomaly detection using geometric transformations. In: Advances in Neural Information Processing Systems, pp. 9758–9769 (2018)
39. Sethuraman, J.: A constructive definition of Dirichlet priors. Statistica Sinica **4**, 639–650 (1994)
40. Nalisnick, E., Smyth, P.: Stick-breaking variational autoencoders. In: International Conference on Learning Representations (ICLR) (2017)
41. Qu, Y., Qi, H., Kwan, C.: Unsupervised sparse Dirichlet-Net for hyperspectral image super-resolution. In: Proceedings of the IEEE Conference on Computer Vision and Pattern Recognition, pp. 2511–2520 (2018)
42. Huang, S., Tran, T.D.: Sparse signal recovery via generalized entropy functions minimization. IEEE Trans. Sig. Process. **67**(5), 1322–1337 (2018)
43. Kingma, D.P., Ba, J.: Adam: a method for stochastic optimization. arXiv preprint arXiv:1412.6980 (2014)
44. Scheirer, W.J., de Rezende Rocha, A., Sapkota, A., Boult, T.E.: Toward open set recognition. IEEE Trans. Pattern Anal. Mach. Intell. **35**(7), 1757–1772 (2012)
45. Krizhevsky, A., Hinton, G., et al.: Learning multiple layers of features from tiny images (2009)
46. Le, Y., Yang, X.: Tiny ImageNet visual recognition challenge. CS **231N**, 7 (2015)

Structure-Aware Human-Action Generation

Ping Yu[1]([✉]), Yang Zhao[1], Chunyuan Li[2], Junsong Yuan[1],
and Changyou Chen[1]

[1] The State University of New York at Buffalo, Buffalo, USA
{pingyu,yzhao63,jsyuan,changyou}@buffalo.edu
[2] Microsoft Research, Redmond, USA
chunyl@microsoft.com

Abstract. Generating long-range skeleton-based human actions has been a challenging problem since small deviations of one frame can cause a malformed action sequence. Most existing methods borrow ideas from video generation, which naively treat skeleton nodes/joints as pixels of images without considering the rich inter-frame and intra-frame structure information, leading to potential distorted actions. Graph convolutional networks (GCNs) is a promising way to leverage structure information to learn structure representations. However, directly adopting GCNs to tackle such continuous action sequences both in spatial and temporal spaces is challenging as the action graph could be huge. To overcome this issue, we propose a variant of GCNs (**SA-GCNs**) to leverage the powerful self-attention mechanism to adaptively sparsify a complete action graph in the temporal space. Our method could dynamically attend to important past frames and construct a sparse graph to apply in the GCN framework, well-capturing the structure information in action sequences. Extensive experimental results demonstrate the superiority of our method on two standard human action datasets compared with existing methods.

Keywords: Action generation · Graph convolutional network · Self-attention · Generative adversarial networks (GAN)

1 Introduction

Recent years have witnessed the development of skeleton-based action generation, which has been applied in a variety of applications, such as action classification [10,17,19,29,44], action prediction [2,24,39] and human-centric video generation [37,45]. Action generation is still a challenging problem since small deviations in one frame can cause confusion in the entire sequence.

Electronic supplementary material The online version of this chapter (https://doi.org/10.1007/978-3-030-58577-8_2) contains supplementary material, which is available to authorized users.

A. Vedaldi et al. (Eds.): ECCV 2020, LNCS 12375, pp. 18–34, 2020.
https://doi.org/10.1007/978-3-030-58577-8_2

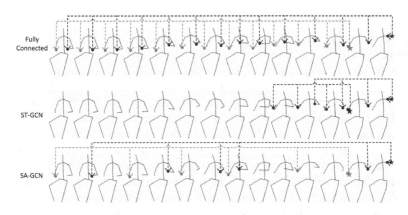

Fig. 1. Comparisons of the construction of action graphs with our proposed method (3rd tow) and two standard methods (1st and 2nd rows) to encode temporal information. First row (*full connection*): the left-hand joint gather information from all left-hand joint of past frames; similar to the right-hand joint. Second row (*ST-GCN*): a 1D convolution of kernel size k is used to encode temporal information. Both the left and right hands could encode information from past k frames with share weights. Third row (*SA-GCN*): both the left- and right-hand joints learn to encode information from a selected left-hand joints based on the attention scores.

One of the most successful methods for skeleton-based action generation considers skeleton-based action generation as a standard video generation problem [7,13,40]. Specifically, the method naively treats skeleton joints as image pixels and sequential actions as videos, without considering the rich structure information among both joints and action frames. The video-generation based methods may produce distorted actions when applied to skeleton generation, if prior structure knowledge is not well leveraged. A first step to consider structure information into action generation is to represent a skeleton as a graph structure to characterize the spatial relations between joints in each frame based on graph convolution networks (GCN) [6,21,50]. However, most existing GCN methods do not have the flexibility to process continuous sequential graphs data. This poses a new challenge: *how to construct a representation to effectively incorporate both temporal and spatial structures into action generation?*

Generally speaking, there are two classes of methods with GCN to model action structure information: (*i*) *Full connection:* an entire action sequence is $7890 = -098765$ considered as a graph. Each node of the current frame is connected with the corresponding nodes in all the past frames. This construction, however, is computationally very inefficient (if ever possible at all). Moreover, the model could be highly redundant since many frames are similar to each other. (*ii*) *Spatial-temporal graph convolutional networks* [44]*:* a graph convolution is first applied to intra-frame skeletons, whose extracted features are then applied with a 1D convolution layer to capture temporal information.

This method typically requires weight sharing among all nodes, and the ability to model temporal information is somewhat weak.

We advocate that a better solution should be proposed to leverage skeleton structures and gather information from action sequences more efficiently. In this paper, we propose *Self-Attention based Graph Convolutional Networks (SA-GCN)* to build generic representations for skeleton sequences. Our *SA-GCN* aims at building a sparse global graph for each action sequence to achieve both computational efficiency and modelling efficacy. Specifically, for a given frame, the proposed *SA-GCN* first calculates self-attention scores for other frames. Based on the attention scores, top k past frames with the most significant scores are selected to be connected to the current frame to construct inter-frame connections. Within each frame, the joints are connected as the original skeleton representation. To demonstrate the differences between our construction and the aforementioned two constructions, Fig. 1 illustrates a sequence of samples in terms of every three consecutive frames on the Human 3.6m dataset *Sitting-Down* sequence. As illustrated in the figure, our method can be considered as an adaptive scheme to construct an action graph, with each node assigning a trainable weight instead of a shared weight as in other methods.

The major contributions of this work are summarized in three aspects:

- We propose *SA-GC* layer, a generic graph-based formulation to encode structure information into action modelling efficiently. Our method is the first sparse and adaptive scheme to encode past frame information for action generation.
- By efficiently leveraging action structure information, our model can generate high-quality long-range action sequences with pure Gaussian noise and provided labels as inputs without pretraining.
- Our model is evaluated on two standard large datasets for skeleton-based action generation, achieving superior and stable performance compared with previous models.

2   Preliminaries and Related Work

2.1   Attention Model

Attention models have become increasingly popular in capturing long-term global dependencies [1,8]. In particular, self-attention [5,33,46] mimics human visual attention, allowing a model to focus on crucial regions and to learn the correlation among elements in the same sequence. [38] proposes a non-local operation as a kind of attention on capturing long-range dependencies in videos. [33] develops the transformer model, which is solely based on attention and achieves state-of-the-art on machine translation. Thus, self-attention can typically lead to a better representation learning. One key advance of our proposed model compared with previous ones is that we adopt self attention to efficiently encode frame-wise correlations by inheriting all merits of the self-attention mechanism.

2.2 Skeleton-Based Action Generation

The task of action generation differs from action prediction [3] in that no past intermediate sub-sequence is provided. Directly generating human actions from noise is considered more challenging. The problem has been well studied in early works [4,27,28], which applied switching linear models to generate stochastic human motions. These models, however, required a large amount of data to fit a model and are difficult to find an appropriate number of switching states. Later on, the Restricted Boltzmann Machine [30] and Gaussian-process latent variable models [32,35,36] were applied. But they still can not scale to massive amounts of data. The rapid development of deep generative models has brought the idea of recurrent-neural-network (RNN) based Variational Autoencoder (VAE) [20] and Generative Adversarial Net (GAN) models [7,11,18,40–42,48]. These models are scalable and usually can generate actions with better quality.

The aforementioned methods still have some limitations, which mainly lie in two aspects. Firstly, spatial relationships among body joints and temporal dynamics along continuous frames have not been well explored. Secondly, these models often require an expensive pre-training phase to capture intra-frame constraints, including the two most recent state-of-the-art works [7,40]. By contrast, Our work moves beyond these limitations and can be trained from scratch to generate high-quality motions.

2.3 Graph Convolutional Network

GCNs have been achieving encouraging results [44]. In general, they can be categorized into two types: spectral approaches [6,21] and spatial approaches [26,50]. The spectral GCN operates on the Fourier domain (locality) with convolution to produce a spectral representation of graphs. The spatial GCN, by contrast, directly applies convolution on the spatially distributed nodes. This work is in the spirit of spatial GCNs and incorporates new ideas of GCNs to fit the task. In particular, to model long-term dependent motion dynamics, we are aware of ideas from graph pruning [47] and jump connection [43], which respectively allows one to extract structure representation more efficiently and to build deeper graph convolutional layers. In terms of GCN-based human motion modelling,

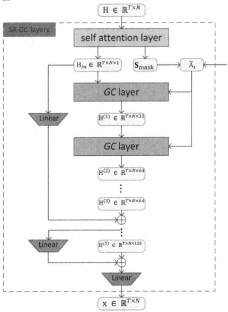

Fig. 2. An illustration of the SA-GC layer. $\tilde{\mathbf{A}}$ and $\tilde{\mathbf{A}}_s$ are two adjacency matrices detailed in Sect. 3.2.

the most related work is ST-GCN [44], which applies a spatial GCN to a different task of action recognition. This method applied a GCN layer for intra-frame skeletons and then used 1D convolution layer for gathering information in temporal space. All nodes in a frame share weights on the temporal space and could only attend limited range of information, depending on the kernel size of the 1D convolution layer. We will compare our method with ST-GCN (for action generation) in Sect. 4.6.

3 Structure-Aware Human-Action Generation

Different from the video-generation task, the skeleton-based action generation contains huge amounts of structure information, $e.g.$, intra-frame structural joints information and inter-frame motion dynamics. Directly treating skeleton frames as images will lose most of these structure information, leading to the distortion of some skeletal frames. Moreover, in the context of skeleton-based actions, where only limited positional information is provided, differences between two continuous frames are virtually impossible to be observed. To address these issues, we propose to incorporate GCNs to encode the rich structural information, with additional consideration to reduce computational burden by using self-attention to automatically learn a sparse action graph.

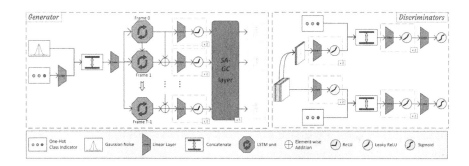

Fig. 3. The overall framework of the proposed method.

3.1 An Overview of the SA-GCN

Figure 3 illustrates the overall framework of our model for action generation. It follows the GAN framework of video generation [9,31], which consists of an action generator $\mathcal{G}$ and a dual discriminator: one video-based discriminator $\mathcal{D}_V$ and one frame-based discriminator $\mathcal{D}_F$.

Generator. For simplicity, we assume the sequence length to be T. Our action generator starts with a RNN with an input at each time as the concatenation of a Gaussian random noise z and an embedded class representation of a label y. The outputs of the RNN layer are denoted as $[o_0, o_1, o_2, ..., o_{T-1}]$. Following [7,40], we consider outputting residuals instead of the exact coordinates of different joints, i.e., $c_0 = o_0$, $c_1 = o_1 + c_0$, ..., $c_{T-1} = o_{T-1} + c_{T-2}$. The output of the RNN will go through three linear transformations before being fed as the input of the newly proposed SA-GC layer, which will be detailed in Sect. 3.2.

The SA-GC Layer. The key component of our framework is a newly defined self-attention based graph convolutional layers (SA-GC layers), as illustrated in Fig. 2. Specifically, we denote the input of the SA-GC layers as a feature vector $\mathbf{H} \in \mathbb{R}^{T \times N}$. Through a self attention layer [33], the output are a new representation $\mathbf{H}_{in} \in \mathbb{R}^{T \times N \times 1}$ and a learned masked attention score matrix $\mathbf{S}_{mask} \in \mathbb{R}^{T \times T}$. This self attention layer is followed by 5 GC layers. Each GC layer takes last layer's hidden state vector and masked adjacency matrix $\tilde{\mathbf{A}}_\mathbf{s}$ as the input. The hidden states, which are outputs of the 5 GC layers, are defined respectively as $\mathbf{H}^{(1)} \in \mathbb{R}^{T \times N \times 32}$, $\mathbf{H}^{(2)} \in \mathbb{R}^{T \times N \times 64}$, $\mathbf{H}^{(3)} \in \mathbb{R}^{T \times N \times 64}$, $\mathbf{H}^{(4)} \in \mathbb{R}^{T \times N \times 128}$ and $\mathbf{H}^{(5)} \in \mathbb{R}^{T \times N \times 128}$. Furthermore, the ResNet mechanism [15] is applied on each two SA-GC layers, i.e., we add the output of the first SA-GC layer to the third SA-GC layer, and the output of the third SA-GC layer to the final output. Detailed operations of the SA-GC layer are described in Sect. 3.2.

Dual Discriminator. The video-based discriminator $\mathcal{D}_V$ takes a sequence of actions and the corresponding labels as the input. The frame-based discriminator $\mathcal{D}_F$ randomly selects k_{frame} frames of an input sequence and the corresponding labels as the input. Both discriminators output either real or fake. In this paper, we apply the conditional GAN objective formulation [11,22,25]:

$$\mathcal{L} = \min_{\mathcal{G}} \max_{\mathcal{D}_F, \mathcal{D}_V} \mathbb{E}_{x \sim p(x)}[\log \mathcal{D}_F(x|y)] + \mathbb{E}_{z \sim p(z)}[\log(1 - \mathcal{D}_F(\mathcal{G}(z|y)))]$$
$$+ \mathbb{E}_{x \sim p(x)}[\log \mathcal{D}_V(x|y)] + \mathbb{E}_{z \sim p(z)}[\log(1 - \mathcal{D}_V(\mathcal{G}(z|y)))] \tag{1}$$

where $p(x)$ defines the ground truth distribution, $p(z)$ is the standard Gaussian distribution and y is the one-hot class indicator.

3.2 Action Graph Construction

In this section, we describe detailed construction of the action graph, which is used in our SA-GCN module. Note a skeleton sequence is usually represented by 2D or 3D coordinates of human joints in each frame. The inter-frame action is the static skeleton in spatial space, and the inter-frame action is the movement in temporal space. To capture the temporal correlation, previous work has applied 1D convolution for learning skeleton representing by concatenating coordinate vectors of all joints in one frame. In our framework,

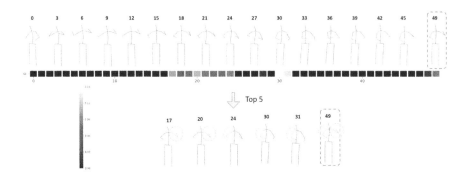

Fig. 4. The pipeline of *SA-GC* layer. The top line shows frames out of every three consecutive frames from Human 3.6 *Direction* class. The heat map under these samples represent the corresponding attention scores for the 49*th* frame. The bottom line shows the top 5 frames with the highest attention scores. Green circles and orange circles show similarity between selected frames and our target frame (the 49*th* frame). (Color figure online)

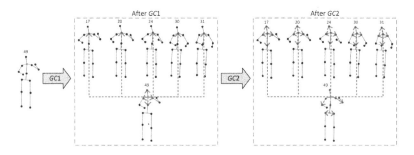

Fig. 5. Information passing through *SA-GC* layers at the node "neck".

as stated before, we propose to construct a connected graph for a whole action sequence, and learn a sparse inter-frame connection by adopting self-attention learning. Particularly, we construct an undirected graph $\mathbf{G} = (\mathbf{V}, \mathbf{E})$ on a whole action sequence of T frames, each consists of N joints. Here, the node set $V = \{v_{ti}|t = 0, \ldots, T-1, i = 1, \ldots N\}$ includes all joints of a skeleton sequence.

Explanation of the *SA-GC* Layer. Figure 2 shows the detailed implementation of the *SA-GC* layer. Our *SA-GC* layer consists of one self attention layer and 5 graph convolution (*GC*) layers. To explain the construction, we detail the pipeline of the construction with an example illustrated in Fig. 4.

The Self Attention Layer. Similar to standard self attention [33], our self-attention layer takes a feature vector $\mathbf{H} \in \mathbb{R}^{T \times N}$ as input, and outputs a self-attention matrix $\mathbf{S}_{mask}$, representing how much influence of the past frames on the current frame. Figure 4 shows one of our generated *Direction* sequences and its corresponding attention score vector's heat map for the last frame (the 49*th* frame). After the self attention layer, we select top 5 past frames with the highest attention scores (only keep 6 elements in each row of the $\mathbf{S}_{mask}$ matrix). As we could see from the example in Fig. 4, the selected 5 past frames have the highest similarity with the $49th$ frame. The skeleton in the $49th$ frame keeps the red arm up and keeps the blue arm bent down. Looking back to past frames, frames before the $21st$ lift up its red arm. Frames between the $24th$ to the $31st$ frame have the similar blue arm pose as the $49th$ frame. Our attention identifies frames $17th$, $20th$, $24th$, $30th$ and $31st$ as the most relevant frames to be attended according to the learned attention matrix $\mathbf{S}_{mask}$.

The GC Layers. As illustrated, the self-attention layer is followed by 5 *GC* layers. After selection, we will connect each node of the $49th$ frame with the corresponding node in the selected 5 frames and assign edge weights with the corresponding self-attention scores. Figure 5 shows information passing path through our *SA-GC* layer at the node neck. The left plot of Fig. 5 shows that after one *GC* layer, the $49th$ neck node can gather information from neck nodes of five selected frames and four neighbor nodes in its own frame. The right plot of Fig. 5 shows that after the two *SA-GC* layers, the $49th$ neck node can gather information for the five nodes of the selected past four frames and seven nodes of its own frame. It is worth noting that nodes in different frame will have distinct attention score for edges in both spatial space and temporal space, thus they will have their particular edge weights through our *SA-GC* layer.

Implementing Self-attention Based GCN. In our case, we consider all joints in an action sequence, ending up with a 2D adjacent matrix with both row size and column size $N * T$. To this end, we first use $\mathbf{A} \in \mathbb{R}^{N \times N}$ to denote the adjacent matrix of intra-frame, which is constructed by strictly following the structure of a skeleton, *e.g.*, the "head" node is connected to the "neck" node. After adding self connections $\mathbf{I}$, the intra-frame adjacency matrix will be $\bar{\mathbf{A}} = \mathbf{A} + \mathbf{I}$. We then define an initial adjacency matrix of a whole sequence as:

$$\tilde{\mathbf{A}} = \begin{pmatrix} \bar{\mathbf{A}} & \mathbf{I} & \cdots & \mathbf{I} \\ \mathbf{I} & \bar{\mathbf{A}} & \cdots & \mathbf{I} \\ \vdots & \vdots & & \vdots \\ \mathbf{I} & \mathbf{I} & \cdots & \bar{\mathbf{A}} \end{pmatrix}_{(N*T) \times (N*T)}, \tag{2}$$

where $\mathbf{I}$ is used to represent connecting each node with all of the corresponding nodes in the temporal space, $(N*T) \times (N*T)$ means $\tilde{\mathbf{A}}$ is a 2D matrix with both row size and column size $N * T$, both N and T are numbers, * means multiply operation. The adjacency matrix $\tilde{\mathbf{A}}$ essentially means each node in one frame is connected to the corresponding node in the temporal space. At the same time, it also connects to the neighboring nodes in spatial space encoded by $\bar{\mathbf{A}}$.

Next, we propose to use self-attention to prune the action graph. The idea is to learn a set of attention scores encoding the relevance of each frame w.r.t. the current frame, and only choose the top-K frames in the temporal space. Specifically, we adopt a similar implementation of the scaled dot-product attention as in [33]. The input of the self-attention layer is represented as $\mathbf{H} \triangleq \{h_0, h_1, \cdots, h_{T-1}\}$, where $h_t \in \mathbb{R}^N$ represents the hidden state vector at time t with N nodes. Following the self-attention in [33], $\mathbf{Q}$, $\mathbf{K}$ and $\mathbf{V}$ are given as:

$$\mathbf{Q} = \mathbf{W}_q \mathbf{H}, \ \mathbf{K} = \mathbf{W}_k \mathbf{H}, \ \mathbf{V} = \mathbf{W}_v \mathbf{H}, \tag{3}$$

where $\mathbf{W}_q$, $\mathbf{W}_k$ and $\mathbf{W}_v$ are projection weights. The attention score $\mathbf{S} \in \mathbb{R}^{T \times T}$ and the attention layer's output $\mathbf{H}_{in}$ are calculated as:

$$\mathbf{S} = \text{softmax}\left(\mathbf{Q}\mathbf{K}^{\mathsf{T}}\right); \ \mathbf{H}_{in} = \mathbf{S}\mathbf{V} \tag{4}$$

In the task of action generation, we need to modify $\mathbf{S}$ as a masked attention $\mathbf{S}_{mask}$ which prevents current frame from attending to subsequent frames

$$\mathbf{S}_{mask} = \begin{pmatrix} s_{0,0} & 0 & \cdots & 0 \\ s_{1,0} & s_{1,1} & \cdots & 0 \\ \vdots & \vdots & & \vdots \\ s_{T-1,0} & s_{T-1,1} & \cdots & s_{T-1,T-1} \end{pmatrix}_{T \times T}, \tag{5}$$

where the element $s_{m,n}$ denotes the n-th frame's influence on the m-th frame and values in the upper triangle are all equal to 0. To enforce the pruning, we further select the top K scores in each row of the $\mathbf{S}_{mask}$ and set the other elements to be 0. Note that, if the number of non-zero elements in some rows is less than K, we will keep all the non-zero elements. Finally, the adjacent matrix is constructed as

$$\tilde{\mathbf{A}}_s = \mathbf{S}_{mask} \odot \tilde{\mathbf{A}} \triangleq \begin{pmatrix} s_{0,0} * \bar{\mathbf{A}} & 0 & \cdots & 0 \\ s_{1,0} * \mathbf{I} & s_{1,1} * \bar{\mathbf{A}} & \cdots & 0 \\ \vdots & \vdots & & \vdots \\ s_{T-1,0} * \mathbf{I} & s_{T-1,1} * \mathbf{I} & \cdots & s_{T-1,T-1} * \bar{\mathbf{A}} \end{pmatrix}_{(N*T) \times (N*T)} \tag{6}$$

Consequently, the output (before activation) of the self-attention based graph convolutional layer becomes:

$$\mathbf{H}^{(1)} = \mathbf{D}^{-1} \tilde{\mathbf{A}}_s \mathbf{H}_{in} \mathbf{W}, \tag{7}$$

where $\mathbf{D}^{ii} = \sum_j \tilde{\mathbf{A}}_s^{ij}$ represents diagonal node degree matrix for normalizing $\tilde{\mathbf{A}}_s$, $\mathbf{H}^{(1)}$ is the hidden state after first $\mathbf{GC}$ layer in Fig. 2. We will conduct graph convolution operation in Eq. 7 using same $\tilde{\mathbf{A}}_s$ for five times. The output of the fifth GC layer in $SA\text{-}GC$ layer is $\mathbf{H}^{(5)}$. After a linear layer, we get the output of the generator, which is a generated sequence $x \in \mathbb{R}^{T \times N}$.

4   Experiments

We perform experiments to evaluate the proposed method on two standard skeleton-based human-action benchmarks, the Human-3.6m dataset [16] and the

NTU RGB+D dataset [29]. Several state-of-the-art methods are used for comparison, including [7,13,40,42]. Following [40], the Maximum Mean Discrepancy (MMD) [12] is adopted to measure the quality of generated actions. Further, we pre-train a classifier on training set to test the recognition accuracy of generated actions. We also conduct human evaluation on the Amazon Mechanical Turk (AMT) to access the perceptual quality of generated sequences. To examine the functionality of each component of the proposed model, we also perform detailed ablation studies on the Human-3.6m dataset.

4.1 Datasets

Human-3.6m. Following the same pre-processing procedure in [7,40], 50 Hz video frames are down-sampled 16 Hz to obtain representative and larger variation 2D human motions. The joint information consists of 2-D locations of 15 major body joints. Ten distinctive classes of actions are used in the following experiments, including *sitting, sitting down, discussion, walking, greeting, direction, phoning, eating, smoking* and *posing*.

NTU RGB+D. This dataset contains 56,000 video clips on 60 classes performed by 40 subjects and recorded with 3 different camera views. Compared with Human-3.6m, it can provide more samples in each class and much more intra-class variations. We select ten classes of motions and obtain their 2-D coordinates of 25 body joints following the same setting in [40], including *drinking water, jump up, make phone call, hand waving, standing up, wear jacket, sitting down, throw, cross hand in front* and *kicking something*. We then apply two commonly used benchmarks for a further evaluation in the ten classes: (*i*) *cross-view:* the training set contains actions captured by two cameras and remaining data are left for testing. (*ii*) *cross-subject:* action clips performed by 20 subjects are randomly picked for training and another 20 subjects are reserved for testing.

4.2 Training Details

Following [40], we set the action sequence length for both datasets to be 50. The image discriminator randomly selects 20 frames from every generated sequence and training sequences as the input. The *SA-GC* layer selects top 5 past frames to construct an adjacency matrix $\tilde{\mathbf{A}}_\mathbf{s}$. We set batch size for training to be 100, for testing to be 1000, and the learning rate to be 0.0002.

4.3 Evaluation Metrics

Maximum Mean Discrepancy. The MMD metric is based on a two-sample test to measure the similarity between two distributions $\mathcal{P}(x)$ and $\mathcal{Q}(y)$, based on samples $x \sim \mathcal{P}(x)$ and $y \sim \mathcal{Q}(y)$. It is widely used to measure the quality of generated samples compared with real data in deep generative model [49] and Bayesian sampling [14]. The metric has also been applied to evaluate the similarity between generated actions and the ground truth in [34,40], which has been proved consistent with human evaluation. As motion dynamics are in the form of sequential data points, we denote MMD_{avg} as the average MMD over each frame and MMD_{seq} to denote the MMD over whole sequences.

Recognition Accuracy. Apart from using MMD to evaluate the model performance, we also pre-train a recognition network on the training data to compute the classification accuracy of generated samples. The recognition network exactly follows the video discriminator except for the last *softmax* layer. This evaluation metrics can examine whether the conditional generated samples are actually residing in the same manifold as the ground truth and can be correctly recognized. Details are given in the Appendix.

4.4   Baselines

We compare our method with six baselines. We first consider the model in [42], which can be used to generate long-term skeleton-based actions in an end-to-end manner. This includes three training alternatives: end-to-end (*E2E*), E2E prediction with visual analogy network (*EPVA*) and EPVA with adversarial loss (*adv-EPVA*). The second baseline [13] is based on VAE, called the *SkeletonVAE*, which improves previous motion generation methods significantly. Finally, two most recent strong baselines are considered, including the previous state-of-the-art method [7] and an improved version [40] with an auxiliary classifier. The latter utilizes a *Single Pose Training* stage and a *Pose Sequence Generation* stage to produce high-quality motions. These two baselines are respectively referred to as *SkeletonGAN* and *c-SkeletonGAN*.

4.5   Detailed Results

Quantitative Results. Our *SA-GCN* model shows superior quantitative results in terms of both MMD and recognition accuracies on the two datasets, compared with related baseline models.

Human-3.6m. Table 1 shows MMD results of our model and the baselines on Human-3.6m. With structure information considered, our model achieves significant performance gains over all baselines, which even without the need of an inefficient pre-training stage. The recognition accuracies are reported in Table 2. Similarly, our model consistently outperforms three baselines by a large margin. Please note none information of the generated actions are used in the pretrained classifier, thus we advocate that the relatively low recognition accuracies are indeed reasonable. On the other hand, this also indicates that existing action generation models are still far from satisfactory.

NTU RGB+D. This dataset is more challenging, which contains more body joints and action variations. In the experiments, we find that three models (E2E, EPVA, adv-EPVA [42]) fail to generate any interpretable action sequences. As a result, we only present MMD results for the other three baselines in Table 3. Again, the proposed method performs the best among all models under *cross-view* and *cross-subject* settings.

Fig. 6. Randomly selected samples on NTU RGB+D dataset. Top: *sitting down* from *cross-subject*, Bottom: *phoning* from *cross-view*.

Table 1. Model comparisons in terms of MMD on Human-3.6m.

Models	Pretrain	MMD$_{avg}$ ↓	MMD$_{seq}$ ↓
E2E [42]	No	0.991	0.805
EPVA [42]	No	0.996	0.806
adv-EPVA [42]	No	0.977	0.792
SkeletonVAE [13]	No	0.452	0.467
SkeletonGAN [7]	Yes	0.419	0.436
c-SkeletonGAN [40]	Yes	0.195	0.218
Ours	No	**0.146**	**0.134**

Table 2. Action recognition accuracy on the generated actions on the Human-3.6m.

Models	Direct	Discuss	Eat	Greet	Phone	Pose	Sit	SitD	Smoke	Walk	Average
SkeletonVAE	0.37	0.01	0.51	0.47	0.10	0.03	0.17	0.33	0.01	0.01	0.201
SkeletonGAN	0.35	0.29	0.72	0.66	0.46	0.09	0.32	0.71	0.14	0.02	0.376
c-SkeletonGAN	0.34	0.44	0.57	0.56	0.52	0.25	0.67	1.00	0.50	0.03	0.488
SA-GCN	0.42	0.40	0.78	0.55	0.72	0.61	0.95	0.79	0.52	0.18	**0.593**

Table 3. Model comparisons in terms of MMD on NTU RGB+D.

Models	Cross-view		Cross-subject	
	MMD$_{avg}$ ↓	MMD$_{seq}$ ↓	MMD$_{avg}$ ↓	MMD$_{seq}$ ↓
SkeletonVAE [13]	1.079	1.205	0.992	1.136
SkeletonGAN [7]	0.999	1.311	0.698	0.788
c-SkeletonGAN [40]	0.371	0.398	0.338	0.402
SA-GCN	**0.316**	**0.335**	**0.285**	**0.299**

Qualitative Results. We present some generated actions in Human-3.6m dataset and NTU RGB+D dataset in Fig. 7(first and third row) and Fig. 6 respectively. It is easy to see that our model can generate very realistic and easily recognizable actions. We also plot action trajectories on a projected space by t-SNE [23] for each generated action class on the Human-3.6m dataset in Fig. 9. It is observed that a group of actions, *i.e.*, *directions, discussion, greeting*, are close to each other, and so is the group *sitting, sitting down, eating*; while actions *smoking* and *sitting down* are far away. These are consistent with what we have observed in the ground truth.

Smooth Action Generation. Humans are capable of switching two actions very smoothly and naturally. For instance, a person can show others directions and walking at the same time. In this part, we verify that our model is expressive enough to perform such transitions as humans do. We use (8) to produce a smooth action transition between action classes y_1 and y_2 with a smoothing parameter $\lambda \in [0,1]$. We generate 100 video clips with every mix and apply t-SNE [23] to project the averaged sequences to a 1D manifold. The histogram of various mixed actions is shown in Fig. 8. As we decrease λ, the mode (action) gradually moves from *directions* towards *walking*, meaning that our model can produce very smooth transitions when interpolating between the two actions. Figure 7 illustrates as randomly selected samples.

$$y_{mix} = \lambda y_1 + (1 - \lambda)y_2; \quad x_{mix} = G(z; y_{mix}), \ z \sim \mathcal{N}(0,1) \tag{8}$$

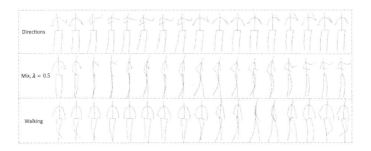

Fig. 7. Generated sequences of *directions, walking* and a mixed action with $\lambda = 0.5$.

4.6   Ablation Study

Our key innovation in our model is the *SA-GC* layer. As a result, we conduct detailed experiments to verify the effectiveness and usefulness of our self-attention based graph convolutional layer on Human 3.6m dataset. Since the self-attention layer has already been proved to be effective for sequential data, we keep the self-attention layer for all the following baselines. Without special mentioning, we keep all the other parts of the model to be the same.

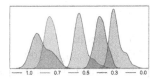

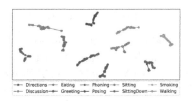

Fig. 8. Histogram of mixed actions where each mode represents an action with a smoothing term λ.

Fig. 9. Action trajectories on Human-3.6m.

Baseline 1: Replace GCN Layers with CNN Layers. We replace 5 GCN layers with 5 CNN layers using the same hidden dimension and kernel size.

Baseline 2: Without the Inter-Frame A Matrix. Based on our model, we drop the attention connections to past frames. That setting is the same as setting our top **k** to be 0 in our *SA-GC* layer. Under this Baseline, each frame in the sequence will be an independent graph for graph convolutional layer.

Baseline 3: Replace Self-Attention Based GCN Layers with the ST-GC Layers [44]. The *ST-GC* layer leverage graph convolution for skeleton-based action recognition. Each *ST-GC* layer combines one graph convolution layer for learning intra-frame skeleton and one 1D convolutional layer for feature aggregation in the temporal space.

The *Fully Connected* model described in Fig. 1 is not applicable and can not scale to long sequences because it demands excessive amount of memory and computational resources. The results of above three baselines are shown in Table 4. Comparing with baseline2 and baseline3, we can see that adding the adjacency matrix makes the model harder to train compared with CNN. However, our proposed self-attention can mitigate the difficulties and surpass standard CNN method on the skeleton based action generation task with much lower MMD scores.

Table 4. Ablation study results.

Baselines	MMD$_{avg}$ ↓	MMD$_{seq}$ ↓
Baseline 1	0.240	0.222
Baseline 2	0.915	0.922
Baseline 3	0.580	0.595
Ours	**0.152**	**0.142**

Table 5. AMT evaluations

Models	Evaluation Score ↑
SkeletonVAE	2.401
SkeletonGAN	2.731
c-SkeletonGAN	3.157
SA-GCN	**3.925**

4.7 Human Evaluation

We finally conduct perceptual human evaluations in the AMT platform. Four models are trained on the Human-3.6m dataset, including *SkeletonVAE*, *Skeleton-GAN*, *c-SkeletonGAN* and our *SA-GCN*. We then sample 100 action clips for each of the 10 action classes; 140 workers were asked to evaluate the quality of the generated sequences and score them in a range from 1 to 5. A higher score indicates a more realistic action clip. We only inform them of the action class and one real action clip to ensure proper judgements. The design detail is given in the Appendix. Table 5 demonstrates that our model is significantly better than other baselines in human evaluation.

5 Conclusions

In this paper, we have presented the self-attention graph convolutional network (*SA-GCN*) to efficiently encode structure information into skeleton-based human action generation. Self-attention can capture long-range dependencies in continuous action sequences and learn to prune the dense action graph for efficient training. Further, the graph convolution is applied to seamlessly encode both spatial joints information and temporal dynamics information into the model. Based on these ideas, our model directly transforms noises to high-quality action sequences and can be trained end-to-end. On two standard human action datasets, we observe a significant improvement of generation quality in terms of both quantitative and qualitative evaluations.

References

1. Bahdanau, D., Cho, K., Bengio, Y.: Neural machine translation by jointly learning to align and translate. arXiv preprint arXiv:1409.0473 (2014)
2. Barsoum, E., Kender, J., Liu, Z.: HP-GAN: probabilistic 3D human motion prediction via GAN (2017). https://arxiv.org/abs/1711.09561
3. Barsoum, E., Kender, J., Liu, Z.: HP-GAN: probabilistic 3D human motion prediction via GAN. In: Proceedings of the IEEE Conference on Computer Vision and Pattern Recognition Workshops, pp. 1418–1427 (2018)
4. Bissacco, A., Soatto, S.: Hybrid dynamical models of human motion for the recognition of human gaits. Int. J. Comput. Vis. **85**(1), 101–114 (2009). https://doi.org/10.1007/s11263-009-0248-7
5. Brock, A., Donahue, J., Simonyan, K.: Large scale GAN training for high fidelity natural image synthesis. arXiv preprint arXiv:1809.11096 (2018)
6. Bruna, J., Zaremba, W., Szlam, A., LeCun, Y.: Spectral networks and locally connected networks on graphs. arXiv preprint arXiv:1312.6203 (2013)
7. Cai, H., Bai, C., Tai, Y.-W., Tang, C.-K.: Deep video generation, prediction and completion of human action sequences. In: Ferrari, V., Hebert, M., Sminchisescu, C., Weiss, Y. (eds.) ECCV 2018. LNCS, vol. 11206, pp. 374–390. Springer, Cham (2018). https://doi.org/10.1007/978-3-030-01216-8_23
8. Chen, X., Mishra, N., Rohaninejad, M., Abbeel, P.: PixelSNAIL: an improved autoregressive generative model. arXiv preprint arXiv:1712.09763 (2017)

9. Clark, A., Donahue, J., Simonyan, K.: Efficient video generation on complex datasets. arXiv preprint arXiv:1907.06571 (2019)
10. Du, Y., Wang, W., Wang, L.: Hierarchical recurrent neural network for skeleton based action recognition. In: CVPR (2015)
11. Goodfellow, I., et al.: Generative adversarial nets. In: Advances in Neural Information Processing Systems, pp. 2672–2680 (2014)
12. Gretton, A., Borgwardt, K.M., Rasch, M.J., Schölkopf, B., Smola, A.: A kernel two-sample test. J. Mach. Learn. Res. **13**(Mar), 723–773 (2012)
13. Habibie, I., Holden, D., Schwarz, J., Yearsley, J., Komura, T., et al.: A recurrent variational autoencoder for human motion synthesis (2017)
14. Han, J., Liu, Q.: Stein variational gradient descent without gradient. arXiv preprint arXiv:1806.02775 (2018)
15. He, K., Zhang, X., Ren, S., Sun, J.: Deep residual learning for image recognition. In: Proceedings of the IEEE Conference on Computer Vision and Pattern Recognition, pp. 770–778 (2016)
16. Ionescu, C., Papava, D., Olaru, V., Sminchisescu, C.: Human 3.6m: large scale datasets and predictive methods for 3D human sensing in natural environments. IEEE Trans. Pattern Anal. Mach. Intell. **36**(7), 1325–1339 (2013)
17. Ke, Q., Bennamoun, M., An, S., Sohel, F., Boussaid, F.: A new representation of skeleton sequences for 3D action recognition. In: CVPR (2017)
18. Kiasari, M.A., Moirangthem, D.S., Lee, M.: Human action generation with generative adversarial networks. arXiv preprint arXiv:1805.10416 (2018)
19. Kim, T.S., Reiter, A.: Interpretable 3D human action analysis with temporal convolutional networks. In: BNMW CVPR (2017)
20. Kingma, D.P., Welling, M.: Auto-encoding variational bayes. arXiv preprint arXiv:1312.6114 (2013)
21. Kipf, T.N., Welling, M.: Semi-supervised classification with graph convolutional networks. arXiv preprint arXiv:1609.02907 (2016)
22. Li, C., et al.: Alice: towards understanding adversarial learning for joint distribution matching. In: Advances in Neural Information Processing Systems, pp. 5495–5503 (2017)
23. van der Maaten, L., Hinton, G.: Visualizing data using t-SNE. J. Mach. Learn. Res. **9**(Nov), 2579–2605 (2008)
24. Martinez, J., Black, M.J., Romero, J.: On human motion prediction using recurrent neural networks. In: CVPR (2017)
25. Mirza, M., Osindero, S.: Conditional generative adversarial nets. arXiv preprint arXiv:1411.1784 (2014)
26. Niepert, M., Ahmed, M., Kutzkov, K.: Learning convolutional neural networks for graphs. In: International Conference on Machine Learning, pp. 2014–2023 (2016)
27. Oh, S.M., Rehg, J.M., Balch, T., Dellaert, F.: Learning and inference in parametric switching linear dynamic systems. In: Tenth IEEE International Conference on Computer Vision, ICCV 2005 Volume 1, vol. 2, pp. 1161–1168. IEEE (2005)
28. Pavlovic, V., Rehg, J.M., MacCormick, J.: Learning switching linear models of human motion. In: Advances in Neural Information Processing Systems, pp. 981–987 (2001)
29. Shahroudy, A., Liu, J., Ng, T.T., Wang, G.: NTU RGB+ D: a large scale dataset for 3D human activity analysis. In: Proceedings of the IEEE Conference on Computer Vision and Pattern Recognition, pp. 1010–1019 (2016)
30. Taylor, G.W., Hinton, G.E., Roweis, S.T.: Two distributed-state models for generating high-dimensional time series. J. Mach. Learn. Res. **12**(Mar), 1025–1068 (2011)

31. Tulyakov, S., Liu, M.Y., Yang, X., Kautz, J.: MoCoGAN: decomposing motion and content for video generation. In: Proceedings of the IEEE Conference on Computer Vision and Pattern Recognition, pp. 1526–1535 (2018)
32. Urtasun, R., Fleet, D.J., Geiger, A., Popović, J., Darrell, T.J., Lawrence, N.D.: Topologically-constrained latent variable models. In: Proceedings of the 25th International Conference on Machine Learning, pp. 1080–1087 (2008)
33. Vaswani, A., et al.: Attention is all you need. In: Advances in Neural Information Processing Systems, pp. 5998–6008 (2017)
34. Walker, J., Marino, K., Gupta, A., Hebert, M.: The pose knows: video forecasting by generating pose futures. In: Proceedings of the IEEE International Conference on Computer Vision, pp. 3332–3341 (2017)
35. Wang, J.M., Fleet, D.J., Hertzmann, A.: Optimizing walking controllers. In: ACM SIGGRAPH Asia 2009 Papers, pp. 1–8 (2009)
36. Wang, J.M., Fleet, D.J., Hertzmann, A.: Optimizing walking controllers for uncertain inputs and environments. ACM Trans. Graph. (TOG) **29**(4), 1–8 (2010)
37. Wang, T.C., et al.: Video-to-video synthesis. In: NeurIPS (2018)
38. Wang, X., Girshick, R., Gupta, A., He, K.: Non-local neural networks. In: Proceedings of the IEEE Conference on Computer Vision and Pattern Recognition, pp. 7794–7803 (2018)
39. Gui, L.-Y., Wang, Y.-X., Liang, X., Moura, J.M.F.: Adversarial geometry-aware human motion prediction. In: Ferrari, V., Hebert, M., Sminchisescu, C., Weiss, Y. (eds.) ECCV 2018. LNCS, vol. 11208, pp. 823–842. Springer, Cham (2018). https://doi.org/10.1007/978-3-030-01225-0_48
40. Wang, Z., et al.: Learning diverse stochastic human-action generators by learning smooth latent transitions. arXiv preprint arXiv:1912.10150 (2019)
41. Wang, Z., Chai, J., Xia, S.: Combining recurrent neural networks and adversarial training for human motion synthesis and control. IEEE Trans. Vis. Comput. Graph. (2019)
42. Wichers, N., Villegas, R., Erhan, D., Lee, H.: Hierarchical long-term video prediction without supervision. arXiv preprint arXiv:1806.04768 (2018)
43. Xu, K., Li, C., Tian, Y., Sonobe, T., Kawarabayashi, K., Jegelka, S.: Representation learning on graphs with jumping knowledge networks. arXiv preprint arXiv:1806.03536 (2018)
44. Yan, S., Xiong, Y., Lin, D.: Spatial temporal graph convolutional networks for skeleton-based action recognition. In: Thirty-Second AAAI Conference on Artificial Intelligence (2018)
45. Yan, Y., Xu, J., Ni, B., Zhang, W., Yang, X.: Skeleton-aided articulated motion generation. In: Proceedings of the 25th ACM International Conference on Multimedia, pp. 199–207 (2017)
46. Zhang, H., Goodfellow, I., Metaxas, D., Odena, A.: Self-attention generative adversarial networks. In: International Conference on Machine Learning, pp. 7354–7363 (2019)
47. Zhang, Y., Qi, P., Manning, C.D.: Graph convolution over pruned dependency trees improves relation extraction. arXiv preprint arXiv:1809.10185 (2018)
48. Zhao, Y., Li, C., Yu, P., Gao, J., Chen, C.: Feature quantization improves GAN training. In: ICML (2020)
49. Zhao, Y., Zhang, J., Chen, C.: Self-adversarially learned Bayesian sampling. Proc. AAAI Conf. Artif. Intell. **33**, 5893–5900 (2019)
50. Zhou, J., et al.: Graph neural networks: a review of methods and applications. arXiv preprint arXiv:1812.08434 (2018)

Towards Efficient Coarse-to-Fine Networks for Action and Gesture Recognition

Niamul Quader[✉], Juwei Lu, Peng Dai, and Wei Li

Huawei Noah's Ark Lab, Toronto, Canada
{niamul.quader1,juwei.lu,peng.dai,wei.li.crc}@huawei.com

Abstract. State-of-the-art approaches to video-based action and gesture recognition often employ two key concepts: First, they employ multistream processing; second, they use an ensemble of convolutional networks. We improve and extend both aspects. First, we systematically yield enhanced receptive fields for complementary feature extraction via coarse-to-fine decomposition of input imagery along the spatial and temporal dimensions, and adaptively focus on training important feature pathways using a reparameterized fully connected layer. Second, we develop a 'use when needed' scheme with a 'coarse-exit' strategy that allows selective use of expensive high-resolution processing in a data-dependent fashion to retain accuracy while reducing computation cost. Our C2F learning approach builds ensemble networks that outperform most competing methods in terms of both reduced computation cost and improved accuracy on the Something-Something V1, V2, and Jester datasets, while also remaining competitive on the Kinetics-400 dataset. Uniquely, our C2F ensemble networks can operate at varying computation budget constraints.

Keywords: Action and gesture recognition · Spatiotemporal coarse to fine decomposition · Weight reparameterization · Budgeted computation

1 Introduction

The human visual system appears to process information across multiple spatial and temporal scales in a coarse-to-fine (C2F) fashion [22]. A key advantage of such a strategy is that perception can be achieved without reliance on the most expensive high resolution processing, as analysis can exit when the needed level of precision has been achieved. Additionally, if all actions do not require equal resolutions, there is no reason to use equal input resolutions to identify all actions.

Electronic supplementary material The online version of this chapter (https://doi.org/10.1007/978-3-030-58577-8_3) contains supplementary material, which is available to authorized users.

A. Vedaldi et al. (Eds.): ECCV 2020, LNCS 12375, pp. 35–51, 2020.
https://doi.org/10.1007/978-3-030-58577-8_3

Although deep learning approaches have achieved enormous success in reliably recognizing action/gesture using ConvNets, e.g. [3,4,9,17,28,34,36,38,42,44] and at very low latency, e.g. [13,17,36,44], all of these approaches still require large computation costs (FLOPs) and energy at runtime. For instance, one of the most efficient and high performing deep Convolutional neural network (ConvNets) for action/gesture recognition [17] required 33 GFLOPs of computations for a single recognition. With this method, if we estimate a 6 GFLOPs per Joule [1] energy consumption, then gesture/action recognition at every 0.1 s interval for around 15 min will completely deplete a powerful mobile battery with a capacity of 50,000 Joules. Additionally, such ConvNets cannot adapt to varying FLOPs constraints (e.g. mobile switching from normal mode to battery saver mode) - a problem that prior works attempt to address by retraining a family of ConvNets with the same baseline (e.g. TSM-ResNet50 family [17] - each of the ConvNets in the family operates on different discrete FLOPs). Overall, there is a need for action/gesture recognizing ConvNets that can operate efficiently and accurately at continuously varying computational cost constraints.

Based on this motivation, we present the first C2F cascaded ensemble network for action/gesture (Fig. 1). At inference, the C2F network starts with attempting to recognize action/gesture using a coarse resolution version of the input video clip. This coarse stream acts as a fast recognition pathway requiring a small number of computations. Only when the coarse pathway is not confident at recognizing an action/gesture, does the next finer resolution pathway gets invoked in a C2F manner (Fig. 1).

While computation costs can be decreased with a C2F scheme built on a baseline ConvNet (e.g. baseline TSM [17] used in each of the C2F pathways), identifying a C2F feature fusion scheme and a multi-pathway loss formulation that enable the C2F ensemble to have better accuracy than the baseline ConvNet is challenging. This is likely due to the following two reasons: First, in contrast to prior multi-scale ConvNets on action/gesture recognition (e.g. SlowFast [4]), the coarse pathways of the C2F are downsampled input videos containing only a subset of information compared to the input at finest pathway - so, it is more difficult to extract complementary information from the coarser inputs; second, the choices on C2F feature fusion strategy and multi-pathway loss formulation can cause some pathways dominating over the rest in the C2F ensemble (e.g. bidirectional feature fusion leading to spatial stream dominating motion stream in [5]). To address the above challenges, we propose the following:

1. A sub-network that fuses features from the ends of each of the C2F pathways using a reparameterized fully connected (FC) layer that adaptively excites gradient flow along the more important feature pathways during training.
2. An end-to-end differentiable multi-loss formulation to train the C2F network.
3. Another multi-loss formulation to train the C2F network when the baseline finest pathway ConvNet is already trained.

Using the above technical contributions, this paper shows the potency of C2F networks in action/gesture recognition by providing the first set of C2F benchmarks that considerably improve baseline ConvNet performances (i.e., improved accuracy and reduced computation cost). Additionally, we implement a coarse-exit scheme that is controllable by a hyperparameter, and implement a controller that adjusts the hyperparameter such that the C2F network operates at a computation budget assigned externally. The ability to operate continuously on a cost-accuracy curve (e.g. Fig. 2) has considerable technical benefits - we no longer need to retrain different models to fit varying computational budget constraints.

2 Related Work

Multi-stream Processing in Action/Gesture Recognition: Multi-stream learning is frequently used in state-of-the-art ConvNets for action/gesture recognition (e.g. [3,5,12,14,27]). For example, the first two-stream network architecture [27] and some of its variants [3,5,6] have achieved competitive results by combining ConvNet activations that are generated from a spatial stream and a temporal stream. The spatial stream usually has a single crop of the RGB video with a small number of frames, whereas the temporal stream has the corresponding optical flow [3,5,27] or a large number of RGB frames with low resolution [4]. In contrast to multiple streams having complementary inputs, a 3D C2F decomposition of the input video results in coarse pathway inputs that are spatiotemporal downsampled signals of the finest pathway input; therefore, a C2F decomposition lacks complementary information at different streams compared to other multi-stream structures [3,4,14,27]. Despite this lack of complementary information at multi-stream inputs in a C2F network, we show that C2F schemes can be very potent in action/gesture recognition by providing a number of benefits: improved accuracy, reduced computation cost, and budgeted inference.

C2F Recognition to Increase Accuracy and Reduce FLOPs: C2F-guided processing has a long history in spatial visual processing [15,16,24,25] - collecting features from coarser pathways computed at a cheaper cost and fusing them with finer pathways can increase accuracy without adding much computational burden. These potential improvements in performance come from extracting features at multiple scales and establishing a hierarchy of structural features extracted at those multiple scales [15,16]. In action/gesture recognition, many research investigated use of multiple temporal and/or spatial resolutions for increasing accuracy [4,14,17,36,42]; however, none have looked into the 3D spatiotemporal decomposition of the input video for C2F feature extraction. This could be likely due to the lack of complementary information in C2F pathways and due to the difficulty in formulating multi-loss joint optimization for multi-stream pathways consisting of greater than two streams. Also, use of C2F prediction cascades [11,29,35,39] and dynamic inference approaches [18,40] to reduce computational complexity is common in object classification/detection tasks; however, their use in action/gesture recognition is rare. Overall, the utility of combining considerably cheaper 3D spatio-temporal coarse pathways and expensive finer scale pathways in a cascade setting and the implementation of budgeted inference are missing in action/gesture recognition tasks.

3   Technical Approach

We begin by documenting the layout for a generic C2F network (Fig. 1). Each of the streams in the C2F ensemble network can be any ConvNet (e.g. TSM [17], R3D [33], etc.) that works on a clip of video with either dense sampling [33,41] or strided sampling [17,36,42] and recognizes actions/gestures. The C2F ensemble can be composed of an arbitrarily large number of C2F pathways (N). In this paper, we use $N = 3$ and we call the coarsest pathway C, the finer pathway M and the finest pathway F (Fig. 1). We decompose input video spatiotemporally that compensates for decaying effective receptive field of convolution kernels [20] (Sect. 3.1), fuse features for complementary processing between the C2F pathways (Sect. 3.2), and formulate two joint optimization schemes for the alternative scenerio where the baseline ConvNet (i.e. the ConvNet at the finest pathway) is either pre-trained or not pre-trained for the action/gesture recognition task

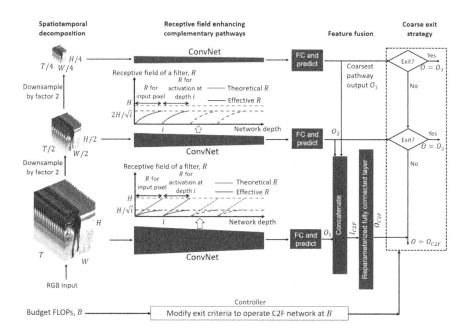

Fig. 1. The layout of our C2F network. Sampled frames of an input video are down-sampled in both the spatial and temporal domains. A filter kernel at a depth i has a progressively larger receptive field for coarser pathways relative to the original input video size. This is demonstrated here with insert graphs that assume the ConvNets used in F, M and C all have same-sized filter sizes (size >1) and that the size of the activation maps does not change at different network depth. C2F features are fused using a reparemterized FC layer. During inference, a coarse-exit scheme encourages recognition output (O) at coarser scales. Additionally, for a given computation budget B, a controller modifies the coarse-exit decision criteria such that C2F operates close to the given budget.

at hand (Sect. 3.3 and 3.4). For C2F inference, we implement a C2F coarse-exit scheme with a controllable hyperparameter and a controller that can operate the C2F network in a given budget computation cost (Sect. 3.5).

3.1 Enhanced Receptive Fields via Spatiotemporal Decomposition

A recent study showed that using the full image rather than center cropping improves accuracy [17] suggesting that pixels in a frame that are located far away from the center could provide valuable information as well and that enhancing receptive field of convolution kernels can be important for action/gesture recognition. Therefore, we propose to systematically enhance receptive fields by spatiotemporally downsizing the input video clip in a C2F manner while compensating for the decreased resolution in the coarser pathways by complementary processing of higher resolution inputs at the finer pathways (Fig. 1). This enhanced receptive field also partially compensates for the $1/\sqrt{(i)}$ decaying effective receptive field in ConvNets [20] where i is the number of layers between the input and the convolution kernel. An illustration of compensating for the decaying receptive field is shown in Fig. 1.

3.2 Feature Fusion Using Reparameterized FC Layer

We concatenate the pre-softmax outputs from each of the C2F pathways (Fig. 1), and fuse these accumulated features (I_{C2F}) through a reparameterized FC layer that adaptively excites gradient flow along the more important features of I_{C2F} during training. Intuitively, for a particular output node of the C2F ensemble (e.g. output node representing 'moving hand from right to left' gesture), there are likely some I_{C2F} features that are more important than others (e.g., corresponding nodes for 'moving hand', 'right to left', 'left to right' in each of the C2F pathways more important than the other nodes). Adaptively exciting gradient flow along these important nodes leads to larger gradient backpropagation along all the learnable ConvNet parameters that contributed to these nodes. We achieve this by modifying each of the weights (w) of the FC layer such that:

$$w_r = 0.5 \times [((2 - \beta) \times w)^{\circ 1} + (\beta \times w)^{\circ 3}], \tag{1}$$

$$\nabla_{w_{new}} = 0.5 \times [(2 - \beta) \times \nabla_w) + 2\beta^3(w^{\circ 2} \times \nabla_w)], \tag{2}$$

where $\circ$ denotes Hadamard power, ∇_w is the backpropagated gradient on w if the above reparameterization was not applied and β is a hyperparameter. $\beta \in \mathbb{R} : \beta \in [0, 2]$, and $\nabla_{w_{new}}$ is the backpropagated gradient of w. For any $\beta > 0$, w_r will have relatively larger magnitude values for larger magnitude w values, and the backpropagated gradient ($\nabla_{w_{new}}$) for higher valued weights will also be higher. Higher values of β will further encourage this asymmetrical gain in magnitude, and at $\beta = 0$ this asymmetrical gain in magnitude disappears. Notably, since this weight reparameterization is only done during training, no computation cost is added during inference compared to a normal FC layer.

3.3 Multi-loss Paradigm of C2F

The C2F network need to be trained such that each pathway becomes reliable for action/gesture recognition by itself and also provides complementary features for use by the finer pathways. We do this by formulating a joint optimization formulation with the multi-loss function,

$$L = \sum_{n=1}^{N} \alpha L_n + (1 - \alpha)L_{C2F}, \tag{3}$$

where L_n and L_{C2F} are the softmax cross-entropy losses comparing ground truth $\hat{O}$ with O_n and O_{C2F}, respectively, O_n is the output at pathway n, O_{C2F} is the output after the reparameterized FC layer, and $\alpha \in \mathbb{R} : \alpha \in [0, 1]$. A high value of α will have the C2F ensemble network focus only on optimizing each of the pathways, whereas a low value of α will have the C2F network focus more on extracting complementary information for improving O_{C2F}. Since we are motivated to improve performances of coarser pathways so we can exit early and save computation costs, we use a high value of $\alpha = 0.9$. Also, note that Eq. (3) is differentiable, so our joint optimization method is end-to-end differentiable and can be trained together.

3.4 Multi-loss Paradigm of C2F with Pre-trained F

With availability of high performing networks that are already trained on large datasets such as the Kinetics dataset [3], we propose extending the student-teacher learning paradigm [10] to a classroom learning paradigm - here, F (i.e. the finest pathway ConvNet) teaches the coarser pathways, and additionally a classroom (the reparameterized FC layer layer of C2F) learns both from the students and the teacher to perform better than F. Similar to a student-teacher learning, the F sub-network is no longer trained and is only used for teaching the students (i.e., the coarser pathways). To optimize all C pathways and the reparameterized FC layer, we train the C2F network by minimizing the following multi-loss function:

$$L_d = \sum_{n=1}^{N} \alpha L_{n,KLD} + \sum_{n=1}^{N} (1 - \alpha/2)L_n + (1 - \alpha/2)L_{C2F}, \tag{4}$$

where $L_{n,KLD}$ is the Kullback-Leibler divergence between the distributions of p_n/τ and p_F/τ, p_n is the softmax output of the nth scale, p_F is the softmax output of F, τ is a temperature parameter that softens the distributions between p_F and p_n [10], and α is a hyperparameter [10]. The primary difference in L_d from the original knowledge distillation scheme [10] is the $(1 - \alpha/2)L_{C2F}$ term that encourages each of the coarse scales to provide some complementary information to F that may help in improving overall performance of O_{C2F}.

3.5 Coarse-Exit and Budgeted Inference

For fast inference, C2F includes a coarse-exit decision stage that encourages using low average computation costs while retaining high accuracy. Inference starts with forward propagation along the coarsest stream (Fig. 1). To ensure that recognition at this coarsest stream is performed accurately, action/gesture recognition is done only when the softmax output $s_N > T$, where s_N is the maximum value in p_N and T is a hyper-parameter controlled externally. However, since softmax outputs tend to be too confident in their predictions and since they are not well-calibrated for uncertainty measures [8], we adjust s_N using a global training accuracy context as follows:

$$s_N^C = 1 - (1 - s_N) \times e_N/e_1, \tag{5}$$

where e_N is the training misclassification rate for the coarsest scale and e_1 is the training misclassification rate at the end of the C2F ensemble. This coarse-exit strategy is repeated and finer streams in the ensemble are only invoked when the coarse-exit in the coarser stream fails (Fig. 1).

Inference stage C2F can also have a budget FLOPs ($B \in \mathbb{R} : B \in [f_C, f_{C2F}]$) as input, where f_C is the FLOPs of C pathway and f_{C2F} is the computation cost of the C2F ensemble for a single recognition. We implement a controller that continuously modifies T as:

$$T = T_{av} + (B - f_{av}) * (f_{av} - f_{C2F})/(T_{av} - 1.0), \tag{6}$$

where f_{av} is the running average FLOPs and T_{av} is the average of previous r recognitions (default $r = 100$). Only when C2F is operating at desired budget (i.e. $B - f_{av} = 0$), T does not update from T_{av}.

3.6 Protocols

Our C2F architecture is generic and can be applied to different baseline ConvNets. In this work, we use C2F with TSM [17] as its baseline ConvNet. We chose TSM since it has been dominating the leaderboard in the Something-Something V1 and V2 datasets [31,32], and is considerably more efficient compared to most competitive ConvNets for action/gesture recognition [17]. We investigate improvements achieved by extending TSM baseline to a C2F architecture on four standard action/gesture recognition datasets, and compare them against the improvements achieved by applying the popular non-local blocks (NL) [37] with TSM baseline. Additionally, we compare C2F-extended TSM with other competitive ConvNets on Jester and Something-Something V1, V2 datasets.

Datasets: For gesture recognition, we evaluate C2F ensemble networks on the Jester dataset [30], which is currently the largest publicly available dataset for gesture recognition. It has ∼119K training videos, ∼15K validation videos, and a total of 27 different gesture and no-gesture classes. For action recognition, we train and evaluate C2F on the Something-Something V1 and V2 datasets

[7], and the Kinetics-400 dataset [3]. The Something-Something V1 dataset has ~86K training videos and ~12K validation videos of humans interacting with everyday objects in some pre-defined action-sets (e.g. pushing some object from left to right, etc.). The V2 dataset expands on the V1 dataset with a total of ~169K training videos and ~25K validation videos. Activity recognition in these videos is challenging since it requires identifying objects as well as the sets of movements resulting in a particular interaction. We also evaluate C2F on the Kinetics-400 [3], which is another large-scale dataset with 400 classes of actions that are less sensitive to temporal movements of objects [17].

Training and Testing: F in our C2F ensemble networks are instantiated with TSM-ResNet50 [17]. We denote a trained C2F that has not used action-pretrained F (i.e., not pretrained on the action recognition task at hand) at initialization as $C2F_A$, and we denote a trained C2F that used action-pretrained F and trained using the classroom learning paradigm as $C2F_C$. For the latter, F is initialized with ImageNet pretraining using only 224×224 resolution. All other pathways also use the same ImageNet pretraining, so multiple ImageNet pretraining at different scales is not necessary. Coarse pathways cause small FLOPs and memory burden on the overall $C2F_C$ architecture (see Table 1). Additionally, instantiating a coarse pathway with a heavier network causes less computational overhead compared to instantiating F with a heavier network. Due to this relatively smaller computation overhead, we also investigate heavier networks (TSM-ResNet101) at the coarser pathways while still using the pretrained TSM-ResNet50 as F in our $C2F_C$ architecture. For convenience, we will call this modified architecture $C2F_{C+}$. For all ConvNet results we report in this section and subsequent sections, we use subscript En to denote use of the entire ensemble network during inference and we use subscript Ex to denote use of coarse-exit scheme during inference.

We sample 16 strided frames for the Something-Something and Jester datasets, and use 8 densely sampled frames for the Kinetics dataset similar to [17]. For the coarser pathways we use 8 frames. We do not decompose the spatiotemporal video to less than 8 frames, since we found empirically that using 4 frames or less can deteriorate C pathway performances. We used data augmentation similar to [36], used a batch size of 64, and optimized with stochastic gradient descent [23] having momentum 0.9 and a weight decay of 5e-4. For the classroom learning, F is not trained any further from its pretrained form. Training was done for 50 epochs - here, learning rate was initialized at 0.02 and reduced thrice with a factor of 10^{-1} after 20th, 40th and 45th epochs. To prevent overfitting, we also added extra dropout layers with dropout rate of 0.5 before each of the final FC layers in O_n. To compare with TSM baseline (Table 1), we use the single center-crop evaluation followed in [17] - here, the center crop has 224/256 of the shorter side of a frame. To compare with other methods (Table 2), we follow [13,37] where the center-crop has the entire shorter side which is then resize to 224×224 for Something-Something V1/V2 and Kinetics datasets and resized to 128×128 for the Jester dataset. We note that single center-crop action/gesture recognition accuracy are likely to approximate accuracy in practical settings that will likely not use multiple crops to ensure efficiency.

We also report multi-crop evaluations using settings used in [17] for each of the datasets. On all experiments, we use $\beta = 1.0$ for the reparameterized FC layer, and $\alpha = 0.1$ and $\tau = 6.0$ for the classroom learning paradigm based on empirical observations on the Jester dataset.

Table 1. Effect of extending TSM baseline with NL [37] or our C2F architectures. Memory usage was computed based on a inference batch size of 1. The C2F ensembles perform better than TSM with NL, except on the Kinetics dataset where using C2F with TSM-NL has the highest accuracy.

Method	Inference costs and accuracy						
Baseline	Module added	Param.	Memory	FLOPs	Top-1 Acc. (%)	Top-5 Acc. (%)	$\triangle$Top-1
Something-something V1							
TSM-ResNet50 (16F-224) [17]	None	1.0×	1.0×	1.0×	47.0	76.9	+0.0
	NL [37]	1.3×	1.24×	1.2×	41.3	72.1	−5.7
	$C2F_{A,En}$	3.0×	1.19×	1.16×	47.9	77.5	0.8
	$C2F_{C,En}$	2.0×	1.12×	1.08×	**48.4**	**78.4**	**+1.4**
	$C2F_{C,Ex}$ (T = .65)	2.0×	1.12×	**0.8×**	47.4	77.7	**+0.4**
Something-something V2							
TSM-ResNet50 (16F-224) [17]	None	1.0×	1.0×	1.0×	61.5	87.5	+0.0
	NL [37]	1.3×	1.24×	1.2×	57.2	84.0	−4.3
	$C2F_{A,En}$	3.0×	1.19×	1.16×	62.1	**88.1**	+0.6
	$C2F_{C,En}$	2.0×	1.12×	1.08×	**62.4**	88.0	**+0.9**
	$C2F_{C,Ex}$ (T = .80)	2.0×	1.12×	**0.7×**	61.8	87.6	**+0.3**
Kinetics							
TSM-ResNet50 (Dense 8F-224) [17]	None	1.0×	1.0×	1.0×	69.8	88.3	+0.0
	NL [37]	1.3×	1.24×	1.2×	70.9	89.3	+1.1
	$C2F_{C,En}$	2.0×	1.12×	1.08×	70.5	88.9	+0.7
	$C2F_{C,En}$ + NL	2.3×	1.39×	1.28×	**71.4**	**90.0**	**+1.6**
	$C2F_{C,Ex}$ + NL (T =.5)	2.3×	1.39×	**0.8×**	70.1	89.0	+0.3
TSM-ResNet50 (Strided 16F-224) [17]	None	1.0×	1.0×	1.0×	72.4	90.8	+0.0
	$C2F_{C,En}$	2.0×	1.12×	1.28×	**73.5**	**91.4**	**+1.1**
	$C2F_{C,Ex}$ (T =.7)	2.0×	1.12×	**0.8×**	72.6	90.9	+0.2
Jester							
TSM-ResNet50 (16F-128) [17]	None	1.0×	1.0×	1.0×	96.3	99.7	+0.0
	NL [37]	1.3×	1.24×	1.2×	96.4	99.8	+0.1
	$C2F_{A,En}$	3.0×	1.19×	1.16×	**96.5**	**99.8**	**+0.2**
	$C2F_{C,En}$	2.0×	1.12×	1.08×	**96.5**	**99.8**	**+0.2**
	$C2F_{C,Ex}$ (T =.99)	2.0×	1.12×	**0.5×**	96.4	**99.8**	+0.1

Table 2. Comparison of $C2F_{C+,En}$ with competitive approaches that use single-crop RGB for evaluation on the Something-Something and Jester datasets. Our $C2F_{C+,En}$ and its coarse-exit scheme outperforms all competitive methods that are at FLOPs $< 50G$ and at FLOPs $> 50G$. For methods with *, we use our training settings on official implementations of the methods.

Model	Jester		Something V1			Something V2		
	FLOP	Val1 (%)	FLOP	Val1 (%)	Val5 (%)	FLOP	Val1 (%)	Val5 (%)
FLOPS < 50G								
TRN-Multiscale-BNInception [42]	-	-	16	34.4	-	16	48.8	77.6
TSN-ResNet50 [36]	-	-	33	19.7	46.6	33	30.0	60.5
ECO [44]	-	-	32	39.6	-	-	-	-
$C2F_{C+,Ex}$	**25**	**96.8**	**15**	**42.1**	**72.0**	**11**	**54.2**	**82.0**
FLOPS > 50G								
GST-ResNet50 [19]	-	-	59	48.6	77.9	59	62.6	87.9
TSM-ResNet50* [17]	65	96.4	65	47.1	77.0	65	61.6	87.7
STM-ResNet50 [13]	-	-	67	49.2	**79.3**	-	-	-
S3D-BNInception-G [41]	-	-	71	48.2	78.7	-	-	-
TSM-ResNet50$_{En}$* [17]	-	-	98	49.0	78.9	-	-	-
ABM-AC-in_{En}-ResNet50 [43]	-	-	106	46.8	-	61.2	-	
TSN-STD-ResNet50[21]	-	-	-	50.1	-	-	-	-
$C2F_{C+,Ex}$	-	-	59	**49.3**	79.1	56	**62.8**	**88.0**
$C2F_{C+,En}$	-	-	85	**50.2**	**79.9**	85	**63.8**	**88.8**

3.7 Results

Performance Improvements Over Baseline: We find consistent improvements with our C2F ensemble networks ($C2F_{A,En}$ and $C2F_{C,En}$ (Table 1), and $C2F_{C+,En}$ (Table 2)) over TSM baseline (Table 1). The accuracy improvements achieved over baseline ($\triangle Top1$) with $C2F_{C,En}$ are larger compared to $\triangle Top1$ achieved by adding NL [37] to baseline TSM (except for the Kinetics dataset, see Table 1), and the improvements come at less FLOPs and memory usage overhead. When 'TSM-ResNet50 with NL added' is used as F in the $C2F_C$ learning on the Kinetics dataset, the accuracy of the ensemble output (i.e., $C2F_{C,En}$ with NL added in F) improves further, suggesting that the beneficial effects of adding the NL block is complementary to extending a base network to C2F. We also find that the smallest $\triangle Top1$ is on the Jester dataset (with the highest baseline TSM accuracy) and the largest $\triangle Top1$ is on the Something-Something V1 dataset (with the lowest baseline TSM accuracy) - this suggests that extending a baseline network to C2F may provide more accuracy improvements for difficult datasets.

Extra FLOPs and Memory Usage: The extra FLOPs of our ensemble networks are considerably reduced when our coarse-exit (Ex) scheme is invoked - e.g. using $C2F_{C,Ex}$ with $T = 0.8$ on the Something-Something V2 dataset reduces FLOPs of $C2F_{C,Ex}$ 30% below the baseline TSM while still retaining a non-trivial accuracy improvement (Table 1). Interestingly, in contrast to the

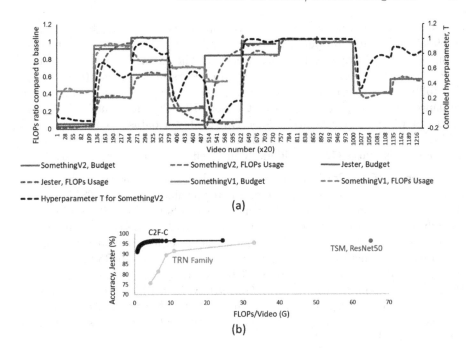

Fig. 2. (a) FLOPs of $C2F_C$ network at inference (dotted blue, green and red lines) follows closely the randomly defined budget FLOPs B that is assigned randomly at every 2500th videos for Jester and Something-Something V1 and V2 datasets (solid blue, green and red lines). The dotted black line represents the modified T values for Something-Something V2 dataset to enable the $C2F_C$ network to operate close to B. (b) $C2F_C$ network's cost-accuracy curve for Jester dataset has considerably better cost-accuracy tradeoff compared to baseline TSM and the more efficient TRN family. The various cost-accuracy operating points on the $C2F_C$ are obtained simply by adjusting the T values ($T \in \mathbb{R} : T \in [1 - e_N/e_1, 1]$) without retraining the entire model. (Color figure online)

trend in accuracy improvement vs. base accuracy, we see an opposite trend for computation cost reduction - easier datasets enjoy larger computation cost reductions (Table 1). Finally, the small additional memory usage of C2F networks are justifiable considering their ability to flexibly and continuously operate on the cost-accuracy curve, and their ability to operate at very low FLOPs when availability of computational resources are limited (Fig. 2). The flexibility to operate continuously on the cost-accuracy curve has enormous technical benefits. We no longer need to retrain different models to fit varying computational budget constraints - we are now able to flexibly control the C2F network to operate in varying cost constraints by simply modifying the parameter T (Fig. 2).

Single-crop RGB Performances: Table 2 shows single-crop RGB performance of C2F networks compared to competitive ConvNets on Jester and Something-Something V1, V2 datasets. While TSM [17] does not exhibit the best single-crop results among competitive methods, $C2F_{C+,En}$ built on TSM baseline considerably outperforms the rest (Table 2); thus, a different baseline to our C2F could bring even higher accuracy. The extra computation costs of $C2F_{C+,En}$ can be flexibly reduced using the coarse-exit scheme ($C2F_{C+,Ex}$). Controlling the hyperparameter T in $C2F_{C+,Ex}$ yields considerably higher performances than competing methods at various FLOPs constraints (e.g. FLOPs $< 50G$ and FLOPs $> 50G$ in Table 2). On the Jester dataset, C2F uses inputs having spatial dimension of 128×128 in contrast to 224×224 used in other methods - this makes C2F considerably more efficient for Jester dataset.

Multi-crop RGB Performances: We find similar accuracy improvements with $C2F_{C,En}$ implementation on multi-crop evaluations over TSM baseline (improvement from 63.1% to 64.1% on Something-Something V2 and improvement from 75.6% to 76.0% on Kinetics). Compared to the competitive multi-resolution approach SlowFast [4], $C2F_{C,En}$ has better accuracy at lower computation cost on Kinetics - 75.6% accuracy with SlowFast4 $\times$ 16R50 at computation cost of 1083 GFLOPs/video vs. 76.0% accuracy with $C2F_{C,En,NL}$ at less than half the computation cost (460GFLOPs/video). Finally, $C2F_{C+,En}$ on a 30-crops softmax-averaged evaluation on the Jester dataset yields a top-1 test accuracy of 97.09%, beating the closest competitor [26] with a misclassification rate reduction by around 14%.

3.8 Ablation Studies

We performed a series of ablation studies, and compare results of all experiments using single-crop validation accuracy. First, we compare performances of the baseline ConvNet (TSM) with alternative multi-stream fusion schemes - (i) late fusion with ensemble of classification scores similar to [4,14,17,27,36], (ii) slow fusion with lateral connections similar to [4]. Next, we separately investigate effectiveness of - (i) the C2F decomposition compared to all fine pathways, (ii) the reparameterized FC layer compared to a vanilla FC layer, (iii) the class-learning method compared to the student-teacher [10] learning method, and (iv) the effectiveness of our generic C2F on other popular baseline ConvNets for action/gesture recognition.

Effectiveness of Multi-loss Schemes: To investigate effectiveness of multi-loss schemes we compare C2F results against simple late fusion schemes with ensemble of classification scores. We investigate three simple C2F ensembling schemes: First, summing up softmax scores of each C2F pathway; second, multiplying softmax scores with training accuracy of each pathway separately and then summing up the weighted scores; third, using a product of softmax scores of each C2F pathway. On the Jester dataset, all of the above results in poorer accuracy (95.3% with additive ensemble, 96.0% with weighted additive ensemble and 95.3% with multiplicative ensemble) compared to baseline TSM with accuracy 96.3% - this could be because softmax outputs tend to be too confident in their predictions [8] such that misclassifications from coarser pathways start disabling correct recognitions at finest pathway. Additionally, end-to-end learning with multi-stream loss added and backpropagated does not improve accuracy over baseline (96.2% with such multi-loss optimization vs. 96.3% with baseline TSM).

Slow Fusion with Lateral Connections: To encourage feature reuse of each convolutional kernel, we experiment with lateral connections between adjacent C2F streams. We only investigate lateral connections sourced at coarser streams and ending at finer streams, so we can still enable coarse-exit at inference. We laterally connect activation maps from the coarse pathway (A_C) to the activation maps in the adjacent finer pathway (A_F) in a manner similar to [5], but find deteriorating accuracy by 0.05% on the Jester dataset. This failure in improving performance is perhaps due to incorporation of noisy activation maps from the coarser pathways to the finer pathways, which outweights the benefits of lateral connections. This could perhaps be resolved with concatenation of activation maps rather than naive arithmetic lateral connections; however, since we are more concerned with keeping overall network computation efficient, we defer that experiment for later.

C2F Decomposition vs. All-Fine Pathways: When the coarser pathways of $C2F_C$ are fed with non-downsampled input rather than downsampled coarse input, the improvement in accuracy compared to baseline TSM ($\triangle Top1 = 1.1$) was slightly lower than with the original $C2F_C$ ($\triangle Top1 = 1.4$), suggesting that a C2F architecture not only reduces computation burden compared to a non-downsampled scheme, but also provides beneficial complementary information for the C2F ensemble.

Utility of the Reparameterized FC Layer: Replacing our reparameterized FC layer with a regular FC layer in the $C2F_C$ ensemble networks deteriorates accuracy - the reduction in top-1 accuracy(%) is around 0.1 for Jester, around 0.3 for both Something-Something V1 and V2 datasets, and a larger 0.6 for the Kinetics dataset. Since the Kinetics dataset has the highest number of classes among the datasets we used (400 compared to 174 of Something-Something and 27 of Jester), the number of connections in the FC layer are much larger for a $C2F_C$ instantiation on Kinetics. The reparameterized FC layer is particularly effective in such instances of large numbers of FC connections since it is able to focus more on the important FC connections. For datasets with higher number of classes (e.g. Kinetics-700 [2]), the reparameterized FC layer or a variant of the reparameterized FC layer that focuses on important FC connections will likely be crucial for a successful C2F implementation.

Comparison with Student-Teacher [10] Learning Method: In contrast to a student-teacher learning method [10], our classroom learning approach enables $C2F_{C,En}$ to perform significantly better than the teacher network F (Table 1). In terms of utility in teaching the student networks (i.e., the coarser networks of $C2F_C$), we find no significant difference in the coarsest pathway accuracy when trained with the student-teacher method (using $\alpha = 0.1$ and $\tau = 6.0$) [10] and our classroom learning method on the Jester dataset (90.4% with student-teacher vs. 90.5% with $C2F_C$) and Something-Something V1 dataset (31.0%

with student-teacher vs. 30.7% with $C2F_C$). Both the classroom learning and the student-teacher learning outperforms a coarsest pathway trained separately (e.g. 90.4% with student-teacher and 90.5% with $C2F_C$ vs. 87.8% with no knowledge-distillation).

C2F Effectiveness on Baseline Other than TSM: Our C2F architectures are generic and can be easily implemented with different baseline ConvNets. While TSM uses 2D convolution-based networks with shift operations added to infuse temporal information, many ConvNets for action/gesture recognition rely on 3D convolution-based networks [3,4,34,41]. Therefore, we investigate effectiveness of C2F extension on the 3D convolution-based R3D-ResNet18 and R2plus1D-ResNet18 ConvNets [34]. $C2F_C$ with these baseline ConvNets instantiated on all three pathways outperforms the baseline ConvNets (94.2% with $C2F_C$-R3D-ResNet18 vs. 93.6% with R3D-ResNet18, 94.6% with $C2F_C$-R2plus1D-ResNet18 vs. 93.9% with R2plus1D-ResNet18).

4   Discussion

Our ablation studies showed that much of the prior popular multi-stream techniques (e.g. slow fusion with lateral connection [4], late fusion using ensemble of classification scores or end-to-end learning with multi-stream loss added and backpropagated) fail to work on C2F multi-stream ConvNets for action/gesture recognition. We conjecture that this ineffectiveness of prior multi-stream techniques primarily arise from the relative lack of complementary information at inputs of the C2F pathways compared to prior multi-stream ConvNets (e.g. complementary spatial and temporal stream in [27]). With much of the prior techniques being ineffective for C2F ConvNets on action/gesture recognition, our proposed baseline C2F ConvNet along with its effective training strategies are significant for future C2F ConvNet research on action/gesture recognition.

To discuss the importance of future C2F ConvNet research on action/gesture recognition, we review the technical benefits achieved with our proposed $C2F_C$ ConvNet. $C2F_C$ on TSM baseline can operate at unprecedented low FLOPs (<1% GFLOPs) and remain competitive with methods such as [42]. In fact, some classes of action (e.g. 'Playing squash or racquetball' in the Kinetics dataset [3]) can be recognized at near-perfect accuracy at such low FLOPs (more details in the supplementary document). When varying FLOPs constraints are imposed, rather than re-designing and training different ConvNets, our proposed $C2F_C$ can operate flexibly at the desired FLOPs, and at a better cost-accuracy trade-off than its baseline ConvNet. Finally, $C2F_C$ is generic and can be built on different ConvNets, so the benefits of $C2F_C$ remain complementary to emerging standalone ConvNets for action/gesture recognition. For all of these advantages, C2F extension approaches on action/gesture recognition are likely to become valuable techniques.

5   Conclusion

'Use when needed' is an extremely effective philosophy for improving efficiency. We introduce this philosophy in the field of action recognition using C2F ConvNets. We demonstrate that fusing complementary feature in a C2F approach with a coarse-exit scheme can significantly improve ConvNet performances. While our C2F ensembles can outperform previous competing methods in terms of both increased accuracy and reduced computation costs, the C2F ensembles are particularly potent in improving accuracy for difficult datasets, in reducing computational costs for easier datasets and in operating efficiently and flexibly at a large range of computational cost constraints.

Acknowledgement. We are very grateful to Professor Richard P. Wildes (York University) who has generously supported this work by providing critical feedback and guidance during concept formulation, early experimentations and writing of earlier drafts.

References

1. Green 500 List for June 2016. https://www.top500.org/green500/lists/2016/06/. Accessed 8 Nov 2019
2. Carreira, J., Noland, E., Hillier, C., Zisserman, A.: A short note on the kinetics-700 human action dataset. arXiv preprint arXiv:1907.06987 (2019)
3. Carreira, J., Zisserman, A.: Quo vadis, action recognition? A new model and the kinetics dataset. In: Proceedings of the IEEE Conference on Computer Vision and Pattern Recognition, pp. 6299–6308 (2017)
4. Feichtenhofer, C., Fan, H., Malik, J., He, K.: Slowfast networks for video recognition. In: Proceedings of the IEEE International Conference on Computer Vision, pp. 6202–6211 (2019)
5. Feichtenhofer, C., Pinz, A., Wildes, R.P.: Spatiotemporal multiplier networks for video action recognition. In: Proceedings of the IEEE Conference on Computer Vision and Pattern Recognition, pp. 4768–4777 (2017)
6. Feichtenhofer, C., Pinz, A., Zisserman, A.: Convolutional two-stream network fusion for video action recognition. In: Proceedings of the IEEE Conference on Computer Vision and Pattern Recognition, pp. 1933–1941 (2016)
7. Goyal, R., et al.: The "something something" video database for learning and evaluating visual common sense. In: ICCV, vol. 1, p. 3 (2017)
8. Guo, C., Pleiss, G., Sun, Y., Weinberger, K.Q.: On calibration of modern neural networks. In: Proceedings of the 34th International Conference on Machine Learning-Volume 70, pp. 1321–1330. JMLR.org (2017)
9. He, D., et al.: Stnet: local and global spatial-temporal modeling for action recognition. arXiv preprint arXiv:1811.01549 (2018)
10. Hinton, G., Vinyals, O., Dean, J.: Distilling the knowledge in a neural network. arXiv preprint arXiv:1503.02531 (2015)
11. Huang, G., Chen, D., Li, T., Wu, F., Van Der Maaten, L., Weinberger, K.Q.: Multi-scale dense convolutional networks for efficient prediction. arXiv preprint arXiv:1703.09844 (2017)

12. Huang, W., et al.: Toward efficient action recognition: principal backpropagation for training two-stream networks. IEEE Trans. Image Process. **28**(4), 1773–1782 (2018)
13. Jiang, B., Wang, M., Gan, W., Wu, W., Yan, J.: STM: spatiotemporal and motion encoding for action recognition. In: Proceedings of the IEEE International Conference on Computer Vision, pp. 2000–2009 (2019)
14. Karpathy, A., Toderici, G., Shetty, S., Leung, T., Sukthankar, R., Fei-Fei, L.: Large-scale video classification with convolutional neural networks. In: Proceedings of the IEEE Conference on Computer Vision and Pattern Recognition, pp. 1725–1732 (2014)
15. Koenderink, J.J.: The structure of images. Biol. Cybern. **50**(5), 363–370 (1984)
16. Koenderink, J.J.: Scale-time. Biol. Cybern. **58**(3), 159–162 (1988)
17. Lin, J., Gan, C., Han, S.: TSM: temporal shift module for efficient video understanding. In: Proceedings of the IEEE International Conference on Computer Vision (2019)
18. Lin, J., Rao, Y., Lu, J., Zhou, J.: Runtime neural pruning. In: Advances in Neural Information Processing Systems, pp. 2181–2191 (2017)
19. Luo, C., Yuille, A.L.: Grouped spatial-temporal aggregation for efficient action recognition. In: Proceedings of the IEEE International Conference on Computer Vision, pp. 5512–5521 (2019)
20. Luo, W., Li, Y., Urtasun, R., Zemel, R.: Understanding the effective receptive field in deep convolutional neural networks. In: Advances in Neural Information Processing Systems, pp. 4898–4906 (2016)
21. Martinez, B., Modolo, D., Xiong, Y., Tighe, J.: Action recognition with spatial-temporal discriminative filter banks. In: Proceedings of the IEEE International Conference on Computer Vision, pp. 5482–5491 (2019)
22. Mermillod, M., Guyader, N., Chauvin, A.: The coarse-to-fine hypothesis revisited: evidence from neuro-computational modeling. Brain Cogn. **57**(2), 151–157 (2005)
23. Robbins, H., Monro, S.: A stochastic approximation method. Ann. Math. Stat. **22**, 400–407 (1951)
24. Rodríguez-Sánchez, A.J., Fallah, M., Leonardis, A.: Hierarchical object representations in the visual cortex and computer vision. Front. Comput. Neurosci. **9**, 142 (2015)
25. Rosenfeld, A.: Multiresolution Image Processing and Analysis, vol. 12. Springer, Heidelberg (2013)
26. Shi, L., Zhang, Y., Hu, J., Cheng, J., Lu, H.: Gesture recognition using spatiotemporal deformable convolutional representation. In: 2019 IEEE International Conference on Image Processing (ICIP), pp. 1900–1904. IEEE (2019)
27. Simonyan, K., Zisserman, A.: Two-stream convolutional networks for action recognition in videos. In: Advances in Neural Information Processing Systems, pp. 568–576 (2014)
28. Stroud, J.C., Ross, D.A., Sun, C., Deng, J., Sukthankar, R.: D3D: distilled 3D networks for video action recognition. arXiv preprint arXiv:1812.08249 (2018)
29. Teerapittayanon, S., McDanel, B., Kung, H.T.: Branchynet: fast inference via early exiting from deep neural networks. In: 2016 23rd International Conference on Pattern Recognition (ICPR), pp. 2464–2469 (2016)
30. The 20BN-jester dataset V1: The 20BN-jester Dataset V1. Accessed 8 Nov 2019
31. The 20BN-something-something Dataset V1: The 20BN-something-something Dataset V1. Accessed 8 Nov 2019
32. The 20BN-something-something Dataset V2: The 20BN-something-something Dataset V2. Accessed 8 Nov 2019

33. Tran, D., Bourdev, L., Fergus, R., Torresani, L., Paluri, M.: Learning spatiotemporal features with 3D convolutional networks. In: Proceedings of the IEEE International Conference on Computer Vision, pp. 4489–4497 (2015)
34. Tran, D., Wang, H., Torresani, L., Ray, J., LeCun, Y., Paluri, M.: A closer look at spatiotemporal convolutions for action recognition. In: Proceedings of the IEEE Conference on Computer Vision and Pattern Recognition, pp. 6450–6459 (2018)
35. Viola, P., Jones, M.J.: Robust real-time face detection. Int. J. Comput. Vision **57**(2), 137–154 (2004)
36. Wang, L., et al.: Temporal segment networks: towards good practices for deep action recognition. In: Leibe, B., Matas, J., Sebe, N., Welling, M. (eds.) ECCV 2016. LNCS, vol. 9912, pp. 20–36. Springer, Cham (2016). https://doi.org/10.1007/978-3-319-46484-8_2
37. Wang, X., Girshick, R., Gupta, A., He, K.: Non-local neural networks. In: Proceedings of the IEEE Conference on Computer Vision and Pattern Recognition, pp. 7794–7803 (2018)
38. Wang, X., Gupta, A.: Videos as space-time region graphs. In: Proceedings of the European Conference on Computer Vision (ECCV), pp. 399–417 (2018)
39. Wang, X., Luo, Y., Crankshaw, D., Tumanov, A., Yu, F., Gonzalez, J.E.: IDK cascades: fast deep learning by learning not to overthink. arXiv preprint arXiv:1706.00885 (2017)
40. Wang, X., Yu, F., Dou, Z.Y., Darrell, T., Gonzalez, J.E.: Skipnet: learning dynamic routing in convolutional networks. In: Proceedings of the European Conference on Computer Vision (ECCV), pp. 409–424 (2018)
41. Xie, S., Sun, C., Huang, J., Tu, Z., Murphy, K.: Rethinking spatiotemporal feature learning: speed-accuracy trade-offs in video classification. In: Proceedings of the European Conference on Computer Vision (ECCV), pp. 305–321 (2018)
42. Zhou, B., Andonian, A., Oliva, A., Torralba, A.: Temporal relational reasoning in videos. In: Proceedings of the European Conference on Computer Vision (ECCV), pp. 803–818 (2018)
43. Zhu, X., Xu, C., Hui, L., Lu, C., Tao, D.: Approximated bilinear modules for temporal modeling. In: Proceedings of the IEEE International Conference on Computer Vision, pp. 3494–3503 (2019)
44. Zolfaghari, M., Singh, K., Brox, T.: Eco: efficient convolutional network for online video understanding. In: Proceedings of the European Conference on Computer Vision (ECCV), pp. 695–712 (2018)

S^3Net: Semantic-Aware Self-supervised Depth Estimation with Monocular Videos and Synthetic Data

Bin Cheng[1]([✉]), Inderjot Singh Saggu[2,5], Raunak Shah[3], Gaurav Bansal[4], and Dinesh Bharadia[5]

[1] Rutgers University, New Brunswick, USA
cb3974@winlab.rutgers.edu
[2] Plus.ai, Cupertino, USA
inderjot.saggu@plus.ai
[3] Indian Institute of Technology, Kanpur, India
raunaks@iitk.ac.in
[4] Blue River Technology, Sunnyvale, USA
gaurav.bansal@bluerivert.com
[5] University of California, San Diego, USA
dineshb@ucsd.edu

Abstract. Solving depth estimation with monocular cameras enables the possibility of widespread use of cameras as low-cost depth estimation sensors in applications such as autonomous driving and robotics. However, learning such a scalable depth estimation model would require a lot of labeled data which is expensive to collect. There are two popular existing approaches which do not require annotated depth maps: (i) using labeled synthetic and unlabeled real data in an adversarial framework to predict more accurate depth, and (ii) self-supervised models which exploit geometric structure across space and time in monocular video frames. Ideally, we would like to leverage features provided by both approaches as they complement each other; however, existing methods do not adequately exploit these additive benefits. We present S^3Net, a self-supervised framework which combines these complementary features: we use synthetic and real-world images for training while exploiting geometric, temporal, as well as semantic constraints. Our novel consolidated architecture provides a new state-of-the-art in self-supervised depth estimation using monocular videos. We present a unique way to train this self-supervised framework, and achieve (i) more than 15% improvement over previous synthetic supervised approaches that use domain adaptation and (ii) more than 10% improvement over previous self-supervised approaches which exploit geometric constraints from the real-world data.

Work done at UCSD.

Electronic supplementary material The online version of this chapter (https://doi.org/10.1007/978-3-030-58577-8_4) contains supplementary material, which is available to authorized users.

© Springer Nature Switzerland AG 2020
A. Vedaldi et al. (Eds.): ECCV 2020, LNCS 12375, pp. 52–69, 2020.
https://doi.org/10.1007/978-3-030-58577-8_4

Keywords: Monocular depth prediction · Self-supervised learning · Domain adaptation · Synthetic data · GANs · Semantic-aware

1 Introduction

Depth estimation is a fundamental component of 3D scene understanding, with applications in fields such as autonomous driving, robotics and space exploration. There has been considerable progress in estimating depth through monocular camera images in the last few years, as monocular cameras are inexpensive and widely deployed on many robots. However, building supervised depth estimation algorithms using monocular cameras is challenging, primarily because collecting ground-truth depth maps for training requires a carefully calibrated setup. As an example, many vehicles currently sold in the market have monocular cameras deployed, but there is no trivial way to obtain ground-truth depth information from the images collected from these cameras. Thus, supervised methods for depth estimation suffer due to the unavailability of extensive training labels.

To overcome the lack of depth annotation for monocular camera data, existing work has explored two areas of research: either designing self-supervised/semi-supervised approaches which require minimal labeling, or leveraging labeled synthetic data. Most self-supervised approaches rely on geometric and spatial constraints [43], and have succeeded in reducing the impact of this issue, however they don't always perform well in challenging environments with conditions like limited visibility, object motion, etc. This is because they lack strong training signal from supervision which lets them learn from and generalize to these conditions. In contrast, some effort has been undertaken to use realistic simulated environments to obtain additional synthetic depth data which can be used to compute a supervised training loss.

Synthetic data can be easily generated in different settings with depth labels - for example by varying the lighting conditions, changing the weather, varying object motion, etc. Simply training the original model on synthetic data, however, does not work well in practice as the model does not generalize well to a real-world dataset. To bridge this domain gap between the real-world and synthetic datasets, many domain adaptation techniques have been proposed. Recent works, like [28,45], have found success in using adversarial approaches to address this issue. These solutions typically involve using an adversarial transformation network to align the domains of the synthetic and real-world images, followed by a task network that is responsible for predicting depth. Naturally, we pose the question - can we build a depth estimation network that combines the benefits of information conveyed through real-world as well as synthetic data?

We present a novel framework S^3Net that trains the depth network by exploiting these self-supervised constraints (derived from real-world sequential images) and supervised constraints (derived from synthetic data and the respective ground-truth depths). This framework is implemented through several integrated stages which are described below: First, as shown in Fig. 1, we present a novel Generative Adversarial Network (GAN)-based domain adaptation network which exploits geometric constraints across space and time, as well as the

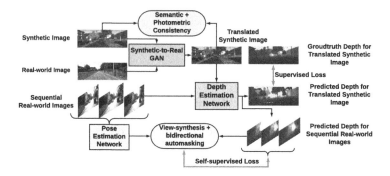

Fig. 1. Overview of our proposed framework: integrating supervised learning on translated synthetic data and self-supervised learning on real-world videos frames while imposing spatial, temporal and semantic constraints

semantic consistency between original synthetic images and translated images. These constraints encode additional latent information and thus enhance the quality of domain adaptation. Next, to leverage 'synthetic' supervised cues and 'real' self-supervised cues, we present a novel training approach - weights of the depth estimation network are updated *alternatively* based on the supervised and self-supervised losses. Finally, to impose explicit constraints on object geometry we augment the input RGB images with semantic labels, and utilize a bi-directional auto-masking technique to limit pixels which would violate rigid motion constraints.

Novel Adversarial Framework: The key idea of our GAN structure is to utilize flow-based photometric consistency and semantic consistency to better guide the image translation and reduce the domain gap. By utilizing the flow and the sequential translated images, the frame at t can be used to reconstruct the frame at $t+1$. The photometric differences between the reconstructed frame $t+1$ and the original frame $t+1$ are primarily due to imperfect image translation. Moreover, the semantic information should remain consistent before and after the image translation. Therefore, we add both a photometric consistency and a semantic consistency loss to create a novel adversarial framework. These offer additional constraints on the domain adaptation and further improve the image translation performance. They also help increase robustness and reduce undesired artifacts in translated images when compared with traditional approaches.

Semantics and Bi-directional Auto-masking: Inspired by the auto-masking technique proposed in [17], we propose a novel bi-directional auto-masking technique for sequential real-world images, which can filter out the pixels violating the fundamental rigid motion assumption for self-supervised depth learning. The key difference from a single direction mask is that the bi-directional technique fuses the masks learned by reconstructing frame $t+1$ from frame t and vice versa, which can substantially increases the accuracy of the proposed mask. Moreover, we augment the input images of our model with semantic labels.

The semantic labels can provide explicit geometry constraints, which can be beneficial to further boost the performance of the image translation and the depth estimation.

The challenges of training our depth model fall under two major categories: the GAN networks are unstable during training due to the presence of supervised synthetic losses and self-supervised losses, which results in lack of convergence. We address the convergence issue by proposing a two-phase training strategy. In the first phase, we train the image translation network and depth estimation network with synthetic supervised losses to stabilize the GAN-based image translation network. In the second phase, we freeze the weights of the image translation network and further train the depth estimation network with both supervised and self-supervised losses.

We evaluate our framework on two challenging datasets, i.e., KITTI [16,27] and Make3D [36]. The evaluation results show that our proposed model can outperform the state-of-the-art approaches in all evaluated metrics. In particular, we show that S^3Net can outperform both the state-of-the-art synthetic supervised domain adaptation approaches [28] by ∼15% and self-supervised approaches [17] by ∼10%. Moreover, we only require the depth estimation network during inference, so our inference compute requirements are comparable to previous state-of-the-art approaches.

2   Related Work

2.1   Supervised Depth Estimation

Eigen *et al.* [10] proposed the first supervised learning architecture that models multi-scale information for pixel-level depth estimation using direct regression. Inspired by this work, many follow-up works have extended supervised depth estimation in various directions [4,8,11,20,23,32,33,39–41,44]. However, acquiring these ground-truth depths is prohibitively expensive and thus it is unlikely to obtain a large amount of labelled training data covering various road conditions, weather conditions, etc., which indicates that these approaches may not generalize well.

One promising approach that reduces the labeling cost is to utilize synthetically generated data. However, models trained on synthetic data typically perform poorly on real-world images due to a large domain gap. Domain adaptation aims to minimize this gap. Recently, GAN-based approaches show promising performance in domain adaptation [3,13,14,34,35]. Atapour *et al.* [1] proposed a CycleGAN-based translator [48] to translate real-world images into the synthetic domain, and then train a depth prediction network using the synthetic labeled data. Zheng *et al.* [45] propose a novel architecture (T^2Net) where the style translator and the depth estimation network are optimized jointly so that they can improve each other. Despite promising performance, these approaches inherently suffer from mode collapse and semantic distortion due to imperfect image translation. Various constraints and techniques have been proposed to improve the quality of the translated images, but image translation[1] still remains a challenging task.

2.2 Self-supervised Depth Estimation

In addition to supervised solutions, various approaches have been studied to predict depths by extracting disparity and depth cues from stereo image pairs or monocular videos. Garg *et al.* [15] introduced a warping loss based on Taylor expansion. An image reconstruction loss with a spatial smoothness constraint was introduced in [21,31,47] to learn depth and camera motion. Recent works [17,18,25,37,47] aim to improve depth estimation by further exploiting geometry constraints. In particular, Godard *et al.* [18] employed epipolar geometry constraints between stereo image pairs and enforced a left-right consistency constraint in training the network. Yin *et al.*[43] proposed GeoNet, which also used depth and pose networks in order to compute rigid flow between sequential images in a video. More specifically, they introduced a temporal, flow-based photometric loss to predict depth for monocular videos in an unsupervised setting. Bian *et al.* [2] used a similar approach along with a self discovered mask to handle dynamic and occluded objects. Gordon *et al.* [19] also addresses these issues using a pure geometric approach. Casser *et al.* [5] adapts a similar framework with an additional online refinement model during inference. Xu *et al.* [42] proposed region deformer networks along with the earlier constraints to handle rigid and non-rigid motion. Zhou *et al.* [46] exploited a dual network attention based model which processes low and high resolution images separately. Godard *et al.* [17] also presented another unsupervised approach which built on their earlier model [18] by modifying the implementation of the unsupervised constraints.

Another set of recently adopted approaches involves using semantic information, which provides additional constraints on object geometry that can potentially boost the accuracy of depth estimation [7,26,29,30]. Meng *et al.* [26] built on top of [43], and proposed several ways to implement a semantic aided network which helped improve performance. Ranjan *et al.* [30] proposed a competitive collaboration framework to leverage segmentation maps, pose and flow for depth estimation. However, even with various constraints, the self-supervised approaches predict depth primarily based on indirect and weak-supervision depth cues, which can be easily affected by undesired artifacts, such as motion blurring and low visibility.

Our model architecture is influenced by various previous work, e.g. approaches in [9,17,28,45]. But, compared to these approaches, our S^3Net cooperatively combines both supervised depth prediction on synthetic data and self-supervised depth prediction on real-world sequential images, such that the two strategies can complement each other in a mutually beneficial setting.

---

[1] "domain adaptation" and "image translation" are used interchangeably.

3   Proposed Methods

We propose a joint framework for monocular depth estimation that is trained on translated synthetic images in a supervised manner and sequences of real-world images in a self-supervised fashion. Our proposed framework can be broken down into two main components: a) Synthetic-to-Real translation and supervised depth estimation ($G_{S \to R}$, D_R, f_T), and, b) View-synthesis guided self-supervised depth estimation ($Pose$, f_T)

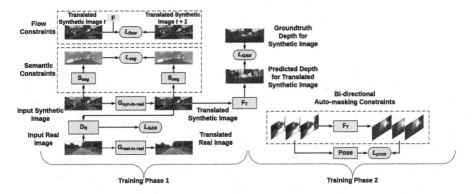

Fig. 2. Our detailed architecture, a) Semantic and photometric consistent GAN for synthetic supervised depth estimation, b) self-supervised architecture trained on sequence of real-world images with warping view-synthesis

Novel GAN Architecture. Models trained on synthetic data do not generalize well to real-world data because of domain gap. To address this problem, we build upon the work of T^2Net [45] for supervised depth estimation on translated synthetic images.

Adversarial Constraints. The goal of our generator is to translate a synthetic image X_s to the real-world domain X_r. To achieve this, a discriminator D_R and a transformer architecture $G_{S \to R}$ are trained jointly such that the discriminator tries to predict if the image is real or synthetic. This accounts for our GAN loss, $\mathcal{L}_{GAN}$ as shown in Fig. 2:

$$\mathcal{L}_{GAN} = \mathbb{E}_{x_r \sim X_r}[\log D_R(x_r)] + \mathbb{E}_{x_s \sim X_s}[\log(1 - D_R(G_{S \to R}(x_s)))] \qquad (1)$$

Identity Constraints. Following T^2Net, we also impose an identity constraint such that if a real-world image x_r is given as input, the generator network ($G_{S \to R}$)'s output should be identical to the input real-world image x_r. This additional constraint is incorporated as an identity loss $\mathcal{L}_r$ in Fig. 2:

$$\mathcal{L}_r = ||G_{S \to R}(x_r) - x_r||_1 \qquad (2)$$

Semantic Consistency. While the identity constraint improves upon a vanilla GAN architecture, we observe artifacts in the translated images, as shown in Fig. 4, which can subsequently compromise the accuracy of depth prediction. To address this we introduce a semantic consistency loss $\mathcal{L}_{seg}$ (Fig. 2) based on the idea that given a semantic segmentation model S_{seg} trained on the source domain, x_s and $G_{S \rightarrow R}(x_s)$ should have identical semantic segmentation maps. This is intuitive as domain translation shouldn't affect the semantic structure of the image. We enforce this by treating $S_{seg}(x_s)$ as a ground-truth label for pixel-wise prediction scores $S_{seg}(G_{S \rightarrow R}(x_s))$. These are used to compute a cross-entropy loss function over semantic labels:

$$\mathcal{L}_{seg} = - \sum_{pixels} \sum_{labels} S_{seg}(x_s) \log(S_{seg}(G_{S \rightarrow R}(x_s))) \tag{3}$$

However, because of domain gap we cannot expect S_{seg} trained on synthetic images to generalize well to the translated image domain, and hence we also fine-tune S_{seg} while training our GAN architecture so that it can learn features that are generalized to both domains.

Photometric Consistency with Ground-Truth Flow. In addition to semantic constraints we introduce a flow guided photometric loss [43] to exploit the temporal structure in translated image sequences. By applying the ground-truth flow, a frame t can be used to reconstructed the frame $t + 1$. We represent this as a transformation $\mathcal{F}$. In Eq. 4 below, $G_{S \rightarrow R}(x_{s,t})$ represents the translated image of a synthetic frame at time t and $\mathcal{F}(G_{S \rightarrow R}(x_{s,t}))$ indicates the reconstructed frame $t + 1$ based on frame t. $L_{pe}(*)$ computes the photometric differences between the reconstructed frame and the original frame. This photometric loss provides an indirect supervision on the synthetic-to-real image translation.

$$\mathcal{L}_{flow} = L_{pe}(\mathcal{F}(G_{S \rightarrow R}(x_{s,t})), G_{S \rightarrow R}(x_{s,t+1})) \tag{4}$$

Incorporating the above constraints in our GAN framework results in an improved domain translator that is largely devoid of artifacts and preserves semantic structure as shown in Fig. 4.

3.1   Combining Supervised and Self-supervised Depth Estimation

Supervised Depth Estimation on Synthetic Data. With the ground-truth depth labels for synthetic data, we formulate the depth estimation on synthetic data as a regression problem. In Eq. 5 below, $f_T(G_{S \rightarrow R}(x_{s,t}))$ is the estimated depth map for frame t and $y_{s,t}$ is the correspond ground-truth label for synthetic frame t.

$$\mathcal{L}_{task} = ||f_T(G_{S \rightarrow R}(x_{s,t})) - y_{s,t}||_1 \tag{5}$$

In accordance with our base network (T^2Net) we add an edge consistency/awareness loss which penalizes discontinuity (or inconsistency) in the edges between the image x_s and its depth map $f_T(x_s)$.

$$\mathcal{L}_s = |\partial_x f_T(x_r)|e^{-|\partial_x x_r|} + |\partial_y f_T(x_r)|e^{-|\partial_y x_r|} \tag{6}$$

Our training is divided into two phases. In the first phase, we train a GAN-based image transfer network ($(G_{S \to R}$ and $D_R)$) and the depth estimation network (f_T). The first loss objective is a weighted combination of the above constraints:

$$\mathcal{L}_{phase1} = \mathcal{L}_{GAN} + \alpha_r \mathcal{L}_r + \alpha_{seg} \mathcal{L}_{seg} + \alpha_{flow} \mathcal{L}_{flow} + \alpha_{task} \mathcal{L}_{task} + \alpha_s \mathcal{L}_s \quad (7)$$

where α_r, α_{seg}, α_{flow}, α_{task}, and α_s are hyper-parameters.

Self-supervised Depth Estimation on Monocular Videos. In addition to supervised depth prediction on translated synthetic images, we also perform self-supervised depth estimation on monocular videos. The corresponding pixel coordinates of one rigid object in two consecutive frames follows the relationship

$$p_{t+1} = \mathbf{K} \mathbf{T}_{t \to t+1} \mathbf{f}_T(p_t) \mathbf{K}^{-1} p_t \qquad\qquad p_{t+1} = \mathcal{W}^{-1} p_t \qquad (8)$$

where p_t and p_{t+1} are the positions of a pixel in frames at time t and $t+1$, $\mathbf{K}$ denotes the camera intrinsic parameters, $\mathbf{T}_{t \to t+1}$ represents the relative camera pose from frame t to from $t+1$ and $\mathcal{W}$ is the equivalent warping transformation. By sampling these pixels at p_{t+1} in frame $t+1$, one can construct frame t from frame $t+1$. The photometric difference (denoted by L_{pe}) between the constructed image and the true image of frame t provides self-supervision for both the depth network (f_T) and pose estimation networks (Fig. 2).

$$\mathcal{L}_{pose} = L_{pe}(\mathcal{W}(x_{r,t+1}), \; x_{r,t}) \qquad (9)$$

Bi-directional Auto-masking. Inspired by the auto-masking method proposed in [17], we compute the photometric loss for different sequential image pairs, e.g., from frame $t+1$ to frame t and from frame $t-1$ to frame t, and then aggregate these photometric losses by extracting their minimum value, i.e., $\min_{t' \in \{...t-1,t+1...\}} L_{pe}(\mathcal{W}(x_{r,t'}), x_{r,t})$. Additionally, the pixels satisfying

$$mask = \min_{t' \in \{...t-1,t+1...\}} L_{pe}(x_{r,t'}, \; x_{r,t}) > \min_{t' \in \{...t-1,t+1...\}} L_{pe}(\mathcal{W}(x_{r,t'}), \; x_{r,t})$$

are selected for further loss computation. It is because the discarded pixels are more likely to belong to moving objects with a similar moving speed as the moving camera. For a more complete loss computation, we consider a bi-directional warping transformation, i.e., from frame t to frame t' as well as from frame t' to frame t.

$$\mathcal{L}_{mask} = mask_{t' \to t} \circ \min_{t' \in \{...t-1,t+1...\}} L_{pe}(\mathcal{W}(x_{r,t'}), \; x_{r,t})$$
$$+ mask_{t \to t'} \circ \min_{t' \in \{...t-1,t+1...\}} L_{pe}(\mathcal{W}(x_{r,t}), \; x_{r,t'})) \qquad (10)$$

In the second phase of our training, we train the depth and pose networks (f_T and $Pose$) with a combination of sequential real-world images and GAN translated synthetic images. The total loss which includes training the depth network f_T with supervised loss $\mathcal{L}_{task}$ from synthetic data, can be written as:

$$\mathcal{L}_{phase2} = \alpha_{pose} \mathcal{L}_{pose} + \alpha_{mask} \mathcal{L}_{mask} + \alpha_{task} \mathcal{L}_{task} \qquad (11)$$

where α_{pose}, α_{mask} and α_{task} are hyper-parameters.

3.2 Semantic Augmentation

Semantic labels provide important information about object shape and geometry. We believe such information helps improve the accuracy of depth estimation by imposing additional constraints. For example, on 2D images, the pixels on the object boundaries can have very different depths. Semantic information can help regulate the pixels belonging to certain objects and facilitate the learning process of depth estimation. In this work, to utilize semantic information, we augment the input RGB images with additional semantic labels. We also experimented with augmenting RGB images with semantic labels during synthetic-to-real image translation and obtained substantial improvements in the quality of our translated images.

4 Experiments

In this section, we first present the implementation details of our framework and then compare our framework with other state-of-the-art work on the KITTI and Make3D benchmarks. Finally, we study the importance of each component in our framework through various ablation experiments.

4.1 Implementation Details

Network Implementation. Our framework mainly consists of two main sub-modules: (i) the synthetic-to-real image translation network which translates synthetic images to real-style images, and (ii) the depth estimation network which predicts depth maps for both translated synthetic images and real-world images. For the synthetic-to-real image translation network, we build our network on top of the T^2Net architecture [45], with added constraints that use the synthetic ground-truth labels for semantic and optical flow. For the semantic consistency loss we apply DeepLab v3+ with a MobileNet backbone as our S_{seg} model. We pre-trained the model on the vKITTI dataset, achieving a median IoU of 0.898 on the validation set. We then tested the U-Net, VGGNet, and ResNet50 architectures for the depth estimation network and selected the U-Net architecture due to its superior performance. A VGG-based architecture was used to estimate the relative camera poses between sequential images. These depth maps and camera poses are subsequently exploited to compute the self-supervised loss.

Data Pre-processing. We select vKITTI [12] and KITTI [27] as the synthetic dataset and the real-world dataset, respectively, while training the synthetic-to-real image translation network. The training dataset consists of 20470 images from vKITTI and 41740 images from KITTI. The training images of the KITTI dataset are further divided into small sequences. We use 697 images from KITTI as our test dataset as in the eigen split [10]. The input images are resized to 640×192 (width $\times$ height) during both training and testing. The ground-truth depth information, semantic labels, and optical flow information from the synthetic vKITTI dataset are also used during training. The input RGB images are augmented by the corresponding semantic labels, which are generated by DeepLab v3+ model [6].

Model Training. Training our model has two major challenges: (i) the training of GAN-based networks is well-known to be unstable; (ii) the depth estimation in our model consists of two components and thus the weights of the depth estimation network are updated by two separate loss functions, which can lead to a convergence issue. To tackle these challenges, we design a two-phase training strategy. In the first phase, we pre-train the synthetic-to-real image translation network along with synthetic supervised depth estimation to provide a stable initialization for the image translation network. In phase 2, we freeze the weights of the image translation network and train the depth estimation network using both synthetic supervised and self-supervised losses. We primarily tested two training methods to harmonize the two sources of losses: 1) *weighted sum training*: updating the weights of the depth estimation network based on a weighted sum of the two sources of losses; 2) *alternating training*: alternatively updating the weights of the depth estimation networks by the two sources of losses. We find that the two training methods are resulting in comparable evaluation results but the alternating training provides another control knob to optimize the model training, and generalize well to both the data sources and therefore to unseen datasets. Due to space limitation, we show the results using alternating training only in this paper.

Further, we use the Adam optimizer [22], with initial learning rate of $2e^{-5}$ for the image translation network, $5e^{-5}$ for the depth estimation network, and $5e^{-5}$ for the camera pose estimation network.

Our network was trained on a RTX 2080Ti GPU. On average, our depth estimation network can process 33 frames per second during inference.

4.2 Monocular Depth Estimation on KITTI Dataset

We follow the procedure defined in [45] when evaluating on the KITTI dataset. The evaluation results are listed in Table 1. As shown in the table, our framework shows the best performance across all metrics. We believe this is because our framework can combine the benefits of both synthetic supervised depth estimation and self-supervised depth estimation. Typical supervised synthetic approaches train models by learning from low-cost synthetic ground-truth depths, but these approaches also suffer from unstable and inconsistent image translation, leading to less accurate translated images with low resolution. On the other hand, self-supervised approaches can learn the depth from high resolution sequential images; however, these depths are learned from indirect cues which are sensitive to in-view object movements, blockages, etc. Training the model with modified supervised and self-supervised constraints in our consolidated framework ensures that we exploit the best of both worlds, which ultimately leads to better prediction results.

Table 1. Monocular depth estimation on KITTI dataset with Eigen *et al.* [10] split. The highlighted scores mark the best performance among selected models. In "Dataset" column, "K" and "V" stands for the KITTI and the vKITTI dataset, respectively.

Method	Dataset	Error-related metrics				Accuracy-related metrics		
		Abs Rel	Sq Rel	RMSE	RMSE log	$\delta < 1.25$	$\delta < 1.25^2$	$\delta < 1.25^3$
depth capped at 80 m								
Zhou *et al.* [47]	K	0.183	1.595	6.709	0.270	0.734	0.902	0.959
Yin *et al.* [43]	K	0.155	1.296	5.857	0.233	0.793	0.931	0.973
Wang *et al.* [38]	K	0.151	1.257	5.583	0.228	0.810	0.936	0.974
Ramirez *et al.* [29]	K	0.143	2.161	6.526	0.222	0.850	0.939	0.972
Casser *et al.* [5]	K	0.141	1.026	5.290	0.215	0.816	0.945	0.979
Ranjan *et al.* [30]	K	0.140	1.070	5.326	0.217	0.826	0.941	0.975
Xu *et al.* [42]	K	0.138	1.016	5.352	0.216	0.823	0.943	0.976
Meng *et al.* [26]	K	0.133	0.905	5.181	0.208	0.825	0.947	0.981
Godard *et al.* [17] [a]	K	0.132	1.044	5.142	0.210	0.845	0.948	0.977
Zheng *et al.* [45]	K + V	0.174	1.410	6.046	0.253	0.754	0.916	0.966
Mou *et al.* [28]	K + V	0.145	1.058	5.291	0.215	0.816	0.941	0.977
Ours	K + V	**0.124**	**0.826**	**4.981**	**0.200**	**0.846**	**0.955**	**0.982**
depth capped at 50 m								
Yin *et al.* [43]	K	0.147	0.936	4.348	0.218	0.810	0.941	0.977
Zheng *et al.* [45]	K +V	0.168	1.199	4.674	0.243	0.772	0.912	0.966
Mou *et al.* [28]	K + V	0.139	0.814	3.995	0.203	0.830	0.949	0.980
Ours	K + V	**0.118**	**0.615**	**3.710**	**0.187**	**0.862**	**0.962**	**0.984**

[a] For fair comparison, we selected the results for the model without pre-training on the ImageNet dataset

In Fig. 3 we compare qualitative depth estimation results of purely self-supervised GeoNet [43], purely synthetic supervised T^2Net [45] and our proposed framework. Purely self-supervised approaches results in depth maps which are blurred and do not model depth discontinuity at object boundaries well. On the other hand purely synthetic supervised approach results in sharper depth maps but because of imperfect domain translation it fails to predict depth for surfaces with multiple textures. For example, in the first row of Fig. 3, T^2Net predicts incorrect depth values for the wall on the right because of the window on the wall adding additional texture. These defects severely limit the real-world application of purely self-supervised and synthetic supervised techniques. Our S^3Net on the other hand generates sharper depth maps than GeoNet and doesn't suffer from the problems discussed for T^2Net depth, further proving our point about combining best features from both.

In Fig. 4, we compare synthetic-to-real translated images for the GAN in T^2Net [45] and the GAN in our proposed framework. Similar to the GAN in T^2Net, our GAN is also tied to a task loss. Without the presence of a specific task loss, e.g., a depth estimation loss, the resulting objective drives the image translator to generate a realistic interpretation of synthetic images. However, when a task loss is introduced the main objective is shifted to project synthetic images to a space that is optimized for the task. Therefore, some of these differences might not be visually perceivable but can lead to a large gain in performance. Comparing to the GAN in T^2Net, our approach significantly reduces artifacts and successfully retains the semantic structure across synthetic and translated images.

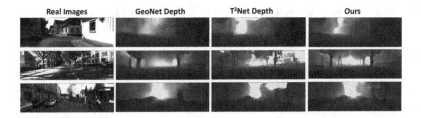

Fig. 3. Qualitative Depth Prediction Results: Column (a) real-world images from KITTI, Column (b), (c), (d) are results for GeoNet [43], T^2Net [45], and our S^3Net framework, respectively.

Fig. 4. Translated images: Column (a) input synthetic images; Column (b) T^2Net; Column (c) our S^3Net GAN with semantic constraints.

We also evaluate our model on the KITTI odometry dataset. We use three sequential images as input and follow the same evaluation strategy as given in [47]. As shown in Table 2, our model outperforms the state-of-the-art approaches by a convincing margin.

4.3   Generalization Study on Make3D Dataset

To show the generalization capability, we also test our framework on the Make3D dataset [36]. We use the model trained on the KITTI dateset and the vKITTI dataset, and evaluate the model on the Make3D test dataset, following the evaluation strategy in [18]. Table 3 shows that our framework outperforms other self-supervised approaches by a considerable margin. It is because our framework gains knowledge from both synthetic and real-world datasets and thus sees more scenarios, which further leads to a better generalization performance.

4.4   Ablation Study

In this subsection, we perform a set of ablation experiments on the KITTI dataset to discuss how each individual component in our framework contributes to the final performance. The evaluation results are reported in Table 4.

Table 2. Absolute Trajectory Error (ATE) on the KITTI odometry dataset.

Method	# of snippets	Seq.09	Seq.10
ORB-SLAM (full)	5	0.014 ± 0.008	0.012 ± 0.011
ORB-SLAM (short)	5	0.064 ± 0.141	0.064 ± 0.130
DDVO (Wang *et al.* [38])	3	0.045 ± 0.108	0.033 ± 0.074
SfmLearner (Zhou *et al.* [47])	5	0.021 ± 0.017	0.020 ± 0.015
SfmLearner [47] updated	5	0.016 ± 0.009	0.013 ± 0.009
GeoNet (Yin *et al.* [43])	5	0.012 ± 0.007	0.012 ± 0.009
MonoDepth2* (Godard *et al.* [17])	2	0.017 ± 0.008	0.015 ± 0.010
EPC++ (Luo *et al.* [24])	3	0.013 ± 0.007	0.012 ± 0.008
Ours	3	**0.0097 ± 0.0046**	**0.0099 ± 0.0071**

Table 3. Error metrics for depth estimation on the Make3D dataset.

Method	Error-related metrics			
	Abs Rel	Sq Rel	RMSE	RMSE log
MonoDepth (Godard *et al.* [18])	0.544	10.940	11.760	0.193
SfmLearner (Zhou *et al.* [47])	0.383	5.321	10.47	0.478
T^2Net (Zheng *et al.* [45])	0.508	6.589	8.935	0.574
MonoDepth2 (Godard *et al.* [17])	0.322	3.589	7.417	0.163
TCDA (Mou *et al.* [28])	0.384	3.885	7.645	0.181
Ours (no semantic augmentation)	0.372	5.699	7.844	0.176
Ours (with semantic augmentation)	**0.322**	**3.238**	**7.187**	**0.164**

Synthetic Supervised Depth Estimation. Due to a large domain gap between the synthetic and the real domain, a model that is only trained on synthetic data typically generates unacceptable depth predictions on the real-world data. Synthetic-to-real image translation is one of the most effective remedies for this issue. Even with a native image translation network as proposed in [45], the depth predictions on the real-world data can be improved by about 40%. Additionally, the flow-guided photometric consistency and semantic consistency constraints further regulates the image translation and improves our depth prediction accuracy by another 8.4% and 8%, respectively. Continuing training S_{seg} gives better performance compared to freezing network parameters. This is because S_{seg} is trained on synthetic semantic labels and cannot generalize to translated domain if not trained further.

Synthetic Supervised + Semantic Augmentation. We investigate the importance of semantic augmentation to our model by (i) only augmenting the input images for the depth estimation network, but keeping RGB images for the image translation network, and (ii) augmenting the input images for both the depth estimation network and the image translation networks. Compared with

Table 4. Performance gain in depth estimation from different model components. The predicted depth is capped at 50 m

Method	Error-related metrics				Accuracy-related metrics		
	Abs Rel	Sq Rel	RMSE	RMSE log	$\delta < 1.25$	$\delta < 1.25^2$	$\delta < 1.25^3$
Synthetic Translated Supervised							
Without synthetic-to-real image translation	0.278	3.216	6.268	0.322	0.681	0.854	0.929
Native synthetic-to-real image translation	0.168	1.199	4.674	0.243	0.772	0.912	0.966
With flow-guided photometric consistency	0.1539	0.993	4.4492	0.2241	0.7986	0.9356	0.9752
With semantic consistency (frozen S_{seg})	0.1555	0.9680	4.7412	0.2324	0.7773	0.9245	0.9721
With semantic consistency	0.1544	0.9633	4.7422	0.2322	0.7786	0.9241	0.9727
Synthetic Translated Supervised with Semantic Augmentation							
Semantic augmentation for depth estimation network input only	0.1532	0.9631	4.3872	0.2275	0.7945	0.9325	0.9738
Semantic augmentation for both the image translation networks & the depth estimation networks	0.1455	0.8869	4.2177	0.2154	0.8133	0.9411	0.9773
Synthetic Translated Supervised + Real-world Self-Supervised							
With self-supervised depth estimation	0.1292	0.6969	3.8399	0.1964	0.8428	0.9554	0.9826
With auto-masking	0.1198	0.6671	3.7696	0.1921	0.8637	0.9583	0.9819
With semantic augmentation	**0.1183**	**0.6150**	**3.7105**	**0.1876**	**0.8620**	**0.9622**	**0.9844**

the first augmentation strategy, the second strategy introduces a larger improvement. It is because the semantic information can impose additional constraints on object geometry and these constraints are useful for regulating the shape of objects and determining the depth prediction on the boundary of objects.

Synthetic Supervised + Self-supervised. By adding photometric losses to real-world sequential images and jointly training the synthetic supervised and self-supervised depth estimation, the depth estimation accuracy on real-world dataset is further improved by 11%. The real-world sequential images are typically clear and accurate, which can compensate for the shortcomings in the imperfect translated images. However, the photometric losses for real-world sequential images are computed based on an assumption that the displacement of pixels is purely caused by movement of the camera. Such an assumption does not always hold. The direct supervision on the synthetic translated images can help alleviate the negative effect of violating this assumption. But, our study indicates that by selecting valid pixels and filtering out the pixels that violate the assumption, our model is further improved by a noticeable margin.

5   Conclusion and Next Steps

In this paper, we present a framework for monocular depth estimation which combines the features of both synthetic images and real-world video frames in a novel semantic-aware, self-supervised setting. The complexity of our model does not affect its scalability, as we only require a depth network during inference time. We outperform all existing approaches on the KITTI benchmark as well as on our generalization to the new Make3D dataset. These factors contribute to the increased accuracy, scalability, and robustness of our framework as compared to other existing approaches. Our framework extends typical dataset-specific models to improve generalization performance, making it more relevant for real-world applications. In the future, we plan to explore strategies which can apply similar frameworks to other related tasks in visual perception.

References

1. Atapour-Abarghouei, A., Breckon, T.P.: Real-time monocular depth estimation using synthetic data with domain adaptation via image style transfer. In: Proceedings of the IEEE Conference on Computer Vision and Pattern Recognition, pp. 2800–2810 (2018)
2. Bian, J., et al.: Unsupervised scale-consistent depth and ego-motion learning from monocular video. In: Advances in Neural Information Processing Systems, pp. 35–45 (2019)
3. Bousmalis, K., Silberman, N., Dohan, D., Erhan, D., Krishnan, D.: Unsupervised pixel-level domain adaptation with generative adversarial networks. In: Proceedings of the IEEE Conference on Computer Vision and Pattern Recognition, pp. 3722–3731 (2017)
4. Cao, Y., Wu, Z., Shen, C.: Estimating depth from monocular images as classification using deep fully convolutional residual networks. IEEE Trans. Circ. Syst. Video Technol. **28**(11), 3174–3182 (2017)
5. Casser, V., Pirk, S., Mahjourian, R., Angelova, A.: Depth prediction without the sensors: leveraging structure for unsupervised learning from monocular videos (2018)
6. Chen, L.-C., Zhu, Y., Papandreou, G., Schroff, F., Adam, H.: Encoder-decoder with atrous separable convolution for semantic image segmentation. In: Ferrari, V., Hebert, M., Sminchisescu, C., Weiss, Y. (eds.) ECCV 2018. LNCS, vol. 11211, pp. 833–851. Springer, Cham (2018). https://doi.org/10.1007/978-3-030-01234-2_49
7. Chen, P.Y., Liu, A.H., Liu, Y.C., Wang, Y.C.F.: Towards scene understanding: unsupervised monocular depth estimation with semantic-aware representation. In: Proceedings of the IEEE Conference on Computer Vision and Pattern Recognition, pp. 2624–2632 (2019)
8. Chen, W., Fu, Z., Yang, D., Deng, J.: Single-image depth perception in the wild. In: Advances in Neural Information Processing Systems, pp. 730–738 (2016)
9. Cherian, A., Sullivan, A.: Sem-gan: semantically-consistent image-to-image translation. CoRR abs/1807.04409 (2018). http://arxiv.org/abs/1807.04409
10. Eigen, D., Puhrsch, C., Fergus, R.: Depth map prediction from a single image using a multi-scale deep network. In: Advances in Neural Information Processing Systems, pp. 2366–2374 (2014)
11. Fu, H., Gong, M., Wang, C., Batmanghelich, K., Tao, D.: Deep ordinal regression network for monocular depth estimation. In: The IEEE Conference on Computer Vision and Pattern Recognition (CVPR), June 2018

12. Gaidon, A., Wang, Q., Cabon, Y., Vig, E.: Virtual worlds as proxy for multi-object tracking analysis. In: The IEEE Conference on Computer Vision and Pattern Recognition (CVPR), pp. 4340–4349 (2016)

13. Ganin, Y., Lempitsky, V.: Unsupervised domain adaptation by backpropagation. arXiv preprint arXiv:1409.7495 (2014)

14. Ganin, Y., Ustinova, E., Ajakan, H., Germain, P., Larochelle, H., Laviolette, F., Marchand, M., Lempitsky, V.: Domain-adversarial training of neural networks. J. Mach. Learn. Res. **17**(1), 2030–2096 (2016)

15. Garg, R., B.G., V.K., Carneiro, G., Reid, I.: Unsupervised CNN for single view depth estimation: geometry to the rescue. In: Leibe, B., Matas, J., Sebe, N., Welling, M. (eds.) ECCV 2016. LNCS, vol. 9912, pp. 740–756. Springer, Cham (2016). https://doi.org/10.1007/978-3-319-46484-8_45

16. Geiger, A., Lenz, P., Urtasun, R.: Are we ready for autonomous driving? the kitti vision benchmark suite. In: Conference on Computer Vision and Pattern Recognition (CVPR) (2012)

17. Godard, C., Aodha, O.M., Firman, M., Brostow, G.J.: Digging into self-supervised monocular depth estimation. In: Proceedings of the IEEE International Conference on Computer Vision, pp. 3828–3838 (2019)

18. Godard, C., Aodha, O.M., Brostow, G.J.: Unsupervised monocular depth estimation with left-right consistency. In: CVPR (2017)

19. Gordon, A., Li, H., Jonschkowski, R., Angelova, A.: Depth from videos in the wild: unsupervised monocular depth learning from unknown cameras (2019)

20. He, L., Wang, G., Hu, Z.: Learning depth from single images with deep neural network embedding focal length. IEEE Trans. Image Process. **27**(9), 4676–4689 (2018)

21. Yu, J.J., Harley, A.W., Derpanis, K.G.: Back to basics: unsupervised learning of optical flow via brightness constancy and motion smoothness. In: Hua, G., Jégou, H. (eds.) ECCV 2016. LNCS, vol. 9915, pp. 3–10. Springer, Cham (2016). https://doi.org/10.1007/978-3-319-49409-8_1

22. Kingma, D.P., Ba, J.: Adam: a method for stochastic optimization. arXiv preprint arXiv:1412.6980 (2014)

23. Liu, F., Shen, C., Lin, G., Reid, I.: Learning depth from single monocular images using deep convolutional neural fields. IEEE Trans. Pattern Anal. Mach. Intell. **38**(10), 2024–2039 (2016)

24. Luo, C., Yang, Z., Wang, P., Wang, Y., Xu, W., Nevatia, R., Yuille, A.: Every pixel counts++: Joint learning of geometry and motion with 3D holistic understanding. arXiv preprint arXiv:1810.06125 (2018)

25. Mahjourian, R., Wicke, M., Angelova, A.: Unsupervised learning of depth and ego-motion from monocular video using 3D geometric constraints. In: The IEEE Conference on Computer Vision and Pattern Recognition (CVPR), June 2018

26. Meng, Y., et al.: Signet: semantic instance aided unsupervised 3D geometry perception. In: Proceedings of the IEEE Conference on Computer Vision and Pattern Recognition, pp. 9810–9820 (2019)

27. Menze, M., Geiger, A.: Object scene flow for autonomous vehicles. In: Proceedings of the IEEE Conference on Computer Vision and Pattern Recognition, pp. 3061–3070 (2015)

28. Mou, Y., Gong, M., Fu, H., Batmanghelich, K., Zhang, K., Tao, D.: Learning depth from monocular videos using synthetic data: a temporally-consistent domain adaptation approach. arXiv preprint arXiv:1907.06882 (2019)

29. Zama Ramirez, P., Poggi, M., Tosi, F., Mattoccia, S., Di Stefano, L.: Geometry meets semantics for semi-supervised monocular depth estimation. In: Jawahar, C.V., Li, H., Mori, G., Schindler, K. (eds.) ACCV 2018. LNCS, vol. 11363, pp. 298–313. Springer, Cham (2019). https://doi.org/10.1007/978-3-030-20893-6_19

30. Ranjan, A., Jampani, V., Balles, L., Kim, K., Sun, D., Wulff, J., Black, M.J.: Competitive collaboration: Joint unsupervised learning of depth, camera motion, optical flow and motion segmentation (2018)

31. Ren, Z., Yan, J., Ni, B., Liu, B., Yang, X., Zha, H.: Unsupervised deep learning for optical flow estimation. In: AAAI, vol. 3 (2017)

32. Repala, V.K., Dubey, S.R.: Dual CNN models for unsupervised monocular depth estimation. arXiv preprint arXiv:1804.06324 (2018)

33. Roy, A., Todorovic, S.: Monocular depth estimation using neural regression forest. In: The IEEE Conference on Computer Vision and Pattern Recognition (CVPR), pp. 5506–5514 (2016)

34. Sankaranarayanan, S., Balaji, Y., Castillo, C.D., Chellappa, R.: Generate to adapt: aligning domains using generative adversarial networks. In: Proceedings of the IEEE Conference on Computer Vision and Pattern Recognition, pp. 8503–8512 (2018)

35. Sankaranarayanan, S., Balaji, Y., Jain, A., Nam Lim, S., Chellappa, R.: Learning from synthetic data: Addressing domain shift for semantic segmentation. In: The IEEE Conference on Computer Vision and Pattern Recognition (CVPR), June 2018

36. Saxena, A., Sun, M., Ng, A.Y.: Make3D: learning 3D scene structure from a single still image. IEEE Trans. Pattern Anal. Mach. Intell. **31**(5), 824–840 (2008)

37. Vijayanarasimhan, S., Ricco, S., Schmid, C., Sukthankar, R., Fragkiadaki, K.: SFM-net: Learning of structure and motion from video. (2017), preprint

38. Wang, C., Miguel Buenaposada, J., Zhu, R., Lucey, S.: Learning depth from monocular videos using direct methods. In: Proceedings of the IEEE Conference on Computer Vision and Pattern Recognition. pp. 2022–2030 (2018)

39. Xian, K., et al.: Monocular relative depth perception with web stereo data supervision. In: The IEEE Conference on Computer Vision and Pattern Recognition (CVPR), June 2018

40. Xu, D., Ricci, E., Ouyang, W., Wang, X., Sebe, N.: Multiscale continuous CRFs as sequential deep networks for monocular depth estimation. In: The IEEE Conference on Computer Vision and Pattern Recognition (CVPR), July 2017

41. Xu, D., Wang, W., Tang, H., Liu, H., Sebe, N., Ricci, E.: Structured attention guided convolutional neural fields for monocular depth estimation. In: Proceedings of the IEEE Conference on Computer Vision and Pattern Recognition, pp. 3917–3925 (2018)

42. Xu, H., Zheng, J., Cai, J., Zhang, J.: Region deformer networks for unsupervised depth estimation from unconstrained monocular videos (2019)

43. Yin, Z., Shi, J.: Geonet: Unsupervised learning of dense depth, optical flow and camera pose. In: Proceedings of the IEEE Conference on Computer Vision and Pattern Recognition, pp. 1983–1992 (2018)

44. Zhang, Z., Schwing, A.G., Fidler, S., Urtasun, R.: Monocular object instance segmentation and depth ordering with CNNs. In: 2015 IEEE International Conference on Computer Vision (ICCV), pp. 2614–2622 (2015)

45. Zheng, C., Cham, T.J., Cai, J.: T2net: synthetic-to-realistic translation for solving single-image depth estimation tasks. In: Proceedings of the European Conference on Computer Vision (ECCV), pp. 767–783 (2018)

46. Zhou, J., Wang, Y., Qin, K., Zeng, W.: Unsupervised high-resolution depth learning from videos with dual networks (2019)
47. Zhou, T., Brown, M., Snavely, N., Lowe, D.G.: Unsupervised learning of depth and ego-motion from video. In: Conference, I. (ed.) IEEE Conference on Computer Vision and Pattern Recognition (CVPR), p. 7 (2017)
48. Zhu, J.Y., Park, T., Isola, P., Efros, A.A.: Unpaired image-to-image translation using cycle-consistent adversarial networks. In: Proceedings of the IEEE International Conference on Computer Vision, pp. 2223–2232 (2017)

Leveraging Seen and Unseen Semantic Relationships for Generative Zero-Shot Learning

Maunil R. Vyas$^{(\boxtimes)}$ ⓘ, Hemanth Venkateswara ⓘ,
and Sethuraman Panchanathan ⓘ

Arizona State University, Tempe, AZ 85281, USA
{mrvyas,hemanthv,panch}@asu.edu

Abstract. Zero-shot learning (ZSL) addresses the unseen class recognition problem by leveraging semantic information to transfer knowledge from seen classes to unseen classes. Generative models synthesize the unseen visual features and convert ZSL into a classical supervised learning problem. These generative models are trained using the seen classes and are expected to implicitly transfer the knowledge from seen to unseen classes. However, their performance is stymied by overfitting, which leads to substandard performance on Generalized Zero-Shot learning (GZSL). To address this concern, we propose the novel LsrGAN, a generative model that Leverages the Semantic Relationship between seen and unseen categories and explicitly performs knowledge transfer by incorporating a novel Semantic Regularized Loss (SR-Loss). The SR-loss guides the LsrGAN to generate visual features that mirror the semantic relationships between seen and unseen classes. Experiments on seven benchmark datasets, including the challenging Wikipedia text-based CUB and NABirds splits, and Attribute-based AWA, CUB, and SUN, demonstrates the superiority of the LsrGAN compared to previous state-of-the-art approaches under both ZSL and GZSL. Code is available at https://github.com/Maunil/LsrGAN.

Keywords: Generalized zero-shot learning · Generative Modeling (GANs) · Seen and unseen relationship

1 Introduction

Consider the following discussion between a kindergarten teacher and her student.

Electronic supplementary material The online version of this chapter (https://doi.org/10.1007/978-3-030-58577-8_5) contains supplementary material, which is available to authorized users.

© Springer Nature Switzerland AG 2020
A. Vedaldi et al. (Eds.): ECCV 2020, LNCS 12375, pp. 70–86, 2020.
https://doi.org/10.1007/978-3-030-58577-8_5

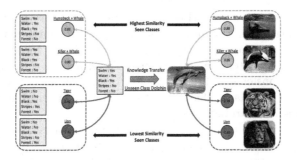

Fig. 1. Driving motivation behind leveraging the semantic relationship between seen and unseen classes to infer the visual characteristics of unseen classes. Notice that though the feature representations are different, the class similarity values are almost the same. e.g. "Dolphin" has almost identical similarity values in visual and semantic space with other seen classes. The similarity values are mentioned in the circles, and computed using the cosine distance.

Teacher: Today we will learn about a new animal that roams the grasslands of Africa. It is called the Zebra.
Student: What does a Zebra look like?
Teacher: It looks like a short white horse but has black stripes like a tiger.

That description is nearly enough for the student to recognize a zebra the next time she sees it. The student is able to take the verbal (textual) description and relate it to the visual understanding of a horse and a tiger and generate a zebra in her mind. In this paper, we propose a zero-shot learning model that transfers knowledge from text to the visual domain to learn and recognize previously unseen image categories.

Collecting and curating large labeled datasets for training deep neural networks is both labor-intensive and nearly impossible for many of the classification tasks, especially for the fine-grained categories in specific domains. Hence, it is desirable to create models that can mitigate these difficulties and learn not to rely on large labeled training sets. Inspired by the human ability to recognize object categories solely based on class descriptions and previous visual knowledge, the research community has extensively pursued the area of "Zero-shot learning" (ZSL) [11,22,23,29,37,38]. Zero-shot learning aims to recognize objects that are not seen during the training process of the model. It leverages textual descriptions/attributes to transfer knowledge from seen to unseen classes.

Generative models are the most popular approach to solve zero-shot learning. Despite the recent progress, generative models for zero-shot learning still have some key limitations. These models show a large quality gap between the synthesized and the actual unseen image features [15,24,31,35,43]. As a result, the performance of generalized zero-shot learning (GZSL) suffers, since many synthesized features of unseen classes are classified as seen classes. The second major concern behind the current generative model based approaches is the assumption that the semantic features are available in the desired form for a class category, e.g., clean attributes. However, in reality, they are hard to get. Getting clean semantic features requires a domain expert to annotate the attributes manually. Moreover, collecting a large number of attributes for all the class categories is again labor-intensive and expensive. Considering this, our generative model learns to transfer semantic knowledge from both noisy text descriptions (like Wikipedia articles) as well as semantic attributes for zero-shot learning and generalized zero-shot learning.

In this paper we propose a novel generative model called the LsrGAN. The LsrGAN leverages semantic relationships between seen and unseen categories and transfers the same to the generated image features. We implement this transfer through a unique semantic regularization framework called the Semantic Regularized Loss (SR-Loss). In Fig. 1, "Dolphin", an unseen class, has a high semantic similarity with classes such as "Killer whale" and "Humpback whale" from the seen class set. These two seen classes are the potential neighbors of the Dolphin in the visual space. Therefore, when we do not have the real visual features for the Dolphin class, we can use these neighbors to form indirect visual references for the Dolphin class. This illustrates the intuition behind the proposed SR-loss. The LsrGAN also trains a classifier to guide the feature generation. The main contributions of our work are:

1. A generative model leveraging the semantic relationship (LsrGAN) between seen and unseen classes overcoming the overfitting concern towards seen classes.
2. A novel semantic regularization framework that enables knowledge transfer across the visual and semantic domain. The framework can be easily integrated into any generalized zero-shot learning model.
3. We have conducted extensive experiments on seven widely used standard benchmark datasets to demonstrate that our model outperforms the previous state-of-the-art approaches.

2 Related Work

Earlier work on ZSL was focused on learning the mapping between visual and semantic space in order to compensate for the lack of visual representation of the unseen classes. These approaches are known as *Embedding methods*. The initial work was focused on two-stage approaches where the attributes of an input image are estimated in the first stage, and the category label is determined in the second phase using highest attribute similarity. DAP and IAP [22] are examples of such an approach. Later, the use of bi-linear compatibility function led to promising ZSL results. Among these, ALE [1] and DEVISE [16] use ranking loss to learn the bilinear compatibility function. ESZSL [30] applies the square loss and explicitly regularizes the objective w.r.t. the Frobenius norm. Unlike standard embedding methods, [8, 40] propose reverse mapping from the semantic to the visual space and perform nearest neighbor classification in the visual space. The hybrid models such as CONSE [26], SSE [41], and SYNC [7], discuss the idea of embedding both visual and semantic features into another intermediate

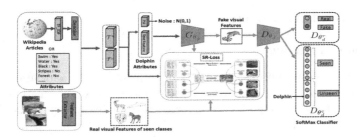

Fig. 2. Conceptual illustration of our LsrGAN model. The basis of our model is a conditional WGAN. The novel SR-Loss is introduced to help the G_{θ_g} to understand the semantic relationship between classes and guide it for applying the same during visual feature generation. The G_{θ_g} will use T^s and T^u to generate visual features. The D_{θ_d} has two branches used to perform real/fake game and classification. Notice that when we train the G_{θ_g} using T^u, only the classification branch remains active in D_{θ_d} as the unseen visual features are not available.

space. These methods perform well in the ZSL setting. However, in GZSL, they show a high bias towards seen classes.

Most of the mentioned embedding methods have a bias towards seen classes leading to substandard GZSL performance. Recently, *Generative methods* [10,15,18,24,31,33,35,43] have attempted to address this concern by synthesizing unseen class features, leading to state of the art ZSL and GZSL performance. Among these, [10,15,24,35,43] are based on Generative Adversarial Networks (GAN), while [5,31] use Variational Autoencoders (VAE) for the ZSL. F-GAN [35] uses a Wasserstein GAN [4,17] to synthesize samples based on the class conditioned attribute. LisGAN [24] proposes a *soul-sample* regularizer to guide the sample generation process to stay close to a generic example of the seen class. Inspired by the cycle consistency loss[42], Felix et al. [15], propose a generative model that maps the generated visual features back to the original semantic feature, thus addressing the unpaired training issue during generation. These generative methods use the annotated attribute as semantic information for the feature generation. However, in reality, such a desired form of the semantic feature representation is hard to obtain. Hence, [13] suggests the use of Wikipedia descriptions for ZSL and GZSL. GAZSL [43] proposes a very first generative model that handles the Wikipedia description to generate features.

Although generative methods have been quite successful in ZSL, the unseen feature generation is biased towards seen classes leading to poor generalization in GZSL. For better generalization, we have proposed a novel SR-loss to leverage the inter-class relationships between seen and unseen classes in GANs. The idea of utilizing inter-class relationships has been addressed before with Triplet loss-based approaches [3,6] and using contrastive networks [20]. In this work, we propose a related approach using generative models. To best of our knowledge, such an approach using GANs has not been investigated.

3 Proposed Approach

3.1 Problem Settings

The zero-shot learning problem consists of seen (observed) and unseen (unobserved) categories of images and their corresponding text information. Images belonging to the seen categories are passed through a feature extractor (for e.g., ResNet-101) to yield features $\{x_i^s\}_{i=1}^{n_s}$, where $x \in \mathcal{X}^s$. The corresponding labels for these features are $\{y_i^s\}_{i=1}^{n_s}$, where $y^s \in \mathcal{Y}^s = \{1, \ldots, C_s\}$ with C_s seen categories. The image features for the unseen categories are denoted as $\{x_i^u\}_{i=1}^{n_u}$, where $x \in \mathcal{X}^u$, and the space of all image features is $\mathcal{X} := \mathcal{X}^s \cup \mathcal{X}^u$. As the name indicates, unseen categories are not observed, and the zero-shot learning model attempts to hallucinate these features with the rest of the information provided. Although we do not have the image features for the unseen categories, we are privy to the C_u unseen categories, where the corresponding labels for the unseen image features would be $\{y_i^u\}_{i=1}^{n_u}$, with $y^u \in \mathcal{Y}^u = \{C_s + 1, \ldots, C\}$, where $C = C_s + C_u$. From the text domain, we have the semantic features for all categories which are either binary attribute vectors or Term-Frequency-Inverse-Document-Frequency (TF-IDF) vectors. The category-wise semantic features are denoted as $\{t_c^s\}_{c=1}^{C_s}$, for seen categories and $\{t_c^u\}_{c=C_s+1}^{C}$ with $t \in \mathcal{T} := \mathcal{T}^s \cup \mathcal{T}^u$. The goal of zero-shot learning is to build a classifier $\mathcal{F}_{zsl} : \mathcal{X}^u \rightarrow \mathcal{Y}^u$, mapping image features to unseen categories, and the goal of the more difficult problem of generalized zero-shot learning is to build a classifier $\mathcal{F}_{gzsl} : \mathcal{X} \rightarrow \mathcal{Y} := \mathcal{Y}^s \cup \mathcal{Y}^u$, mapping image features to seen and unseen categories.

3.2 Adversarial Image Feature Generation

Using the image features of the seen categories and the semantic features of the seen and unseen categories, we propose a generative adversarial network to hallucinate the unseen image features for each of the unseen categories. We apply a conditional Wasserstein Generative Adversarial Network (WGAN) to generate image features for the unseen categories using semantic features as input [4]. The WGAN aligns the real and generated image feature distributions. In addition, we have a feature classifier that is trained to classify image features into C categories of seen and unseen classes. The components of the WGAN are described in the following.

Feature Generator: The conditional generator in the WGAN has parameters θ_g and is represented as $G_{\theta_g} : \mathcal{Z} \times \mathcal{T} \rightarrow \mathcal{X}$, where $\mathcal{Z}$ is the space of random normal vectors $\mathcal{N}(0, I)$ of $|\mathcal{Z}|$ dimensions. Since the TF-IDF features from the Wikipedia articles may contain repetitive and non-discriminating feature information, we apply a denoising transformation upon the TF-IDF vector using a fully-connected neural network layer as proposed by [43]. The WGAN takes as input a random noise vector $z \in \mathcal{Z}$ concatenated with the semantic feature vector t_c for a category c, and generates an image feature $\tilde{x}_c \leftarrow G_{\theta_g}(z, t_c)$. The generator is trained to generate image features for both seen categories

$(\tilde{\boldsymbol{x}}_c^s \leftarrow G_{\theta_g}(\boldsymbol{z}, \boldsymbol{t}_c^s))$ and unseen categories $(\tilde{\boldsymbol{x}}_c^u \leftarrow G_{\theta_g}(\boldsymbol{z}, \boldsymbol{t}_c^u))$. In order to generate image features that are structurally similar to the real image features, we implement visual pivot regularization, $\mathcal{L}_{vp}$ that aligns the cluster centers of the real image features with the cluster centers of the generated image features for each of the C_s categories [43]. This is implemented only for the seen categories where we have real image features.

$$\mathcal{L}_{vp} = \min_{\theta_g} \frac{1}{C_s} \sum_{c=1}^{C_s} \left|\left| \mathbb{E}_{(x,y=c)\sim(\mathcal{X}^s,\mathcal{Y}^s)}[\boldsymbol{x}] - \mathbb{E}_{(z,t_c^s)\sim(\mathcal{Z},\mathcal{T}^s)}[G_{\theta_g}(\boldsymbol{z}, \boldsymbol{t}_c^s)]\right|\right|. \quad (1)$$

Feature Discriminator: We train the WGAN with an adversarial discriminator having two branches to perform the real/fake game and classification. The discriminator has parameters θ_d^r and θ_d^c for two branches respectively and is denoted as D_{θ_d}. The real/fake branch of the discriminator learns a mapping $D_{\theta_d^r} : \mathcal{X} \rightarrow \mathbb{R}$ using the generated and real image features to output $D_{\theta_d^r}(\tilde{\boldsymbol{x}}^s)$ and $D_{\theta_d^r}(\boldsymbol{x}^s)$ that are used to estimate the objective term $\mathcal{L}_d$. The objective $\mathcal{L}_d$ is maximized w.r.t. the discriminator parameters θ_d^r and minimized w.r.t. the generator parameters θ_g.

$$\mathcal{L}_d = \min_{\theta_g} \max_{\theta_d} \mathbb{E}_{\boldsymbol{x}\sim\mathcal{X}^s}\left[D_{\theta_d}(\boldsymbol{x})\right] - \mathbb{E}_{(z,t)\sim(\mathcal{Z},\mathcal{T}^s)}\left[D_{\theta_d}(G_{\theta_g}(\boldsymbol{z},\boldsymbol{t}))\right] - \lambda_{gp}\mathcal{L}_{gp} \quad (2)$$

where, the first two terms control the alignment of real image feature and the generated image feature distributions. The third term is the gradient penalty to enforce the Lipschitz constraint with $\mathcal{L}_{gp} = (||\nabla_{\boldsymbol{x}} D_{\theta_d}(\boldsymbol{x})||_2 - 1)^2$ where, input $\boldsymbol{x}$ are real image features, generated image features and random samples on a straight line connecting real image features and generated image features [17]. The parameter λ_{gp} controls the importance of the Lipschitz constraint. The discriminator parameters are trained using only seen category image features since $\boldsymbol{x}^u$ are unavailable.

Feature Classifier: The category classifier has parameters θ_d^c and is denoted as $D_{\theta_d^c}$. It is a softmax cross-entropy classifier for the generated and real image features $\boldsymbol{x}^s, \tilde{\boldsymbol{x}}^s, \tilde{\boldsymbol{x}}^u$ and is trained to minimize the loss $\mathcal{L}_c$. For ease of notation we represent the real/generated image features as $\boldsymbol{x}$ and corresponding labels y

$$\mathcal{L}_c = \min_{\theta_g, \theta_d^c} -\mathbb{E}_{(x,y)\sim(\mathcal{X},\mathcal{Y})}\left[\sum_{c=1}^{C} 1(y = c)\log(D_{\theta_d^c}(\boldsymbol{x}))_c\right], \quad (3)$$

where $(D_{\theta_d^c}(\boldsymbol{x}))_c$ is the c-th component of the C-dimension softmax output and $1(y = c)$ is the indicator function. While the discriminator performs a marginal alignment of real and generated features, the classifier performs category based conditional alignment.

So far, in the proposed WGAN model the *Generator* generates image features from seen and unseen categories. The *Discriminator* aligns the image feature distributions and the *Classifier* performs image classification. The LsrGAN model is illustrated in Fig. 2. In the following we introduce a novel regularization technique that transfers knowledge across the semantic and image feature spaces for the unseen and seen categories respectively.

3.3   Semantic Relationship Regularization

Conventional generative zero-shot learning approaches [24,35,43], have poor generalization performance in GZSL since the generated visual features are biased towards the seen classes. We address this issue by proposing a novel regularization procedure that will explicitly transfer knowledge of unseen classes from the semantic domain to guide the model in generating distinct seen and unseen image features. We term this the "Semantic Regularized Loss (SR-Loss)".

We argue that the visual and semantic feature spaces share a common underlying latent space that generates the visual and semantic features. We propose to exploit this relationship by transferring knowledge from the semantic space to the visual space to generate image features. Knowing the inter-class relationships in the semantic space can help us impose the same relationship constraints among the generated visual features. This is the idea behind the SR-Loss in the WGAN where we transfer inter-class relationships from the semantic domain to the visual domain. Figure 1 illustrates this concept. The visual similarity between class c_i and c_j is represented as $\mathcal{X}_{sim}(\boldsymbol{\mu}_{c_i}, \boldsymbol{\mu}_{c_j})$, where $\boldsymbol{\mu}_c$ is the mean of the image features of class c. Note that for visual similarity we are considering the relationship between the class centers and not between individual image features. Likewise, the semantic similarity between class c_i and c_j is represented as $\mathcal{T}_{sim}(\boldsymbol{t}_{c_i}, \boldsymbol{t}_{c_j})$. For semantic similarity, we do not have the mean, since we only have one semantic vector $\boldsymbol{t}_c$ for every category, although the proposed approach can be extended to include multiple semantic feature vectors. We impose the following semantic relationship constraint for the image features,

$$\mathcal{T}_{sim}(\boldsymbol{t}_{c_i}, \boldsymbol{t}_{c_j}) - \epsilon_{ij} \ \leq \ \mathcal{X}_{sim}(\boldsymbol{\mu}_{c_i}, \boldsymbol{\mu}_{c_j}) \ \leq \ \mathcal{T}_{sim}(\boldsymbol{t}_{c_i}, \boldsymbol{t}_{c_j}) + \epsilon_{ij}, \qquad (4)$$

where, hyper-parameter $\epsilon_{ij} \geq 0$ is a soft margin enforcing the similarity between semantic and image features for classes i and j. Large values of ϵ_{ij} allow more deviation between semantic similarities and visual similarities. We incorporate the constraint into the objective by applying the penalty method [25],

$$p_{c_{ij}} \big[|| \max(0, \mathcal{X}_{sim}(\boldsymbol{\mu}_{c_i}, \boldsymbol{\mu}_{c_j}) - (\mathcal{T}_{sim}(\boldsymbol{t}_{c_i}, \boldsymbol{t}_{c_j}) + \epsilon_{ij}))||^2$$
$$+ || \max(0, (\mathcal{T}_{sim}(\boldsymbol{t}_{c_i}, \boldsymbol{t}_{c_j}) - \epsilon_{ij}) - \mathcal{X}_{sim}(\boldsymbol{\mu}_{c_i}, \boldsymbol{\mu}_{c_j}))||^2 \big], \qquad (5)$$

where, $p_{c_{ij}}$ is the penalty for violating the constraint. The penalty becomes zero when the constraints are satisfied and is non-zero otherwise. We use cosine distance to compute the similarities.

We intend to transfer semantic inter-class relationships to enhance the image feature representations that are output from the generator. Consider a seen class c_i. We estimate its semantic similarity $\mathcal{T}_{sim}(\boldsymbol{t}_{c_i}, \boldsymbol{t}_{c_j})$ with all the other seen semantic features $\boldsymbol{t}_{c_j}$ where $j \in \{1, \ldots, C_s\} \wedge j \neq i$. Not all the similarities are important and for the ease of implementation we select the highest n_c similarities that we wish to transfer. Let I_{c_i} represent the set of n_c seen categories with the highest semantic similarity with c_i. We apply Eq. 5 to train the generator to output image features that satisfy the semantic similarity constraints against the top n_c similarities from the seen categories. For the seen image categories, the objective function is,

$$\mathcal{L}_{sr}^s = \min_{\theta_g} \frac{1}{C_s} \sum_{i=1}^{C_s} \sum_{j \in I_{c_i}} [\| \max(0, \mathcal{X}_{sim}(\boldsymbol{\mu}_{c_j}^s, \tilde{\boldsymbol{\mu}}_{c_i}^s) - (\mathcal{T}_{sim}(t_{c_j}^s, t_{c_i}^s) + \epsilon))\|^2$$
$$+ \| \max(0, (\mathcal{T}_{sim}(t_{c_j}^s, t_{c_i}^s) - \epsilon) - \mathcal{X}_{sim}(\boldsymbol{\mu}_{c_j}^s, \tilde{\boldsymbol{\mu}}_{c_i}^s))\|^2], \qquad (6)$$

where, penalty $p_{ij} = 1$, and $\boldsymbol{\mu}_{c_j}^s := \mathbb{E}_{(x,y=c_j) \sim (\mathcal{X}^s, \mathcal{Y}^s)}[\boldsymbol{x}]$ is the mean of the image features for seen class c_j, and $\tilde{\boldsymbol{\mu}}_{c_i}^s := \mathbb{E}_{\boldsymbol{z} \sim \mathcal{Z}}[G_{\theta_g}(\boldsymbol{z}, t_{c_i}^s)]$ is the mean of the generated image features of seen class c_i. We use a constant value of ϵ as the soft margin to simplify the solution. Similarly, the objective function for the unseen categories is,

$$\mathcal{L}_{sr}^u = \min_{\theta_g} \frac{1}{C_u} \sum_{i=C_s+1}^{C} \sum_{j \in I_{c_i}} [\| \max(0, \mathcal{X}_{sim}(\boldsymbol{\mu}_{c_j}^s, \tilde{\boldsymbol{\mu}}_{c_i}^u) - (\mathcal{T}_{sim}(t_{c_j}^s, t_{c_i}^u) + \epsilon_{ij}))\|^2$$
$$+ \| \max(0, (\mathcal{T}_{sim}(t_{c_j}^s, t_{c_i}^u) - \epsilon_{ij}) - \mathcal{X}_{sim}(\boldsymbol{\mu}_{c_j}^s, \tilde{\boldsymbol{\mu}}_{c_i}^u))\|^2], \qquad (7)$$

where, $\tilde{\boldsymbol{\mu}}_{c_i}^u := \mathbb{E}_{\boldsymbol{z} \sim \mathcal{Z}}[G_{\theta_g}(\boldsymbol{z}, t_{c_i}^u)]$ is the mean of the generated image features of unseen class c_i. The time complexity of the proposed SR-loss is $\mathcal{O}(Bn_c|\mathcal{X}|E) + \mathcal{O}(C^2|\mathcal{T}|E) + \mathcal{O}(C^2 \log n_c E)$ where B, n_c, $|\mathcal{X}|$, E, C and $|\mathcal{T}|$ denote the batch size, neighbor size, number of visual features, epoch, number of total classes and the size of the semantic features respectively. This shows the computation cost is manageable.

3.4   LsrGAN Objective Function

The LsrGAN leverages the semantic relationship between seen and unseen categories to generate robust image features for unseen categories using the objective function defined in Eq. 6 and 7. The model generates robust seen image features conditioned by the regularizer in Eq. 1. The LsrGAN trains a classifier over all the C categories as outlined in Eq. 3. The LsrGAN is based on a WGAN model that aligns the image feature distributions using the objective function defined in Eq. 2. The overall objective function of the LsrGAN model is given by,

$$\lambda_c \mathcal{L}_c + \mathcal{L}_d + \lambda_{vp} \mathcal{L}_{vp} + \lambda_{sr}(\mathcal{L}_{sr}^s + \mathcal{L}_{sr}^u) \qquad (8)$$

where, λ_c, λ_{vp} and λ_{sr} are hyper parameters controlling the importance of each of the loss terms. Unlike standard zero-shot learning models that generate image features and then have to train a supervised classifier [24,35,43], the LsrGAN model has an inbuilt classifier that can also be used for evaluating zero-shot learning and generalized zero-shot learning.

4   Experiments

4.1   Datasets

Attribute-Based Datasets: We conduct experiments on three widely used attribute-based datasets: *Animal with Attributes* (AWA) [22], *Caltech-UCSD-Birds 200-2011* (CUB) [34] and Scene UNderstanding (SUN) [27]. AWA is a medium scale coarse-grained animal dataset having 50 animal classes with 85 attributes annotated. CUB is a fine-grained, medium-scale dataset having 200 bird classes annotated with 312 attributes. SUN is a medium scale dataset having 717 types of scenes with 102 annotated attributes. We followed the split mentioned in [36] to have a fair comparison with existing approaches.

Wikipedia Descriptions-Based Datasets: In order to address a more challenging ZSL problem with Wikipedia descriptions as auxiliary information, we have used two common fine-grained datasets with textual descriptions: CUB and *North America Birds* (NAB) [32]. The NAB dataset is larger compared to CUB having 1011 classes in total. We have used two splits, suggested by [14] in our experiments to have a fair comparison with other methods. The splits are termed as *Super-Category-Shared* (SCS, Easy split) and *Super-Category-Exclusive* (SCE, Hard split). These splits represent the similarity between seen and unseen classes. The SCS-split has at least one seen class for every unseen class belonging to the same parent. For example, "Harris's Hawk" in the unseen set and "Cooper's Hawk" in the seen set belong to the same parent category, "Hawks." On the other hand, in the SCE-split, the parent categories are disjoint for the seen and unseen classes. Therefore, SCS and SCE splits are considered as Easy and Hard splits. The details for each dataset and class splits are given in Table 1.

Table 1. Dataset information. For the attribute-based datasets, the (number) in seen classes denotes the number of classes used for test in GZSL.

	Attribute-based			Wikipedia descriptions			
	AWA	CUB	SUN	CUB (Easy)	CUB (Hard)	NAB (Easy)	NAB (Hard)
No. of Samples	30,475	11,788	14,340	11,788	11,788	48,562	48,562
No. of Features	85	312	102	7551	7551	13217	13217
No. of Seen classes	40(13)	150(50)	645(65)	150	160	323	323
No. of Unseen classes	10	50	72	50	40	81	81

4.2   Implementation Details and Performance Metrics

The 2048-dimensional ResNet-101 [19] features are considered as a real visual feature for attribute-based datasets, and part-based features (e.g., belly, leg, wing, etc.) from VPDE-net [39] are used for the Wikipedia based datasets, as suggested by [35,43]. We have utilized the TF-IDF to extract the features from the Wikipedia descriptions. For a fair comparison, all of our experiment settings are kept the same as reported in [35,36,43].

Table 2. ZSL and GZSL results on AWA, CUB, and SUN with attributes as semantic information. T1 indicates the Top-1% accuracy in the ZSL setting. On the other hand, "U", "S" and "H" denotes the Top-1% accuracy for the unseen, seen, and Harmonic mean (seen + unseen).

Methods	Zero-shot learning			Generalized zero-shot learning								
	AWA	CUB	SUN	AwA			CUB			SUN		
	T1	T1	T1	U	S	H	U	S	H	U	S	H
DAP [22]	44.1	40.0	39.9	0.0	88.7	0.0	1.7	67.9	3.3	4.2	25.2	7.2
CONSE [26]	45.6	34.3	38.8	0.4	88.6	0.8	1.6	72.2	3.1	6.8	39.9	11.6
SSE [41]	60.1	43.9	51.5	7.0	80.5	12.9	8.5	46.9	14.4	2.1	36.4	4.0
DeViSE [16]	54.2	50.0	56.5	13.4	68.7	22.4	23.8	53.0	32.8	16.9	27.4	20.9
SJE [2]	65.6	53.9	53.7	11.3	74.6	19.6	23.5	59.2	33.6	14.7	30.5	19.8
ESZSL [30]	58.2	53.9	54.5	5.9	77.8	11.0	2.4	70.1	4.6	11.0	27.9	15.8
ALE [1]	59.9	54.9	58.1	14.0	81.8	23.9	4.6	73.7	8.7	21.8	33.1	26.3
SYNC [7]	54.0	55.6	56.3	10.0	**90.5**	18.0	7.4	66.3	13.3	7.9	**43.3**	13.4
SAE [21]	53.0	33.3	40.3	1.1	82.2	2.2	0.4	**80.9**	0.9	8.8	18.0	11.8
DEM [40]	68.4	51.7	61.9	30.5	86.4	45.1	11.1	75.1	19.4	20.5	34.3	25.6
PSR [3]	63.8	56.0	61.4	20.7	73.8	32.3	24.6	54.3	33.9	20.8	37.2	26.7
TCN [20]	70.3	59.5	61.5	49.4	76.5	60.0	52.6	52.0	52.3	31.2	37.3	34.0
GAZSL [43]	68.2	55.8	61.3	19.2	86.5	31.4	23.9	60.6	34.3	21.7	34.5	26.7
F-GAN [35]	68.2	57.3	60.8	**57.9**	61.4	59.6	43.7	57.7	49.7	42.6	36.6	39.4
cycle-CLSWGAN [15]	66.3	58.4	60.0	56.9	64.0	60.2	45.7	61.0	52.3	**49.4**	33.6	40.0
LisGAN [24]	**70.6**	58.8	61.7	52.6	76.3	62.3	46.5	57.9	51.6	42.9	37.8	40.2
LsrGAN [ours]	66.4	**60.3**	**62.5**	54.6	74.6	**63.0**	48.1	59.1	**53.0**	44.8	37.7	**40.9**

The base block of our model is GAN, which is implemented using a multi-layer perceptron. Specifically, the feature generator G_{θ_g} has one hidden unit having 4096 neurons and LeakyReLU as an activation function. For attribute-based datasets, we intend to get the top max-pooling units of ResNet - 101 (visual features). Hence, the output layer has ReLU activation in the feature generator. On the other hand, for the Wikipedia based datasets, we have used Tanh as an output activation for the feature generator since the VPDE-net feature varies from -1 to 1. $\mathcal{Z}$ is sampled from the normal Gaussian distribution. To perform the denoising and dimensionality reduction from Wikipedia descriptions, we have employed a fully connected layer with a feature generator. Also, notice that the semantic similarity for the SR-Loss is computed using the denoiser's output in Wikipedia based datasets. We will discard this layer when dealing with attribute-based datasets. In our model, the discriminator D_{θ_d} has two branches. One is used to play the real/fake game, and the other performs the classification on the generated/real visual feature. The discriminator also has 4096 units in the hidden layer with ReLU as an activation. Since the cosine distance is less prone to the curse of dimensionality when the features are sparse (semantic features), we have considered it in the SR-loss.

To perform the zero-shot recognition we have used nearest neighbor prediction on datasets having Wikipedia descriptions, and the classifier attached to the discriminator for the attributes based recognition. Notice that the classifier

Table 3. ZSL and GZSL results on CUB and NAB datasets with Wikipedia descriptions as semantic information on the two-split setting. We have used Top-1% accuracy and Seen-Unseen AUC (%) for ZSL and GZSL, respectively.

Methods	Zero-shot Learning				Generalized zero-shot Learning			
	CUB		NAB		CUB		NAB	
	Easy	Hard	Easy	Hard	Easy	Hard	Easy	Hard
WAC-Linear [13]	27.0	5.0	-	-	23.9	4.9	23.5	-
WAC-Kernal [12]	33.5	7.7	11.4	6.0	14.7	4.4	9.3	2.3
ESZSL [30]	28.5	7.4	24.3	6.3	18.5	4.5	9.2	2.9
ZSLNS [28]	29.1	7.3	24.5	6.8	14.7	4.4	9.3	2.3
Sync-fast [7]	28.0	8.6	18.4	3.8	13.1	4.0	2.7	3.5
ZSLPP [14]	37.2	9.7	30.3	8.1	30.4	6.1	12.6	3.5
GAZSL [43]	43.7	10.3	35.6	8.6	35.4	8.7	20.4	5.8
LsrGAN [ours]	**45.2**	**14.2**	**36.4**	**9.04**	**39.5**	**12.1**	**23.2**	**6.4**

is not re-trained for the recognition part. The Top-1 accuracy is used to assess the ZSL setting. Furthermore, to capture the more realistic scenario, we have examine the Generalized zero-shot recognition as well. As suggested by [9], we report the area under the seen and unseen curve (AUC score) as GZSL performance metric for Wikipedia based datasets, and the harmonic mean of the seen and unseen Top-1 accuracies for attribute based dataset. Notice that the choice of these measures and predictions models is to make a fair comparison with existing methods.

4.3   ZSL and GZSL Performance

The results for the ZSL are provided in the left part of Table 2 and 3. It can be seen that our LsrGAN achieves superior performance in both attribute and Wikipedia based datasets compared to the previous state of the art models, especially with generative models GAZSL, F-GAN, cycle-CLSWGAN, and LisGAN. It is worth noticing that all the mentioned generative models have the same base architecture. e.g., WGAN. Hence, the superiority of our model suggests that our motivation is realistic, and our experiments are effective. In summary, we achieve, 1.5%, 3.9%, 0.8%, 0.44% improvement on CUB (Easy), CUB (Hard), NAB (Easy) and NAB (Hard) respectively for the Wikipedia based datasets under ZSL. On the other hand, for the attribute-based ZSL, we attain 1.5% and 0.8% improvement on CUB and SUN, respectively. ZSL under performs on AWA probably due to the high feature correlation between similar unseen classes having a common neighbor among the seen classes, e.g. Dolphin and Blue whale. The availability of similar unseen classes slightly affects the prediction capability of the classifier for ZSL as the SR-loss tends to cluster them together due to high semantic similarity with common seen classes. However, it is worth noticing that the GZSL result for the same dataset is superior.

Our primary focus is to elevate the GZSL performance in this work, which is apparent from the right side of Table 2 and 3. Following [35, 36, 43], we report the harmonic mean and AUC score for the attribute and Wikipedia based datasets, respectively. The mentioned metrics help us to showcase the approach's generalizability as the harmonic mean and AUC scores are only high when the performance on seen and unseen classes is high. From the results, we can see that LsrGAN outperforms the previous state of the art for the GZSL. In terms of numbers, we achieve, 4.1%, 3.4% 2.8% and 0.6% gain on CUB (Easy), CUB (Hard), NAB (Easy) and NAB (Hard) respectively for the Wikipedia based datasets and 0.7%, 1.4% and 0.7% improvement on attribute-based AWA, CUB and SUN respectively. It is worth noticing that the LsrGAN improves the unseen Top-1 performance in the GZSL setting for the attribute-based CUB and SUN by 1.6% and 1.9% with the previous state of the art LisGAN [24].

The majority of the conventional approaches, including the generative models overfit the seen classes, which results in lower GZSL performance. Notice that most of the approaches mentioned in Table 2 achieve very high performance on seen classes compared to the unseen classes in the GZSL setting. For example, SYNC [7] has around 90% recognition capability on the seen classes, and it drops to only 10% (80% difference) for the unseen classes on AWA dataset. It is also evident from Tables 2 and 3 that the performance on the unseen categories drops drastically when the search space includes both seen and unseen classes in the GZSL setting. For instance, DAP [22] drops from 40% to 1.7%, GAZSL [43] drops from 55.8% to 23.9% and F-GAN [35] drops from 57.3% to 43.7% on attribute-based CUB dataset. This indicates that the previous approaches are easy to overfit the seen classes. Although the generative models have achieved significant progress compared to the previous embedding methods, they still possess the overfitting issue towards seen classes by having substandard GZSL performance. On the contrary, our model incorporates the novel SR-Loss that enables the explicit knowledge transfer from the similar seen classes to the unseen classes. Therefore, the proposed LsrGAN alleviates the overfitting concern and helps us to achieve a state of the art GZSL performance. It is worth noticing that our model not only outperforms the generative zero-shot models having single GAN but also proves its worth against cycle-GAN [15] in ZSL and GZSL setting. Lastly, to have fair comparisons, we have taken the performance numbers from [35, 36, 43].

4.4 Effectiveness of SR-Loss

Utilizing the semantic relationship between seen and unseen classes to infer the visual characteristics of an unseen class is at the heart of the proposed SR-Loss. Contrary to other generative approaches, it enables explicit knowledge transfer in the generative model to make it learn from the unseen classes together with seen classes during the training process itself. As a result, our LsrGAN will become more robust towards the unseen classes leading to address the seen class overfitting concern. To demonstrate such ability, we have computed the average class confidence score (Average Softmax probabilities) of the classifier

Fig. 3. Average class confidence score (Average SoftMax Probability) comparison for classifier trained with F-GAN and LsrGAN. Top 3 average guesses are mentioned here. The red marked label showcase the top 1 average guess. The class names with underline represent seen classes.

trained with the generated features from the LsrGAN or F-GAN model. Since the confidence scores are computed under the GZSL, the classifier's training set contains real visual features from the seen classes and generated unseen features from LsrGAN or F-GAN. To have a fair comparison, we have used the same F-GAN's Softmax classifier in LsrGAN. Since LsrGAN learns from the seen and unseen classes during the training process itself, the Softmax classifier associated with it is not trained in an offline fashion like in the F-GAN model.

Figure 3 depicts the confidence results on the AWA dataset under the GZSL setting. We have taken mainly four confusing seen and unseen classes for the comparison with classifier's top 3 guesses. It is evident from the figure that the classifier trained with the F-GAN has lower confidence for the unseen classes, and it mainly showcases very high confidence towards the similar seen classes even if the test image comes from the unseen classes. Also, during the seen class classification, the classifier's confidence values are mainly distributed among the seen classes in F-GAN. For instance, "mouse" has its confidence spread between mouse and hamster only. On the other hand, LsrGAN showcases decent confidence values for the seen and unseen, both leading to better GZSL performance. It is worth noticing that the LsrGAN fails for the "mouse" classification. However, the confidence is well spread among all the three categories showing it has considered an unseen class "rat" with other seen classes "mouse" and "Hamster". These observations reflect the fact that F-GAN has an overfitting issue towards the seen classes. On the other hand, the balanced performance of LsrGAN manifests that explicit knowledge transfer from SR-loss helps it to overcome the overfitting concern towards the seen classes. To bolster our claim, we have also computed the average class confidence across all the seen and unseen classes for these two models on attribute-based AWA, CUB, and SUN datasets. Table 4 reports average confidence values for seen and unseen classes. Clearly, it shows the superiority of LsrGAN in terms of generalizability compared to F-GAN.

Table 4. Comparison of average class confidence score across all seen or unseen classes (SoftMax Probability) between F-GAN and LsrGAN for attribute-based AWA, CUB and SUN

| | F-GAN [35] | | LrsGAN [ours] | |
	Unseen	Seen	Unseen	Seen
AWA	0.29	0.86	0.69	0.83
CUB	0.33	0.65	0.60	0.64
SUN	0.32	0.35	0.65	0.39

4.5 Model Analysis

Parameter Sensitivity: We have tuned our parameters following the conventional grid search approach. We mainly tune the SR-Loss parameters ϵ, λ_{sr} and n_c. For the fair comparison, we have adopted other parameters λ_{vp}, λ_{gp} from [35,43], also the λ_c is considered between $(0,1]$, specifically, 0.01 for the majority of our experiments. Figure 4(b)–(d) show the parameter sensitivity for the SR-Loss. Notice that we estimate the ϵ value from the seen class visual and semantic relations. It can be seen that a lower and higher value of ϵ affects the performance. On the other hand, λ_{sr} and n_c maintain consistent performance after reaching a certain threshold value.

(a) Ablation study   (b) ϵ (CUB E)   (c) λ_{sr} (AWA)   (d) n_c (CUB E)

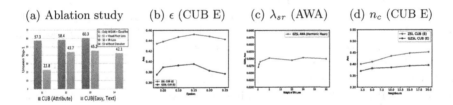

Fig. 4. Ablation study under ZSL(a), Parameter sensitivity for the SR-loss (b-d).

(a) ZSL Text   (b) GZSL Text   (c) ZSL-GZSL Attribute

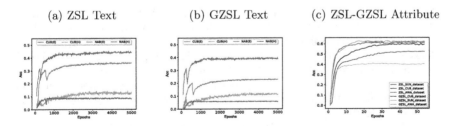

Fig. 5. Training stability (a-c) across all datasets under ZSL and GZSL

Training Stability: Since GANs are notoriously hard to train, and our proposed LsrGAN model not only uses GAN but also optimizes the similarity constraints from the SR-loss. Therefore, we also report the training stability for ZSL and GZSL for attribute and Wikipedia based datasets. Specifically, for the ZSL, we report the unseen Top-1 accuracy and Epoch behavior in Fig. 5(a) and Fig. 5(c). We have used harmonic mean of the seen and unseen Top-1 accuracy and Epoch to showcase the GZSL stability, mentioned in Fig. 5(b) and Fig. 5(c). Mainly we see the stable performance across all the datasets.

Ablation Study: We have reported the Ablation study of our model in Fig. 4(a) under ZSL for both attribute and Wikipedia based CUB. Primarily, we have used CUB (Easy) split for the Wikipedia based dataset. The $S1$ - WGAN with a classifier is considered as a baseline model. $S2$ reflects the effect of the visual pivot regularizer in our model. Finally, $S3$ showcase the performance of a complete LsrGAN with SR-loss. To highlight the effect of the denoiser used to process the Wikipedia based features, we have also reported the LsrGAN without denoiser in $S4$. In summary, Fig. 4 (a) showcases the importance of each component used in our model.

5   Conclusions

In this paper, we have proposed a novel generative zero-shot model named Lsr-GAN that Leverages the Semantic Relationship between seen and unseen classes to address the seen class overfitting concern in GZSL. Mainly, LsrGAN employs a novel Semantic Regularized Loss (SR-Loss) to perform explicit knowledge transfer from seen classes to unseen ones. The SR-Loss explores the semantic relationships between seen and unseen classes to guide the LsrGAN to generate visual features that mirror the same relationship. Extensive experiments on seven benchmarks, including attribute and Wikipedia description based datasets, verify that our LsrGAN effectively addresses the overfitting concern and demonstrates a superior ZSL and GZSL performance.

References

1. Akata, Z., Perronnin, F., Harchaoui, Z., Schmid, C.: Label-embedding for image classification. IEEE Trans. Pattern Anal. Mach. Intell. **38**(7), 1425–1438 (2015)
2. Akata, Z., Reed, S., Walter, D., Lee, H., Schiele, B.: Evaluation of output embeddings for fine-grained image classification. In: Proceedings of the IEEE Conference on Computer Vision and Pattern Recognition, pp. 2927–2936 (2015)
3. Annadani, Y., Biswas, S.: Preserving semantic relations for zero-shot learning. In: Proceedings of the IEEE Conference on Computer Vision and Pattern Recognition, pp. 7603–7612 (2018)
4. Arjovsky, M., Chintala, S., Bottou, L.: Wasserstein GAN. arXiv preprint arXiv:1701.07875 (2017)
5. Bucher, M., Herbin, S., Jurie, F.: Generating visual representations for zero-shot classification. In: Proceedings of the IEEE International Conference on Computer Vision Workshops, pp. 2666–2673 (2017)

6. Cacheux, Y.L., Borgne, H.L., Crucianu, M.: Modeling inter and intra-class relations in the triplet loss for zero-shot learning. In: Proceedings of the IEEE International Conference on Computer Vision, pp. 10333–10342 (2019)
7. Changpinyo, S., Chao, W.L., Gong, B., Sha., F.: Synthesized classifiers for zero-shot learning. In: CVPR, pp. 5327–5336 (2016)
8. Changpinyo, S., Chao, W.L., Sha, F.: Predicting visual exemplars of unseen classes for zero-shot learning. In: Proceedings of the IEEE International Conference on Computer Vision, pp. 3476–3485 (2017)
9. Chao, W.-L., Changpinyo, S., Gong, B., Sha, F.: An empirical study and analysis of generalized zero-shot learning for object recognition in the wild. In: Leibe, B., Matas, J., Sebe, N., Welling, M. (eds.) ECCV 2016. LNCS, vol. 9906, pp. 52–68. Springer, Cham (2016). https://doi.org/10.1007/978-3-319-46475-6_4
10. Chen, L., Zhang, H., Xiao, J., Liu, W., Chang, S.F.: Zero-shot visual recognition using semantics-preserving adversarial embedding networks. In: Proceedings of the IEEE Conference on Computer Vision and Pattern Recognition, pp. 1043–1052 (2018)
11. Ding, Z., Shao, M., Fu, Y.: Low-rank embedded ensemble semantic dictionary for zero-shot learning. In: Proceedings of the IEEE Conference on Computer Vision and Pattern Recognition, pp. 2050–2058 (2017)
12. Elhoseiny, M., Elgammal, A., Saleh, B.: Write a classifier: predicting visual classifiers from unstructured text. In: PAMI (2016)
13. Elhoseiny, M., Saleh, B., Elgammal, A.: Write a classifier: Zero-shot learning using purely textual descriptions. In: ICCV (2013)
14. Elhoseiny, M., Zhu, Y., Zhang, H., Elgammal, A.: Link the head to the "beak": Zero shot learning from noisy text description at part precision. In: CVPR, July 2017
15. Felix, R., Kumar, V., Reid, I., Carnerio, G.: Multi-model cycle-consistent generalized zero-shot leanring. In: ECCV (2018)
16. Frome, A., Corrado, G.S., Shlens, J., Bengio, S., Dean, J., Mikolov, T., et al.: Devise: a deep visual-semantic embedding model. In: NIPS, pp. 2121–2129 (2013)
17. Gulrajani, I., Ahmed, F., Arjovsky, M., Dumoulin, V., Courville, A.C.: Improved training of wasserstein GANs. In: Advances in Neural Information Processing Systems, pp. 5767–5777 (2017)
18. Guo, Y., Ding, G., Han, J., Gao, Y.: Synthesizing samples FRO zero-shot learning. IJCAI (2017)
19. He, K., Zhang, X., Ren, S., Sun, J.: Deep residual learning for image recognition. In: Proceedings of the IEEE Conference on Computer Vision and Pattern Recognition, pp. 770–778 (2016)
20. Jiang, H., Wang, R., Shan, S., Chen, X.: Transferable contrastive network for generalized zero-shot learning. In: Proceedings of the IEEE International Conference on Computer Vision, pp. 9765–9774 (2019)
21. Kodirov, E., Xiang, T., Gong, S.: Semantic autoencoder for zero-shot learning. In: Proceedings of the IEEE Conference on Computer Vision and Pattern Recognition, pp. 3174–3183 (2017)
22. Lampert, C.H., Nickisch, H., Harmeling, S.: Attribute-based classification for zero-shot visual object categorization. IEEE Trans. Pattern Anal. Mach. Intell. 36(3), 453–465 (2013)
23. Larochelle, H., Erhan, D., Bengio, Y.: Zero-data learning of new tasks. In: AAAI (2008)
24. Li, J., Jin, M., Lu, K., Ding, Z., Zhu, L., Huang, Z.: Leveraging the invariant side of generative zero-shot learning. In: CVPR (2019)

25. Lillo, W.E., Loh, M.H., Hui, S., Zak, S.H.: On solving constrained optimization problems with neural networks: a penalty method approach. IEEE Trans. Neural Networks **4**(6), 931–940 (1993)
26. Norouzi, M., et al.: Zero-shot learning by convex combination of semantic embeddings. arXiv preprint arXiv:1312.5650 (2013)
27. Patterson, G., Hays, J.: Sun attribute database: discovering, annotating, and recognizing scene attributes. In: CVPR (2012)
28. Qiao, R., Liu, L., Shen, C., Hengel, A.V.D.: Less is more: zero-shot learning from online textual documents with noise suppression. In: CVPR, June 2016
29. Rohrbach, M., Stark, M., Schiele, B.: Evaluating knowledge transfer and zero-shot learning in a large-scale setting. In: CVPR 2011, pp. 1641–1648. IEEE (2011)
30. Romera-Paredes, B., Torr, P.: An embarrassingly simple approach to zero-shot learning. In: ICML, pp. 2152–2161 (2015)
31. Verma, V., Arora, G., A.M., Rai, P.: Generalized zero-shot learning via synthesized examples. In: CVPR (2018)
32. Van Horn, G., et al.: Building a bird recognition app and large scale dataset with citizen scientists: the fine print in fine-grained dataset collection. In: Proceedings of the IEEE Conference on Computer Vision and Pattern Recognition, pp. 595–604 (2015)
33. Verma, V.K., Rai, P.: A simple exponential family framework for zero-shot learning. In: Ceci, M., Hollmén, J., Todorovski, L., Vens, C., Džeroski, S. (eds.) ECML PKDD 2017. LNCS (LNAI), vol. 10535, pp. 792–808. Springer, Cham (2017). https://doi.org/10.1007/978-3-319-71246-8_48
34. Welinder, P., et al.: Caltech-UCSD birds 200 (2010)
35. Xian, Y., Lorenz, T., Schiele, B., Akata, Z.: Feature generating networks for zero shot learning. In: CVPR (2018)
36. Xian, Y., Lampert, C.H., Schiele, B., Akata, Z.: Zero-shot learning - a comprehensive evaluation of the good, the bad and the ugly. In: TPAMI (2018)
37. Xu, X., Shen, F., Yang, Y., Zhang, D., Tao Shen, H., Song, J.: Matrix trifactorization with manifold regularizations for zero-shot learning. In: Proceedings of the IEEE Conference on Computer Vision and Pattern Recognition, pp. 3798–3807 (2017)
38. Yu, X., Aloimonos, Y.: Attribute-based transfer learning for object categorization with zero/one training example. In: Daniilidis, K., Maragos, P., Paragios, N. (eds.) ECCV 2010. LNCS, vol. 6315, pp. 127–140. Springer, Heidelberg (2010). https://doi.org/10.1007/978-3-642-15555-0_10
39. Zhang, H., et al.: SPDA-CNN: unifying semantic part detection and abstraction for fine-grained recognition. In: Proceedings of the IEEE Conference on Computer Vision and Pattern Recognition, pp. 1143–1152 (2016)
40. Zhang, L., Xiang, T., Gong, S.: Learning a deep embedding model for zero-shot learning. In: Proceedings of the IEEE Conference on Computer Vision and Pattern Recognition, pp. 2021–2030 (2017)
41. Zhang, Z., Saligrama, V.: Zero-shot learning via semantic similarity embedding. In: ICCV, pp. 4166–4174 (2015)
42. Zhu, J.Y., Park, T., Isola, P., Efros, A.A.: Unpaired image-to-image translation using cycle-consistent adversarial networks. In: Proceedings of the IEEE International Conference on Computer Vision, pp. 2223–2232 (2017)
43. Zhu, Y., Elhoseiny, M., Liu, B., Peng, X., Elgammal, A.: A generative adversarial approach for zero-shot learning from noisy texts. In: Proceedings of the IEEE Conference on Computer Vision and Pattern Recognition, pp. 1004–1013 (2018)

Weight Excitation: Built-in Attention Mechanisms in Convolutional Neural Networks

Niamul Quader$^{(\boxtimes)}$, Md Mafijul Islam Bhuiyan, Juwei Lu, Peng Dai, and Wei Li

Huawei Noah's Ark Lab, Montreal, Canada
{niamul.quader1,juwei.lu,peng.dai,wei.li.crc}@huawei.com,
mbhuiyan@ualberta.ca

Abstract. We propose novel approaches for simultaneously identifying important weights of a convolutional neural network (ConvNet) and providing more attention to the important weights during training. More formally, we identify two characteristics of a weight, its magnitude and its location, which can be linked with the importance of the weight. By targeting these characteristics of a weight during training, we develop two separate weight excitation (WE) mechanisms via weight reparameterization-based backpropagation modifications. We demonstrate significant improvements over popular baseline ConvNets on multiple computer vision applications using WE (e.g. 1.3% accuracy improvement over ResNet50 baseline on ImageNet image classification, etc.). These improvements come at no extra computational cost or ConvNet structural change during inference. Additionally, including WE methods in a convolution block is straightforward, requiring few lines of extra code. Lastly, WE mechanisms can provide complementary benefits when used with external attention mechanisms such as the popular Squeeze-and-Excitation attention block.

Keywords: Convolutional neural network · Convolution filter weights · Weight reparameterization · Attention mechanism

1 Introduction

Convolutional neural networks (ConvNets) are extremely powerful in analyzing visual imagery and have brought tremendous success in many computer vision tasks, e.g. image classification [9,29,40], video action and gesture recognition [14,34], etc. A ConvNet is made up of convolution blocks, each of which consists of a number of learnable parameters including convolution filter weights. The effectiveness of training these weights in convolution blocks can depend on the

Electronic supplementary material The online version of this chapter (https://doi.org/10.1007/978-3-030-58577-8_6) contains supplementary material, which is available to authorized users.

© Springer Nature Switzerland AG 2020
A. Vedaldi et al. (Eds.): ECCV 2020, LNCS 12375, pp. 87–103, 2020.
https://doi.org/10.1007/978-3-030-58577-8_6

ConvNet design, the training data available, choice of optimizer and many other factors [10,15,19,26,37].

As a ConvNet is trained, some of these weights become less important than others [18], and removing the less important weights in a ConvNet can often lead to compressed ConvNets without compromising on accuracy considerably [6–8,18,22,25]. Recently, Frankle & Carbin [4] hypothesized that an iteratively pruned network can be retrained from scratch to similar accuracy using its initial ConvNet parameters, suggesting that some weights may, in fact, start as being more important than others after random initialization of weights.

Assuming that some convolution kernel weights of a ConvNet are more important than others during the ConvNet training, we hypothesize that providing more attention to optimizing the more important weights during ConvNet training can improve ConvNet performances without the addition of computation costs or modifications in ConvNet structure during inference. To test this hypothesis, we implement two novel schemes that simultaneously identify the importance of weights and provide more attention to training the more important weights, and investigate their effects on ConvNet performances. The highlights of our contributions in this paper are:

1. We investigate two characteristics of weights (i.e., their magnitude and location in a ConvNet) that could provide indications on how important each of the weights is. Note that 'location' of a weight is defined by the layer the weight is in, and the input and output channel that the weight is connecting.
2. We propose two novel weight reparameterization mechanisms that target the identified characteristics of weights in (1) and modify the weights in a way that enables more attention to optimizing the more important weights via adjusting the backpropagated gradients. We broadly term such training mechanisms as Weight Excitation (WE)-based training.
3. We conduct several experiments on image classification (ImageNet [30], CIFAR100 [16]), semantic segmentation (PASCAL VOC [3], Cityscapes [2]), gesture recognition (Jester [33]) and action recognition (mini Kinetics [41]). The ConvNets studied for these tasks have varying types of convolutions (e.g. 2D convolution, 3D convolution, shift-based 3D convolution [20], etc.). Additionally, we experiment with ConvNets of different structures and model sizes, hyperparameter settings, normalization approaches, optimizers, and schedulers.

In all our experiments, WE-based training demonstrates considerable improvements over popular baseline ConvNets. The accuracy gains with using WE can be similar to that of adding the popular squeeze-and-excitation (SE) based attention to a baseline ConvNet [12]. Notably, in contrast to SE, WE-based training does not require any ConvNet structural changes or added costs at inference. Additionally, WE-based training complements prior attention mechanisms [12,27,35] in terms of improving accuracy. Finally, WE-based training provides considerably better acceleration and accuracy gains compared to the other weight reparameterization approaches [24,28,31].

2 Related Works

Identifying Important ConvNet Parameters: Identifying important parameters and then removing unimportant parameters is common in ConvNet pruning approaches [6–8,18,25]. Two common criteria for identifying important ConvNet parameters are - (1) higher minimal increase in training error when a parameter is removed corresponding to the parameter being of higher importance [8,18], and (2) higher magnitude parameter corresponding to higher importance (e.g. [22,25]). A less explored characteristic of a weight that can be linked to its importance is its location in a ConvNet. One evidence that the location of a weight could be important is in an ablation study in [12]. The ablation study shows that earlier layers in SE-added ConvNets tend to put more attention or importance on some activation map output channels than others regardless of the inputs to those layers [12], suggesting that the convolution filter weights involved in calculating those activation map output channels are more important than others. In this work, we use novel weight reparameterization approaches to provide more attention to training important weights identified using their magnitude and location characteristics.

Applying Attention on Weights: In prior works, attention mechanisms in ConvNets have been applied to activation maps [11,12,27,35]. Such activation map-based attention methods do not have fine-grained control on providing more attention to a particular weight in a convolution filter kernel - for example, in SE ConvNets [12], during a backpropagation through an excited activation map channel, attention is provided to all the weights that contributed to generating that activation map channel. In contrast, weight reparameterization-based WE methods can provide fine-grained weight-wise attention to the weights during backpropagation.

3 Technical Approach

We begin by investigating the relationship between the importance of a weight and its magnitude/location properties (Fig. 1a, Fig. 2) (Sect. 3.1). Next, we implement a weight reparameterization approach that simultaneously identifies important weights via location characteristics and provides more attention to them (Fig. 1c, Fig. 3a) (Sect. 3.2). We call this method location-based WE (LWE). We then implement another WE method that targets magnitude property of a weight for identifying its importance (Fig. 1b, Fig. 3b) (Sect. 3.3). We call this second method magnitude-based WE (MWE).

3.1 Investigating the Importance of Weights

To investigate the importance of weights, we systematically decimate the effects of weights (i.e., by making them zeroes) that exhibit certain characteristics in a ConvNet and investigate the resulting decrease in ConvNet performance. This approach is similar to finding the importance of a ConvNet parameter in [8,18].

We study two separate characteristics of a weight in a ConvNet - the magnitude and the location of a weight. For these investigations, we use a ResNet50 [9] ConvNet pretrained on the ImageNet [30] since it is popularly and variously used in computer vision research (e.g. [12,21]).

Weight Magnitude and Importance: To study this relationship, we do the following: (1) sort weights in each convolution layer in ascending order of absolute values of weights, (2) make the bth percentile position ($b \in \{1,2,3,4, \ldots, 100\}$) weight of the sorted absolute weights of each layer zero and record the decrease in performance D (Fig. 1a). Figure 1a shows our findings on decrease in performance D when a weight at position b is zeroed. For higher b, D progressively increases (Fig. 1a), suggesting that importance increases with magnitude.

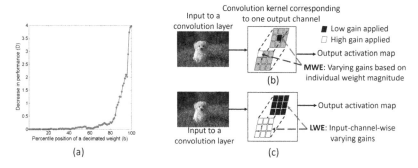

Fig. 1. (a) Higher magnitude weights correspond to larger D, and therefore are more important. To focus on optimizing more important weights, higher gain is provided to the more important weights via weight reparameterization. This reparameterization can be based on magnitude-importance relationship of each weight (b) (Sect. 3.3) or can be based on location-importance relationship of all weights along each of the input channels (c) (Sect. 3.2).

Weight Location and Importance: To study this relationship, we do the following: (1) select all the L convolutional blocks having 3×3 filters in the pretrained ResNet50 ($L = 16$ for ResNet50), (2) for each selected 3×3 convolution block (e.g. Fig. 3) (S_l, $l \in \{1, \ldots, L\}$), select N_1 output channels ($S_{l,O_j}, j \in \{1, \ldots, N_1\}$) equidistant from each other, (3) for each S_{l,O_j}, select N_2 input channels ($S_{l,O_j,I_i}, i \in \{1, \ldots, N_2\}$) equidistant from each other, and then (4) zero each of the weights in S_{l,O_j,I_i} and record the decrease in performance $D_{S_{l,O_j,I_i}}$. In this paper, we consider $N_1 = N_2 = 5$. This results in 5×5 D values for each layer of the pretrained ResNet50 (Fig. 2). Higher D values correspond to 3×3 filters that are more important to retaining the performances of the ResNet50. We notice larger fluctuations in D in the early layers, and the fluctuations in D tend to fade away in deeper layers, though some variations in D still remain at the deepest layer (Fig. 2). This suggests that the importance of weights in different locations of a ConvNet can be different and that the difference in importance is more obvious in the ConvNet's earlier layers.

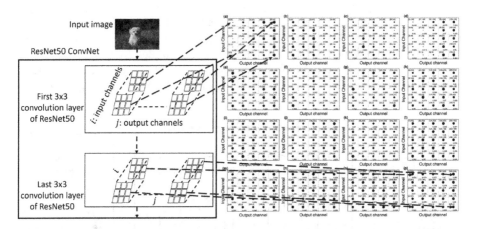

Fig. 2. Decrease in performance map (or importance map) of sampled 3×3 convolution kernels in various layers of the pretrained ResNet50. Here, (a) corresponds to the first convolution layer in the ConvNet that consists of 3 × 3 convolution kernels, and (b) to (p) correspond to progressively deeper layers containing 3 × 3 kernels. Early layers exhibit larger variation in 3 × 3 kernel importance than deeper layers.

3.2   Location-Based Weight Excitation

As we see in Sect. 3.1, importance of weight kernels can vary depending on where they are located (Fig. 2). Therefore, for weights W having dimensions $Out \times In \times h \times w$ in a convolutional layer that connects In input channels to Out output channels (e.g. Fig. 3a where $h = 3$ and $w = 3$), importance of each $h \times w$ weight kernels can vary. To provide separate attention to each of the $Out \times In$ weight kernels, an $Out \times In$-sized location-importance multiplier map ($m \in \mathbb{R}$: $m \in [0,1]$) can be multiplied to each of the $h \times w$ weight kernels separately. This will result in larger backpropagated gradients through weight kernels that were multiplied with larger m values. To find the multiplier map m, a simple approach could be to instantiate $Out \times In$-sized learnable parameters that are dynamically trained during training; however, this results in drastically increased ConvNet parameters (e.g. around 60% increase in parameters for ResNet50 [9]). Instead, we propose to use a simple subnetwork that takes in $In \times h \times w$ weights and generates In sized importance map values. The same subnetwork is re-used for all the Out pathways in W to generate m. While this subnetwork could have many different structures, we chose the SE block [12] for its efficient design - a small artificial neural network comprising of only two fully connected layers and some activation functions. The subnetwork (Fig. 3a) is defined as:

$$m_j = A_2(FC_2(A_1(FC_1(Avg(W_j))))) \tag{1}$$

where W_j is the weights across the jth output channel [28], Avg is an average pooling operation that averages every $h \times w$ to one averaged value, A_1 and A_2 are two activation functions (instantiated as ReLU and Sigmoid, respectively), FC_1

and FC_2 are two fully connected layers. To extend this formulation to 1D and 3D convolution layer, Avg is modified to average over w for a 1D convolution and over $t \times h \times w$ in a 3D convolution. Finally, m_j is expanded by value replication to a $In \times h \times w$, and multiplied elementwise to W_j (Fig. 3a). Repeating this across all output channels result in LWE transformed weights W_{LWE} (Fig. 3a).

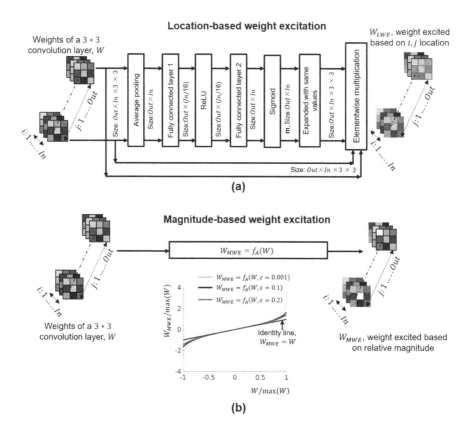

Fig. 3. (a) Outline of our LWE method that identifies location-wise importance map m, and then provides more attention to important locations via elementwise multiplication. (b) Outline of our MWE method that uses the f_A activation function to provide more attention to higher magnitude weights. This behavior is seen for all f_A under different ϵ_A values, with lower ϵ_A increasing this effect of attention. As ϵ_A value is increased, f_A behaves closer to an identity line (blue line). (Color figure online)

3.3 Magnitude-Based Weight Excitation

Our magnitude-based weight mechanism (MWE) is a novel activation function $f_A(\omega)$ that takes in a weight ω of W, and provides relatively larger gains to ω if it has a relatively larger magnitude among all weights of W. Additionally,

$f_A(\omega)$ is differentiable and avoids vanishing and exploding gradient problems. We define $f_A(\omega)$ which transforms ω to ω_{MWE} as:

$$\omega_{MWE} = f_A(W) = M_A \times 0.5 \times ln\frac{1+\omega/M_A}{1-\omega/M_A}, \tag{2}$$

where $M_A = (1 + \epsilon_A) \times M$, M is the maximum magnitude of a weight in W and ϵ_A is a hyperparameter with a small value greater than 0 (e.g. $0 < \epsilon_A < 0.2$). The gradient for ω (i.e., ∇_ω) becomes:

$$\nabla_\omega = M_A{}^2/(M_A{}^2 - \omega^2) \times \nabla_{\omega_{MWE}}. \tag{3}$$

For $\epsilon_A > 0$, $\omega < M_A$ and the denominator of $M_A{}^2/(M_A{}^2 - \omega^2)$ remains numerically stable. Minimum value of $M_A{}^2/(M_A{}^2 - \omega^2)$ is 1 when ω has a zero magnitude. As magnitude of ω increases, $M_A{}^2/(M_A{}^2 - \omega^2)$ progressively increases upto a maximum value defined by ϵ_A (e.g. maximum of 5.76 when $\epsilon_A = 0.1$) for ω that has the highest magnitude in W. Thus, more attention or gain is provided to backpropagated gradients corresponding to higher magnitude weights. For larger values of ϵ_A, f_A becomes closer to an identity line, whereas smaller values of ϵ_A increases the level of attention provided via MWE (Fig. 3b).

Note that, before feeding W as input for LWE or MWE, we can normalize W similar to normalizing an input before being fed to a neural network. We do this by standardizing W across each jth output channel similar to [28]. Using normalized weight W_n instead of W as input results in additional small performance improvements on most ConvNets.

4   Experiments and Results

4.1   Experimental Setup

We test effectiveness of our proposed WE methods on image classification (ImageNet [30], CIFAR100 [16]), and semantic segmentation (PASCAL VOC [3]). Additionally, we test WE methods on a standard 3D convolution-based ConvNet [34] for action recognition on the Mini-Kinetics dataset [41], and on a temporal shift convolution-based 3D ConvNet [20] for gesture recognition on Jester dataset [33]. In all experiments, we modify all convolution blocks such that WE methods are added to the convolution blocks during training only, so no ConvNet structural change or additional cost is required at inference. We use normalized weights as inputs for WE methods, except when WE methods are used on depthwise convolutions. This exception is because ConvNet performances deteriorate when weights are normalized in depthwise convolutions [39].

4.2   ImageNet Image Classification

ImageNet is a large-scale standard dataset containing around 1.28 million training and 50K validation images [30]. For experiments on ImageNet, we used a standard data augmentation and training recipe as described in [42]. We adopt the data augmentation used in [9], and adopt the standard single-crop evaluation for testing [12,35]. We used a batch size of 128 for ResNet152 [9] and a batch size of 256 for all other ConvNets, and optimized with stochastic gradient descent (SGD) with backpropagation [17] having momentum 0.9 and a decay of 10^{-4}. Each of the ConvNets were trained for 100 epochs. Learning rate was initialized at 0.1 and reduced by a factor of 10^{-1} at 30th, 60th and 90th epochs. We used MWE with $\epsilon_A = 0.1$ in all convolution blocks. Since our LWE outperforms our MWE method, we primarily investigate efficacy of LWE on different ConvNets (Table 1).

Additionally, we conduct a series of ablation studies on the ImageNet image classification dataset [30]: (1) compare performance gains achieved with LWE against performance gains with activation map-based attention (or external attention) based approaches [12,27,35], (2) compare convergence speed with LWE-based training against convergence speeds with popular weight normalizing reparameterization approaches [24,28,31], (3) study effectiveness of LWE for different normalization approaches, optimizers and schedulers, (4) compare utility of LWE against that of MWE, (5) check validity of learned attention multiplier in LWE, and (6) study hyperparameter sensitivity.

WE Improves Accuracy on Baseline ConvNets: LWE-based training on popular baseline ConvNets of varying parameter sizes (e.g. MobileNetV2 [32], ResNet50 [9], ResNeXt50 (32×4d) [40], ResNet152-SE [12], Wide ResNet50-2 [43]) provides significant accuracy gains over baseline ConvNets (Table 1, Fig. 4a). MWE-based training also provides considerable accuracy gain on ResNet50 [9]; however, this accuracy gain is much lower than that with LWE. To provide perspective on the accuracy gains achieved with WE-based training, we compare them with the accuracy gains obtained using the popular SE blocks [12]. We find the two sets of accuracy gains to be similar (Table 1), except for MobileNetV2 [32] where SE [12] provides better accuracy gain. This exception is likely due to LWE's inability in providing excitations on depthwise convolution blocks that have a size of $Out \times In \times h \times w$ where $In = 1$. In such a case, we conjecture that any excitation provided by LWE's importance map m of sized $Out \times 1$ likely gets absorbed by the following batch normalization (BN) that also has a $Out \times 1$ learnable scaling paramter. This limitation of LWE in depthwise convolution is not shared by MWE - adding MWE on depthwise convolutions of MobileNetV2 provides improved performance gain (MobileNetV2-WE in Table 1).

Comparing LWE with Activation Map-Based Attention Mechanisms: LWE-based built-in weight attention mechanism brings large accuracy gains on baseline ConvNets similar to popular attention methods such as SE [12] (Table 1); however, in contrast to prior attention approaches [12,27,35] LWE does not burden the baseline ConvNets with any structural changes or any added computation costs at inference. Also, training costs with LWE methods are lower than prior attention approaches such as SE [12] since LWE is applied on weights whereas prior attention approaches are applied on considerably larger activation maps (e.g. lower training memory with LWE, see Table 1). Finally, LWE-based attention is applied on each of the In input filter kernels separately for each output channel j (Fig. 3), which is complementary to prior attention methods that provide attention to only the output channels separately [12,27,35]. We conjecture that, due to this complementary attention mechanism, consistent accuracy gains (Table 1) and speed in convergence (Fig. 4b) are observed when LWE is used with other attention methods [12,35]. Overall, the utility of LWE is in its ability to provide considerable accuracy gains that are comparable to SE [12] with lower training costs than SE and without addition of inference costs, and in LWE's complementary nature to prior attention approaches.

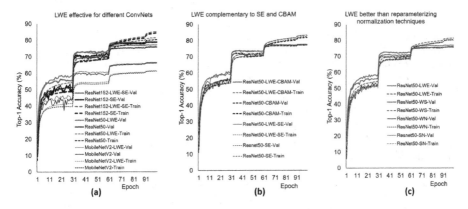

Fig. 4. (a) When LWE-based training (red lines) is applied on baseline ConvNets, consistent gains in performance are sustained throughout the training process. (b) Using LWE with other baseline activation map-based attention methods provide further performance gains. (c) Compared with weight normalizing reparameterization approaches that can speed up training convergence, LWE demonstrates faster convergence and better accuracy gains. (Color figure online)

Comparing LWE with Weight Normalizing Approaches: Weight normalizing reparameterization approaches have shown promise in speeding ConvNet training and in providing accuracy gains to baseline ConvNets [24,28,31]. We compare accuracy gains and speed of convergence of LWE-based training with three weight normalizing reparameterization approaches [24,28,31] for the ResNet50 [9] ConvNet. We find that LWE-based training outperforms all weight normalizing reparameterizing approaches by a considerable margin in terms of accuracy gain (Table 2) and speed of convergence (Fig. 4c). Taking training time into consideration, since LWE-based training adds around 7% training time compared to baseline ResNet50, we conservatively reduce total training epochs of LWE from 100 to 90 (with learning rate decays at 27th, 54th and 81st epochs), and still find improvements over other approaches (Table 2).

Applicability to Group Normalization, AdamW and Cosine Learning Rate: As a preliminary study to identify whether LWE retains its beneficial

Table 1. Validation accuracy of different baseline ConvNet architectures without and with WE-based training on the ImageNet image classification task. Memory requirements for training and inference are shown for batch size of 128. LWE consistently provides considerable accuracy gains that are comparable to accuracy gains achieved when SE blocks [12] are used with baseline ConvNet. In contrast to SE [12], WE methods do not add inference costs or ConvNet structural changes. Also, training with WE in baseline ConvNets require relatively small extra training memory compared to the extra memory requirement of SE (extra memory with WE around 1/3 to 2/3 of the extra memory with SE). LWE also complements different attention methods [12,27,35] in improving accuracy.

Method	Train Param. (M)	Train Mem. (GB)	Inference Param. (M)	Inference Mem. (GB)	GFLOPs	Top-1 Acc. (%)	Top-5 Acc. (%)
Small sized models							
MobileNetV2 [32]	3.5	9.5	3.5	2.9	0.317	66.2	87.1
MobileNetV2-**LWE**	4.1	9.6	3.5	2.9	0.317	66.5	87.3
MobileNetV2-**WE**	4.1	**9.6**	**3.5**	**2.9**	**0.317**	**67.0**	**87.5**
MobileNetV2-SE	3.6	9.8	3.6	3.0	0.320	67.3	87.8
MobileNetV2-**WE**-SE	4.2	9.9	**3.6**	**3.0**	**0.320**	**68.1**	**88.2**
ResNet18	11.7	4.9	11.7	2.2	1.80	70.6	89.5
ResNet18-**LWE**	11.9	**5.1**	**11.7**	**2.2**	**1.80**	**71.0**	**90.0**
ResNet18-SE	11.8	5.4	11.8	2.2	1.81	71.0	90.1
ResNet18-**LWE**-SE	11.9	5.6	**11.8**	**2.3**	**1.81**	**71.7**	**90.5**
Medium sized models							
ResNet50	25.6	14.0	25.6	2.6	3.86	75.8	92.8
ResNet50-BAM	25.9	15.8	25.9	2.8	3.94	76.0	92.8
ResNet50-**LWE**	28.1	**14.8**	**25.6**	**2.6**	**3.86**	**77.1**	**93.5**
ResNet50-SE	28.1	16.7	28.1	2.9	3.87	77.1	93.5
ResNet50-CBAM	28.1	19.3	28.1	3.4	3.87	77.2	93.7
ResNet50-**LWE**-SE	30.6	17.6	28.1	**2.9**	3.87	77.5	93.8
ResNet50-**LWE**-BAM	28.4	16.7	25.9	2.8	3.94	77.4	93.7
ResNet50-**LWE**-CBAM	30.6	20.2	**28.1**	3.4	**3.87**	**77.7**	**93.9**
ResNeXt50	25.0	16.8	25.0	2.4	4.24	77.2	93.4
ResNeXt50-**LWE**	27.9	**17.2**	**25.0**	**2.4**	**4.24**	**77.7**	**93.8**
ResNeXt50-SE	27.5	19.5	27.5	2.9	4.25	77.8	93.9
ResNeXt50-**LWE**-SE	30.4	20.0	**27.5**	**2.9**	**4.25**	**78.1**	**94.1**
Large sized models							
WideResNet50-2	68.9	18.3	68.9	5.2	11.46	77.3	93.5
WideResNet50-2-**LWE**	72.3	**20.0**	**68.9**	**5.2**	**11.46**	**78.3**	**94.2**
WideResNet50-2-SE	71.4	21.2	71.4	5.29	11.48	78.3	94.2
WideResNet50-2-**LWE**-SE	74.9	22.8	**71.4**	**5.29**	**11.48**	**78.6**	**94.3**
ResNet152	60.2	25.8	60.2	3.27	11.30	77.9	93.8
ResNet152-**LWE**	67.3	**29.7**	**60.2**	**3.27**	**11.30**	**78.6**	**94.3**
ResNet152-SE	66.8	31.4	66.8	3.46	11.32	78.7	94.3
ResNet152-**LWE**-SE	73.9	35.4	**66.8**	**3.46**	**11.32**	**79.0**	**94.6**

properties when different normalizations, optimizers and schedulers are used, we study the following on ResNet50 [9] baseline - (a) effectiveness of LWE when used with BN [13] and group normalization (GN) [38] (Fig. 5), and (b) effectiveness of LWE when using AdamW optimzer [23] with initial learning rate of 10^{-3} and weight decay of 0.1, and a cosine learning rate with initial restart period set as 5 epochs and a restart period multiplier of 1.2 [23], and run for four cosine annealing cycles (Fig. 5). We find that LWE-based training provides similar accuracy gains on ResNet50 [9] ConvNets with both BN [13] and GN [38] (Table 4) (Fig. 5). We also find that ResNet50-LWE continually shows better training and validation accuracy compared to baseline ResNet50 and ResNet50-SE at every epoch for the AdamW and cosine learning rate optimizer/scheduler setting (accuracy at last epoch: baseline ResNet50 70.9% vs. ResNet50-SE 71.2% vs. ResNet50-LWE 72.0%) (Fig. 5). Notably, LWE performs considerably better than SE [12] in this optimizer/scheduler setting. These results provide preliminary evidence that LWE can be effective in various normalization, optimizer and learning rate settings.

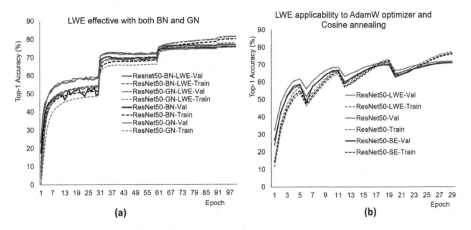

Fig. 5. (a) LWE-based training improves on baseline ResNet50 with both GN [38] and BN [13] normalizations. (b) LWE-based training on ResNet50 consistently performs better than baseline ResNet50 or ResNet50-SE for AdamW optimizer and Cosine annealing scheduler.

Utility of LWE and MWE Methods: While MWE provides a substantial improvement over baseline ResNet50, LWE causes an even larger improvement (Table 3). This improved utility of LWE is perhaps because it is based on a data-dependant learning approach and also because it provides a sense of relative structural importance (e.g. grouping weights in a 3×3 filter together, and comparing relative importance with other groups). Combining LWE and MWE together did not result in any noticeable accuracy gain (77.07% with LWE vs. 77.09% with LWE+MWE).

While our LWE method outperforms our MWE method, MWE has some advantages over LWE. First, as explained earlier, MWE is effective on depthwise convolutions whereas LWE is not (Table 1). Also, since LWE has an effect of suppressing some input channels, it is detrimental for shift-based convolutions [20, 36] - this is because suppressing an input channel can result in consistently

Table 2. WE-based weight reparameterization provides considerable accuracy gains compared to weight reparameterization approaches

Method	Top-1 Acc.(%)	Top-5 Acc.(%)	$\triangle$Top-1
ResNet50	75.8	92.8	+0.0
+WS [28]	76.1	92.9	+0.3
+WN [31]	76.0	92.9	+0.2
+SN [24]	75.7	92.8	-0.1
+MWE	76.5	93.1	+0.7
+LWE,90Epoch	**76.9**	**93.4**	+1.1
+LWE	**77.1**	**93.5**	+1.3

Table 3. Effectiveness of each WE processes in improving accuracy from ResNet50 baseline ($\triangle$Top-1).

Method	Top-1 Acc.(%)	Top-5 Acc.(%)	$\triangle$Top-1
ResNet50	75.83	92.8	+0.00
+MWE	76.52	93.1	+0.69
+LWE	77.07	93.5	+1.24
+LWE+MWE	77.09	93.5	+1.26

Table 4. LWE provides considerable accuracy gains with both batch normalization- [13] and group normalization-based [38] ResNet50.

Method	Top-1 Acc.(%)	Top-5 Acc.(%)	$\triangle$Top-1
ResNet50-GN	75.5	92.7	+0.0
ResNet50-GN-WN	75.9	92.9	+0.4
ResNet50-GN-WS	76.0	92.9	+0.5
ResNet50-GN-LWE	**76.8**	**93.3**	**+1.3**
ResNet50-BN	75.8	92.8	+0.0
ResNet50-BN-LWE	**77.1**	**93.5**	**+1.3**

Table 5. Effectiveness of MWE in improving accuracy over baseline ResNet50 for different ϵ_A.

Method	Top-1 Acc.(%)	Top-5 Acc.(%)	$\triangle$Top-1
ResNet50	75.83	92.8	+0.0
+MWE, $\epsilon_A = 10^{-3}$	76.38	93.0	+0.55
+MWE, $\epsilon_A = 0.1$	76.52	93.1	+0.69
+MWE, $\epsilon_A = 0.2$	76.43	93.0	+0.50

ignoring some parts of the input signal. MWE remains effective for shift-based convolutions (Sect. 4.6).

Validity of Attention Multiplier of LWE: To verify that LWE's learned attention multiplier m provides excitation to the more important channel locations, we study decrease in performance after channels with the lowest LWE multiplier are decimated (D_L) in each layer and decrease in performance after channels with the highest LWE multiplier are decimated (D_H). As expected, we find $D_H > D_L$ ($D_H - D_L = 8.15$), meaning that the channels with high importance multiplier are more important and loss of such channels are more detrimental to the network. Interestingly, we find $D_L = -0.03$ meaning that removing the channels with the lowest LWE multiplier in each layer contributes to small improvement in accuracy, though the difference in accuracy is small and could turn out to be insignificant after repeated experiments. Investigating whether some of the channels in a ConvNet are adversarial or detrimental to the ConvNet's performance is interesting future work.

Sensitivity to Hyperparameter: We have used $\epsilon_A = 0.1$ in MWE. A large ϵ_A makes f_A close to an identity line thereby negating any effect that f_A may have, while ϵ_A close to 0 can amplify the maximum value in W_L towards infinity thereby causing exploding gradients. We investigate three values of ϵ_A: 10^{-3}, 0.1 and 0.2 (Table 5), and find consistent improvements over baseline in all settings, suggesting that use of ϵ_A within a reasonable range (e.g. between 10^{-3} and 0.2) may provide consistent performance gains. We also used reduction ratio of 16 in LWE (Fig. 3a). Similar to [12], reducing this ratio to 8 improved ImageNet accuracy slightly by 0.05%.

4.3 CIFAR-100 Image Classification

Continuing on our ablation studies on image classification task, we investigate LWE method applied on some other popular ConvNet architectures, VGG19 [29] and ResNeXt-29-16x64d [40]), on the standard CIFAR-100 dataset [16]. We used a batch size of 256, and optimized with SGD with backpropagation [17] having momentum 0.9 and a decay of 5×10^{-4}. Learning rate was initialized at 0.1 and reduced with a factor of 10^{-1} at 81 and 122 epochs, and training was completed at 164 epochs. We see consistent improvements when LWE is used with base ConvNets. LWE improves top-1 accuracy of VGG19 from 73.2% to 73.8%, and yields a larger top-1 accuracy improvement on ResNeXt-29-16x64d (80.5% with ResNeXt-29-16x64d vs. 81.5% with ResNeXt-29-16x64d-LWE) (Table 6).

Table 6. Efficacy of WE on different tasks and convolution operations. Superscript * denotes average result from 5 repeated experiments.

Baseline ConvNet	Convolution type	Performance metric	Baseline performance	WE type	△ Performance with WE added
CIFAR100 image classification					
VGG19	2D	Accuracy	73.2%	LWE	+0.6
ResNeXt-29-16x64d	2D	Accuracy	80.5%	LWE	+1.0
PASCAL VOC semantic segmentation					
DeepLabv3 (ResNet50)	2D	IOU	74.4%*	MWE	+0.8
DeepLabv3 (ResNet50)	2D	IOU	74.4%*	LWE	+1.3
CityScapes VOC semantic segmentation					
DeepLabv3 (ResNet50)	2D	IOU	76.0%	LWE	+0.7
Mini-Kinetics action recognition					
R3D-ResNet18	3D	Accuracy	43.7%	LWE	+0.6
Gesture recognition on Jester dataset					
TSM-ResNet50	Shift-based 3D	Accuracy	96.3%	MWE	+0.3

4.4 PASCAL VOC and CityScapes Semantic Segmentation

To explore applicability of WE methods for tasks other than image classification, we investigate semantic segmentation on the PASCAL VOC 2012 [3] and CityScapes [2] datasets. For PASCAL VOC, we use the training recipe in [28] and use DeepLabv3 with ResNet50 as backbone [1] for its competitive performance, and find that MWE improves mean Intersection over Union (IOU) from 74.4% to 75.2% (+0.8% improvement) and LWE improves IOU to 75.7% (+1.3% improvement) (Table 6). For CityScapes, we also use DeepLabv3 with ResNet50 as backbone [1] and use the training recipe in [5].

4.5   Mini-Kinetics Action Recognition

To show the general applicability of WE mechanisms to improve training of 3D convolution kernels, we use LWE in the 3D convolution blocks of the R3D-ResNet18 ConvNet [34] and experiment on the Mini-Kinetics action recognition dataset [41]. We train R3D-ResNet18 on 16 densely sampled frames per video for 50 epochs using an initial learning rate of 0.02 which later gets reduced twice with a factor of 10^{-1} after 20th and 40th epochs. The weights of the network were randomly initialized, and trained using SGD [17] with momentum of 0.9 and a weight decay of 5×10^{-4}. A batch size of 64 was used. Evaluating on standard single center-crop [20], we find non-trivial accuracy improvement when LWE is used in R3D-ResNet18 (top-1 accuracy: R3D-ResNet18 at 43.7% vs. R3D-ResNet18-LWE at 44.3%, Table 6), suggesting LWE can be effective for ConvNets that use 3D convolution, without adding any computation cost at inference.

4.6   Gesture Recognition on Jester Dataset

To show applicability in other recognition tasks on video, we experiment with the Jester dataset for gesture recognition [33]. Here, we choose the TSM [20] ConvNet as baseline for two reasons: First, TSM [20] uses an efficient shift-based operation [36] that can be used to reduce complexity of a 3D convolution, and we want to evaluate effectiveness of using WE in such shift-based convolutions; second, TSM outperforms most other ConvNets in gesture recognition on the Jester dataset [20].

TSM's temporal shift-based convolution operation assigns different temporal information to different input channels. To train TSM, we use the same training recipe described in Sect. 4.5 except we use a strided sampling [20]. Implementing MWE in shift-based convolutions of TSM-ResNet50 [20] yields improved single center-crop accuracy (96.3% for TSM only vs. 96.6% with TSM+MWE), with no additional inference cost.

5   Conclusion

In this paper, we proposed novel weight reparameterization mechanisms that simultaneously identify the relative importance of weights in a convolution block in a ConvNet, and provide more attention to training the more important weights. Training a baseline ConvNet using such WE mechanisms can provide strong performance gains, without adding any additional computational costs or

ConvNet structural change at inference. We demonstrated this potency of our WE mechanisms in diverse settings (e.g. varying computer vision tasks, different baseline ConvNets with and without activation map-based attention methods, multiple normalization methods, etc.). Finally, WE mechanisms can easily be implemented in a convolution block with minimal effort, requiring few lines of extra code. Overall, the diversity on applicability, the complementarity with previous attention mechanisms and the simplicity in implementation make our WE mechanisms valuable in variously and popularly used ConvNets in computer vision research.

References

1. Chen, L.C., Papandreou, G., Schroff, F., Adam, H.: Rethinking Atrous convolution for semantic image segmentation. arXiv preprint arXiv:1706.05587 (2017)
2. Cordts, M., et al.: The cityscapes dataset for semantic urban scene understanding. In: Proceedings of the IEEE Conference on Computer Vision and Pattern Recognition, pp. 3213–3223 (2016)
3. Everingham, M., Eslami, S.A., Van Gool, L., Williams, C.K., Winn, J., Zisserman, A.: The pascal visual object classes challenge: a retrospective. Int. J. Comput. Vision 111(1), 98–136 (2015)
4. Frankle, J., Carbin, M.: The lottery ticket hypothesis: finding sparse, trainable neural networks. In: International Conference on Learning Representations (2019)
5. Gongfan, F.: pytorch-classification (2020). https://github.com/VainF/DeepLab V3Plus-Pytorch
6. Han, S., Mao, H., Dally, W.J.: Deep compression: compressing deep neural networks with pruning, trained quantization and huffman coding. arXiv preprint arXiv:1510.00149 (2015)
7. Han, S., Pool, J., Tran, J., Dally, W.: Learning both weights and connections for efficient neural network. In: Advances in Neural Information Processing Systems, pp. 1135–1143 (2015)
8. Hassibi, B., Stork, D.G.: Second order derivatives for network pruning: Optimal brain surgeon. In: Advances in Neural Information Processing Systems, pp. 164–171 (1993)
9. He, K., Zhang, X., Ren, S., Sun, J.: Deep residual learning for image recognition. In: Proceedings of the IEEE Conference on Computer Vision and Pattern Recognition, pp. 770–778 (2016)
10. He, T., Zhang, Z., Zhang, H., Zhang, Z., Xie, J., Li, M.: Bag of tricks for image classification with convolutional neural networks. In: Proceedings of the IEEE Conference on Computer Vision and Pattern Recognition, pp. 558–567 (2019)
11. Hu, J., Shen, L., Albanie, S., Sun, G., Vedaldi, A.: Gather-excite: exploiting feature context in convolutional neural networks. In: Advances in Neural Information Processing Systems, pp. 9401–9411 (2018)
12. Hu, J., Shen, L., Sun, G.: Squeeze-and-excitation networks. In: Proceedings of the IEEE Conference on Computer Vision and Pattern Recognition, pp. 7132–7141 (2018)
13. Ioffe, S., Szegedy, C.: Batch normalization: accelerating deep network training by reducing internal covariate shift. arXiv preprint arXiv:1502.03167 (2015)
14. Kay, W., et al.: The kinetics human action video dataset. arXiv preprint arXiv:1705.06950 (2017)

15. Kingma, D.P., Ba, J.: Adam: a method for stochastic optimization. arXiv preprint arXiv:1412.6980 (2014)
16. Krizhevsky, A., Hinton, G., et al.: Learning multiple layers of features from tiny images. Technical report, Citeseer (2009)
17. LeCun, Y., et al.: Backpropagation applied to handwritten zip code recognition. Neural Comput. 1(4), 541–551 (1989)
18. LeCun, Y., Denker, J.S., Solla, S.A.: Optimal brain damage. In: Advances in Neural Information Processing Systems, pp. 598–605 (1990)
19. Li, H., Xu, Z., Taylor, G., Studer, C., Goldstein, T.: Visualizing the loss landscape of neural nets. In: Advances in Neural Information Processing Systems, pp. 6389–6399 (2018)
20. Lin, J., Gan, C., Han, S.: Temporal shift module for efficient video understanding. arXiv preprint arXiv:1811.08383 (2018)
21. Lin, T.Y., Goyal, P., Girshick, R., He, K., Dollár, P.: Focal loss for dense object detection. In: Proceedings of the IEEE International Conference on Computer Vision, pp. 2980–2988 (2017)
22. Liu, J., Xu, Z., Shi, R., Cheung, R.C.C., So, H.K.: Dynamic sparse training: find efficient sparse network from scratch with trainable masked layers. In: International Conference on Learning Representations (2020). https://openreview.net/forum?id=SJlbGJrtDB
23. Loshchilov, I., Hutter, F.: Decoupled weight decay regularization (2018)
24. Miyato, T., Kataoka, T., Koyama, M., Yoshida, Y.: Spectral normalization for generative adversarial networks. arXiv preprint arXiv:1802.05957 (2018)
25. Molchanov, P., Tyree, S., Karras, T., Aila, T., Kautz, J.: Pruning convolutional neural networks for resource efficient inference. arXiv preprint arXiv:1611.06440 (2016)
26. Nesterov, Y.: A method of solving a convex programming problem with convergence rate. In: Soviet Math. Dokl. vol. 27
27. Park, J., Woo, S., Lee, J.Y., Kweon, I.S.: Bam: Bottleneck attention module. arXiv preprint arXiv:1807.06514 (2018)
28. Qiao, S., Wang, H., Liu, C., Shen, W., Yuille, A.: Weight standardization. arXiv preprint arXiv:1903.10520 (2019)
29. Ren, S., He, K., Girshick, R., Sun, J.: Faster R-CNN: towards real-time object detection with region proposal networks. In: Advances in Neural Information Processing Systems, pp. 91–99 (2015)
30. Russakovsky, O., et al.: Imagenet large scale visual recognition challenge. Int. J. Comput. Vision 115(3), 211–252 (2015)
31. Salimans, T., Kingma, D.P.: Weight normalization: a simple reparameterization to accelerate training of deep neural networks. In: Advances in Neural Information Processing Systems, pp. 901–909 (2016)
32. Sandler, M., Howard, A., Zhu, M., Zhmoginov, A., Chen, L.C.: Mobilenetv 2: inverted residuals and linear bottlenecks. In: Proceedings of the IEEE Conference on Computer Vision and Pattern Recognition, pp. 4510–4520 (2018)
33. The 20bn-jester dataset v1: The 20BN-jester Dataset V1. Accessed 16 July 2020
34. Tran, D., Wang, H., Torresani, L., Ray, J., LeCun, Y., Paluri, M.: A closer look at spatiotemporal convolutions for action recognition. In: Proceedings of the IEEE conference on Computer Vision and Pattern Recognition, pp. 6450–6459 (2018)
35. Woo, S., Park, J., Lee, J.-Y., Kweon, I.S.: CBAM: convolutional block attention module. In: Ferrari, V., Hebert, M., Sminchisescu, C., Weiss, Y. (eds.) ECCV 2018. LNCS, vol. 11211, pp. 3–19. Springer, Cham (2018). https://doi.org/10.1007/978-3-030-01234-2_1

36. Wu, B., et al.: Shift: a zero flop, zero parameter alternative to spatial convolutions. In: Proceedings of the IEEE Conference on Computer Vision and Pattern Recognition, pp. 9127–9135 (2018)

37. Wu, L., Zhu, Z., et al.: Towards understanding generalization of deep learning: perspective of loss landscapes. arXiv preprint arXiv:1706.10239 (2017)

38. Wu, Y., He, K.: Group normalization. In: Ferrari, V., Hebert, M., Sminchisescu, C., Weiss, Y. (eds.) ECCV 2018. LNCS, vol. 11217, pp. 3–19. Springer, Cham (2018). https://doi.org/10.1007/978-3-030-01261-8_1

39. Xiang, L., Shuo, C., Yan, X., Jian, Y.: Understanding the disharmony between weight normalization family and weight decay: $epsilon-$ shifted l_2 regularizer. arXiv preprint arXiv:1911.05920 (2019)

40. Xie, S., Girshick, R., Dollár, P., Tu, Z., He, K.: Aggregated residual transformations for deep neural networks. In: Proceedings of the IEEE Conference on Computer Vision and Pattern Recognition, pp. 1492–1500 (2017)

41. Xie, S., Sun, C., Huang, J., Tu, Z., Murphy, K.: Rethinking spatiotemporal feature learning: speed-accuracy trade-offs in video classification. In: Ferrari, V., Hebert, M., Sminchisescu, C., Weiss, Y. (eds.) ECCV 2018. LNCS, vol. 11219, pp. 318–335. Springer, Cham (2018). https://doi.org/10.1007/978-3-030-01267-0_19

42. Yang, W.: pytorch-classification (2017). https://github.com/bearpaw/pytorch-classification

43. Zagoruyko, S., Komodakis, N.: Wide residual networks. arXiv preprint arXiv:1605.07146 (2016)

UNITER: UNiversal Image-TExt Representation Learning

Yen-Chun Chen[✉], Linjie Li, Licheng Yu, Ahmed El Kholy, Faisal Ahmed,
Zhe Gan, Yu Cheng, and Jingjing Liu

Microsoft Dynamics 365 AI Research, Redmond, USA
{yen-chun.chen,lindsey.li,licheng.yu,ahmed.elkholy,fiahmed,
zhe.gan,yu.cheng,jingjl}@microsoft.com

Abstract. Joint image-text embedding is the bedrock for most Vision-and-Language (V+L) tasks, where multimodality inputs are simultaneously processed for joint visual and textual understanding. In this paper, we introduce UNITER, a UNiversal Image-TExt Representation, learned through large-scale pre-training over four image-text datasets (COCO, Visual Genome, Conceptual Captions, and SBU Captions), which can power heterogeneous downstream V+L tasks with joint multimodal embeddings. We design four pre-training tasks: Masked Language Modeling (MLM), Masked Region Modeling (MRM, with three variants), Image-Text Matching (ITM), and Word-Region Alignment (WRA). Different from previous work that applies joint random masking to both modalities, we use conditional masking on pre-training tasks (*i.e.*, masked language/region modeling is conditioned on full observation of image/text). In addition to ITM for global image-text alignment, we also propose WRA via the use of Optimal Transport (OT) to *explicitly* encourage fine-grained alignment between words and image regions during pre-training. Comprehensive analysis shows that both conditional masking and OT-based WRA contribute to better pre-training. We also conduct a thorough ablation study to find an optimal combination of pre-training tasks. Extensive experiments show that UNITER achieves new state of the art across six V+L tasks (over nine datasets), including Visual Question Answering, Image-Text Retrieval, Referring Expression Comprehension, Visual Commonsense Reasoning, Visual Entailment, and NLVR² (Code is available at https://github.com/ChenRocks/UNITER.).

1 Introduction

Most Vision-and-Language (V+L) tasks rely on joint multimodal embeddings to bridge the semantic gap between visual and textual clues in images and text,

---

Y.-C. Chen, L. Li and L. Yu—Equal contribution.

Electronic supplementary material The online version of this chapter (https://doi.org/10.1007/978-3-030-58577-8_7) contains supplementary material, which is available to authorized users.

A. Vedaldi et al. (Eds.): ECCV 2020, LNCS 12375, pp. 104–120, 2020.
https://doi.org/10.1007/978-3-030-58577-8_7

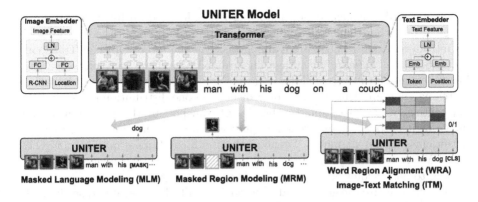

Fig. 1. Overview of the proposed UNITER model (best viewed in color), consisting of an Image Embedder, a Text Embedder and a multi-layer Transformer, learned through four pre-training tasks (Color figure online)

although such representations are usually tailored for specific tasks. For example, MCB [8], BAN [14] and DFAF [10] proposed advanced multimodal fusion methods for Visual Question Answering (VQA) [3]. SCAN [18] and MAttNet [45] studied learning latent alignment between words and image regions for Image-Text Retrieval [40] and Referring Expression Comprehension [13]. While each of these models has pushed the state of the art on respective benchmarks, their architectures are diverse and the learned representations are highly task-specific, preventing them from being generalizable to other tasks. This raises a million-dollar question: can we learn a universal image-text representation for all V+L tasks?

In this spirit, we introduce **UN**iversal **I**mage-**TE**xt **R**epresentation (**UNITER**), a large-scale pre-trained model for joint multimodal embedding. We adopt Transformer [39] as the core of our model, to leverage its elegant self-attention mechanism designed for learning contextualized representations. Inspired by BERT [6], which has successfully applied Transformer to NLP tasks through large-scale language modeling, we pre-train UNITER through four pre-training tasks: (*i*) Masked Language Modeling (MLM) *conditioned on image*; (*ii*) Masked Region Modeling (MRM) *conditioned on text*; (*iii*) Image-Text Matching (ITM); and (*iv*) Word-Region Alignment (WRA). To further investigate the effectiveness of MRM, we propose three MRM variants: (*i*) Masked Region Classification (MRC); (*ii*) Masked Region Feature Regression (MRFR); and (*iii*) Masked Region Classification with KL-divergence (MRC-kl).

As shown in Fig. 1, UNITER first encodes image regions (visual features and bounding box features) and textual words (tokens and positions) into a common embedding space with Image Embedder and Text Embedder. Then, a Transformer module is applied to learn generalizable contextualized embeddings for each region and each word through well-designed pre-training tasks. Compared with previous work on multimodal pre-training [1,19,20,23,33,37,50]:

(i) our masked language/region modeling is conditioned on full observation of image/text, rather than applying joint random masking to both modalities; (ii) we introduce a novel WRA pre-training task via the use of Optimal Transport (OT) [5,29] to *explicitly* encourage fine-grained alignment between words and image regions. Intuitively, OT-based learning aims to optimize for distribution matching via minimizing the cost of transporting one distribution to another. In our context, we aim to minimize the cost of transporting the embeddings from image regions to words in a sentence (and vice versa), thus optimizing towards better cross-modal alignment. We show that both conditional masking and OT-based WRA can successfully ease the misalignment between images and text, leading to better joint embeddings for downstream tasks.

To demonstrate the generalizable power of UNITER, we evaluate on six V+L tasks across nine datasets, including: (i) VQA; (ii) Visual Commonsense Reasoning (VCR) [48]; (iii) NLVR^2 [34]; (iv) Visual Entailment [42]; (v) Image-Text Retrieval (including zero-shot setting) [18]; and (vi) Referring Expression Comprehension [46]. Our UNITER model is trained on a large-scale V+L dataset composed of four subsets: (i) COCO [21]; (ii) Visual Genome (VG) [16]; (iii) Conceptual Captions (CC) [32]; and (iv) SBU Captions [26]. Experiments show that UNITER achieves new state of the art with significant performance boost across all nine downstream datasets. Moreover, training on additional CC and SBU data (containing unseen images/text in downstream tasks) further boosts model performance over training on COCO and VG only.

Our contributions are summarized as follows: (i) We introduce UNITER, a powerful UNiversal Image-TExt Representation for V+L tasks. (ii) We present Conditional Masking for masked language/region modeling, and propose a novel Optimal-Transport-based Word-Region Alignment task for pre-training. (iii) We achieve new state of the art on a wide range of V+L benchmarks, outperforming existing multimodal pre-training methods by a large margin. We also present extensive experiments and analysis to provide useful insights on the effectiveness of each pre-training task/dataset for multimodal encoder training.

2   Related Work

Self-supervised learning utilizes original data as its own source of supervision, which has been applied to many Computer Vision tasks, such as image colorization [49], solving jigsaw puzzles [25,38], inpainting [27], rotation prediction [11], and relative location prediction [7]. Recently, pre-trained language models, such as ELMo [28], BERT [6], GPT2 [31], XLNet [44], RoBERTa [22] and ALBERT [17], have pushed great advances for NLP tasks. There are two keys to their success: effective pre-training tasks over large language corpus, and the use of Transformer [39] for learning contextualized text representations.

More recently, there has been a surging interest in self-supervised learning for multimodal tasks, by pre-training on large-scale image/video and text pairs, then finetuning on downstream tasks. For example, VideoBERT [36] and CBT [35] applied BERT to learn a joint distribution over video frame features and linguistic tokens from video-text pairs. ViLBERT [23] and LXMERT [37] introduced the two-stream architecture, where two Transformers are applied to images and text independently, which is fused by a third Transformer in a later stage. On the other hand, B2T2 [1], VisualBERT [20], Unicoder-VL [19] and VL-BERT [33] proposed the single-stream architecture, where a single Transformer is applied to both images and text. VLP [50] applied pre-trained models to both image captioning and VQA. More recently, multi-task learning [24] and adversarial training [9] were used to further boost the performance. VALUE [4] developed a set of probing tasks to understand pre-trained models.

Our Contributions. The key differences between our UNITER model and the other methods are two-fold: (*i*) UNITER uses conditional masking on MLM and MRM, *i.e.*, masking only one modality while keeping the other untainted; and (*ii*) a novel Word-Region Alignment pre-training task via the use of Optimal Transport, while in previous work such alignment is only implicitly enforced by task-specific losses. In addition, we examine the best combination of pre-training tasks through a thorough ablation study, and achieve new state of the art on multiple V+L datasets, often outperforming prior work by a large margin.

3   UNiversal Image-TExt Representation

In this section, we first introduce the model architecture of UNITER (Sect. 3.1), then describe the designed pre-training tasks and V+L datasets used for pre-training (Sect. 3.2 and 3.3).

3.1   Model Overview

The model architecture of UNITER is illustrated in Fig. 1. Given a pair of image and sentence, UNITER takes the visual regions of the image and textual tokens of the sentence as inputs. We design an Image Embedder and a Text Embedder to extract their respective embeddings. These embeddings are then fed into a multi-layer Transformer to learn a cross-modality contextualized embedding across visual regions and textual tokens. Note that the self-attention mechanism in Transformer is order-less, thus it is necessary to explicitly encode the positions of tokens and the locations of regions as additional inputs.

Specifically, in *Image Embedder*, we first use Faster R-CNN[1] to extract the visual features (pooled ROI features) for each region. We also encode the location features for each region via a 7-dimensional vector.[2] Both visual and location features are then fed through a fully-connected (FC) layer, to be projected into the same embedding space. The final visual embedding for each region is obtained by summing up the two FC outputs and then passing through a layer normalization (LN) layer. For *Text Embedder*, we follow BERT [6] and tokenize the input sentence into WordPieces [41]. The final representation for each sub-word

---

[1] Our Faster R-CNN was pre-trained on Visual Genome object+attribute data [2].

[2] $[x_1, y_1, x_2, y_2, w, h, w * h]$ (normalized top/left/bottom/right coordinates, width, height, and area.).

token[3] is obtained via summing up its word embedding and position embedding, followed by another LN layer.[4]

We introduce four main tasks to pre-train our model: Masked Language Modeling *conditioned on image regions* (MLM), Masked Region Modeling *conditioned on input text* (with three variants) (MRM), Image-Text Matching (ITM), and Word-Region Alignment (WRA). As shown in Fig. 1, our MRM and MLM are in analogy to BERT, where we randomly mask some words or regions from the input and learn to recover the words or regions as the output of Transformer. Specifically, word masking is realized by replacing the token with a special token [MASK], and region masking is implemented by replacing the visual feature vector with all zeros. Note that each time we only mask one modality while keeping the other modality intact, instead of randomly masking both modalities as used in other pre-training methods. This prevents potential misalignment when a masked region happens to be described by a masked word (detailed in Sect. 4.2).

We also learn an instance-level alignment between the whole image and the sentence via ITM. During training, we sample both positive and negative image-sentence pairs and learn their matching scores. Furthermore, in order to provide a more fine-grained alignment between word tokens and image regions, we propose WRA via the use of Optimal Transport, which effectively calculates the minimum cost of transporting the contextualized image embeddings to word embeddings (and vice versa). The inferred transport plan thus serves as a propeller for better cross-modal alignment. Empirically, we show that both conditional masking and WRA contributes to performance improvement (in Sect. 4.2). To pre-train UNITER with these tasks, we randomly sample one task for each mini-batch, and train on only one objective per SGD update.

3.2 Pre-training Tasks

Masked Language Modeling (MLM). We denote the image regions as $\mathbf{v} = \{\mathbf{v}_1, ..., \mathbf{v}_K\}$, the input words as $\mathbf{w} = \{\mathbf{w}_1, ..., \mathbf{w}_T\}$, and the mask indices as $\mathbf{m} \in \mathbb{N}^M$.[5] In MLM, we randomly mask out the input words with probability of 15%, and replace the masked ones $\mathbf{w}_\mathbf{m}$ with special token [MASK].[6] The goal is to predict these masked words based on the observation of their surrounding words $\mathbf{w}_{\backslash \mathbf{m}}$ and all image regions $\mathbf{v}$, by minimizing the negative log-likelihood:

$$\mathcal{L}_{\text{MLM}}(\theta) = -\mathbb{E}_{(\mathbf{w}, \mathbf{v}) \sim D} \log P_\theta(\mathbf{w}_\mathbf{m} | \mathbf{w}_{\backslash \mathbf{m}}, \mathbf{v}), \qquad (1)$$

---

[3] We use word/sub-word and token interchangeably throughout the rest of the paper.

[4] We also use a special modality embedding to help the model distinguish between textual and visual input, which is similar to the 'segment embedding' in BERT. This embedding is also summed before the LN layer in each embedder. For simplicity, this modality embedding is omitted in Fig. 1.

[5] $\mathbb{N}$ is the natural numbers, M is the number of masked tokens, and $\mathbf{m}$ is the set of masked indices.

[6] Following BERT, we decompose this 15% into 10% random words, 10% unchanged, and 80% [MASK].

where θ is the trainable parameters. Each pair $(\mathbf{w}, \mathbf{v})$ is sampled from the whole training set D.

Image-Text Matching (ITM). In ITM, an additional special token [CLS] is fed into our model, which indicates the fused representation of both modalities. The inputs to ITM are a sentence and a set of image regions, and the output is a binary label $y \in \{0, 1\}$, indicating if the sampled pair is a match. We extract the representation of [CLS] token as the joint representation of the input image-text pair, then feed it into an FC layer and a sigmoid function to predict a score between 0 and 1. We denote the output score as $s_\theta(\mathbf{w}, \mathbf{v})$. The ITM supervision is over the [CLS] token.[7] During training, we sample a positive or negative pair $(\mathbf{w}, \mathbf{v})$ from the dataset D at each step. The negative pair is created by replacing the image or text in a paired sample with a randomly-selected one from other samples. We apply the binary cross-entropy loss for optimization:

$$\mathcal{L}_{\text{ITM}}(\theta) = -\mathbb{E}_{(\mathbf{w}, \mathbf{v}) \sim D}[y \log s_\theta(\mathbf{w}, \mathbf{v}) + (1 - y) \log(1 - s_\theta(\mathbf{w}, \mathbf{v}))]). \tag{2}$$

Word-Region Alignment (WRA). We use Optimal Transport (OT) for WRA, where a transport plan $\mathbf{T} \in \mathbb{R}^{T \times K}$ is learned to optimize the alignment between $\mathbf{w}$ and $\mathbf{v}$. OT possesses several idiosyncratic characteristics that make it a good choice for WRA: (*i*) *Self-normalization*: all the elements of $\mathbf{T}$ sum to 1 [29]. (*ii*) *Sparsity*: when solved exactly, OT yields a sparse solution $\mathbf{T}$ containing $(2r - 1)$ non-zero elements at most, where $r = \max(K, T)$, leading to a more interpretable and robust alignment [29]. (*iii*) *Efficiency*: compared with conventional linear programming solvers, our solution can be readily obtained using iterative procedures that only require matrix-vector products [43], hence readily applicable to large-scale model pre-training.

Specifically, $(\mathbf{w}, \mathbf{v})$ can be considered as two discrete distributions μ, ν, formulated as $\mu = \sum_{i=1}^{T} \mathbf{a}_i \delta_{\mathbf{w}_i}$ and $\nu = \sum_{j=1}^{K} \mathbf{b}_j \delta_{\mathbf{v}_j}$, with $\delta_{\mathbf{w}_i}$ as the Dirac function centered on $\mathbf{w}_i$. The weight vectors $\mathbf{a} = \{\mathbf{a}_i\}_{i=1}^{T} \in \Delta_T$ and $\mathbf{b} = \{\mathbf{b}_j\}_{j=1}^{K} \in \Delta_K$ belong to the T- and K-dimensional simplex, respectively (*i.e.*, $\sum_{i=1}^{T} \mathbf{a}_i = \sum_{j=1}^{K} \mathbf{b}_j = 1$), as both μ and ν are probability distributions. The OT distance between μ and ν (thus also the alignment loss for the $(\mathbf{w}, \mathbf{v})$ pair) is defined as:

$$\mathcal{L}_{\text{WRA}}(\theta) = \mathcal{D}_{ot}(\mu, \nu) = \min_{\mathbf{T} \in \Pi(\mathbf{a}, \mathbf{b})} \sum_{i=1}^{T} \sum_{j=1}^{K} \mathbf{T}_{ij} \cdot c(\mathbf{w}_i, \mathbf{v}_j), \tag{3}$$

where $\Pi(\mathbf{a}, \mathbf{b}) = \{\mathbf{T} \in \mathbb{R}_+^{T \times K} | \mathbf{T} \mathbf{1}_m = \mathbf{a}, \mathbf{T}^\top \mathbf{1}_n = \mathbf{b}\}$, $\mathbf{1}_n$ denotes an n-dimensional all-one vector, and $c(\mathbf{w}_i, \mathbf{v}_j)$ is the cost function evaluating the distance between $\mathbf{w}_i$ and $\mathbf{v}_j$. In experiments, the cosine distance $c(\mathbf{w}_i, \mathbf{v}_j) = 1 - \frac{\mathbf{w}_i^\top \mathbf{v}_j}{||\mathbf{w}_i||_2 ||\mathbf{v}_j||_2}$ is used. The matrix $\mathbf{T}$ is denoted as the transport plan, interpreting the alignment between two modalities. Unfortunately, the exact minimization

---

[7] Performing this during pre-training also alleviates the mismatch problem between pre-training and downstream finetuning tasks, since most of the downstream tasks take the representation of the [CLS] token as the joint representation.

over $\mathbf{T}$ is computational intractable, and we consider the IPOT algorithm [43] to approximate the OT distance (details are provided in the supplementary file). After solving $\mathbf{T}$, the OT distance serves as the WRA loss that can be used to update the parameters θ.

Masked Region Modeling (MRM). Similar to MLM, we also sample image regions and mask their visual features with a probability of 15%. The model is trained to reconstruct the masked regions $\mathbf{v_m}$ given the remaining regions $\mathbf{v_{\setminus m}}$ and all the words $\mathbf{w}$. The visual features of the masked region are replaced by zeros. Unlike textual tokens that are represented as discrete labels, visual features are high-dimensional and continuous, thus cannot be supervised via class likelihood. Instead, we propose three variants for MRM, which share the same objective base:

$$\mathcal{L}_{\mathrm{MRM}}(\theta) = \mathbb{E}_{(\mathbf{w},\mathbf{v}) \sim D} f_\theta(\mathbf{v_m} | \mathbf{v_{\setminus m}}, \mathbf{w}). \tag{4}$$

1) **Masked Region Feature Regression (MRFR)** MRFR learns to regress the Transformer output of each masked region $\mathbf{v_m}^{(i)}$ to its visual features. Specifically, we apply an FC layer to convert its Transformer output into a vector $h_\theta(\mathbf{v_m}^{(i)})$ of same dimension as the input ROI pooled feature $r(\mathbf{v_m}^{(i)})$. Then we apply L2 regression between the two: $f_\theta(\mathbf{v_m} | \mathbf{v_{\setminus m}}, \mathbf{w}) = \sum_{i=1}^{M} \|h_\theta(\mathbf{v_m}^{(i)}) - r(\mathbf{v_m}^{(i)})\|_2^2$.

2) **Masked Region Classification (MRC)** MRC learns to predict the object semantic class for each masked region. We first feed the Transformer output of the masked region $\mathbf{v_m}^{(i)}$ into an FC layer to predict the scores of K object classes, which further goes through a softmax function to be transformed into a normalized distribution $g_\theta(\mathbf{v_m}^{(i)}) \in \mathbb{R}^K$. Note that there is no ground-truth label, as the object categories are not provided. Thus, we use the object detection output from Faster R-CNN, and take the detected object category (with the highest confidence score) as the label of the masked region, which will be converted into a one-hot vector $c(\mathbf{v_m}^{(i)}) \in \mathbb{R}^K$. The final objective minimizes the cross-entropy (CE) loss: $f_\theta(\mathbf{v_m} | \mathbf{v_{\setminus m}}, \mathbf{w}) = \sum_{i=1}^{M} \mathrm{CE}(c(\mathbf{v_m}^{(i)}), g_\theta(\mathbf{v_m}^{(i)}))$.

3) **Masked Region Classification with KL-Divergence (MRC-kl)** MRC takes the most likely object class from the object detection model as the hard label (w.p. 0 or 1), assuming the detected object class is the ground-truth label for the region. However, this may not be true, as no ground-truth label is available. Thus, in MRC-kl, we avoid this assumption by using soft label as supervision signal, which is the raw output from the detector (*i.e.*, a distribution of object classes $\tilde{c}(\mathbf{v_m}^{(i)})$). MRC-kl aims to distill such knowledge into UNITER as [12], by minimizing the KL divergence between two distributions: $f_\theta(\mathbf{v_m} | \mathbf{v_{\setminus m}}, \mathbf{w}) = \sum_{i=1}^{M} D_{KL}(\tilde{c}(\mathbf{v_m}^{(i)}) \| g_\theta(\mathbf{v_m}^{(i)}))$.

3.3 Pre-training Datasets

We construct our pre-training dataset based on four existing V+L datasets: COCO [21], Visual Genome (VG) [16], Conceptual Captions (CC) [32], and SBU

Table 1. Statistics on the datasets used for pre-training. Each cell shows #image-text pairs (#images)

Split	In-domain		Out-of-domain	
	COCO captions	VG dense captions	Conceptual captions	SBU captions
train	533K (106K)	5.06M (101K)	3.0M (3.0M)	990K (990K)
val	25K (5K)	106K (2.1K)	14K (14K)	10K (10K)

Captions [26]. Only image and sentence pairs are used for pre-training, which makes the model framework more scalable, as additional image-sentence pairs are easy to harvest for further pre-training.

To study the effects of different datasets on pre-training, we divide the four datasets into two categories. The first one consists of image captioning data from COCO and dense captioning data from VG. We call it "In-domain" data, as most V+L tasks are built on top of these two datasets. To obtain a "fair" data split, we merge the raw training and validation splits from COCO, and exclude all validation and test images that appear in downstream tasks. We also exclude all co-occurring Flickr30K [30] images via URL matching, as both COCO and Flickr30K images were crawled from Flickr and may have overlaps.[8] The same rule was applied to Visual Genome as well. In this way, we obtain 5.6M image-text pairs for training and 131K image-text pairs for our internal validation, which is half the size of the dataset used in LXMERT [37], due to the filtering of overlapping images and the use of image-text pairs only. We also use additional Out-of-domain data from Conceptual Captions [32] and SBU Captions [26] for model training.[9] The statistics on the cleaned splits are provided in Table 1.

4   Experiments

We evaluate UNITER on six V+L tasks[10] by transferring the pre-trained model to each target task and finetuning through end-to-end training. We report experimental results on two model sizes: UNITER-base with 12 layers and UNITER-large with 24 layers.[11]

---

[8] A total of 222 images were eliminated through this process.

[9] We apply the same URL matching method, excluding 109 images from training.

[10] VQA, VCR, NLVR^2, Visual Entailment, Image-Text Retrieval, and Referring Expression Comprehension. Details about the tasks are listed in the supplementary.

[11] UNITER-base: $L = 12$, $H = 768$, $A = 12$, Total Parameters $= 86M$. UNITER-large: $L = 24$, $H = 1024$, $A = 16$, Total Parameters $= 303M$ (L: number of stacked Transformer blocks; H: hidden activation dimension; A: number of attention heads). 882 and 3645 V100 GPU hours were used for pre-training UNITER-base and UNITER-large.

Table 2. Evaluation on pre-training tasks and datasets using VQA, Image-Text Retrieval on Flickr30K, NLVR^2, and RefCOCO+ as benchmarks. All results are obtained from UNITER-base. Averages of R@1, R@5 and R@10 on Flickr30K for Image Retrieval (IR) and Text Retrieval (TR) are reported. Dark and light grey colors highlight the top and second best results across all the tasks trained with In-domain data

Pre-training Data	Pre-training Tasks	Meta-Sum	VQA	IR (Flickr)	TR (Flickr)	NLVR^2	Ref-COCO+
			test-dev	val	val	dev	val^d
None	1 None	314.34	67.03	61.74	65.55	51.02	68.73
Wikipedia + BookCorpus	2 MLM (text only)	346.24	69.39	73.92	83.27	50.86	68.80
In-domain (COCO+VG)	3 MRFR	344.66	69.02	72.10	82.91	52.16	68.47
	4 ITM	385.29	70.04	78.93	89.91	74.08	72.33
	5 MLM	386.10	71.29	77.88	89.25	74.79	72.89
	6 MLM + ITM	393.04	71.55	81.64	91.12	75.98	72.75
	7 MLM + ITM + MRC	393.97	71.46	81.39	91.45	76.18	73.49
	8 MLM + ITM + MRFR	396.24	71.73	81.76	92.31	76.21	74.23
	9 MLM + ITM + MRC-kl	397.09	71.63	82.10	92.57	76.28	74.51
	10 MLM + ITM + MRC-kl + MRFR	399.97	71.92	83.73	92.87	76.93	74.52
	11 MLM + ITM + MRC-kl + MRFR + WRA	400.93	72.47	83.72	93.03	76.91	74.80
	12 MLM + ITM + MRC-kl + MRFR (w/o cond. mask)	396.51	71.68	82.31	92.08	76.15	74.29
Out-of-domain (SBU+CC)	13 MLM + ITM + MRC-kl + MRFR + WRA	396.91	71.56	84.34	92.57	75.66	72.78
In-domain + Out-of-domain	14 MLM + ITM + MRC-kl + MRFR + WRA	**405.24**	**72.70**	**85.77**	**94.28**	**77.18**	**75.31**

4.1 Downstream Tasks

In VQA, VCR and NLVR^2 tasks, given an input image (or a pair of images) and a natural language question (or description), the model predicts an answer (or judges the correctness of the description) based on the visual content in the image. For Visual Entailment, we evaluate on the SNLI-VE dataset. The goal is to predict whether a given image semantically entails an input sentence. Classification accuracy over three classes ("Entailment", "Neutral" and "Contradiction") is used to measure model performance. For Image-Text Retrieval, we consider two datasets (COCO and Flickr30K) and evaluate the model in two settings: Image Retrieval (IR) and Text Retrieval (TR). Referring Expression (RE) Comprehension requires the model to select the target from a set of image region proposals given the query description. Models are evaluated on both ground-truth objects and detected proposals[12] (MAttNet [45]).

For VQA, VCR, NLVR^2, Visual Entailment and Image-Text Retrieval, we extract the joint embedding of the input image-text pairs via a multi-layer perceptron (MLP) from the representation of the [CLS] token. For RE Comprehension, we use the MLP to compute the region-wise alignment scores. These MLP layers are learned during the finetuning stage. Specifically, we formulate VQA, VCR, NLVR^2, Visual Entailment and RE Comprehension as classification prob-

---

[12] The evaluation splits of RE comprehension using detected proposals are denoted as val^d, test^d, etc.

lems and minimize the cross-entropy over the ground-truth answers/responses. For Image-Text Retrieval, we formulate it as a ranking problem. During finetuning, we sample three pairs of image and text, one positive pair from the dataset and two negative pairs by randomly replacing its sentence/image with others. We compute the similarity scores (based on the joint embedding) for both positive and negative pairs, and maximize the margin between them through triplet loss.

4.2   Evaluation on Pre-training Tasks

We analyze the effectiveness of different pre-training settings through ablation studies over VQA, NLVR^2, Flickr30K and RefCOCO+ as representative V+L benchmarks. In addition to standard metrics[13] for each benchmark , we also use Meta-Sum (sum of all the scores across all the benchmarks) as a global metric.

Firstly, we establish two baselines: Line 1 (L1) in Table 2 indicates no pre-training is involved, and L2 shows the results from MLM initialized with pre-trained weights from [6]. Although MLM trained on text only did not absorb any image information during pre-training, we see a gain of approximately +30 on Meta-Sum over L1. Hence, we use the pre-trained weights in L2 to initialize our model for the following experiments.

Secondly, we validate the effectiveness of each pre-training task through a thorough ablation study. Comparing L2 and L3, MRFR (L3) achieves better results than MLM (L2) only on NLVR^2. On the other hand, when pre-trained on ITM (L4) or MLM (L5) only, we observe a significant improvement across all the tasks over L1 and L2 baselines. When combining different pre-training tasks, MLM + ITM (L6) improves over single ITM (L4) or MLM (L5). When MLM, ITM and MRM are jointly trained (L7–L10), we observe consistent performance gain across all the benchmarks. Among the three variants of MRM (L7–L9), we observe that MRC-kl (L9) achieves the best performance (397.09) when combined with MLM + ITM, while MRC (L7) the worst (393.97). When combining MRC-kl and MRFR together with MLM and ITM (L10), we find that they are complimentary to each other, which leads to the second highest Meta-Sum score. The highest Meta-Sum Score is achieved by MLM + ITM + MRC-kl + MRFR + WRA (L11). We observe significant performance improvements from adding WRA, especially on VQA and RefCOCO+. It indicates the fine-grained alignment between words and regions learned through WRA during pre-training benefits the downstream tasks involving region-level recognition or reasoning. We use this optimal pre-training setting for the further experiments.

Additionally, we validate the contributions of conditional masking through a comparison study. When we perform random masking on both modalities simultaneously during pre-training, i.e., w/o conditional masking (L12), we observe a decrease in Meta-Sum score (396.51) compared to that with conditional masking (399.97). This indicates that the conditional masking strategy enables the model to learn better joint image-text representations effectively.

---

[13] Details about the metrics are listed in the supplementary.

Lastly, we study the effects of pre-training datasets. Our experiments so far have been focused on In-domain data. In this study, we pre-train our model on Out-of-domain data (Conceptual Captions + SBU Captions). A performance drop (396.91 in L13) from the model trained on In-domain data (COCO + Visual Genome) (400.93 in L11) shows that although Out-of-domain data contain more images, the model still benefits more from being exposed to similar downstream images during pre-training. We further pre-train our model on both In-domain and Out-of-domain data. With doubled data size, the model continues to improve (405.24 in L14).

4.3   Results on Downstream Tasks

Table 3 presents the results of UNITER on all downstream tasks. Both our base and large models are pre-trained on In-domain+Out-of-domain datasets, with the optimal pre-training setting: MLM+ITM+MRC-kl+MRFR+WRA. The implementation details of each task are provided in the supplementary file. We compare with both task-specific models and other pre-trained models on each downstream task. SOTA task-specific models include: MCAN [47] for VQA, MaxEnt [34] for NLVR^2, B2T2 [1] for VCR, SCAN [18] for Image-Text Retrieval, EVE-Image [42] for SNLI-VE, and MAttNet for RE Comprehension (RefCOCO, RefCOCO+ and RefCOCOg).[14] Other pre-trained models include: ViLBERT [23], LXMERT [37], Unicoder-VL [19], VisualBERT [20] and VLBERT [33].

Results show that our UNITER-large model achieves new state of the art across all the benchmarks. UNITER-base model also outperforms the others by a large margin across all tasks except VQA. Specifically, our UNITER-base model outperforms SOTA by approximately +2.8% for VCR on Q→AR, +2.5% for NLVR^2, +7% for SNLI-VE, +4% on R@1 for Image-Text Retrieval (+15% for zero-shot setting), and +2% for RE Comprehension.

Note that LXMERT pre-trains with downstream VQA (+VG+GQA) data, which may help adapt the model to VQA task. However, when evaluated on unseen tasks such as NLVR^2, UNITER-base achieves 3% gain over LXMERT. In addition, among all the models pre-trained on image-text pairs only, our UNITER-base outperforms the others by >1.5% on VQA.

It is also worth mentioning that both VilBERT and LXMERT observed two-stream model outperforms single-stream model, while our results show empirically that with our pre-training setting, single-stream model can achieve new state-of-the-art results, with much fewer parameters (UNITER-base: 86M, LXMERT: 183M, VilBERT: 221M).[15]

For VCR, we propose a two-stage pre-training approach: (*i*) pre-train on standard pre-training datasets; and then (*ii*) pre-train on downstream VCR dataset. Interestingly, while VLBERT and B2T2 observed that pre-training is not very

---

[14] MAttNet results are updated using the same features as the others. More details are provided in the supplementary file.

[15] The word embedding layer contains excessive rare words, thus excluded from the parameter counts.

Table 3. Results on downstream V+L tasks from UNITER model, compared with task-specific state-of-the-art (SOTA) and previous pre-trained models. ZS: Zero-Shot, IR: Image Retrieval and TR: Text Retrieval

Tasks		SOTA	ViLBERT	VLBERT (Large)	Unicoder-VL	VisualBERT	LXMERT	UNITER Base	Large
VQA	test-dev	70.63	70.55	71.79	-	70.80	72.42	72.70	**73.82**
	test-std	70.90	70.92	72.22	-	71.00	72.54	72.91	**74.02**
VCR	Q→A	72.60	73.30	75.80	-	71.60	-	75.00	**77.30**
	QA→R	75.70	74.60	78.40	-	73.20	-	77.20	**80.80**
	Q→AR	55.00	54.80	59.70	-	52.40	-	58.20	**62.80**
NLVR²	dev	54.80	-	-	-	67.40	74.90	77.18	**79.12**
	test-P	53.50	-	-	-	67.00	74.50	77.85	**79.98**
SNLI-VE	val	71.56	-	-	-	-	-	78.59	**79.39**
	test	71.16	-	-	-	-	-	78.28	**79.38**
ZS IR (Flickr)	R@1	-	31.86	-	48.40	-	-	66.16	**68.74**
	R@5	-	61.12	-	76.00	-	-	88.40	**89.20**
	R@10	-	72.80	-	85.20	-	-	92.94	**93.86**
IR (Flickr)	R@1	48.60	58.20	-	71.50	-	-	72.52	**75.56**
	R@5	77.70	84.90	-	91.20	-	-	92.36	**94.08**
	R@10	85.20	91.52	-	95.20	-	-	96.08	**96.76**
IR (COCO)	R@1	38.60	-	-	48.40	-	-	50.33	**52.93**
	R@5	69.30	-	-	76.70	-	-	78.52	**79.93**
	R@10	80.40	-	-	85.90	-	-	87.16	**87.95**
ZS TR (Flickr)	R@1	-	-	-	64.30	-	-	80.70	**83.60**
	R@5	-	-	-	85.80	-	-	**95.70**	95.70
	R@10	-	-	-	92.30	-	-	**98.00**	97.70
TR (Flickr)	R@1	67.90	-	-	86.20	-	-	85.90	**87.30**
	R@5	90.30	-	-	96.30	-	-	97.10	**98.00**
	R@10	95.80	-	-	99.00	-	-	98.80	**99.20**
TR (COCO)	R@1	50.40	-	-	62.30	-	-	64.40	**65.68**
	R@5	82.20	-	-	87.10	-	-	87.40	**88.56**
	R@10	90.00	-	-	92.80	-	-	93.08	**93.76**
Ref-COCO	val	87.51	-	-	-	-	-	91.64	**91.84**
	testA	89.02	-	-	-	-	-	92.26	**92.65**
	testB	87.05	-	-	-	-	-	90.46	**91.19**
	val^d	77.48	-	-	-	-	-	81.24	**81.41**
	testA^d	83.37	-	-	-	-	-	86.48	**87.04**
	testB^d	70.32	-	-	-	-	-	73.94	**74.17**
Ref-COCO+	val	75.38	-	80.31	-	-	-	83.66	**84.25**
	testA	80.04	-	83.62	-	-	-	86.19	**86.34**
	testB	69.30	-	75.45	-	-	-	78.89	**79.75**
	val^d	68.19	72.34	72.59	-	-	-	75.31	**75.90**
	testA^d	75.97	78.52	78.57	-	-	-	81.30	**81.45**
	testB^d	57.52	62.61	62.30	-	-	-	65.58	**66.70**
Ref-COCOg	val	81.76	-	-	-	-	-	86.52	**87.85**
	test	81.75	-	-	-	-	-	86.52	**87.73**
	val^d	68.22	-	-	-	-	-	74.31	**74.86**
	test^d	69.46	-	-	-	-	-	74.51	**75.77**

Table 4. Experiments on two-stage pre-training for VCR. Results are from UNITER-base on VCR val split. Stage I and Stage II denote first-stage and second-stage pre-training

Stage I	Stage II	Q→A	QA→ R	Q → AR
N	N	72.44	73.71	53.52
N	Y	73.52	75.34	55.6
Y	N	72.83	75.25	54.94
Y	Y	**74.56**	**77.03**	**57.76**

Table 5. Experiments on three modified settings for NLVR^2. All models use pre-trained UNITER-base

Setting	dev	test-P
Triplet	73.03	73.89
Pair	75.85	75.80
Pair-biattn	**77.18**	**77.85**

helpful on VCR, we find that the second-stage pre-training can significantly boost model performance, while the first-stage pre-training still helps but with limited effects (results shown in Table 4). This indicates that the proposed two-stage approach is highly effective in our pre-trained model over new data that are unseen in pre-training datasets.

Different from other tasks, NLVR^2 takes two images as input. Thus, directly finetuning UNITER pre-trained with image-sentence pairs might not lead to optimal performance, as the interactions between paired images are not learned during the pre-training stage. Thus, we experimented with three modified settings on NLVR^2: (*i*) *Triplet*: joint embedding of images pairs and query captions; (*ii*) *Pair*: individual embedding of each image and each query caption; and (*iii*) *Pair-biattn*: a bidirectional attention is added to the *Pair* model to learn the interactions between the paired images.

Comparison results are presented in Table 5. The *Pair* setting achieves better performance than the *Triplet* setting even without cross-attention between the image pairs. We hypothesize that it is due to the fact that our UNITER is pre-trained with image-text pairs. Thus, it is difficult to finetune a pair-based pre-trained model on triplet input. The bidirectional attention mechanism in the *Pair-biattn* setting, however, compensates the lack of cross-attention between images, hence yielding the best performance with a large margin. This show that with minimal surgery on the top layer of UNITER, our pre-trained model can adapt to new tasks that are very different from pre-training tasks.

4.4 Visualization

Similar to [15], we observe several patterns in the attention maps of the UNITER model, as shown in Fig. 2. Note that different from [15], our attention mechanism operates in both inter- and intra-modality manners. For completeness, we briefly discuss each pattern here:

- *Vertical:* attention to special tokens [CLS] or [SEP];
- *Diagonal:* attention to the token/region itself or preceding/following tokens/regions;
- *Vertical + Diagonal:* mixture of vertical and diagonal;

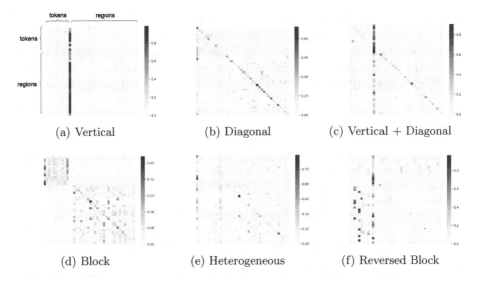

Fig. 2. Visualization of the attention maps learned by the UNITER-base model

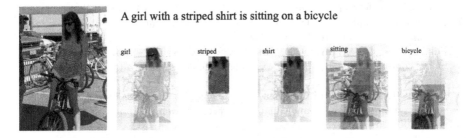

Fig. 3. Text-to-image attention visualization example

- *Block:* intra-modality attention, *i.e.*, textual self-attention and visual self-attention;
- *Heterogeneous:* diverse attentions that cannot be categorized and is highly dependent on actual input;
- *Reversed Block:* inter-modality attention, *i.e.*, text-to-image and image-to-text attention.

Note that *Reversed Block* (Fig. 2f) shows cross-modality alignment between tokens and regions. In Fig. 3, we visualize several examples of text-to-image attention to demonstrate the local cross-modality alignment between regions and tokens.

5   Conclusion

In this paper, we present UNITER, a large-scale pre-trained model providing UNiversal Image-TExt Representations for Vision-and-Language tasks. Four main pre-training tasks are proposed and evaluated through extensive ablation studies. Trained with both in-domain and out-of-domain datasets, UNITER outperforms state-of-the-art models over multiple V+L tasks by a significant margin. Future work includes studying early interaction between raw image pixels and sentence tokens, as well as developing more effective pre-training tasks.

References

1. Alberti, C., Ling, J., Collins, M., Reitter, D.: Fusion of detected objects in text for visual question answering. In: EMNLP (2019)
2. Anderson, P., et al.: Bottom-up and top-down attention for image captioning and visual question answering. In: CVPR (2018)
3. Antol, S., et al.: VQA: visual question answering. In: ICCV (2015)
4. Cao, J., Gan, Z., Cheng, Y., Yu, L., Chen, Y.C., Liu, J.: Behind the scene: revealing the secrets of pre-trained vision-and-language models. arXiv preprint arXiv:2005.07310 (2020)
5. Chen, L., Gan, Z., Cheng, Y., Li, L., Carin, L., Liu, J.: Graph optimal transport for cross-domain alignment. In: ICML (2020)
6. Devlin, J., Chang, M.W., Lee, K., Toutanova, K.: Bert: pre-training of deep bidirectional transformers for language understanding. In: NAACL (2019)
7. Doersch, C., Gupta, A., Efros, A.A.: Unsupervised visual representation learning by context prediction. In: ICCV (2015)
8. Fukui, A., Park, D.H., Yang, D., Rohrbach, A., Darrell, T., Rohrbach, M.: Multimodal compact bilinear pooling for visual question answering and visual grounding. In: EMNLP (2017)
9. Gan, Z., Chen, Y.C., Li, L., Zhu, C., Cheng, Y., Liu, J.: Large-scale adversarial training for vision-and-language representation learning. arXiv preprint arXiv:2006.06195 (2020)
10. Gao, P., et al.: Dynamic fusion with intra-and inter-modality attention flow for visual question answering. In: CVPR (2019)
11. Gidaris, S., Singh, P., Komodakis, N.: Unsupervised representation learning by predicting image rotations. In: ICLR (2018)
12. Hinton, G., Vinyals, O., Dean, J.: Distilling the knowledge in a neural network. arXiv preprint arXiv:1503.02531 (2015)
13. Kazemzadeh, S., Ordonez, V., Matten, M., Berg, T.: ReferItGame: referring to objects in photographs of natural scenes. In: EMNLP (2014)
14. Kim, J.H., Jun, J., Zhang, B.T.: Bilinear attention networks. In: NeurIPS (2018)
15. Kovaleva, O., Romanov, A., Rogers, A., Rumshisky, A.: Revealing the dark secrets of BERT. In: EMNLP (2019)
16. Krishna, R., et al.: Visual genome: connecting language and vision using crowdsourced dense image annotations. Int. J. Comput. Vis. **123**, 32–73 (2017). https://doi.org/10.1007/s11263-016-0981-7
17. Lan, Z., Chen, M., Goodman, S., Gimpel, K., Sharma, P., Soricut, R.: Albert: a lite BERT for self-supervised learning of language representations. In: ICLR (2020)
18. Lee, K.-H., Chen, X., Hua, G., Hu, H., He, X.: Stacked cross attention for image-text matching. In: Ferrari, V., Hebert, M., Sminchisescu, C., Weiss, Y. (eds.) ECCV 2018. LNCS, vol. 11208, pp. 212–228. Springer, Cham (2018). https://doi.org/10.1007/978-3-030-01225-0_13

19. Li, G., Duan, N., Fang, Y., Jiang, D., Zhou, M.: Unicoder-VL: a universal encoder for vision and language by cross-modal pre-training. In: AAAI (2020)
20. Li, L.H., Yatskar, M., Yin, D., Hsieh, C.J., Chang, K.W.: VisualBERT: a simple and performant baseline for vision and language. arXiv preprint arXiv:1908.03557 (2019)
21. Lin, T.-Y., et al.: Microsoft COCO: common objects in context. In: Fleet, D., Pajdla, T., Schiele, B., Tuytelaars, T. (eds.) ECCV 2014. LNCS, vol. 8693, pp. 740–755. Springer, Cham (2014). https://doi.org/10.1007/978-3-319-10602-1_48
22. Liu, Y., et al.: RoBERTa: a robustly optimized BERT pretraining approach. arXiv preprint arXiv:1907.11692 (2019)
23. Lu, J., Batra, D., Parikh, D., Lee, S.: ViLBERT: pretraining task-agnostic visiolinguistic representations for vision-and-language tasks. In: NeurIPS (2019)
24. Lu, J., Goswami, V., Rohrbach, M., Parikh, D., Lee, S.: 12-in-1: multi-task vision and language representation learning. In: CVPR (2020)
25. Noroozi, M., Favaro, P.: Unsupervised learning of visual representations by solving Jigsaw puzzles. In: Leibe, B., Matas, J., Sebe, N., Welling, M. (eds.) ECCV 2016. LNCS, vol. 9910, pp. 69–84. Springer, Cham (2016). https://doi.org/10.1007/978-3-319-46466-4_5
26. Ordonez, V., Kulkarni, G., Berg, T.L.: Im2Text: describing images using 1 million captioned photographs. In: NeurIPS (2011)
27. Pathak, D., Krahenbuhl, P., Donahue, J., Darrell, T., Efros, A.A.: Context encoders: feature learning by inpainting. In: CVPR (2016)
28. Peters, M.E., et al.: Deep contextualized word representations. In: NAACL (2018)
29. Peyré, G., Cuturi, M., et al.: Computational optimal transport. Found. Trends® Mach. Learn. 11(5–6), 355–607 (2019)
30. Plummer, B.A., Wang, L., Cervantes, C.M., Caicedo, J.C., Hockenmaier, J., Lazebnik, S.: Flickr30k entities: collecting region-to-phrase correspondences for richer image-to-sentence models. In: ICCV (2015)
31. Radford, A., Wu, J., Child, R., Luan, D., Amodei, D., Sutskever, I.: Language models are unsupervised multitask learners (2019)
32. Sharma, P., Ding, N., Goodman, S., Soricut, R.: Conceptual captions: a cleaned, hypernymed, image alt-text dataset for automatic image captioning. In: ACL (2018)
33. Su, W., et al.: VL-BERT: pre-training of generic visual-linguistic representations. In: ICLR (2020)
34. Suhr, A., Zhou, S., Zhang, I., Bai, H., Artzi, Y.: A corpus for reasoning about natural language grounded in photographs. In: ACL (2019)
35. Sun, C., Baradel, F., Murphy, K., Schmid, C.: Contrastive bidirectional transformer for temporal representation learning. arXiv preprint arXiv:1906.05743 (2019)
36. Sun, C., Myers, A., Vondrick, C., Murphy, K., Schmid, C.: VideoBERT: a joint model for video and language representation learning. In: ICCV (2019)
37. Tan, H., Bansal, M.: LXMERT: learning cross-modality encoder representations from transformers. In: EMNLP (2019)
38. Trinh, T.H., Luong, M.T., Le, Q.V.: Selfie: self-supervised pretraining for image embedding. arXiv preprint arXiv:1906.02940 (2019)
39. Vaswani, A., et al.: Attention is all you need. In: NeurIPS (2017)
40. Wang, L., Li, Y., Lazebnik, S.: Learning deep structure-preserving image-text embeddings. In: CVPR (2016)
41. Wu, Y., et al.: Google's neural machine translation system: bridging the gap between human and machine translation. arXiv preprint arXiv:1609.08144 (2016)

42. Xie, N., Lai, F., Doran, D., Kadav, A.: Visual entailment: a novel task for fine-grained image understanding. arXiv preprint arXiv:1901.06706 (2019)
43. Xie, Y., Wang, X., Wang, R., Zha, H.: A fast proximal point method for Wasserstein distance. arXiv:1802.04307 (2018)
44. Yang, Z., Dai, Z., Yang, Y., Carbonell, J., Salakhutdinov, R., Le, Q.V.: XLNet: generalized autoregressive pretraining for language understanding. In: NeurIPS (2019)
45. Yu, L., et al.: MAttNet: modular attention network for referring expression comprehension. In: CVPR (2018)
46. Yu, L., Poirson, P., Yang, S., Berg, A.C., Berg, T.L.: Modeling context in referring expressions. In: Leibe, B., Matas, J., Sebe, N., Welling, M. (eds.) ECCV 2016. LNCS, vol. 9906, pp. 69–85. Springer, Cham (2016). https://doi.org/10.1007/978-3-319-46475-6_5
47. Yu, Z., Yu, J., Cui, Y., Tao, D., Tian, Q.: Deep modular co-attention networks for visual question answering. In: CVPR (2019)
48. Zellers, R., Bisk, Y., Farhadi, A., Choi, Y.: From recognition to cognition: visual commonsense reasoning. In: CVPR (2019)
49. Zhang, R., Isola, P., Efros, A.A.: Colorful image colorization. In: Leibe, B., Matas, J., Sebe, N., Welling, M. (eds.) ECCV 2016. LNCS, vol. 9907, pp. 649–666. Springer, Cham (2016). https://doi.org/10.1007/978-3-319-46487-9_40
50. Zhou, L., Palangi, H., Zhang, L., Hu, H., Corso, J.J., Gao, J.: Unified vision-language pre-training for image captioning and VQA. In: AAAI (2020)

Oscar: Object-Semantics Aligned Pre-training for Vision-Language Tasks

Xiujun Li[1,2(✉)], Xi Yin[1], Chunyuan Li[1], Pengchuan Zhang[1], Xiaowei Hu[1], Lei Zhang[1], Lijuan Wang[1], Houdong Hu[1], Li Dong[1], Furu Wei[1], Yejin Choi[2], and Jianfeng Gao[1]

[1] Microsoft Corporation, Redmond, USA
xiul@microsoft.com
[2] University of Washington, Seattle, USA

Abstract. Large-scale pre-training methods of learning cross-modal representations on image-text pairs are becoming popular for vision-language tasks. While existing methods simply concatenate image region features and text features as input to the model to be pre-trained and use self-attention to learn image-text semantic alignments in a brute force manner, in this paper, we propose a new learning method OSCAR (**O**bject-**S**emantics **A**ligned Pre-training), which uses object tags detected in images as *anchor points* to significantly ease the learning of alignments. Our method is motivated by the observation that the salient objects in an image can be accurately detected, and are often mentioned in the paired text. We pre-train an OSCAR model on the public corpus of 6.5 million text-image pairs, and fine-tune it on downstream tasks, creating new state-of-the-arts on six well-established vision-language understanding and generation tasks (The code and pre-trained models are released: https://github.com/microsoft/Oscar).

Keywords: Object semantics · Vision-and-language · Pre-training

1 Introduction

Learning cross-modal representations is fundamental to a wide range of vision-language (V+L) tasks, such as visual question answering, image-text retrieval, image captioning. Recent studies [5,18,19,21,33,36,41] on vision-language pre-training (VLP) have shown that it can effectively learn generic representations from massive image-text pairs, and that fine-tuning VLP models on task-specific data achieves state-of-the-art (SoTA) results on well-established V+L tasks.

These VLP models are based on multi-layer Transformers [37]. To pre-train such models, existing methods simply concatenate image region features and text features as input and resort to the self-attention mechanism to learn semantic

Electronic supplementary material The online version of this chapter (https://doi.org/10.1007/978-3-030-58577-8_8) contains supplementary material, which is available to authorized users.

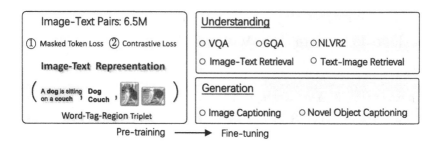

Fig. 1. OSCAR pipeline. The model takes a triple as input, is pre-trained with two losses (a masked token loss over words & tags, and a contrastive loss between tags and others), and fine-tuned for 5 understanding and 2 generation tasks (detailed in Sect. 4).

alignments between image regions and text in a brute force manner. However, the lack of explicit alignment information between the image regions and text poses alignment modeling a weakly-supervised learning task. In addition, visual regions are often over-sampled [2], noisy and ambiguous, which makes the task even more challenging.

In this study, we show that the learning of cross-modal representations can be significantly improved by introducing object tags detected in images as *anchor points* to ease the learning of semantic alignments between images and texts. We propose a new VLP method OSCAR, where we define the training samples as triples, each consisting of a word sequence, a set of object tags, and a set of image region features. Our method is motivated by the observation that the salient objects in an image can be accurately detected by modern object detectors [27], and that these objects are often mentioned in the paired text. For example, on the MS COCO dataset [20], the percentages that an image and its paired text share at least 1, 2, 3 objects are 49.7%, 22.2%, 12.9%, respectively. Our OSCAR model is pre-trained on a large-scale V+L dataset composed of 6.5 million pairs, and is fine-tuned and evaluated on seven V+L understanding and generation tasks. The overall setting is illustrated in Fig. 1.

Although the use of anchor points for alignment modeling has been explored in natural language processing *e.g.,* [3], to the best of our knowledge, this work is the first that explores the idea for VLP. There have been previous works that use object or image tags in V+L tasks for the sake of enhancing the feature representation of image regions, rather than for learning image-text alignments. For example, Zhou *et al.* [41] uses the object prediction probability as a soft label and concatenate it with its corresponding region features. Wu *et al.* [38] and You *et al.* [39] introduce image-level labels or attributes to improve image-level visual representations.

The main contributions of this work can be summarized as follows: (*i*) We introduce OSCAR, a powerful VLP method to learn generic image-text representations for V+L understanding and generation tasks. (*ii*) We have developed an OSCAR model that achieves new SoTA on multiple V+L benchmarks, outperforming existing approaches by a significant margin; (*iii*) We present extensive experiments and analysis to provide insights on the effectiveness of using object tags as anchor points for cross-modal representation learning and downstream tasks.

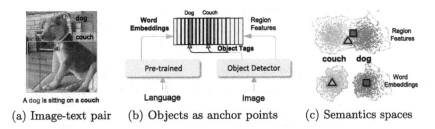

(a) Image-text pair   (b) Objects as anchor points   (c) Semantics spaces

Fig. 2. Illustration on the process that OSCAR represents an image-text pair into semantic space via dictionary look up. (a) An example of input image-text pair (b) The object tags are used as anchor points to align image regions with word embeddings of pre-trained language models. (c) The word semantic space is more representative than image region features. In this example, dog and couch are similar in the visual feature space due to the overlap regions, but distinctive in the word embedding space.

2   Background

The training data for many V+L tasks consists of image-text pairs, as shown in Fig. 2(a). We denote a dataset of size N by $\mathcal{D} = \{(\mathbf{I}_i, \boldsymbol{w}_i)\}_{i=1}^N$, with image $\mathbf{I}$ and text sequence $\boldsymbol{w}$. The goal of pre-training is to learn cross-modal representations of image-text pairs in a self-supervised manner, which can be adapted to serve various down-stream tasks via fine-tuning.

VLP typically employs multi-layer self-attention Transformers [37] to learn cross-modal *contextualized* representations, based on the *singular* embedding of each modality. Hence, the success of VLP fundamentally relies on the quality of the input singular embeddings. Existing VLP methods take visual region features $\boldsymbol{v} = \{v_1, \cdots, v_K\}$ of an image and word embeddings $\boldsymbol{w} = \{w_1, \cdots, w_T\}$ of its paired text as input, and relies on the self-attention mechanism to learn image-text alignments and produce cross-modal contextual representations.

Though intuitive and effective, existing VLP methods suffer from two issues:

(*i*) *Ambiguity.* The visual region features are usually extracted from over-sampled regions [2] via Faster R-CNN object detectors [27], which inevitably results in overlaps among image regions at different positions. This renders ambiguities for the extracted visual embeddings. For example, in Fig. 2(a) the region features for dog and couch are not easily distinguishable, as their regions heavily overlap. (*ii*) *Lack of grounding.* VLP is naturally a weakly-supervised learning problem because there is no explicitly labeled alignments between regions or objects in an image and words or phrases in text. However, we can see that salient objects such as dog and couch are presented in both image and its paired text as in Fig. 2(a), and can be used as anchor points for learning semantic alignments between image regions and textual units as in Fig. 2(b). In this paper we propose a new VLP method that utilizes these anchor points to address the aforementioned issues.

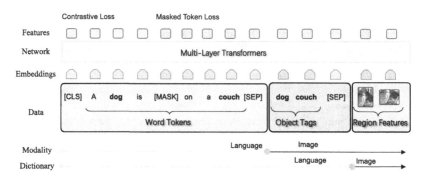

Fig. 3. Illustration of OSCAR. We represent the image-text pair as a triple [word tokens , object tags , region features], where the object tags (*e.g.,* "dog" or "couch") are proposed to align the cross-domain semantics; when removed, OSCAR reduces to previous VLP methods. The input triple can be understood from two perspectives: a *modality* view and a *dictionary* view.

3   Oscar Pre-training

Humans perceive the world through many channels. Even though any individual channel might be incomplete or noisy, important factors are still perceivable since they tend to be shared among multiple channels (*e.g.,* dog can be described visually and verbally, as in Fig. 2). With this motivation, we propose a new VLP method OSCAR to learn representations that capture channel-invariant (or modality-invariant) factors at the semantic level. Oscar differs from existing VLP in the way that the input image-text pairs are represented and the pre-training objective, as outlined in Fig. 3.

Input. OSCAR represents each input image-text pair as a Word-Tag-Image triple $(\boldsymbol{w}, \boldsymbol{q}, \boldsymbol{v})$, where $\boldsymbol{w}$ is the sequence of word embeddings of the text, $\boldsymbol{q}$ is the word embedding sequence of the object tags (in text) detected from the image, and $\boldsymbol{v}$ is the set of region vectors of the image.

Existing VLP methods represent each input pair as $(\boldsymbol{w}, \boldsymbol{v})$. OSCAR introduces $\boldsymbol{q}$ as anchor points to ease the learning of image-text alignment. This is motivated by the observation that in training data, important objects in an image are often also *presented* in the image-paired text, using either the same words as object tags or different but semantically similar or related words. Since the alignments between $\boldsymbol{q}$ and $\boldsymbol{w}$, both in text, are relatively easy to identified by using pre-trained BERT models [6], which are used as initialization for VLP in OSCAR, the image regions from which the object tags are detected are likely to have higher attention weights than other regions, when queried by the semantically related words in the text. This alignment learning process is conceptually illustrated in Fig. 2(b). The process can also be interpreted as learning to ground the image objects, which might be ambiguously represented in the vision space such as dog and couch in Fig. 2(a), in distinctive entities represented in the language space, as illustrated in Fig. 2(c).

Specifically, v and q are generated as follows. Given an image with K regions of objects (normally over-sampled and noisy), Faster R-CNN [27] is used to extract the visual semantics of each region as (v', z), where region feature $v' \in \mathbb{R}^P$ is a P-dimensional vector (*i.e.*, $P = 2048$), and region position z a R-dimensional vector (*i.e.*, $R = 4$ or 6)[1]. We concatenate v' and z to form a position-sensitive region feature vector, which is further transformed into v using a linear projection to ensure that it has the same vector dimension as that of word embeddings. Meanwhile, the same Faster R-CNN is used to detect a set of high precision object tags. q is the sequence of word embeddings of the object tags.

Pre-Training Objective. The OSCAR input can be viewed from two different perspectives as

$$x \triangleq [\underbrace{w}_{\text{language}}, \underbrace{q, v}_{\text{image}}] = [\underbrace{w, q}_{\text{language}}, \underbrace{v}_{\text{image}}] \triangleq x' \tag{1}$$

where x is the *modality* view to distinguish the representations between a text and an image; while x' is the *dictionary* view[2] to distinguish the two different semantic spaces, in which the input is represented. The two-view perspective allows us to design a novel pre-training objective.

A Dictionary View: Masked Token Loss. The use of different dictionaries determines the semantic spaces utilized to represent different sub-sequences. Specifically, the object tags and word tokens share the same linguistic semantic space, while the image region features lie in the visual semantic space. We define the *discrete token sequence* as $h \triangleq [w, q]$, and apply the Masked Token Loss (MTL) for pre-training. At each iteration, we randomly mask each input token in h with probability 15%, and replace the masked one h_i with a special token [MASK]. The goal of training is to predict these masked tokens based on their surrounding tokens $h_{\backslash i}$ and all image features v by minimizing the negative log-likelihood:

$$\mathcal{L}_{\text{MTL}} = -\mathbb{E}_{(v,h) \sim \mathcal{D}} \log p(h_i | h_{\backslash i}, v) \tag{2}$$

This is similar to masked language model used by BERT. The masked word or tag needs to be recovered from its surroundings, with additional image information attended to help ground the learned word embeddings in the vision context.

---

[1] It includes coordinates of top-left & bottom-right corners, and/or height & width.

[2] A semantic space can be viewed a vector space defined by a dictionary, which maps an input to a vector representation in the semantic space. For example, BERT can be viewed as a dictionary that defines a linguistic semantic space. BERT maps an input word or word sequence into a feature vector in the semantic space.

A Modality View: Contrastive Loss. For each input triple, we group $\boldsymbol{h}' \triangleq [\boldsymbol{q}, \boldsymbol{v}]$ to represent the image modality, and consider $\boldsymbol{w}$ as the language modality. We then sample a set of "polluted" image representations by replacing $\boldsymbol{q}$ with probability 50% with a different tag sequence randomly sampled from the dataset $\mathcal{D}$. Since the encoder output on the special token [CLS] is the fused vision-language representation of $(\boldsymbol{h}', \boldsymbol{w})$, we apply a fully-connected (FC) layer on the top of it as a binary classifier $f(.)$ to predict whether the pair contains the original image representation ($y = 1$) or any polluted ones ($y = 0$). The contrastive loss is defined as

$$\mathcal{L}_\text{C} = -\mathbb{E}_{(\boldsymbol{h}', \boldsymbol{w}) \sim \mathcal{D}} \log p(y|f(\boldsymbol{h}', \boldsymbol{w})). \tag{3}$$

During the cross-modal pre-training, we utilize object tags as the proxy of images to adjust the word embedding space of BERT, where a text is similar to its paired image (or more specifically, the object tags detected from the image), and dissimilar to the polluted ones.

The full pre-training objective of OSCAR is:

$$\mathcal{L}_\text{Pre-training} = \mathcal{L}_\text{MTL} + \mathcal{L}_\text{C}. \tag{4}$$

Discussion. Although other loss function designs can be considered as pre-training objectives, we perform experiments with these two losses for two reasons: (*i*) Each loss provides a representative learning signal from its own perspective. We deliberately keep a clear and simple form for the joint loss to study the effectiveness of the proposed dictionary and modality views, respectively. (*ii*) Though the overall loss is much simpler than those of existing VLP methods, it yields superior performance in our experiments.

Pre-training Corpus. We have built the pre-training corpus based on the existing V+L datasets, including COCO [20], Conceptual Captions (CC) [30], SBU captions [25], flicker30k [40], GQA [12] *etc.*. In total, the unique image set is 4.1 million, and the corpus consists of 6.5 million text-tag-image triples. The detail is in Appendix.

Implementation Details. We pre-train two model variants, denoted as OSCAR$_\text{B}$ and OSCAR$_\text{L}$, initialized with parameters $\boldsymbol{\theta}_\text{BERT}$ of BERT base ($H = 768$) and large ($H = 1024$), respectively, where H is the hidden size. To ensure that the image region features have the same input embedding size as BERT, we transform the position-sensitive region features using a linear projection via matrix $\mathbf{W}$. The trainable parameters are $\boldsymbol{\theta} = \{\boldsymbol{\theta}_\text{BERT}, \mathbf{W}\}$. The AdamW Optimizer is used. OSCAR$_\text{B}$ is trained for at least 1.0M steps, with learning rate $5e^{-5}$ and batch size 768. OSCAR$_\text{L}$ is trained for at least 900k steps, with learning rate $1e^{-5}$ and batch size 512. The sequence length of discrete tokens $\boldsymbol{h}$ and region features $\boldsymbol{v}$ are 35 and 50, respectively.

4 Adapting to V+L Tasks

We adapt the pre-trained models to seven downstream V+L tasks, including five understanding tasks and two generation tasks. Each task poses different challenges for adaptation. We introduce the tasks and our fine-tuning strategy in this section, and leave the detailed description of datasets and evaluation metrics to Appendix.

Image-Text Retrieval heavily relies on the joint representations. There are two sub-tasks: *image retrieval* and *text retrieval*, depending on which modality is used as the retrieved target. During training, we formulate it as a binary classification problem. Given an aligned image-text pair, we randomly select a different image or a different caption to form an unaligned pair. The final representation of [CLS] is used as the input to the classifier to predict whether the given pair is aligned or not. We did not use ranking losses [13,17], as we found that the binary classification loss works better, similarly as reported in [26]. In the testing stage, the probability score is used to rank the given image-text pairs of a query. Following [18], we report the top-K retrieval results on both the 1K and 5K COCO test sets.

Image Captioning requires the model to generate a natural language description of the content of an image. To enable sentence generation, we fine-tune OSCAR using the seq2seq objective. The input samples are processed to triples consisting of image region features, captions, and object tags, in the same way as that during the pre-training. We randomly mask out 15% of the caption tokens and use the corresponding output representations to perform classification to predict the token ids. Similar to VLP [41], the self-attention mask is constrained such that a caption token can only attend to the tokens before its position to simulate a uni-directional generation process. Note that all caption tokens will have full attentions to image regions and object tags but not the other way around.

During inference, we first encode the image regions, object tags, and a special token [CLS] as input. Then the model starts the generation by feeding in a [MASK] token and sampling a token from the vocabulary based on the likelihood output. Next, the [MASK] token in the previous input sequence is replaced with the sampled token and a new [MASK] is appended for the next word prediction. The generation process terminates when the model outputs the [STOP] token. We use beam search (*i.e.,* beam size = 5) [2] in our experiments and report our results on the COCO image captioning dataset.

Novel Object Captioning (NoCaps) [1] extends the image captioning task, and provides a benchmark with images from the Open Images dataset [16] to test models' capability of describing novel objects which are not seen in the training corpus. Following the restriction guideline of NoCaps, we use the predicted Visual Genome and Open Images labels to form tag sequences, and train OSCAR on COCO without the initialization of pre-training.

VQA [8] requires the model to answer natural language questions based on an image. Given an image and a question, the task is to select the correct answer from a multi-choice list. Here we conduct experiments on the widely-used VQA v2.0 dataset [8], which is built based on the MSCOCO [20] image corpus. The dataset is split into training (83k images and 444k questions), validation (41k images and 214k questions), and test (81k images and 448k questions) sets. Following [2], for each question, the model picks the corresponding answer from a shared set consisting of 3,129 answers.

When fine-tuning on the VQA task, we construct one input sequence, which contains the concatenation of a given question, object tags and region features, and then the [CLS] output from OSCAR is fed to a task-specific linear classifier for answer prediction. We treat VQA as a multi-label classification problem [2] – assigning a soft target score to each answer based on its relevancy to the human answer responses, and then we fine-tune the model by minimizing the cross-entropy loss computed using the predicted scores and the soft target scores. At inference, we simply use a Softmax function for prediction.

GQA [12] is similar to VQA, except that GQA tests the reasoning capability of the model to answer a question. We conduct experiments on the public GQA dataset [12]. For each question, the model chooses an answer from a shared set of $1,852$ candidate answers. We develop two fine-tuned models using OSCAR_B. One is similar to that of VQA. The other, denoted as OSCAR_B^* in Table 2(d), is first fine-tuned on unbalanced "all-split" for 5 epochs, and then fine-tuned on the "balanced-split" for 2 epochs, as suggested in [4].

Natural Language Visual Reasoning for Real (NLVR2) [34] takes a pair of images and a natural language, and the goal is to determine whether the natural language statement is true about the image pair. When fine-tuning on the NLVR2 task, we first construct two input sequences, each containing the concatenation of the given sentence (the natural language description) and one image, and then two [CLS] outputs from OSCAR are concatenated as the joint input for a binary classifier, implemented by an MLP[3].

5   Experimental Results and Analysis

5.1   Performance Comparison with SoTA

To account for parameter efficiency, we compare OSCAR against three types of SoTA's: (*i*) SoTA$_S$ indicates the best performance achieved by small models prior to the Transformer-based VLP models. (*ii*) SoTA$_B$ indicates the best performance achieved by VLP models of similar size to BERT base. (*iii*) SoTA$_L$

---

[3] This is not necessarily the best fine-tuning choice for NLVR2, please refer to the *Pair-biattn* finetuning in UNITER [5] for a better choice, which introduces a multi-head attention layer to look back the concatenated text-image sequences.

Table 1. Overall results on six tasks. Δ indicates the improvement over SoTA. SoTA with subscript S, B, L indicates performance achieved by small models, VLP of similar size to BERT base and large model, respectively. Most results are from [5], except that image captioning results are from [10,41], NoCaps results are from [1], VQA results are from [36].

Task	Image Retrieval			Text Retrieval			Image Captioning				NoCaps		VQA	NLVR2
	R@1	R@5	R@10	R@1	R@5	R@10	B@4	M	C	S	C	S	test-std	test-P
SoTA$_S$	39.2	68.0	81.3	56.6	84.5	92.0	38.9	29.2	129.8	22.4	61.5	9.2	70.90	53.50
SoTA$_B$	48.4	76.7	85.9	63.3	87.0	93.1	39.5	29.3	129.3	23.2	73.1	11.2	72.54	78.87
SoTA$_L$	51.7	78.4	86.9	66.6	89.4	94.3	–	–	–	–	–	–	73.40	79.50
Oscar$_B$	54.0	80.8	88.5	70.0	91.1	95.5	40.5	29.7	137.6	22.8	78.8	11.7	73.44	78.36
Oscar$_L$	57.5	82.8	89.8	73.5	92.2	96.0	41.7	30.6	140.0	24.5	80.9	11.3	73.82	80.37
Δ	5.8 ↑	4.4 ↑	2.9 ↑	6.9 ↑	2.8 ↑	1.7 ↑	2.2 ↑	1.3 ↑	10.7 ↑	1.3 ↑	7.8 ↑	0.5 ↑	0.42 ↑	0.87 ↑

indicates the best performance yielded by models that have a similar size to BERT large. To the best of our knowledge, UNITER [5] is the only model of BERT large size.

Table 1 summarizes the overall results on all tasks[4]. For all the tables in this paper, **Blue** indicates the best result for a task, and gray background indicates results produced by Oscar. As shown in the table, our base model outperforms previous large models on most tasks, often by a significantly large margin. It demonstrates that the proposed Oscar is highly parameter-efficient, partially because the use of object tags as anchor points significantly eases the learning of semantic alignments between images and texts. Note that Oscar is pre-trained on 6.5 million pairs, which is less than 9.6 million pairs used for UNITER pre-training and 9.18 million pairs for LXMERT.

We report the detailed comparison on each task in Table 2. (*i*) VLP methods dominate empirical performance across many V+L tasks, compared with small models. Oscar outperforms all existing VLP methods on all seven tasks, and achieves new SoTA on six of them. On GQA, neural state machine (NSM) [11] relies on a strong structural prior, which can also be incorporated into Oscar for improvement in the future. (*ii*) 12-in-1 is a recently proposed multi-task learning model [22] for V+L, implemented on BERT base. We see that Oscar$_B$ outperforms 12-in-1 on almost all the tasks, except on Test-P of NLVR2. Given that our method is based on single task fine-tuning, the result demonstrates the effectiveness of our proposed pre-training scheme. (*iii*) overall, Oscar is the best performer on both understanding and generation tasks. On the captioning task, we further fine-tune Oscar with self-critical sequence training (SCST) [29] to improve sequence-level learning. The only comparable VLP method for captioning is [41]. The results in Table 2 (e) show that Oscar yields a much better performance, *e.g.*, improving BLEU@4 and CIDEr by more than 2 and 10 points, respectively. (*iv*) The NoCaps guideline requires to only use the COCO captioning training set. Hence, we initialize with BERT, and train Oscar on the COCO training set. Constrained beam search (CBS) is used. The results in

---

[4] All the (single-model) SoTAs are from the published results.

Table 2. Detailed results on V+L tasks.

Method	Size	Text Retrieval R@1	R@5	R@10	Image Retrieval R@1	R@5	R@10	Text Retrieval R@1	R@5	R@10	Image Retrieval R@1	R@5	R@10
					1K Test Set						5K Test Set		
DVSA [14]	-	38.4	69.9	80.5	27.4	60.2	74.8	-	-	-	-	-	-
VSE++ [7]	-	64.7	-	95.9	52.0	-	92.0	41.3	-	81.2	30.3	-	72.4
DPC [46]	-	65.6	89.8	95.5	47.1	79.9	90.0	41.2	70.5	81.1	25.3	53.4	66.4
CAMP [42]	-	72.3	94.8	98.3	58.5	87.9	95.0	50.1	82.1	89.7	39.0	68.9	80.2
SCAN [18]	-	72.7	94.8	98.4	58.8	88.4	94.8	50.4	82.2	90.0	38.6	69.3	80.4
SCG [33]	-	76.6	96.3	99.2	61.4	88.9	95.1	56.6	84.5	92.0	39.2	68.0	81.3
PFAN [41]	-	76.5	96.3	99.0	61.6	89.6	95.2	-	-	-	-	-	-
Unicoder-VL [19]	B	84.3	97.3	99.3	69.7	93.5	97.2	62.3	87.1	92.8	46.7	76.0	85.3
12-in-1 [24]	B	-	-	-	65.2	91.0	96.2	-	-	-	-	-	-
UNITER [5]	B	-	-	-	-	-	-	63.3	87.0	93.1	48.4	76.7	85.9
UNITER [5]	L	-	-	-	-	-	-	66.6	89.4	94.3	51.7	78.4	86.9
OSCAR	B	88.4	**99.1**	**99.8**	75.7	95.2	98.3	70.0	91.1	95.5	54.0	80.8	88.5
	L	**89.8**	98.8	99.7	**78.2**	**95.8**	**98.3**	**73.5**	**92.2**	**96.0**	**57.5**	**82.8**	**89.8**

(a) Image-text retrieval

Method	ViLBERT	VL-BERT	VisualBERT	LXMERT	12-in-1	UNITER_B	UNITER_L	OSCAR_B	OSCAR_L
Test-dev	70.63	70.50	70.80	72.42	73.15	72.27	**73.24**	73.16	**73.61**
Test-std	70.92	70.83	71.00	72.54	–	72.46	73.40	**73.44**	**73.82**

(b) VQA

Method	MAC	VisualBERT	LXMERT	12-in-1	UNITER_B	UNITER_L	OSCAR_B	OSCAR_L
Dev	50.8	67.40	74.90	–	77.14	**78.40**	78.07	**79.12**
Test-P	51.4	67.00	74.50	78.87	77.87	**79.50**	78.36	**80.37**

(c) NLVR2

Method	Test-dev	Test-std
LXMERT [39]	60.00	60.33
MMN [4]	–	60.83
12-in-1 [24]	–	60.65
NSM [12]	–	63.17
OSCAR_B	61.19	61.23
OSCAR_B*	**61.58**	**61.62**

(d) GQA

Method	cross-entropy optimization B@4	M	C	S	CIDEr optimization B@4	M	C	S
BUTD [2]	36.2	27.0	113.5	20.3	36.3	27.7	120.1	21.4
VLP [47]	36.5	28.4	117.7	21.3	39.5	29.3	129.3	23.2
AoANet [11]	37.2	28.4	119.8	21.3	38.9	29.2	129.8	22.4
OSCAR_B	36.5	**30.3**	123.7	23.1	40.5	29.7	137.6	22.8
OSCAR_L	**37.4**	**30.7**	**127.8**	**23.5**	**41.7**	**30.6**	**140.0**	**24.5**

(e) Image captioning on COCO

Method	in-domain CIDEr	SPICE	near-domain CIDEr	SPICE	out-of-domain CIDEr	SPICE	overall CIDEr	SPICE
UpDown [1]	78.1	11.6	57.7	10.3	31.3	8.3	55.3	10.1
UpDown + CBS [1]	80.0	12.0	73.6	11.3	66.4	9.7	73.1	11.1
UpDown + ELMo + CBS [1]	79.3	**12.4**	73.8	11.4	71.7	9.9	74.3	11.2
OSCAR_B	79.6	12.3	66.1	11.5	45.3	9.7	63.8	11.2
OSCAR_B + CBS	80.0	12.1	80.4	**12.2**	75.3	10.6	79.3	**11.9**
OSCAR_B + SCST + CBS	**83.4**	12.0	**81.6**	12.0	**77.6**	10.6	**81.1**	11.7
OSCAR_L	79.9	**12.4**	68.2	11.8	45.1	9.4	65.2	11.4
OSCAR_L + CBS	78.8	12.2	78.9	**12.1**	77.4	10.5	78.6	**11.8**
OSCAR_L + SCST + CBS	**85.4**	11.9	**84.0**	11.7	**80.3**	10.0	**83.4**	11.4

(f) Evaluation on NoCaps Val. Models are trained on COCO only without pre-training.

Table 2 (f) show that the variants of OSCAR consistently outperform the previous SoTA method UpDown [1]. The gap is much larger on the near-domain or out-of-domain cases, demonstrating the strong generalization ability of OSCAR.

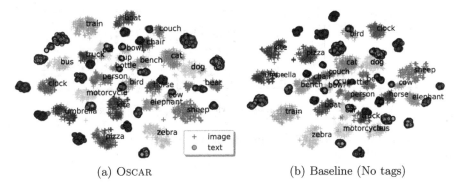

(a) OSCAR   (b) Baseline (No tags)

Fig. 4. 2D visualization using t-SNE. The points from the same object class share the same color. Please refer Appendix for full visualization. (Color figure online)

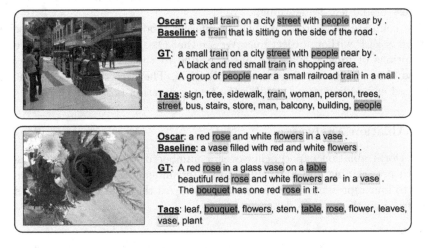

Fig. 5. Examples of image captioning. Objects are colored, based on their appearance against the groud-truth (GT): all , OSCAR & tags , tags only . (Color figure online)

5.2 Qualitative Studies

We visualize the learned semantic feature space of image-text pairs of the COCO test set on a 2D map using t-SNE [23]. For each image region and word token, we pass it through the model, and use its last-layer output as features. Pre-trained models with and without object tags are compared. The results in Fig. 4 reveal some interesting findings. (*i*) *Intra-class.* With the aid of object tags, the distance of the same object between two modalities is substantially reduced. For example, the visual and textual representations for person (or zebra) in OSCAR is much closer than that in the baseline method. (*ii*) *Inter-class.* Object classes of related semantics are getting closer (but still distinguishable) after adding tags, while there are some mixtures in the baseline, such as animal (person,

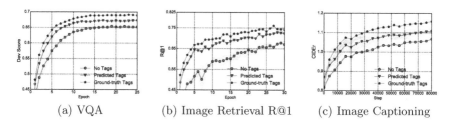

(a) VQA   (b) Image Retrieval R@1   (c) Image Captioning

Fig. 6. The learning curves of fine-tuning downstream tasks with different object tags. Each curve is with 3 runs.

zebra, sheep, bird), furniture (chair, couch, bench), and transportation (bus, train, truck, motorcycle, car). This verifies the importance of object tags in alignment learning: it plays the role of anchor points in linking and regularizing the cross-modal feature learning.

We compare generated captions of different models in Fig. 5. The baseline method is VLP without object tags. We see that OSCAR generates more detailed descriptions of images than the baseline, due to the use of the accurate and diverse object tags detected by Faster R-CNN. They are the anchor points in the word embedding space, guiding the text generation process.

5.3   Ablation Analysis

We perform ablation experiments over a number of design choices of OSCAR in both pre-training and fine-tuning to better understand their relative importance to four representative downstream tasks. All the ablation experiments are conducted on the base model.

The Effect of Object Tags. To study the effect of object tags, we experiment three different settings: (*i*) *Baseline (No Tags)*: this reduces the models to their previous VLP counterparts, where no tag information is exploited. (*ii*) *Predicted Tags*: we use an off-the-shelf object detector (trained on COCO dataset) to predict object tags. (*iii*) *Ground-truth Tags*: The ground-truth tags from COCO dataset are utilized to serve as a performance "upper bound" for our method. The experiments are conducted with the same BERT base model on three representative tasks, including VQA, image retrieval, and image captioning. As shown in Fig. 6, the learning curves for fine-tuning with object tags converges significantly faster and better than the VLP method without tags on all tasks. On the VQA and retrieval tasks, training using tags only takes half of the training time to achieve the final performance of the baseline, showing that OSCAR is a more practical and efficient scheme for VLP. With more accurate object detectors developed in the future, OSCAR can achieve even better performance, closing the gap demonstrated by using the ground-truth tags.

Attention Interaction. To further understand the interaction among the text, object tags and object regions, we conduct fine-tuning experiments by varying the attention masks for image-text retrieval. The default setting uses full attentions across all modalities. We then enable certain part of the attention masks. All models are initialized from BERT base without pre-training. Table 3 reports the performance on the COCO 1K

Table 3. Retrieval results on the COCO 1K test set, with different types of attention interactions.

Attention			Text R.		Image R.	
w-v	*w-q*	*v-q*	R@1	R@5	R@1	R@5
✓	✓	✓	77.3	95.6	65.2	91.5
✓			75.4	94.8	64.2	91.4
		✓	32.3	57.6	25.7	60.1

test set. By comparing the results of using full attention and partial attention *w-v*, we see that it is beneficial to add object tags. Moreover, region features are more informative than object tags (*w-v*, vs. *v-q*) in representing an image. This suggests that tags yield minor improvement when used as features; a more promising way is to use them as anchor points, as done in OSCAR.

Object Tags in Pre-training. To study the impact of different object tag sets in pre-trained models, we pre-train two variants: OSCAR^VG and OSCAR^OI utilizes object tags produced by the object detector trained on the visual genome (VG) dataset [15] and the open images (OI) dataset [16], respectively. In this ablation, all the models are pre-trained for 589k steps. The results are shown in Table 4, where BASELINE (NO TAGS) is also listed for comparison. It is clear that the OSCAR scheme of using object tags as anchor points improves the baseline, regardless of which set of object tags is used. VG tags performs slightly better than OI. We hypothesize that the object detector trained on VG has a more diverse set of objects, although the object detector trained on OI has a higher precision.

Table 4. Results with various pre-training schemes.

Pre-train	VQA	Text retrieval			Image retrieval			Image captioning			
	dev	R@1	R@5	R@10	R@1	R@5	R@10	B@4	M	C	S
BASELINE (NO TAGS)	70.93	84.4	98.1	99.5	73.1	94.5	97.9	34.5	29.1	115.6	21.9
OSCAR^VG	71.70	88.4	99.1	99.8	75.7	95.2	98.3	36.4	30.3	123.4	23.0
OSCAR^OI	71.15	85.9	97.9	99.5	72.9	94.3	97.6	35.3	29.6	119.5	22.6

6   Related Work

Vision-Language Pre-training. There is a growing interest in pre-training generic models to solve a variety of V+L problems, such as visual question-answering (VQA), image-text retrieval and image captioning *etc.* The existing methods [5,9,18,21,33,35,36,41] employ BERT-like objectives [6] to learn cross-modal representations from a concatenated-sequence of visual region features and language token embeddings. They heavily rely on the self-attention mechanism of Transformers to learn joint representations that are appropriately contextualized in both modalities. For example, early efforts such as [21,36] propose a two-stream and three-stream Transformer-based framework with co-attention to fuse the two modalities, respectively. Chen *et al.* [5] conduct comprehensive studies on the effects of different pre-training objectives for the learned generic representations. Zhou *et al.* [41] propose the first unified model to deal with both understanding and generation tasks, using only VQA and image captioning as the downstream tasks. In this paper, the OSCAR models have been applied to a wider range of downstream tasks, including both understanding and generation tasks, and have achieved new SoTA in most of them. Compared to existing VLP methods, the most salient difference of the proposed OSCAR is the use of object tags for aligning elements in two modalities. It alleviates the challenge of VLP models having to figure out the cross-modal semantic alignment from scratch, and thus improves the learning efficiency. In fact, our base model already outperforms the existing large VLP models on most V+L tasks.

Object Tags. Anderson *et al.* [2] introduce the bottom-up mechanism to represent an image as a set of visual regions via Faster R-CNN [27], each with an associated feature vector. It enables attention to be computed at the object level, and has quickly become the de facto standard for fine-grained image understanding tasks. In this paper, we propose to use object tags to align the object-region features in [2] in the pre-trained linguistic semantic space. The idea of utilizing object tags has been explored for image understanding [38,39,41]. Based on grid-wise region features of CNNs, Wu *et al.* [38] employ the predicted object tags only as the input to LSTM for image captioning, while You *et al.* [39] consider both tags and region features. Based on salient regions proposed by object detectors, Zhou *et al.* [41] concatenate the object prediction probability vector with region features as the visual input for VLP. Unfortunately, the tags in these works are not simultaneously associated with both object regions and word embeddings of text, resulting in a lack of grounding. Our construction of object tags with their corresponding region features & word embeddings yields more complete and informative representations for objects, particularly when the linguistic entity embeddings are pre-trained, as described next.

Multimodal Embeddings. It has been shown that V+L tasks can benefit from a shared embedding space to align the inter-modal correspondences between images and text. Early attempts from Socher *et al.* [31] project words and image regions into a common space using kernelized canonical correlation analysis, and achieve good results for annotation and segmentation. Similar ideas are employed for image captioning [13] and text-based image retrieval [28]. In particular, the seminal work DeViSE [7] proposes to identify visual objects using semantic information gleaned from un-annotated text. This semantic information is exploited to make predictions of image labels that are not observed during training, and improves zero-shot predictions dramatically across thousands of novel labels that have never been seen by the vision model. The idea has been extended in [14, 24, 32], showing that leveraging pre-trained linguistic knowledge is highly effective for aligning semantics and improving sample efficiency in cross-modal transfer learning. Inspired by this line of research, we revisit the idea and propose to leverage the rich semantics from the learned word embeddings in the era of neural language model pre-training. Indeed, our results on novel object captioning demonstrate that OSCAR helps improve the generalizability of the pre-trained models.

7   Conclusion

In this paper, we have presented a new pre-training method OSCAR, which uses object tags as anchor points to align the image and language modalities in a shared semantic space. We validate the schema by pre-training OSCAR models on a public corpus with 6.5 million text-image pairs. The pre-trained models archive new state-of-the-arts on six established V+L understanding and generation tasks.

References

1. Agrawal, H., et al.: Nocaps: novel object captioning at scale. In: ICCV (2019)
2. Anderson, P., et al.: Bottom-up and top-down attention for image captioning and visual question answering. In: CVPR (2018)
3. Brown, P.F., Lai, J.C., Mercer, R.L.: Aligning sentences in parallel corpora. In: Proceedings of the 29th Annual Meeting on Association for Computational Linguistics (1991)
4. Chen, W., Gan, Z., Li, L., Cheng, Y., Wang, W., Liu, J.: Meta module network for compositional visual reasoning (2019). arXiv preprint arXiv:1910.03230
5. Chen, Y.C., et al.: Uniter: learning universal image-text representations (2019). arXiv preprint arXiv:1909.11740
6. Devlin, J., Chang, M.W., Lee, K., Toutanova, K.: BERT: Pre-training of deep bidirectional transformers for language understanding. In: NAACL (2019)
7. Frome, A., et al.: DeViSE: a deep visual-semantic embedding model. In: NeurIPS (2013)
8. Goyal, Y., Khot, T., Summers-Stay, D., Batra, D., Parikh, D.: Making the V in VQA matter: elevating the role of image understanding in visual question answering. In: CVPR (2017)
9. Hao, W., Li, C., Li, X., Carin, L., Gao, J.: Towards learning a generic agent for vision-and-language navigation via pre-training. In: CVPR (2020)
10. Huang, L., Wang, W., Chen, J., Wei, X.Y.: Attention on attention for image captioning. In: ICCV (2019)
11. Hudson, D., Manning, C.D.: Learning by abstraction: the neural state machine. In: NeurIPS (2019)
12. Hudson, D.A., Manning, C.D.: GQA: a new dataset for real-world visual reasoning and compositional question answering (2019). arXiv preprint arXiv:1902.09506

13. Karpathy, A., Fei-Fei, L.: Deep visual-semantic alignments for generating image descriptions. In: CVPR (2015)
14. Kiros, R., Salakhutdinov, R., Zemel, R.S.: Unifying visual-semantic embeddings with multimodal neural language models (2014). arXiv preprint arXiv:1411.2539
15. Krishna, R., et al.: Visual genome: connecting language and vision using crowd-sourced dense image annotations. Int. J. Comput. Vis. **123**(1), 32–73 (2017). https://doi.org/10.1007/s11263-016-0981-7
16. Kuznetsova, A., et al.: The open images dataset v4: unified image classification, object detection, and visual relationship detection at scale (2018). arXiv preprint arXiv:1811.00982
17. Lee, K.H., Chen, X., Hua, G., Hu, H., He, X.: Stacked cross attention for image-text matching. In: ECCV (2018)
18. Li, G., Duan, N., Fang, Y., Jiang, D., Zhou, M.: Unicoder-VL: a universal encoder for vision and language by cross-modal pre-training (2019). arXiv preprint arXiv:1908.06066
19. Li, L.H., Yatskar, M., Yin, D., Hsieh, C.J., Chang, K.W.: Visualbert: a simple and performant baseline for vision and language (2019). arXiv preprint arXiv:1908.03557
20. Lin, T.Y., et al.: Microsoft COCO: common objects in context. In: ECCV (2014)
21. Lu, J., Batra, D., Parikh, D., Lee, S.: VilBERT: pretraining task-agnostic visiolinguistic representations for vision-and-language tasks. In: NeurIPS (2019)
22. Lu, J., Goswami, V., Rohrbach, M., Parikh, D., Lee, S.: 12-in-1: Multi-Task vision and language representation learning (2019). arXiv preprint arXiv:1912.02315
23. Maaten, L.V.D., Hinton, G.: Visualizing data using t-SNE. J. Mach. Learn. Res. **9**, 2579–2605 (2008)
24. Norouzi, M., et al.: Zero-shot learning by convex combination of semantic embeddings (2013). arXiv preprint arXiv:1312.5650
25. Ordonez, V., Kulkarni, G., Berg, T.L.: Im2text: describing images using 1 million captioned photographs. In: NeurIPS (2011)
26. Qi, D., Su, L., Song, J., Cui, E., Bharti, T., Sacheti, A.: Imagebert: cross-modal pre-training with large-scale weak-supervised image-text data (2020). arXiv preprint arXiv:2001.07966
27. Ren, S., He, K., Girshick, R., Sun, J.: Faster R-CNN: towards real-time object detection with region proposal networks. In: Advances in Neural Information Processing Systems, pp. 91–99 (2015)
28. Ren, Z., Jin, H., Lin, Z., Fang, C., Yuille, A.: Joint image-text representation by gaussian visual-semantic embedding. In: Multimedia (2016)
29. Rennie, S.J., Marcheret, E., Mroueh, Y., Ross, J., Goel, V.: Self-critical sequence training for image captioning. In: CVPR (2017)
30. Sharma, P., Ding, N., Goodman, S., Soricut, R.: Conceptual captions: a cleaned, hypernymed, image alt-text dataset for automatic image captioning. In: Annual Meeting of the Association for Computational Linguistics (2018)
31. Socher, R., Fei-Fei, L.: Connecting modalities: semi-supervised segmentation and annotation of images using unaligned text corpora. In: CVPR (2010)
32. Socher, R., Ganjoo, M., Manning, C.D., Ng, A.: Zero-shot learning through cross-modal transfer. In: NeurIPS (2013)
33. Su, W., et al.: VL-BERT: pre-training of generic visual-linguistic representations (2019). arXiv preprint arXiv:1908.08530
34. Suhr, A., Zhou, S., Zhang, A., Zhang, I., Bai, H., Artzi, Y.: A corpus for reasoning about natural language grounded in photographs (2018). arXiv preprint arXiv:1811.00491

35. Sun, C., Myers, A., Vondrick, C., Murphy, K., Schmid, C.: VideoBERT: a joint model for video and language representation learning. In: ICCV (2019)
36. Tan, H., Bansal, M.: LXMERT: learning cross-modality encoder representations from transformers. In: EMNLP (2019)
37. Vaswani, A., et al.: Attention is all you need. In: NeurIPS (2017)
38. Wu, Q., Shen, C., Liu, L., Dick, A., Van Den Hengel, A.: What value do explicit high level concepts have in vision to language problems? In: CVPR (2016)
39. You, Q., Jin, H., Wang, Z., Fang, C., Luo, J.: Image captioning with semantic attention. In: CVPR (2016)
40. Young, P., Lai, A., Hodosh, M., Hockenmaier, J.: From image descriptions to visual denotations: new similarity metrics for semantic inference over event descriptions. Trans. Assoc. Comput. Linguist. **2**, 67–78 (2014)
41. Zhou, L., Palangi, H., Zhang, L., Hu, H., Corso, J.J., Gao, J.: Unified vision-language pre-training for image captioning and VQA. In: AAAI (2020)

Improving Face Recognition from Hard Samples via Distribution Distillation Loss

Yuge Huang[1], Pengcheng Shen[1], Ying Tai[1(✉)], Shaoxin Li[1(✉)], Xiaoming Liu[2], Jilin Li[1], Feiyue Huang[1], and Rongrong Ji[3]

[1] Youtu Lab, Tencent, Shanghai, China
{yugehuang,quantshen,yingtai,darwinli,jerolinli,garyhuang}@tencent.com
[2] Michigan State University, East Lansing, USA
liuxm@cse.msu.edu
[3] Xiamen University, Xiamen, China
rrji@xmu.seu.cn

Abstract. Large facial variations are the main challenge in face recognition. To this end, previous variation-specific methods make full use of task-related prior to design special network losses, which are typically not general among different tasks and scenarios. In contrast, the existing generic methods focus on improving the feature discriminability to minimize the intra-class distance while maximizing the inter-class distance, which perform well on easy samples but fail on hard samples. To improve the performance on hard samples, we propose a novel Distribution Distillation Loss to narrow the performance gap between easy and hard samples, which is simple, effective and generic for various types of facial variations. Specifically, we first adopt state-of-the-art classifiers such as Arcface to construct two similarity distributions: a teacher distribution from easy samples and a student distribution from hard samples. Then, we propose a novel distribution-driven loss to constrain the student distribution to approximate the teacher distribution, which thus leads to smaller overlap between the positive and negative pairs in the student distribution. We have conducted extensive experiments on both generic large-scale face benchmarks and benchmarks with diverse variations on race, resolution and pose. The quantitative results demonstrate the superiority of our method over strong baselines, *e.g.*, Arcface and Cosface. Code will be available at https://github.com/HuangYG123/DDL.

Keywords: Face recognition · Loss function · Distribution distillation

---

Y. Huang and P. Shen—Equal contribution.

Electronic supplementary material The online version of this chapter (https://doi.org/10.1007/978-3-030-58577-8_9) contains supplementary material, which is available to authorized users.

1 Introduction

A primary challenge of large-scale face recognition on unconstrained imagery is to handle the diverse variations on pose, resolution, race and illumination, *etc.* While some variations are easy to address, many others are relatively difficult. As in Fig. 1, State-of-the-Art (SotA) facial classifiers like Arcface [6] well address images with small variations with tight groupings in the feature space. We denote these as easy samples. In contrast, images with large variations are usually far away from the easy ones in the feature space, and are much more difficult to tackle. We denote these as hard samples. To better recognize these hard samples, there are usually two schemes: *variation-specific* and *generic* methods.

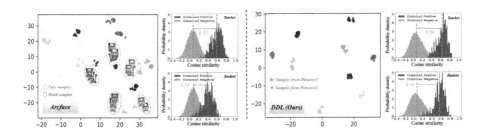

Fig. 1. Comparisons with Arcface [6] on SCface [10] dataset. T-SNE [21] visualizations on features, where the same color indicates samples of the same subject. Distance1 (d_1) and Distance3 (d_3) indicate low-resolution and high-resolution images, which were captured at distances of 4.2 and 1.0 m, respectively. Each method has two distributions from d_3 and d_1, where there are also two distributions from the positive and negative pairs with a margin indicating the difference of their expectations. With our distribution distillation loss between the teacher and student distributions, our method effectively narrows the performance gap between the easy and hard samples, decreasing the expectation margin from **0.21** (0.52–0.31) to **0.07** (0.56–0.49).

Variation-specific methods are usually designed for a specific task. For instance, to achieve pose-invariant face recognition, either handcrafted or learned features are extracted to enhance robustness against pose while remaining discriminative to the identities [33]. Recently, joint face frontalization and disentangled identity preservation are incorporated to facilitate the pose-invariant feature learning [35,49]. To address resolution-invariant face recognition, a unified feature space is learned in [16,27] for mapping Low-Resolution (LR) and High-Resolution (HR) images. The works [4,50] first apply super-resolution on LR images and then perform recognition on the super-resolved images. However, the above methods are specifically designed for the respective variations, therefore their ability to generalize from one variation to another is limited. Yet, it is highly desirable to handle multiple variations in real world recognition systems.

Different from variation-specific methods, generic methods focus on improving the discriminative power of facial features for small intra-class and large interclass distances. Basically, the prior works fall into two categories, *i.e.*, softmax loss-based and triplet loss-based methods. Softmax loss-based methods regard each identity as a unique class to train the classification networks. Since the traditional softmax loss is insufficient to acquire the discriminative features, several variants [6,18,40,43] are proposed to enhance the discriminability. In contrast, triplet loss-based methods [23,26] directly learn a Euclidean space embedding for each face, where faces from the same person form a separate cluster from faces of other people. With large-scale training data and well-designed network structures, both types of methods can obtain promising results.

However, the performance of these methods degrades dramatically on hard samples, such as very large-pose and low-resolution faces. As illustrated in Fig. 1, the features extracted from HR images (*i.e.*, d_3) by the strong face classifier of Arcface [6] are well separated, but the features extracted from LR images (*i.e.*, d_1) cannot be well distinguished. From the perspective of the angle distributions of positive and negative pairs, we can easily observe that Arcface exists more confusion regions on LR face images. It is thereby a natural consequence that such generic methods perform worse on hard samples.

To narrow the performance gap between the easy and hard samples, we propose a novel Distribution Distillation Loss (DDL). By leveraging the best of both the *variation-specific* and *generic* methods, our method is generic and can be applied to diverse variations to improve face recognition in hard samples. Specifically, we first adopt current SotA face classifiers as the baseline (*e.g.*, Arcface) to construct the initial similarity distributions between teacher (*e.g.*, easy samples from d_3 in Fig. 1) and student (*e.g.*, hard samples from d_1 in Fig. 1) according to the difficulties of samples, respectively. Compared to finetuning the baseline models with domain data, our method firstly does not require extra data or inference time (*i.e.*, *simple*); secondly makes full use of hard sample mining and directly optimizes the similarity distributions to improve the performance on hard samples (*i.e.*, *effective*); and finally can be easily applied to address different kinds of large variations in extensive real applications, *e.g.*, women with makeup in fashion stores, surveillance faces in railway stations, and apps looking for missing senior person or children, *etc.*

To sum up, the contributions of this work are three-fold:

- Our method narrows the performance gap between easy and hard samples on *diverse* facial variations, which is simple, effective and general.
- To our best knowledge, it is the first work that adopts similarity distribution distillation loss for face recognition, which provides a new perspective to obtain more discriminative features to better address hard samples.
- Significant gains compared to the SotA Arcface are reported, *e.g.*, 97.0% over 92.7% on SCface, 93.4% over 92.1% on CPLFW, 90.7% over 89.9% (@FAR=1e−4) on IJB-B and 93.1% over 92.1% (@FAR=1e−4) on IJB-C.

2   Related Work

Loss Function in FR. Loss function design is pivotal for large-scale face recognition. Softmax is commonly used for face recognition [30,34,39], which encourages the separability of features but the learned features are not guaranteed to

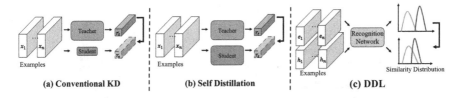

Fig. 2. Comparisons among conventional knowledge distillation, self-distillation and our DDL. The student in KD is usually smaller than the teacher. $\{e\}_1^n$ and $\{h\}_1^n$ indicate the easy and hard samples, respectively.

be discriminative. To address this issue, contrastive [29] and triplet [23,26] losses are proposed to increase the margin in the Euclidean space. However, both contrastive and triplet losses occasionally encounter training instability due to the selection of effective training samples. As a simple alternative, center loss and its variants [7,43,52] are proposed to compress the intra-class variance. More recently, angular margin-based losses [6,13,18,19,38] facilitate feature discrimination, and thus lead to larger angular/cosine separability between learned features. The above loss functions are designed to apply constraints either between samples, or between sample and center of the corresponding subject. In contrast, our proposed loss is *distribution* driven. While being similar to the histogram loss [37] that constrains the overlap between the distributions of positive and negative pairs across the training set, our loss differs in that we first separate the training set into a teacher distribution (easy samples) and student distribution (hard samples), and then constrain the student distribution to approximate the teacher distribution via our novel loss, which narrows the performance gap between easy and hard samples.

Variation-Specific FR. Apart from generic solutions [30,34] for face recognition, there are also many methods designed to handle specific facial variations, such as resolutions, poses, illuminations, expressions and demographics [8]. For example, cross-pose FR [33,35,48,54] is very challenging, and previous methods mainly focus on either face frontalization or pose invariant representations. Low resolution FR is also a difficult task, especially in the surveillance scenario. One common approach is to learn a unified feature space for LR and HR images [11,20,55]. The other way is to perform super resolution [4,31,32] to enhance the facial identity information. Differing from the above methods that mainly deal with one specific variation, our novel loss is a generic approach to improve FR from hard samples, which is applicable to a wide variety of variations.

Knowledge Distillation. Knowledge Distillation (KD) is an emerging topic. Its basic idea is to distill knowledge from a large teacher model into a small one by learning the class distributions provided by the teacher via softened softmax [12]. Typically, Kullback Leibler (KL) divergence [12,53] and Maximum Mean Discrepancy (MMD) [14] can be adopted to minimize the posterior probabilities between teacher and student models. More recently, transferring mutual relations of data examples from the teacher to the student is proposed [22,36]. In particular, RKD [22]

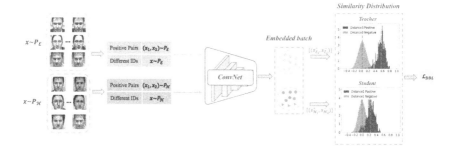

Fig. 3. Illustration of our DDL. We sample b positive pairs (*i.e.*, $2b$ samples) and b samples with different identities, for both the teacher $P_{\mathcal{E}}$ and student $P_{\mathcal{H}}$ distributions, to form one mini-batch (*i.e.*, $6b$ in total). $\{(s_{\mathcal{E}_i}^+, s_{\mathcal{E}_i}^-)|i = 1, ..., b\}$ indicates we construct b positive and negative pairs from $P_{\mathcal{E}}$ via Eqs. 1 and 2 respectively to estimate the teacher distribution. $\{(s_{\mathcal{H}_i}^+, s_{\mathcal{H}_i}^-)|i = 1, ..., b\}$ also indicates we construct b positive and negative pairs from $P_{\mathcal{H}}$ via Eqs. 1 and 2 respectively to estimate the student distribution.

reported that KD can improve the original performance when the student has the same structure as the teacher (*i.e.*, self-distillation).

Compared to the above distillation methods, our DDL differs in several aspects (see Fig. 2): 1) KD has at least *two* networks, a teacher and a student, while DDL only learns *one* network. Although in KD the student may have the same structure as the teacher (*e.g.*, self-distillation), they have *different parameters* in training. 2) KD uses sample-wise, Euclidean distance-wise or angle-wise constraints, while DDL proposes a novel cosine similarity distribution-wise constraint which is specifically designed for face recognition. 3) To our best knowledge, currently no KD methods outperform SotA face classifiers on face benchmarks, while DDL consistently outperforms the SotA Arcface classifier.

3 The Proposed Method

Figure 3 illustrates the framework of our DDL. We separate the training set into two parts, *i.e.*, $\mathcal{E}$ for easy samples and $\mathcal{H}$ for hard samples to form the teacher and student distributions, respectively. In general, for each mini-batch during training, we sample from both parts. To ensure a good teacher distribution, we use the SotA FR model [6] as our initialization. The extracted features are used to construct the positive and negative pairs (Sect. 3.1), which are further utilized to estimate the similarity distributions (Sect. 3.2). Finally, based on the similarity distributions, the proposed DDL is utilized to train the classifier (Sect. 3.3).

3.1 Sampling Strategy from $P_{\mathcal{E}}$ and $P_{\mathcal{H}}$

First, we introduce the details on how we construct the positive and negative pairs in one mini-batch during training. Given two types of input data from both $P_{\mathcal{E}}$ and $P_{\mathcal{H}}$, each mini-batch consists of four parts, two kinds of positive pairs (*i.e.*, (x_1, x_2)

$\sim P_{\mathcal{E}}$ and $(x_1, x_2) \sim P_{\mathcal{H}}$), and two kinds of samples with different identities (*i.e.*, x $\sim P_{\mathcal{E}}$ and $x \sim P_{\mathcal{H}}$). To be specific, we on one hand construct b positive pairs (*i.e.*, $2b$ samples), and on the other hand b samples with different identities from $P_{\mathcal{E}}$ and $P_{\mathcal{H}}$. As the result, there are $6b = (2b + b) * 2$ samples in each mini-batch (see Fig. 3 for more details).

Positive Pairs. The positive pairs are constructed offline in advance, and each pair consist of two samples with the same identity. As shown in Fig. 3, samples of each positive pair are arranged in order. After embedding data into a high-dimensional feature space by a deep network $\mathcal{F}$, the similarity of a positive pair s^+ can be obtained as follows:

$$s_i^+ =< \mathcal{F}(x_{pos_{i1}}), \mathcal{F}(x_{pos_{i2}}) >, i = 1, ..., b \tag{1}$$

where $x_{pos_{i1}}$, $x_{pos_{i2}}$ are the samples of one positive pair. Note that positive pairs with similarity less than 0 are usually outliers, which are deleted as a practical setting since our main goal is not to specifically handle noise.

Negative Pairs. Different from the positive pairs, we construct negative pairs online from the samples with different identities via hard negative mining, which selects negative pairs with the largest similarities. To be specific, the similarity of a negative pair s^- is defined as:

$$s_i^- = \max_j \left(\{s_{ij}^- =< \mathcal{F}(x_{neg_i}), \mathcal{F}(x_{neg_j}) > | j = 1, ..., b \} \right), \tag{2}$$

where x_{neg_i}, x_{neg_j} are from different subjects. Once the similarities of positive and negative pairs are constructed, the corresponding distributions can be estimated, which is described in the next subsection.

3.2 Similarity Distribution Estimation

The process of similarity distribution estimation is similar to [37], which is performed in a simple and piece-wise differentiable manner using 1D histograms with soft assignment. Specifically, two samples x_i, x_j from the same person form a positive pair, and the corresponding label is denoted as $m_{ij} = +1$. In contrast, two samples from different persons form a negative pair, and the label is denoted as $m_{ij} = -1$. Then, we obtain two sample sets $\mathcal{S}^+ = \{s^+ = \langle \mathcal{F}(x_i), \mathcal{F}(x_j) \rangle | m_{ij} = +1\}$ and $\mathcal{S}^- = \{s^- = \langle \mathcal{F}(x_i), \mathcal{F}(x_j) \rangle | m_{ij} = -1\}$ corresponding to the similarities of positive and negative pairs, respectively.

Let p^+ and p^- denote the two probability distributions of $\mathcal{S}^+$ and $\mathcal{S}^-$, respectively. As in cosine distance-based methods [6], the similarity of each pair is bounded to $[-1, 1]$, which is demonstrated to simplify the task [37]. Motivated by the histogram loss, we estimate this type of one-dimensional distribution by fitting simple histograms with uniformly spaced bins. We adopt R-dimensional histograms H^+ and H^-, with the nodes $t_1 = -1, t_2, \cdots, t_R = 1$ uniformly filling $[-1, 1]$ with the step $\triangle = \frac{2}{R-1}$. Then, we estimate the value h_r^+ of the histogram H^+ at each bin as:

$$h_r^+ = \frac{1}{|S^+|} \sum_{(i,j):m_{ij}=+1} \delta_{i,j,r}, \tag{3}$$

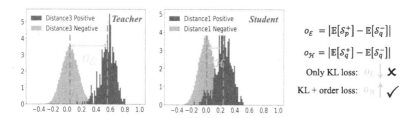

Fig. 4. Illustration of the effects of our order loss. Similarity distributions are constructed by Arcface [6] on SCface, in which we have 2 kinds of order distances formed from both of the teacher and student distributions according to Eq. 6.

where (i, j) spans all the positive pairs. Different from [37], the weights $\delta_{i,j,r}$ are chosen by an exponential function as:

$$\delta_{i,j,r} = exp(-\gamma(s_{ij} - t_r)^2), \tag{4}$$

where γ denotes the spread parameter of Gaussian kernel function, and t_r denotes the rth node of histograms. We adopt the Gaussian kernel function because it is the most commonly used kernel function for density estimation and robust to the small sample size. The estimation of H^- proceeds analogously.

3.3   Distribution Distillation Loss

We make use of SotA face recognition engines like [6], to obtain the similarity distributions from two kinds of samples: easy and hard samples. Here, easy samples indicate that the FR engine performs well, in which the similarity distributions of positive and negative pairs are clearly separated (see the teacher distribution in Fig. 4), while hard samples indicate that the FR engine performs poorly, in which the similarity distributions may be highly overlapped (see the student distribution in Fig. 4).

KL Divergence Loss. To narrow the performance gap between the easy and hard samples, we constrain the similarity distribution of hard samples (*i.e.*, student distribution) to approximate the similarity distribution of easy samples (*i.e.*, teacher distribution). The teacher distribution consists of two similarity distributions of both positive and negative pairs, denoted as P^+ and P^-, respectively. Similarly, the student distribution also consists of two similarity distributions, denoted as Q^+ and Q^-. Motivated by the previous KD methods [12,53], we adopt the KL divergence to constrain the similarity between the student and teacher distributions, which is defined as follows:

$$\mathcal{L}_{KL} = \lambda_1 \mathbb{D}_{KL}(P^+||Q^+) + \lambda_2 \mathbb{D}_{KL}(P^-||Q^-)$$
$$= \lambda_1 \underbrace{\sum_s P^+(s) \log \frac{P^+(s)}{Q^+(s)}}_{KL\,loss\,on\,pos.\,pairs} + \lambda_2 \underbrace{\sum_s P^-(s) \log \frac{P^-(s)}{Q^-(s)}}_{KL\,loss\,on\,neg.\,pairs}, \tag{5}$$

where λ_1, λ_2 are the weight parameters.

Order Loss. However, only using KL loss does not guarantee good performance. In fact, the teacher distribution may choose to approach the student distribution and leads to more confusion regions between the distributions of positive and negative pairs, which is the opposite of our objective (see Fig. 4). To address this problem, we design a simple yet effective term named *order loss*, which minimizes the distances between the expectations of similarity distributions from the negative and positive pairs to control the overlap. Our order loss can be formulated as follows:

$$\mathcal{L}_{order} = -\lambda_3 \sum_{(i,j)\in(p,q)} (\mathbb{E}[\mathcal{S}_i^+] - \mathbb{E}[\mathcal{S}_j^-]), \tag{6}$$

where $\mathcal{S}_p^+$ and $\mathcal{S}_p^-$ denote the similarities of positive and negative pairs of the teacher distribution; $\mathcal{S}_q^+$ and $\mathcal{S}_q^-$ denote the similarities of positive and negative pairs of the student distribution; and λ_3 is the weight parameter.

In summary, the entire formulation of our distribution distillation loss is: $\mathcal{L}_{DDL} = \mathcal{L}_{KL} + \mathcal{L}_{order}$. DDL can be easily extended to multiple student distributions varied from one specific variation as follows:

$$\mathcal{L}_{DDL} = \sum_{i=1}^{K} \mathbb{D}_{KL}(P||Q_i) - \lambda_3 \sum_{i,j\in(p,q_1\cdots q_K)} (\mathbb{E}[\mathcal{S}_i^+] - \mathbb{E}[\mathcal{S}_j^-]), \tag{7}$$

where K is the number of student distributions. Further, to maintain the performance on easy samples, we incorporate the loss function of Arcface [6], and thus the final loss is:

$$\mathcal{L}(\Theta) = \mathcal{L}_{DDL} + \mathcal{L}_{Arcface}, \tag{8}$$

where Θ denotes the parameter set. Note that $\mathcal{L}_{Arcface}$ can be easily replaced by any kind of popular losses in FR.

3.4   Generalization on Various Variations

Next, we discuss the generalization of DDL on various variations, which defines our *application scenarios* and how we select easy/hard samples. Basically, we can distinguish the easy and hard samples according to whether the image contains large facial variations that may hinder the identity information, *e.g.*, low-resolution and large pose variation.

Observation from Different Variations. Our method assumes that two or more distributions, each computed from a subset of training data, have differences among themselves, which is a popular phenomenon in face recognition and is demonstrated in Fig. 5. It shows similarity distributions of normal and challenging samples based on Arcface [6] trained on CASIA except CFP, which is trained on VGGFace2. As we can see, 1) since CASIA is biased to Caucasian, Mongolian samples in COX are more difficult and thus relatively regarded as the hard samples, 2) different variations share a common observation that the similarity distributions of challenging samples are usually different from those of easy samples, 3) variations with different extents may have different similarity distributions (*e.g.*, $\mathcal{H}1$ and $\mathcal{H}2$ in Fig. 5(c)). In summary, when a task satisfies that *the similarity distributions differ between*

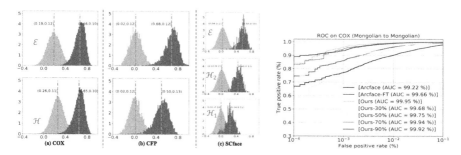

Fig. 5. Similarity distribution differences between easy and hard samples on various variations, including race on COX, pose on CFP, and resolution on SCface respectively. $(\cdot,\cdot)$ indicates the mean and standard deviation.

Fig. 6. Effects of number of training subjects on COX. Compared to Arcface-FT, DDL achieves comparable results with only *half* the number of training subjects.

easy and hard samples, our method is a good solution and one can enjoy the performance improvement by properly constructing the positive and negative pairs, as validated in Sect. 4.3.

Performance Balance Between Easy and Hard Samples. Improving the performance on hard samples while maintaining the performance on easy samples is a trade-off. Two factors in our method help maintain performance on easy samples. First, we incorporate the SotA Arcface loss [6] to maintain feature discriminability on easy samples. Second, our order loss minimizes the distance between the expectations of similarity distributions from the negative and positive pairs, which helps control the overlap between positive and negative pairs.

Discussions on Mixture Variations. As shown in Eq. 7, our method can be easily extended to multiple variations for one task (*e.g.*, low resolution, large pose, *etc*). An alternative is to mix the variations with different extents from one task into *one* student distribution, which, as shown in Sect. 4.2, is not good enough to specifically model the different extents and tends to lead to lower performance. As for different variations from different tasks, one may also construct multiple teacher-student distribution pairs to address the corresponding task respectively, which can be a good future direction.

4   Experiments

4.1   Implementation Details

Datasets. We separately employ SCface [10], COX [15], CASIA-WebFace [47], VGGFace2 [3] and the refined MS1M [6] as our training data to conduct fair comparisons with other methods. We extensively test our method on benchmarks with diverse variations, *i.e.*, COX on race, SCface on resolution, CFP and CPLFW on

Pose, as well as generic large-scale benchmarks IJB-B and IJB-C. For COX, the data are collected from two races: Caucasian and Mongolian. Since no race label is given, we manually label 428 Mongolians and 572 Caucasians to conduct experiments, in which half of both races are used for finetuning and the others for testing. For SCface, following [20], 50 subjects are used for finetuning and 80 subjects are for testing. In the testing stage, we conduct face identification, where HR image is used as the gallery and LR images with three different resolutions form the probe. Specifically, the LR images are captured at three distances: 4.2 m for d_1, 2.6 m for d_2 and 1.0 m for d_3. We split easy and hard samples according to the main variation in each dataset. For race, since the dataset on which the model is pre-trained is biased to Caucasian, Mongolian samples on COX are more difficult and thus relatively regarded as the hard samples. For pose, we estimate the pose of each image [25] on VGGFace2, and images with yaw $< 10°$ and yaw $> 45°$ are used as easy and hard samples respectively. For resolution, images captured under d_3 and d_1/d_2 are used as easy and hard samples respectively on SCface.

Training Setting. We follow [6,40] to generate the normalized faces (112×112) with five landmarks [51]. For the embedding network, we adopt ResNet50 and ResNet100 as in [6]. Our work is implemented in Tensorflow [1]. We train models on 8 NVIDIA Tesla P40 GPUs. On SCface, we set the number of positive/negative pairs as $b = 16$, thus the batch size on one GPU is $3b \times 3 = 144$, including one teacher distribution and two student distributions (see Fig. 5(c)). On other datasets, we set b to 32, thus the batch size per GPU is $3b \times 2 = 192$. The numbers of iterations are 1K, 2K and 20K on SCface, COX and VGGFace2, respectively. The models are trained with SGD, with momentum 0.9 and weight decay $5e^{-4}$. The learning rate is $1e^{-3}$, and is divided by 10 at half of iterations. All of the weight parameters are consistent across all the experiments. λ_1, λ_2 and λ_3 are set to $1e^{-1}$, $2e^{-2}$ and $5e^{-1}$, respectively.

4.2 Ablation Study

Effects of Distance Metric on Distributions. We investigate the effects of several commonly used distribution metrics to constrain the teacher and student distributions in our DDL, including KL divergence, Jensen-Shannon (JS) divergence, and Earth Mover Distance (EMD). Although KL divergence does not qualify as a statistical metric, it is widely used as a measure of how one distribution is different from another. JS divergence is a symmetric version of KL divergence. EMD is another distance function between distributions on a given metric space and has seen success in image synthesis [9]. We incorporate our order loss with the above distance metrics, and report the results in Table 1. We choose KL divergence in our DDL since it achieves the best performance, which shares similar conclusion with [53]. To further investigate the effectiveness of each component in our loss, we train the network with each component separately. As shown in Table 1, only KL or only Order does not guarantee satisfying performance, while using both components leads to better results.

Table 1. Extensive ablation studies on SCface dataset. All methods are trained on CASIA with a ResNet50 backbone. Each color corresponds to a type of ablation study experimental setting.

	EMD	JS	KL	order	Random	hard mining	Mixture	specific	d_1	d_2	d_3	Avg.
Arcface-FT									67.3	93.5	98.0	86.3
Distance metric	✓			✓		✓		✓	78.0	97.8	96.8	90.5
		✓	✓			✓		✓	83.0	**98.3**	**99.0**	93.4
			✓			✓		✓	76.0	94.3	98.5	89.6
				✓		✓		✓	80.8	97.5	**99.0**	92.4
Mining strategy		✓	✓		✓			✓	80.3	96.3	95.3	90.6
Mixture training		✓	✓			✓	✓		81.5	97.0	97.8	92.1
DDL (ours)		✓	✓			✓		✓	**86.8**	**98.3**	98.3	**94.4**

Effects of Random vs. Hard Mining. To investigate the effect of hard sample mining in our method, we train models on SCface with the corresponding strategy (*i.e.*, negative pairs with the largest similarity are selected), and without the strategy by randomly selecting the negative pairs, respectively. The comparative results are reported in Table 1. Comparing with the results of "Random" selecting, it is clear that our hard mining version outperforms the one without.

Effects of Mixture vs. Specific training. As mentioned in Sect. 3.4, we basically construct different student distributions for samples with different extents of variations on SCface. Here, we mix two variations from d_1 and d_2 into *one* student distribution. The comparison between our specific and mixture training is also shown in Table 1. As we expected, the mixture version is worse than the specific version, but is still better than the conventional finetuning (*i.e.*, Avg. being 86.3), which indicates that properly constructing different hard samples for the target tasks may maximize the advantages of our method.

Effects of Number of Training Subjects. Here, we conduct tests on COX dataset to show the effects of using different numbers of training subjects. Specifically, we adopt 10%, 30%, 50%, 70%, 90% and 100% of training subjects, respectively. A pre-trained Arcface on CASIA is used as the baseline. For fair comparison, we also compare our method against Arcface with conventional finetuning (*i.e.*, Arcface-FT). From Fig. 6 we see that: 1) Compared to Arcface-FT, our method clearly boosts the performance on Mongolian-Mongolian verification tests with comparable training data. 2) Our method can have comparable performance with the only *half* of the entire training subjects, which demonstrate the superiority of utilizing the global similarity distributions.

4.3   Comparisons with SotA Methods

Resolution on SCface. SCface mimics the real-world surveillance watch-list problem, where the gallery contains HR faces and the probe consists of LR faces captured from surveillance cameras. We compare our method with SotA low-resolution face recognition methods in Table 2. Most results are directly cited

Table 2. Rank-1 performance (%) of face identification on SCface testing set. '-FT' represents finetuning with training set from SCface.

Distance →	d_1	d_2	d_3	Avg.
LDMDS [46]	62.7	70.7	65.5	66.3
Center Loss [43]	36.3	81.8	94.3	70.8
Arcface (CASIA+R50)	48.0	92.0	99.3	79.8
Arcface (MS1M+R100)	58.9	98.3	**99.5**	85.5
Center Loss-FT	54.8	86.3	95.8	79.0
DCR-FT [20]	73.3	93.5	98.0	88.3
Histogram (CASIA+R50)-FT [37]	74.3	95.0	97.3	88.8
OHEM (CASIA+R50)-FT [28]	82.5	97.3	97.5	92.7
Focal (CASIA+R50)-FT [17]	76.8	95.5	96.8	89.7
Triplet (CASIA+R50)-FT [6]	84.2	97.2	99.2	93.5
Arcface (CASIA+R50)-FT	67.3	93.5	98.0	86.3
Arcface (MS1M+R100)-FT	80.5	98.0	99.5	92.7
Ours (CASIA+R50)	86.8	98.3	98.3	94.4
Ours (MS1M+R100)	**93.2**	**99.2**	98.5	**97.0**

Table 3. Verification comparisons with SotA methods on LFW and two popular pose benchmarks, including CFP-FP and CPLFW.

Methods (%)	LFW	CFP-FP	CPLFW
Triplet Loss (CVPR'15)	98.98	91.90	–
Center Loss (ECCV'16) [43]	98.75	–	77.48
SphereFace (CVPR'17) [18]	99.27	–	81.40
DRGAN (CVPR'17) [35]	–	93.41	–
Peng et al. (ICCV'17) [24]	–	93.76	–
Yin et al. (TIP'17) [48]	98.27	94.39	–
VGGFace2 (FG'18) [3]	99.43	–	84.00
Dream (CVPR'18) [2]	–	93.98	–
Deng et al. (CVPR'18) [5]	99.60	94.05	–
SV-Arc-Softmax (arXiv'19) [42]	99.78	98.28	92.83
CO-Mining (ICCV'19) [41]	–	95.87	87.31
Arcface (MS1M+R100)-Official [6]^a	**99.82**	98.37	92.08
Arcface (MS1M+R100)	99.80	98.29	92.52
Arcface (VGG+R100)	99.62	98.30	93.13
Ours (VGG+R100)	99.68	**98.53**	**93.43**

^{a} Results are from the official model: https://github.com/deepinsight/insightface, which is trained on MS1M and adopts ResNet100 as the backbone.

from [20], while the results of Arcface come from our re-implementation. From Table 2, we have some observations: 1) The baseline Arcface achieves much better results than the other methods without finetuning, especially on the relatively high-resolution images from d_3. 2) Our (CASIA+ResNet50)-FT version already outperforms all of the other methods, including Arcface (MS1M+ResNet100)-FT, which uses a larger model that is trained by a much larger dataset. 3) We achieve significant improvement on d_1 setting, which is the hardest. This demonstrates the effectiveness of our novel loss. 4) Histogram loss performs poorly, which demonstrates the effects of our constraint between teacher and student distributions.

Moreover, different to the prior hard mining methods [17,26,28] where the hard samples are mined based on the loss values during the training process, we *predefine* hard samples according to human prior. Penalizing *individual* samples or triplets as in previous hard mining methods does not leverage sufficient contextual insight of the overall distribution. DDL minimizes the difference of *global* similarity distributions between the easy and hard samples, which is more robust for tackling hard samples and against the noisy samples. The word "global" means our method leverages sufficient contextual insight of the overall distribution in a mini-batch, rather than focusing on a sample.

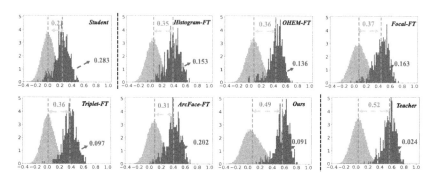

Fig. 7. Illustrations of similarity distributions of different SotA methods, which are all pre-trained by CASIA with ResNet50 and then finetuned on SCface. The leftmost and rightmost are the student and teacher distributions estimated from a pre-trained Arcface model on d_1 and d_3 settings, respectively. The similarity distributions in the middle are obtained by various methods finetuned on SCface. The red number indicates the *histogram intersection* between the estimated similarity distributions from the positive and negative pairs. (Color figure online)

Figure 7 illustrates the estimated similarity distributions of various SotA methods. To quantify the differences among these methods, we introduce two statistics for evaluation, the expectation margin and histogram intersection (*i.e.*, $\sum_{r=1}^{R} \min(h_r^+, h_r^-)$) between the two distributions from positive and negative pairs. Typically, smaller histogram intersection and larger expectation margin indicate better verification/identification performance, since it means more discriminative embeddings are learned [37]. Our DDL achieves the closest statistics to the teacher distribution, and thus obtains the best performance.

Pose on CFP-FP and CPLFW. We compare our method with SotA pose-invariant methods [2,5,24,35,48] and generic solutions [3,6,18,41–43]. Since VGGFace2 includes comprehensive pose variations, we use it to pre-train a ResNet100 with Arcface. Next, we construct teacher and student distributions to finetune the model with our loss. From Table 3, we can see that: 1) Our Arcface re-implementations achieve comparable results against the official version, with similar results on LFW and CFP-FP, as well as better performance on CPLFW. Arcface is also much better than other methods, including those pose-invariant face recognition methods. 2) Our method achieves the best performance on both pose benchmarks, while also maintaining the performance on LFW (*i.e.*, 99.68% vs. 99.62%).

Note that when using the model pre-trained on MS1M, and finetuning it with easy/hard samples from VGGFace2, our method can further push the performance to a higher level (**99.06%** on CFP-FP and **94.20%** on CPLFW), which is the *first* method that exceeds 99.0% on CFP-FP and 94.0% on CPLFW using images cropped by MTCNN. Besides, we also train our DDL on the smaller training set CASIA with a smaller backbone ResNet50. Again, our DDL outperforms the competitors. Please refer to our supplementary material for details.

Table 4. 1:1 **verification TAR** on the IJB-B and IJB-C datasets. All methods are trained on VGGFace2 with ResNet50.

Methods (%)	IJB-B			IJB-C		
	FAR = 1e−5	FAR = 1e−4	FAR = 1e−3	FAR = 1e−5	FAR = 1e−4	FAR = 1e−3
VGGFace2 [3]	67.1	80.0	88.7	74.7	84.1	90.9
MN[45]	70.8	83.1	90.9	77.1	86.2	92.7
DCN [44]	–	84.9	93.7	–	88.5	94.7
Arcface [6]	80.5	89.9	94.5	86.1	92.1	96.0
SP [36]	79.4	89.8	94.9	85.9	92.3	96.2
RKD [22]	78.4	89.6	94.7	85.5	92.1	96.1
Ours	**83.4**	**90.7**	**95.2**	**88.4**	**93.1**	**96.3**

Table 5. 1:N **(mixed media) Identification** on IJB-B/C. All methods are trained on VGGFace2 with ResNet50. VGGFace2 is cited from the paper, and Arcface is from its official released model.

Methods (%)	IJB-B				IJB-C			
	FPIR = 0.01	FPIR = 0.1	Rank 1	Rank 5	FPIR = 0.01	FPIR = 0.1	Rank 1	Rank 5
VGGFace2 [3]	70.6	83.9	90.1	94.5	74.6	84.2	91.2	94.9
Arcface [6]	73.1	88.2	93.6	96.5	79.6	89.5	94.8	96.9
SP [36]	72.4	88.0	93.8	96.6	79.9	89.5	94.7	97.0
RKD [22]	70.6	87.6	93.4	96.5	79.3	89.1	94.6	96.9
Ours	**76.3**	**89.5**	**93.9**	**96.6**	**85.4**	**91.1**	**95.4**	**97.2**

Large-Scale Benchmarks: IJB-B and IJB-C. On IJB-B/C datasets, we employ VGGFace2 with ResNet50 for a fair comparison with recent methods. We first construct the teacher and student distributions according to the pose of each image, and then follow the testing protocol in [6] to take the *average of the image features* as the corresponding template representation without bells and whistles. Tables 4 and 5 show the 1:1 verification and 1:N identification comparisons with the recent SotA methods, respectively. Note that our method is *not* a set-based face recognition method, and the experiments on these two datasets are just to prove that our DDL can obtain more discriminate features than generic methods like Arcface, even on all-variations-included datasets. Please refer to our supplementary material for the detailed analysis.

Comparisons with SotA KD Methods. We further conduct fair comparisons between our DDL and the recent SotA KD/self-distillation methods, *i.e.*, SP [36] and RKD [22]. Note that since both SP and RKD have not reported SOTA results on face recognition tasks, we re-implement the two methods under the same experimental setting on VGGFace2, using their officially released code. Specifically, we first train a ResNet50 with Arcface on VGGFace2 as the teacher model, and then train a student ResNet50 via combining the knowledge distillation method (*e.g.*, SP or RKD) and Arcface loss under the guidance of the teacher model. As in Tables 4 and 5, our DDL outperforms the SotA KD/self-distillation methods, which achieve similar results to vanilla Arcface.

5   Conclusion

In this paper, we propose a novel framework Distribution Distillation Loss (DDL) to improve various variation-*specific* tasks, which comes from the observations that state-of-the-art methods (*e.g.*, Arcface) witness significant performance gaps between easy and hard samples. The key idea of our method is to construct a teacher and a student distribution from easy and hard samples, respectively. Then, the proposed loss drives the student distribution to approximate the teacher distribution to reduce the overlap between the positive and negative pairs. Extensive experiments demonstrate the effectiveness of our DDL on a wide range of recognition tasks compared to the state-of-the-art face recognition methods. In subsequent work, we can try to extend our method to multiple teacher-student distribution pairs for the corresponding task respectively.

References

1. Abadi, M., et al.: TensorFlow: large-scale machine learning on heterogeneous systems (2015). http://tensorflow.org/. software available from tensorflow.org
2. Cao, K., Rong, Y., Li, C., Tang, X., Change Loy, C.: Pose-robust face recognition via deep residual equivariant mapping. In: CVPR, pp. 5187–5196 (2018)
3. Cao, Q., Shen, L., Xie, W., Parkhi, O.M., Zisserman, A.: Vggface2: a dataset for recognising faces across pose and age. In: FG, pp. 67–74. IEEE (2018)
4. Chen, Y., Tai, Y., Liu, X., Shen, C., Yang, J.: FSRNet: end-to-end learning face super-resolution with facial priors. In: CVPR, pp. 2492–2501 (2018)
5. Deng, J., Cheng, S., Xue, N., Zhou, Y., Zafeiriou, S.: UV-GAN: adversarial facial UV map completion for pose-invariant face recognition. In: CVPR, pp. 7093–7102 (2018)
6. Deng, J., Guo, J., Xue, N., Zafeiriou, S.: Arcface: additive angular margin loss for deep face recognition. In: CVPR, pp. 4690–4699 (2019)
7. Deng, J., Zhou, Y., Zafeiriou, S.: Marginal loss for deep face recognition. In: CVPR Workshops, pp. 60–68 (2017)
8. Gong, S., Liu, X., Jain, A.: Jointly de-biasing face recognition and demographic attribute estimation. In: ECCV (2020)
9. Goodfellow, I., et al.: Generative adversarial nets. In: NIPS, pp. 2672–2680 (2014)
10. Grgic, M., Delac, K., Grgic, S.: SCface-surveillance cameras face database. Multimedia Tools Appli. **51**(3), 863–879 (2011). https://doi.org/10.1007/s11042-009-0417-2
11. Hennings-Yeomans, P.H., Baker, S., Kumar, B.V.: Simultaneous super-resolution and feature extraction for recognition of low-resolution faces. In: CVPR, pp. 1–8. IEEE (2008)
12. Hinton, G., Vinyals, O., Dean, J.: Distilling the knowledge in a neural network. In: NIPS Workshop (2014)
13. Huang, Y., et al.: CurricularFace: adaptive curriculum learning loss for deep face recognition. In: CVPR (2020)
14. Huang, Z., Wang, N.: Like what you like: knowledge distill via neuron selectivity transfer (2017). arXiv:1707.01219v2
15. Huang, Z., et al.: A benchmark and comparative study of video-based face recognition on cox face database. IEEE Trans. Image Process. **24**(12), 5967–5981 (2015)

16. Lei, Z., Ahonen, T., Pietikäinen, M., Li, S.Z.: Local frequency descriptor for low-resolution face recognition. In: FG, pp. 161–166. IEEE (2011)
17. Lin, T.Y., Goyal, P., Girshick, R., He, K., Dollár, P.: Focal loss for dense object detection. In: ICCV, pp. 2980–2988 (2017)
18. Liu, W., Wen, Y., Yu, Z., Li, M., Raj, B., Song, L.: Sphereface: deep hypersphere embedding for face recognition. In: CVPR, pp. 212–220 (2017)
19. Liu, W., Wen, Y., Yu, Z., Yang, M.: Large-margin softmax loss for convolutional neural networks. In: ICML, vol. 2, p. 7 (2016)
20. Lu, Z., Jiang, X., Kot, A.: Deep coupled resnet for low-resolution face recognition. IEEE Sig. Process. Lett. 25(4), 526–530 (2018)
21. Maaten, L.V.D., Hinton, G.: Visualizing data using t-SNE. J. Mach. Learn. Res. 9(Nov), 2579–2605 (2008)
22. Park, W., Kim, D., Lu, Y., Cho, M.: Relational knowledge distillation. In: Proceedings of the IEEE Conference on Computer Vision and Pattern Recognition, pp. 3967–3976 (2019)
23. Parkhi, O.M., Vedaldi, A., Zisserman, A., et al.: Deep face recognition. In: BMVC, vol. 1, p. 6 (2015)
24. Peng, X., Yu, X., Sohn, K., Metaxas, D.N., Chandraker, M.: Reconstruction-based disentanglement for pose-invariant face recognition. In: ICCV, pp. 1623–1632 (2017)
25. Ruiz, N., Chong, E., Rehg, J.M.: Fine-grained head pose estimation without keypoints. In: CVPR Workshops, pp. 2074–2083 (2018)
26. Schroff, F., Kalenichenko, D., Philbin, J.: Facenet: a unified embedding for face recognition and clustering. In: CVPR, pp. 815–823 (2015)
27. Shekhar, S., Patel, V.M., Chellappa, R.: Synthesis-based recognition of low resolution faces. In: IJCB, pp. 1–6. IEEE (2011)
28. Shrivastava, A., Gupta, A., Girshick, R.: Training region-based object detectors with online hard example mining. In: CVPR, pp. 761–769 (2016)
29. Sun, Y., Chen, Y., Wang, X., Tang, X.: Deep learning face representation by joint identification-verification. In: NIPS, pp. 1988–1996 (2014)
30. Sun, Y., Wang, X., Tang, X.: Deep learning face representation from predicting 10,000 classes. In: CVPR, pp. 1891–1898 (2014)
31. Tai, Y., Yang, J., Liu, X.: Image super-resolution via deep recursive residual network. In: CVPR, pp. 3147–3155 (2017)
32. Tai, Y., Yang, J., Liu, X., Xu, C.: Memnet: a persistent memory network for image restoration. In: ICCV, pp. 4539–4547 (2017)
33. Tai, Y., Yang, J., Zhang, Y., Luo, L., Qian, J., Chen, Y.: Face recognition with pose variations and misalignment via orthogonal procrustes regression. IEEE Trans. Image Process. 25(6), 2673–2683 (2016)
34. Taigman, Y., Yang, M., Ranzato, M., Wolf, L.: Deepface: closing the gap to human-level performance in face verification. In: CVPR, pp. 1701–1708 (2014)
35. Tran, L., Yin, X., Liu, X.: Disentangled representation learning GAN for pose-invariant face recognition. In: CVPR, pp. 1415–1424 (2017)
36. Tung, F., Mori, G.: Similarity-preserving knowledge distillation. In: Proceedings of the IEEE International Conference on Computer Vision, pp. 1365–1374 (2019)
37. Ustinova, E., Lempitsky, V.: Learning deep embeddings with histogram loss. In: NIPS, pp. 4170–4178 (2016)
38. Wang, F., Cheng, J., Liu, W., Liu, H.: Additive margin softmax for face verification. IEEE Sig. Process. Lett. 25(7), 926–930 (2018)
39. Wang, F., Xiang, X., Cheng, J., Yuille, A.L.: Normface: L2 hypersphere embedding for face verification. In: ACMMM, pp. 1041–1049. ACM (2017)

40. Wang, H., et al.: Cosface: large margin cosine loss for deep face recognition. In: CVPR, pp. 5265–5274 (2018)

41. Wang, X., Wang, S., Wang, J., Shi, H., Mei, T.: Co-mining: deep face recognition with noisy labels. In: ICCV, pp. 9358–9367 (2019)

42. Wang, X., Wang, S., Zhang, S., Fu, T., Shi, H., Mei, T.: Support vector guided softmax loss for face recognition (2018). arXiv:1812.11317

43. Wen, Y., Zhang, K., Li, Z., Qiao, Yu.: A discriminative feature learning approach for deep face recognition. In: Leibe, B., Matas, J., Sebe, N., Welling, M. (eds.) ECCV 2016. LNCS, vol. 9911, pp. 499–515. Springer, Cham (2016). https://doi.org/10.1007/978-3-319-46478-7_31

44. Xie, W., Shen, L., Zisserman, A.: Comparator networks. In: ECCV, pp. 782–797 (2018)

45. Xie, W., Zisserman, A.: Multicolumn networks for face recognition. In: BMVC (2018)

46. Yang, F., Yang, W., Gao, R., Liao, Q.: Discriminative multidimensional scaling for low-resolution face recognition. IEEE Sig. Process. Lett. **25**(3), 388–392 (2017)

47. Yi, D., Lei, Z., Liao, S., Li, S.Z.: Learning face representation from scratch (2014).arXiv:1411.7923

48. Yin, X., Liu, X.: Multi-task convolutional neural network for pose-invariant face recognition. IEEE Trans. Image Process. **27**(2), 964–975 (2017)

49. Yin, X., Yu, X., Sohn, K., Liu, X., Chandraker, M.: Towards large-pose face frontalization in the wild. In: ICCV, pp. 3990–3999 (2017)

50. Zhang, K., et al.: Super-identity convolutional neural network for face hallucination. In: ECCV, pp. 183–198 (2018)

51. Zhang, K., Zhang, Z., Li, Z., Qiao, Y.: Joint face detection and alignment using multitask cascaded convolutional networks. IEEE Sig. Process. Lett. **23**(10), 1499–1503 (2016)

52. Zhang, X., Fang, Z., Wen, Y., Li, Z., Qiao, Y.: Range loss for deep face recognition with long-tailed training data. In: ICCV, pp. 5409–5418 (2017)

53. Zhang, Y., Xiang, T., Hospedales, T.M., Lu, H.: Deep mutual learning. In: CVPR, pp. 4320–4328 (2018)

54. Zhao, J., et al.: 3D-aided deep pose-invariant face recognition. In: IJCAI, vol. 2, p. 11 (2018)

55. Zou, W.W., Yuen, P.C.: Very low resolution face recognition problem. IEEE Trans. Image Process. **21**(1), 327–340 (2011)

Extract and Merge: Superpixel Segmentation with Regional Attributes

Jianqiao An[1,2], Yucheng Shi[1,2], Yahong Han[1,2,3(✉)] (ID), Meijun Sun[1] (ID), and Qi Tian[4] (ID)

[1] College of Intelligence and Computing, Tianjin University, Tianjin, China
{anjianqiao,yucheng,yahong,sunmeijun}@tju.edu.cn
[2] Tianjin Key Lab of Machine Learning, Tianjin University, Tianjin, China
[3] Peng Cheng Laboratory, Shenzhen, China
[4] Noah's Ark Lab, Huawei Technologies, Shenzhen, China
tian.qi1@huawei.com

Abstract. For a certain object in an image, the relationship between its central region and the peripheral region is not well utilized in existing superpixel segmentation methods. In this work, we propose the concept of **regional attribute**, which indicates the location of a certain region in the object. Based on the regional attributes, we propose a novel superpixel method called **Extract and Merge (EAM)**. In the extracting stage, we design square windows with a side length of a power of two, named **power-window**, to extract regional attributes by calculating boundary clearness of objects in the window. The larger windows are for the central regions and the smaller ones correspond to the peripheral regions. In the merging stage, power-windows are merged according to the defined attraction between them. Specifically, we build a graph model and propose an efficient method to make the large windows merge the small ones strategically, regarding power-windows as vertices and the adjacencies between them as edges. We demonstrate that our superpixels have fine boundaries and are superior to the respective state-of-the-art algorithms on multiple benchmarks.

Keywords: Superpixel · Power-window · Boundary clearness · Graph model

1 Introduction

Superpixel segmentation is to divide an image into several fragments without intersecting, as illustrated in Fig. 1. The major advantage of using superpixels is computational efficiency. It can be a helpful pre-process of numerous computer-vision tasks such as image semantic segmentation [5,12], object detection [14,17], salient object detection [18,21], tracking [19] etc. Some work in the field of deep learning that specifically utilizes the features of the superpixels has also been proposed in recent years. SPN [8] employs superpixel segmentation as a pooling

© Springer Nature Switzerland AG 2020
A. Vedaldi et al. (Eds.): ECCV 2020, LNCS 12375, pp. 155–170, 2020.
https://doi.org/10.1007/978-3-030-58577-8_10

Fig. 1. The images in the first and third columns show superpixel boundaries and the ones in the second and fourth columns show superpixels colored by their mean colors. From left to right, the results are generated by SLIC [1] and the proposed EAM, respectively, with both 300 superpixels.

layout to reflect low-level image structure for learning and inferring semantic segmentation. However, this requires the superpixel boundaries generated by algorithms can be as close as possible to the object boundary of the original image under various complicated conditions, so that the semantic information expressed by superpixels is almost equivalent to that expressed by the original image.

In light of fundamental importance of superpixels in the computer vision, various representative superpixel segmentation algorithms have been proposed and widely applied, including unsupervised methods [1,2,13,20] and deep learning based methods [6,16] etc. Deep learning based methods relies on the training set, which is inefficient and lacks generalization ability. Here we list some disadvantages of existing superpixel methods:

1) Pixels are directly merged into superpixels which are not well resistant to noise and complex textures in the images.
2) Most of the algorithms focus on the color and spatial information but ignore the deeper relationship between image pixels.
3) The superpixels cannot be flexibly allocated according to the regional features, thus complicated details cannot be well segmented (see Fig. 1).

Innovatively, we divide the image region of an object into several categories, from the central region to the peripheral region. The concept of **regional attribute** is proposed to describe the labels for each category. Instead of directly merging independent original image pixels, we extract regional attributes of each image region through a proposed structure called Power-Window and merge them into superpixels. Superpixels generated in this way are more disciplined and more consistent with human vision. The contributions of this paper are summarized as follows:

1) **Extract and Merge (EAM)**, a novel method focusing on **regional attributes** of objects in an image is proposed.

2) An ingenious structure called **power-window** is proposed to classify pixels by regional attribute. Moreover, the definition and solution of **boundary clearness** is proposed to help accomplish this extract regional attributes.

3) A graph model is built on power-windows and the solution called **attraction competition** is designed to merge windows into flexible and diverse superpixels.

4) Experimental comparisons with several state-of-the-art superpixel segmentation methods are taken on the BSDS500 dataset. The results show that EAM keeps almost all the details of the original images and performs favorably against the state-of-the-art.

2   Related Work

The superpixel segmentation methods can be roughly divided into two categories: unsupervised methods and deep learning methods. Unsupervised methods include graph-based approach and clustering-based approach.

Graph-Based Methods: Prominent graph-based approaches for image segmenation rely on pixel graphs. Superpixels are generated by minimizing a cost function using a graph model, in which pixels are vertices and pixel-level similarities are treated as edge weights. The normalized cuts (NC) algorithm [7] is one of the representative works of graph-based methods, which creates superpixels by recursively computing normalized cuts for the pixel graph. However, NC suffers from the high computational complexity cost. Felzenszwalb and Huttenlocher (FH) [4] propose a minimum spanning tree based segmentation approach. Their algorithm progressively joins components until a stopping criterion is met, which prevents the spannning tree from covering the whole image. Although the FH method preserves boundaries well, it often generates both extremely large and small segments. The ERS [10] algorithm presents a superpixel segmentation method by maximizing an objective function of entropy rate on graph topology, which generates homogeneous superpixels and adhere to the boundaries. A greedy algorithm is used to obtain solutions. Lazy random walk (LRW) [13] adheres to object boundary well and preserves texture regions by translating input image into a graph. The graph vertex is the image pixel, and then superpixels are initialized and optimized iteratively by an energy function. In general, separate image pixels are treated as vertices in these graph-based methods, causing them susceptible to noise interference.

Clustering-Based Methods: Series of clustering-based superpixel methods are developed based on clustering techniques. SLIC [1] produces superpixels by adopting k-means clustering approach in the five dimensional CIELab color and position feature space to cluster pixels. The LSC [20] method projects the five dimensional features to a ten dimensional space and performs clustering in the projected space. Normalized cuts formulation is adopted in LSC to generate the final superpixels. The SEEDS [2] algorithm uses uniform blocks as initial approximation of superpixels and iteratively exchanges neighboring blocks in a coarse-to-fine manner based on an objective function. However, these clustering-based methods are still in units of a single pixel, affecting the quality of the superpixels.

Deep Learning Methods: Recently, deep learning based methods for superpixel segmentation have been proposed. Two representative deep learning based superpixel segmentation methods are SEAL [16] and SSN [6]. In SEAL, a new loss function is proposed to take the segmentation error into account for affinity learning and Pixel Affinity Net is designed for affinity prediction. SSN presents a differentiable superpixel sampling model which can be integrated into end-to-end trainable networks.

Individual pixels are the basic unit in all of these algorithms, which cause the algorithms to be insensitive to semantic information. Pre-extracting the central and peripheral regions, and then merging these regions (instead of independent pixels) into superpixels is demonstrated by us to be a smarter way.

3   The Proposed Method

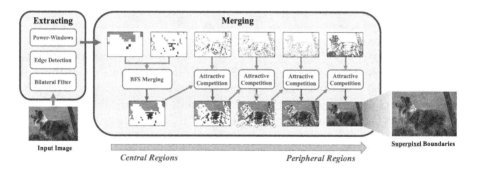

Fig. 2. The EAM pipeline consists of two parts: the extracting process on the left and the merging process on the right. The green arrow in the figure indicates the input and the red arrow indicates the output. Refer to Fig. 3 and Fig. 4 for more details of the extracting stage.

In this section we will describe in detail the EAM method. Our method (See Fig. 2) consists of two stages, extracting and merging. The extracting stage is essentially a process of pixel classification. According to the proposed concept of regional attribute, we use squares with various sizes (power-window) to classify pixels into several categories. The pixels in each window are regarded as a whole, and they have the same regional attribute. The larger the window is, the closer it is to the central region of a certain object, and vice versa. In the merging stage, the classified power-windows are merged according to their regional attributes. The power-windows of the central region are first merged. Then we build graph model to define the attraction between the power-windows

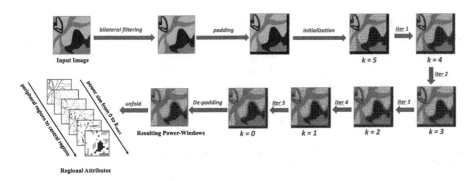

Fig. 3. To show the process of extracting more clearly, we deliberately choose an image with simple and clear color. The size of the input image is 321×321, hence we take the value of k_{max} to be 5 and iterate 5 times in total to get the regional attributes.

and merge the remaining windows according to their regional attribute from central to peripheral. In this way, we can generate superpixels that are not disorganized while at the same time maximize the consistency with the human visual object boundaries. The degree of detail as well as the number of superpixels can be determined by one parameter ϵ_{delta}, under the condition of semi-manual intervention, which will be further explained later.

3.1 Extracting

We perform bilateral filtering [15] on the original image before extracting. The bilateral filtering method can remove the noise in the image while keeping the boundary definition as much as possible, which can greatly help our subsequent extracting process. The method can be easily used by calling the *opencv* package in python.

Power-Window: In order to extract the regional attributes of the pixels in the image, we designed a structure of $(k_{max}+1)$ different sizes of square windows with a side length of $2^0, 2^1, ..., 2^{k_{max}}$ (called **power-window**) to assist the process. k is called the **power size** of a power-window with the side length of 2^k. We define that

$$k_{max} = \lfloor \log_2 \frac{min(h, w)}{10} \rfloor \tag{1}$$

where h is the height of the image and w is the width. Each power-window can be defined as a triple (x, y, k), where (x, y) represents the coordinate of the upper left corner of the window and k represents its power size. The algorithm is executed as follows. First, we initialize the original image (size $h \times w$) into a series of power-windows with power size of k_{max}, $\frac{h}{k_{max}}$ rows and $\frac{w}{k_{max}}$ columns. If h or w are not divisible by k_{max}, we pad black pixels to the bottom and right sides of the original image corresponding to the amount missing (padding) and remove them (de-padding) when the stage is complete. Then we enumerate the windows in order k from k_{max} to 0. For the current window with power size k, if its *boundary clearness* $< \epsilon_{bc}$, the attribute k of this region will be determined. Otherwise it will be divided into four power-windows with power size $k - 1$ for the next iteration and so on, until the power size equals to 1. Boundary clearness is a quantity that we define to determine whether a power-window contains a single object, which will be explained in more detail in the following section. And ϵ_{bc} is the threshold for boundary clearness which is set to 3 in this paper. Refer to the Algorithm 1 and Fig. 3 for more details.

Algorithm 1. Attributes Extracting

Input: Input image I after bilateral filtering and padding
Output: Sets of power-windows with different sizes: $set_0, set_1, ..., set_{k_{max}}$

1: Empty $set_0, set_1, ..., set_{k_{max}}$
2: Split I into power-windows with power size k_{max}
3: Insert the power-windows into $set_{k_{max}}$
4: **for** $k = k_{max} \rightarrow 1$ **do**
5:     **for** each $pw_i \in set_k$ **do**
6:         **if** BoundaryClearness$(pw_i) < \epsilon_{bc}$ **then**
7:             **Erase** pw_i from set_k
8:             **Split** pw_i into 4 smaller power-windows.
9:             **Insert** the 4 power-windows into set_{k-1}
10:         **end if**
11:     **end for**
12: **end for**
13: **return** $set_0, set_1, ..., set_{k_{max}}$

Boundary Clearness: To determine whether a power-window contains only one single object, it is necessary to calculate the boundary clearness of the image within the window. We present a graph-based definition of the concept of boundary clearness and propose a corresponding calculation method. Items in a power-window can be interpreted as a graph $G = (V, E)$, where four corners of each pixel in the window are the vertices and the boundaries between pixels are the edges (see Fig. 4). We define the weight of each edge as the Euclidean distance in [rgb] space of the two pixels with this edge as a dividing line. Our proposed algorithm for calculating boundary clearness can be roughly seen as a process of adding edges to the graph. First, we add to the graph all the outermost edges, which represent the boundaries of the power-window and form a large cycle. Next, we add the remaining edges one by one according to the descending order of their weights, until a new cycle appears in the graph. The weight of the last edge added to the graph is the expected boundary clearness. To explain further, if the weight is large enough, it means that the two regions inside and outside the new cycle have clear boundaries, which can be considered as two objects. Refer to the Algorithm 2 and Fig. 4 for more details.

The resulting power-windows with different power sizes through this stage correspond to different regional attributes. The larger the resulting power-

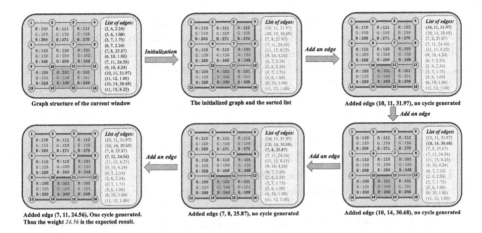

Fig. 4. Computation flow of boundary clearness. The example shows a portion of an image with nine pixels. A weighted graph is set up and the edge list is sorted at first. The edges are added one by one until a new cycle is generated. The cycle encloses three pixels in the right bottom corner that express another object in the image. The last edge weight 24.56 represents the boundary clearness of these two regions. The larger the boundary clearness value is, the less likely we think these two regions belong to the same object and we should try smaller power-windows, and vice versa.

Algorithm 2. Boundary Clearness

Input: Input power-window P

Output: The Boundary Clearness of P

1: Initialize set V
2: V contains corner nodes of pixels in P.
3: Initialize set E
4: E contains edges represented as triples (u, v, w), which $u, v \in V$ and w is defined as the Euclidean distance in [rgb] space of the two pixels which divided by the edge.
5: Initialize set E' as a subset of E containing the outermost edges of P.
6: Initialize $G = (V, E')$
7: List $E \setminus E'$ into L as $[edge_1, edge_2, ...]$
8: Sort L by $edge.w$ in **descending** order.
9: **for** each $edge_i \in L$ **do**
10:     Add $edge_i$ to G
11:     **if** a new cycle in G generated by $edge_i$ **then**
12:         $LastEdge \leftarrow edge_i$
13:         **break**
14:     **end if**
15: **end for**
16: **return** $LastEdge.w$

---

window is, the closer it is to the central region of a certain object, and vice versa. In the graph struture of the method calculating boundary clearness, the number of vertices and edges are both on the same order of magnitude as the number of pixels in the power-window. We use the DSU (disjoint-set-union) to determine if a new cycle is generated, so the time complexity of the method is linear. For the entire extracting method, the number of iterations is at most K_{max}, which is on the order of logarithm. Therefore, the time complexity of extracting method is $O(NlogN)$, where N is the number of pixels in the input image. The proposed method has the following outstanding features:

1) Simple definition of power-windows. Each window is a square structure which could be defined simply by the upper left coordinate (x, y) and the power size k.
2) The square structure with a side length of 2^k can be easily cut into four squares with the same side length of 2^{k-1}, which can be seen as an efficient half-interation process.
3) The proposed definition and solution method of boundary clearness can itelligently ignore the color gradation in the same object and be sensitive to the real object boundaries.

3.2 Merging

Using the resulting power-windows extracted in last stage, we propose a graph-based methods for merging them into superpixels. Since the image pixels in each power-window have the same regional attribute, each power-window is integrally treated as an independent vertex in the graph, and the adjacencies between power-windows are the edges. The two types of power-windows with the largest power size (i.e. k_{max}, $k_{max} - 1$) are first merged through BFS-based (breath-first search) algorithm. For the smaller windows, we enumerate in descending order of power size values and merge them through the proposed Attractive Competive method. Each of the superpixel we ultimately generate is mapped to a label, called **spColor**. The output of the merging stage is a map called *ColorMap*, which records the spColor of each vertices in the graph structure. The initial spColor of all vertices is 0, and the final spColor of them are integer numbers between 1 and the total number of superpixels generated. Our final superpixels are generated according to *ColorMap* by mapping the spColor of the vertices to the corresponding pixel area in the original image. Image pixels with the same spColor belong to the same superpixel.

Graph Structure: To merge power-windows into superpixels, the proposed method represents the image as a graph where vertices are power-windows and the edge weight are the similarities between the adjacent windows (i.e. sharing a boundary). In this work, the [rgb] color of a power-window is defined as the mean color of the pixels inside it and the similarity of two adjacent power-windows is defined as the Euclidean distance in [rgb] space of their colors. It is clear that the number of vertices and edges are both of the same order as the number of pixels of the image, expressed by N. The structure is more clearly shown in Fig. 5.

BFS-Based Merging: The power-windows with the two largest power size k_{max}, $k_{max} - 1$ represent the central regions, which should be merged into con-

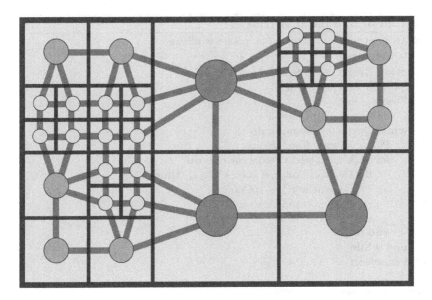

Fig. 5. Graph structure: Three types of power-windows separated by red lines with different sizes are represented in the shown graph by three different types of vertices. The purple lines in the figure are the edges connecting adjacent vertices. (Color figure online)

nected regions first. The algorithm consists of two iterations, while each power-windows with k_{max} and $k_{max} - 1$ power size are traversed respectively to expand into some connected regions. The expansion of vertices has three limitations.

1) Only adjacent vertices can be expanded, just as in general BFS algorithm.
2) Only vertices with power size $\geq k_{max} - 1$ can be expanded.
3) A vertex only expands the vertices which the weight of the edge between them $< \epsilon_{bfs}$. The threshold ϵ_{bfs} is set to 25 in this work.

The limitations are intended to form better initial subject regions, laying the groundwork for the subsequent Attractive Competive algorithm. Refer to the Algorithm 3 and Fig. 2 for more details.

Attraction Competition: In this section, We define the concept of attraction between two vertices as the length of the shortest path between them. Further more, we propose an efficient competition algorithm based on Dijkstra Algorithm [3]. The proposed algorithm goes through $k_{max} - 1$ iterations (i.e. from power size $(k_{max} - 2)$ downto 0). For each iteration, we compute the shortest path from the colored vertices (spColor of which is not 0) to the uncolored ones with Dijkstra algorithm. When one vertex u relaxes another vertex v in the process, we assign the spColor of u to v temporarily. As v may be relaxed again soon by another vertex u' and the spColor of v should change to the spColor of u', the process is just like an attraction competition between the vertices. We do iterations in order of power size from $(k_{max} - 2)$ to 0, so in each iteration only

---

Algorithm 3. BFS-Based Merging

Input: Graph $G = (V, E)$ created on power-windows
Output: $ColorMap$

1: **function** BFSEXTENSION($v, ColorMap, spColor$)
2:     Initialize an empty $Queue$
3:     Push v into $Queue$
4:     $ColorMap[v] \leftarrow spColor$
5:     **while** $Queue$ is not empty **do**
6:         Pop a vertex v from $Queue$
7:         **for** each uncolored v' adjacent to v **do**
8:             **if** Distance($v.color, v'.color$) $\leq \epsilon_{bfs}$ **then**
9:                 $ColorMap[v'] \leftarrow spColor$
10:                Push v' into $Queue$
11:            **end if**
12:        **end for**
13:    **end while**
14: **end function**
15:
16: **for** each uncolored $v \in V$ **do**
17:     **if** $v.power\_size \geq k_{max} - 1$ **then**
18:         $spColor \leftarrow spColor + 1$
19:         BFSEXTENSION($v, ColorMap, spColor$)
20:     **end if**
21: **end for**
22: **return** $ColorMap$

---

the vertices with the current power size are colored. For the verices that are far away from any colored vertices, we take a clustering approach to merge them and form new color regions, similar to SLIC [1]. This is done in the five-dimensional $[rgbxy]$ space, where $[rgb]$ is the color of the power-window represented by a vertex, and $[xy]$ is the middle pixel position of the power-window. The initial cluster centers are selected randomly and the distance measure D is defined as

$$D = d_{rgb} + \alpha d_{xy} \tag{2}$$

where d_{rgb} and d_{xy} is the Euclidean norm in $[rgb]$ space and $[xy]$ space respectively. α is set to 0.02 in this paper. A **binary search** method is taken to minimize the number of new colors to added under the condition that the maximum color difference in one cluster is $\leq \epsilon_{cl}$. The threshold ϵ_{cl} is the only parameter in this work allowing us to control the degree of detail and the number of superpixels. The smaller ϵ_{cl} is set, the finer the superpixels are and the more the superpixel numbers, vice versa. In particular, for vertices with power size = 0, we only let them be attracted to the colored regions but cannot cluster by themselves as they express the details of the image boundaries. Refer to the Algorithm 4 and Fig. 2 for more details of Attraction Competition method.

As the number of vertices and edges in our proposed graph structure are both of the same order as the number of the pixels of the image, expressed

---

Algorithm 4. Attraction Competition

Input: Graph $G = (V, E)$ created on power-windows
Output: $ColorMap$

1: **for** $k = k_{max} - 2 \rightarrow 0$ **do**
2:     $Set_{ori} \leftarrow$ colored vertices
3:     $Set_{ter} \leftarrow$ vertices with power size k
4:     DIJKSTRA($Set_{ori}, Set_{ter}, ColorMap$)
5:     **Update** $ColorMap$
6:     **if** $k > 0$ **then**
7:         $Set' \leftarrow$ uncolored vertices in Set_{ter}
8:         $L \leftarrow 1, R \leftarrow Set'.size$
9:         **while** $L < R$ **do**
10:             $Mid = \lfloor \frac{L+R}{2} \rfloor$
11:             $Dif_{max} \leftarrow$ CLUSTER(Set', Mid)
12:             **if** $Dif_{max} > \epsilon_{cl}$ **then**
13:                 $R \leftarrow Mid$
14:                 **Update** $ColorMap$
15:             **else**
16:                 $L \leftarrow Mid$
17:             **end if**
18:         **end while**
19:     **end if**
20: **end for**
21: **return** $ColorMap$

---

by N, the time complexity of BFS-based Merging is $O(N)$. For the subsequent Attraction Competition method, $k_{max} - 1$ iterations are executed which is on the order of logarithm. Further more, the complexity of Dijkstra algorithm with priority queue is $O(NlogN)$ and the complexity of the Cluster Method with few constant number of iterations is $O(N)$. So the total complexity of this algorithm is $O(Nlog^2N)$.

4   Experiments

In this section, the proposed method is compared with some unsupervised state-of-the-art superpixel segmentation methods, including two representative clustering-based algorithms (SLIC [1] and LSC [20]), a graph-based algorithm (LRW [13]) and an algorithm based on energy optimization (SEEDS [2]) on **BSDS500**[1] dataset. Furthermore, we also compared our method with two latest deep learning based superpixel segmentation methods (SEAL [16] and SSN [6]) to show the great advantage on **EV** [11], which quantifies the variation of the image explained by the superpixels. We also show that our algorithm can retain more details in the visual comparison, and it is consistent with EV score.

---

[1] https://www2.eecs.berkeley.edu/Research/Projects/CS/vision/grouping.

4.1   Datasets and Performance Metrics

For evaluation of the proposed method, we use the BSDS500 dataset which containing 500 images with size 321×481. Achievable Segmentation Accuracy (ASA) [9] is the most widely adopted criteria for measuring superpixel segmentation quality. Moreover, in order to better measure how well the data in the original pixels is represented by superpixels we use Explained Variation (EV) [11]. The detailed definitions of these metrics are as follows:

Achievable Segmentation Accuracy (ASA): ASA [9] quantifies the achievable accuracy for segmentation using superpixels as pre-processing step. The performance of subsequent processing is expected to be unaffected. Then ASA is defined as:

$$ASA(G,S) = \frac{\sum_i \max_j |S_i \cap G_j|}{\sum_j |G_j|} \tag{3}$$

ASA computes the highest achievable accuracy by labeling each superpixel with the label of ground truth segmentation that has the biggest overlap area, i.e. higher is better.

Explained Variation (EV): EV [11] evaluates the superpixel segmentation quality independent of annotated ground truth. As image boundaries tend to exhibit strong change in color and structure, EV assesses boundary adherence independent of human annotions. EV is defined as:

$$EV(S) = \frac{\sum_{S_j} |S_j|(\mu(S_j) - \mu(I))^2}{\sum_{x_n} (I(x_n) - \mu(I))^2} \tag{4}$$

where $\mu(S_j)$ and $\mu(I)$ are the mean color of superpixel S_j and the image I, respectively. As result, EV quantifies the variation of the image explained by the superpixels, i.e. higher is better.

4.2   Compare with State-of-the-art

In all comparative experiments in this section, instead of directly controlling the number of superpixels manually, we use the cluster diameter threshold ϵ_{cl} to control the degree of detail of segmentation since it is difficult for the human eye to determine the appropriate number of superpixels of an image but the degree of detail. We obtained the average number of superpixels by adjusting the value of ϵ_{cl} on the BSDS500 dataset, as shown in Table 1.

Table 1. The correspondence between ϵ_{cl} and the average number of superpixels of EAM in the experiment on the BSDS500 dataset.

Number of superpixels	100	200	300	400	500
ϵ_{cl}		30	20	15	8	5

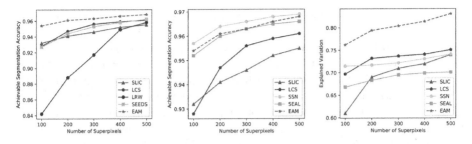

Fig. 6. Comparing state-of-the-art methods for a range of 5 different numbers of superpixels on the BSDS500 datasets, ASA on the left and middle, EV on the right. The leftmost graph shows the performance comparison of the unsupervised methods and the middle one shows the deep learning based methods. For each such value, all methods were initialized, and ended with about the same number of superpixels.

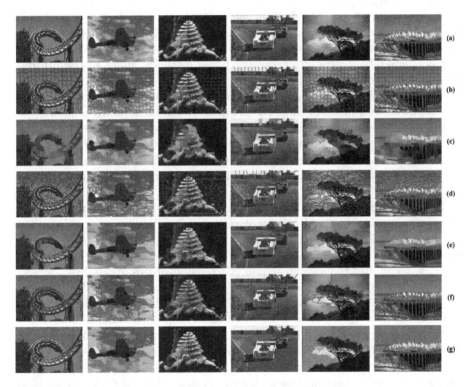

Fig. 7. Visual comparison of superpixel segmentation results, the superpixel boundaries and mean colors. (a) Origional. (b) SLIC (boundaries). (c) SLIC (mean color). (d) SEAL (boundaries). (e) SEAL (mean color). (f) EAM (boundaries). (g) EAM (mean color). The average superpixel numbers in all images is roughly 200. All methods is initialized and ended with about the same number of superpixels. The images with mean color show huge advantage of EAM compared to other methods in term of detail retention, as demonstrated by EV score. (Color figure online)

We compare EAM with four unsupervised state-of-the-art superpixel algorithms, including SLIC [1], LSC [20], LRW [13] and SEEDS [2] on the BSDS500 dataset. For all four algorithms, the implementations are based on publicly available codes. The experiments are performed on the BSDS500 dataset. EAM has achieved higher ASA than other methods as shown in the **leftmost** graph of Fig. 6. It shows that our method can better express the object boundaries.

In order to test the performance of our method in more depth, we also performed comparison experiments with two latest deep learning based methods (SEAL [16] and SSN [6]) on 200 images of the BSDS500 dataset (300 images for training). The results are also presented on Fig. 6. The experimental results of two kinds of unsupervised state-of-the-art methods (SLIC [1], LSC [20]) are also added to the line graph, in order to better form the performance comparison between deep learning based methods and unsupervised methods (including the proposed EAM).

In terms of ASA, compared with unsupervised methods, deep learning based methods has obvious advantages. The performance of EAM ranks second, not as good as the algorithms SSN [6], but EAM is the only one unsupervised methods that is comparable to the deep learning based methods. Refer to the **middle** graph of Fig. 6 for more details.

As the explained variation (EV) metric quantifies the quality of superpixel segmentation **without relying on ground truth**, it is more objective for comparison experiments with deep learning based methods. In terms of EV, the proposed method performs the best and shows the huge advantage against the deep learning based methods (SSN [6] and SEAL [16]) and unsupervised methods (LSC [20] and SLIC [1]). Refer to the **rightmost** graph of Fig. 6 for more details.

Deep learning methods rely too much on the ground truth from training sets thus they score highly on ASA but not very well on EV. The proposed method can achieve performance comparable to deep learning based methods without the need of extensive sample data for model training and has strong universality, which demonstrates the advantage and effectiveness of this approach.

To make more intuitive, we do visual experiment on the BSDS500 dataset, as shown in Fig. 7. Consistent with the EV score in Fig. 6, EAM retains much more details than other methods under the same superpixel numbers.

For time efficiency, EAM has no training overhead and the total complexity of the method is $O(Nlog^2N)$. Compared with most unsupervised methods such as ERS [10] $O(N^2logN)$ and NC [7] $O(N^{\frac{3}{2}})$, EAM has higher computational efficiency, but with some efficient algorithms such as SLIC [1] $O(N)$, we do need to strengthen.

5   Conclusion

In this paper, we focus on the regional attribute of image pixels for the first time in the field of superpixel segmentation. A novel superpixel method called EAM is proposed to generate finer superpixels. Compared with the direct merging of independent pixels, the pre-extraction process can obtain useful prior knowledge (the regional attributes). This allows EAM to smartly divide the detail portion with more superpixels and not waste on the bulk portion.

Acknowledgments. This work is supported by the NSFC (under Grant 61876130, 61932009).

References

1. Achanta, R., Shaji, A., Smith, K., Lucchi, A., Fua, P., Süsstrunk, S.: Slic superpixels compared to state-of-the-art superpixel methods. IEEE Trans. Pattern Anal. Mach. Intell. **34**(11), 2274–2282 (2012). https://doi.org/10.1109/TPAMI.2012.120
2. den Bergh, M.V., Boix, X., Roig, G., Gool, L.V.: SEEDS: superpixels extracted via energy-driven sampling. CoRR abs/1309.3848 (2013). http://arxiv.org/abs/1309.3848
3. Dijkstra, E.W.: A note on two problems in connexion with graphs. Numer. Math. **1**(1), 269–271 (1959). https://doi.org/10.1007/BF01386390
4. Felzenszwalb, P.F., Huttenlocher, D.P.: Efficient graph-based image segmentation. Int. J. Comput. Vis. **59**(2), 167–181 (2004). https://doi.org/10.1023/B:VISI.0000022288.19776.77
5. Gadde, R., Jampani, V., Kiefel, M., Gehler, P.V.: Superpixel convolutional networks using bilateral inceptions. CoRR abs/1511.06739 (2015). http://arxiv.org/abs/1511.06739
6. Jampani, V., Sun, D., Liu, M., Yang, M., Kautz, J.: Superpixel sampling networks. CoRR abs/1807.10174 (2018). http://arxiv.org/abs/1807.10174
7. Shi, J., Malik, J.: Normalized cuts and image segmentation. IEEE Trans. Pattern Anal. Mach. Intell. **22**(8), 888–905 (2000). https://doi.org/10.1109/34.868688
8. Kwak, S., Hong, S., Han, B.: Weakly supervised semantic segmentation using superpixel pooling network. In: AAAI (2017)
9. Liu, M., Tuzel, O., Ramalingam, S., Chellappa, R.: Entropy rate superpixel segmentation. CVPR **2011**, 2097–2104 (2011). https://doi.org/10.1109/CVPR.2011.5995323
10. Liu, M.Y., Tuzel, O., Ramalingam, S., Chellappa, R.: Entropy rate superpixel segmentation. CVPR **2011**, 2097–2104 (2011)
11. Moore, A.P., Prince, S.J.D., Warrell, J., Mohammed, U., Jones, G.: Superpixel lattices. In: 2008 IEEE Conference on Computer Vision and Pattern Recognition, pp. 1–8, June 2008. https://doi.org/10.1109/CVPR.2008.4587471
12. Sharma, A., Tuzel, O., Liu, M.Y.: Recursive context propagation network for semantic scene labeling. In: Ghahramani, Z., Welling, M., Cortes, C., Lawrence, N.D., Weinberger, K.Q. (eds.) Advances in Neural Information Processing Systems 27, pp. 2447–2455. Curran Associates, Inc. (2014). http://papers.nips.cc/paper/5282-recursive-context-propagation-network-for-semantic-scene-labeling.pdf
13. Shen, J., Du, Y., Wang, W., Li, X.: Lazy random walks for superpixel segmentation. IEEE Trans. Image Process. **23**(4), 1451–1462 (2014). https://doi.org/10.1109/TIP.2014.2302892
14. Shu, G., Dehghan, A., Shah, M.: Improving an object detector and extracting regions using superpixels. In: 2013 IEEE Conference on Computer Vision and Pattern Recognition, pp. 3721–3727, June 2013. https://doi.org/10.1109/CVPR.2013.477

15. Tomasi, C., Manduchi, R.: Bilateral filtering for gray and color images. In: Sixth International Conference on Computer Vision (IEEE Cat. No.98CH36271), pp. 839–846, January 1998. https://doi.org/10.1109/ICCV.1998.710815
16. Tu, W.C., et al.: Learning superpixels with segmentation-aware affinity loss. In: IEEE Conference on Computer Vision and Pattern Recognition (CVPR), June 2018
17. Yan, J., Yu, Y., Zhu, X., Lei, Z., Li, S.Z.: Object detection by labeling superpixels. In: 2015 IEEE Conference on Computer Vision and Pattern Recognition (CVPR), pp. 5107–5116, June 2015. https://doi.org/10.1109/CVPR.2015.7299146
18. Yang, C., Zhang, L., Lu, H., Ruan, X., Yang, M.H.: Saliency detection via graph-based manifold ranking. In: 2013 IEEE Conference on Computer Vision and Pattern Recognition, pp. 3166–3173 (2013)
19. Yang, F., Lu, H., Yang, M.H.: Robust superpixel tracking. Trans. Img. Proc. **23**(4), 1639–1651 (2014). https://doi.org/10.1109/TIP.2014.2300823
20. Li, Z., Chen, J.: Superpixel segmentation using linear spectral clustering. In: 2015 IEEE Conference on Computer Vision and Pattern Recognition (CVPR), pp. 1356–1363, June 2015. https://doi.org/10.1109/CVPR.2015.7298741
21. Zhu, W., Liang, S., Wei, Y., Sun, J.: Saliency optimization from robust background detection. In: 2014 IEEE Conference on Computer Vision and Pattern Recognition. pp. 2814–2821, June 2014. https://doi.org/10.1109/CVPR.2014.360

Spatial-Adaptive Network for Single Image Denoising

Meng Chang, Qi Li$^{(\boxtimes)}$, Huajun Feng, and Zhihai Xu

State Key Lab for Modern Optical Instruments, Zhejiang University,
Hangzhou, China
{changm,liqi,fenghj,xuzh}@zju.edu.cn

Abstract. Previous works have shown that convolutional neural networks can achieve good performance in image denoising tasks. However, limited by the local rigid convolutional operation, these methods lead to oversmoothing artifacts. A deeper network structure could alleviate these problems, but at the cost of additional computational overhead. In this paper, we propose a novel spatial-adaptive denoising network (SAD-Net) for efficient single image blind noise removal. To adapt to changes in spatial textures and edges, we design a residual spatial-adaptive block. Deformable convolution is introduced to sample the spatially related features for weighting. An encoder-decoder structure with a context block is introduced to capture multiscale information. By conducting noise removal from coarse to fine, a high-quality noise-free image is obtained. We apply our method to both synthetic and real noisy image datasets. The experimental results demonstrate that our method outperforms the state-of-the-art denoising methods both quantitatively and visually.

Keywords: Image denoising · Image restoration · Image processing

1 Introduction

Image denoising is an important task in computer vision. During image acquisition, noise is often unavoidable due to imaging environment and equipment limitations. Therefore, noise removal is an essential step, not only for visual quality but also for other computer vision tasks. Image denoising has a long history, and many methods have been proposed. Many of the early model-based methods found natural image priors and then applied optimization algorithms to solve the model iteratively [2,23,30,41]. However, these methods are time consuming and cannot effectively remove noise. With the rise of deep learning, convolutional neural networks (CNNs) have been applied to image denoising tasks and have achieved high-quality results.

Electronic supplementary material The online version of this chapter (https://doi.org/10.1007/978-3-030-58577-8_11) contains supplementary material, which is available to authorized users.

A. Vedaldi et al. (Eds.): ECCV 2020, LNCS 12375, pp. 171–187, 2020.
https://doi.org/10.1007/978-3-030-58577-8_11

On the other hand, the early works assumed that noise is independent and identically distributed. Additive white Gaussian noise (AWGN) is often adopted to create synthetic noisy images. People now realize that noise presents in more complicated forms that are spatially variant and channel dependent. Therefore, some recent works have made progress in real image denoising [4,12,26,39].

However, despite numerous advances in image denoising, some issues remain to be resolved. A traditional CNN can use only the features in local fixed-location neighborhoods, but these may be irrelevant or even exclusive to the current location. Due to their inability to adapt to textures and edges, CNN-based methods result in oversmoothing artifacts and some details are lost. In addition, the receptive field of a traditional CNN is relatively small. Many methods deepen the network structure [27] or use a non-local module to expand the receptive field [18,37]. However, these methods lead to high computational memory and time consumption, hence they cannot be applied in practice.

In this paper, we propose a spatial-adaptive denoising network (SADNet) to address the above issues. A residual spatial-adaptive block (RSAB) is designed to adapt to changes in spatial textures and edges. We introduce the modulated deformable convolution in each RSAB to sample the spatially relevant features for weighting. Moreover, we incorporate the RSAB and residual blocks (ResBlock) in an encoder-decoder structure to remove noise from coarse to fine. To further enlarge the receptive field and capture multiscale information, a context block is applied to the coarsest scale. Compared to the state-of-the-art methods, our method can achieve good performance while maintaining a relatively small computational overhead.

In conclusion, the main contributions of our method are as follows:

- We propose a novel spatial-adaptive denoising network for efficient noise removal. The network can capture the relevant features from complex image content, and recover details and textures from heavy noise.
- We propose the residual spatial-adaptive block, which introduces deformable convolution to adapt to spatial textures and edges. In addition, using an encoder-deocder structure with a context block to capture multiscale information, we can estimate offsets and remove noise from coarse to fine.
- We conduct experiments on multiple synthetic image datasets and real noisy datasets. The results demonstrate that our model achieves state-of-the-art performances on both synthetic and real noisy images with a relatively small computational overhead.

2   Related Works

In general, image denoising methods include model-based and learning-based methods. Model-based methods attempt to model the distribution of natural images or noise. Then, using the modeled distribution as the prior, they attempt to obtain clear images with optimization algorithms. The common priors include local smoothing [23,30], sparsity [2,20,33], non-local self-similarity [5,8,9,11,34] and external statistical prior [32,41]. Non-local self-similarity is the notable prior in the image denoising task. This prior assumes that the image information is redundant and that similar structures exist within a single image. Then, self-similar patches are found in the image to remove noise. Many methods have been proposed based on the non-local self-similarity prior including NLM [5], BM3D [8,9], and WNNM [11,34], all of which are currently widely used.

With the popularity of deep neural networks, learning-based denoising methods have developed rapidly. Some works combine natural priors with deep neural networks. TRND [7] introduced the field-of-experts prior into a deep neural network. NLNet [17] combined the non-local self-similarity prior with a CNN. Limited by the designed priors, their performance is often inferior compared to end-to-end CNN methods. DnCNN [35] introduced residual learning and batch normalization to implement end-to-end denoising. FFDNet [36] introduced the noise level map as the input and enhanced the flexibility of the network for non-uniform noise. MemNet [27] proposed a very deep end-to-end persistent memory network for image restoration, which fuses both short-term and long-term memories to capture different levels of information. Inspired by the non-local self-similarity prior, a non-local module [28] was designed for neural networks. NLRN [18] attempted to incorporate non-local modules into a recurrent neural network (RNN) for image restoration. N3Net [26] proposed neural nearest neighbors block to achieve non-local operation. RNAN [37] designed non-local attention blocks to capture global information and pay more attention to the challenging parts. However, non-local operations lead to high memory usage and time consumption.

Recently, the focus of researchers has shifted from AWGN to more realistic noise. Some recent works have made progress on real noisy images. Several real noisy datasets have been established by capturing real noisy scenes [1,3,25], which promotes research into real-image denoising. N3Net [26] demonstrated the significance on real noisy dataset. CBDNet [12] trained two subnets to sequentially estimate noise and perform non-blind denoising. PD [39] applied the pixel-shuffle downsampling strategy to approximate the real noise to AWGN, which can adapt the trained model to real noises. RIDNet [4] proposed a one-stage denoising network with feature attention for real image denoising. However, these methods lack adaptability to image content and result in oversmoothing artifacts.

3   Framework

The architecture of our proposed spatial-adaptive denoising network (SADNet) is shown in Fig. 1. Let x denotes a noisy input image and $\hat{y}$ denotes the corresponding output denoised image. Then our model can be described as follows:

$$\hat{y} = \text{SADNet}(x). \tag{1}$$

We use one convolutional layer to extract the initial features from the noisy input; then those features are input into a multiscale encoder-decoder architecture. In the encoder component, we use ResBlocks [14] to extract features of different

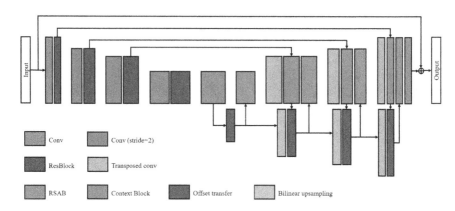

Fig. 1. The framework of our proposed spatial-adaptive denoising network.

scales. However, unlike the original ResBlock, we remove the batch normalization and use leaky ReLU [19] as the activation function. To avoid damaging the image structures, we limit the number of downsampling operations and implement a context block to further enlarge the receptive field and capture multiscale information. Then, in the decoder component, we design residual spatial-adaptive blocks (RSABs) to sample and weight the related features to remove noise and reconstruct the textures. In addition, we estimate the offsets and transfer them from coarse to fine, which is beneficial for obtaining more accurate feature locations. Finally the reconstructed features are fed to the last convolutional layer to restore the denoised image. By using the long residual connection, our network learns only the noise component.

In addition to the network architecture, the loss function is crucial to the performance. Several loss functions, such as L_2 [35–37], L_1 [4], perceptual loss [15], and asymmetric loss [12], have been used in denoising tasks. In general, L_1 and L_2 are the two losses used most commonly in previous works. The L_2 loss has good confidence for Gaussian noise, whereas the L_1 loss has better tolerance for outliers. In our experiment, we use the L_2 loss for training on synthetic image datasets and the L_1 loss for training on real-image noise datasets.

The following subsections focus on the RSAB and context block to provide more detailed explanations.

3.1 Residual Spatial-Adaptive Block

In this section, we first introduce the deformable convolution [10,40] and then propose our RSAB in detail.

Let $x(p)$ denote the features at location p from the input feature map x. Then, for a traditional convolution operation, the corresponding output features $y(p)$ can be obtained by

$$y(p) = \sum_{p_i \in N(p)} w_i \cdot x(p_i), \tag{2}$$

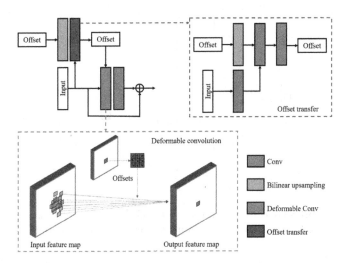

Fig. 2. The architecture of the residual spatial-adaptive block (RSAB). The offset transfer component is shown in the green dashed box. The deformable convolution architecture is shown in the blue dashed box. (Color figure online)

where $N(p)$ denotes the neighborhood of location p, whose size is equal to the size of the convolutional kernel. w_i denotes the weight of location p in the convolutional kernel, and p_i denotes the location in $N(p)$. The traditional convolution operation strictly takes the feature of the fixed location around p when calculating the output feature. Thus, some unwanted or unrelated features can interfere with the output calculation. For example, when the current location is near the edge, the distinct features located outside the object are introduced for weighting, which may smooth the edges and destroy the texture. For the denoising task, we would prefer that only the related or similar features are used for noise removal, similar to the self-similarity weighted denoising methods [5,8,9].

Therefore, we introduce deformable convolution [10,40] to adapt to spatial texture changes. In contrast to traditional convolutional layers, deformable convolution can change the shapes of convolutional kernels. It first learns an offset map for every location and applies the resulting offset map to the feature map, which resamples the corresponding features for weighting. Here, we use modulated deformable convolution [40], which provides another dimension of freedom to adjust its spatial support regions,

$$y(p) = \sum_{p_i \in N(p)} w_i \cdot x(p_i + \Delta p_i) \cdot \Delta m_i, \tag{3}$$

where Δp_i is the learnable offset for location p_i, and Δm_i is the learnable modulation scalar, which lies in the range $[0,1]$. It reflects the degree of correlation between the sampled features $x(p_i)$ and the features in the current location. Thus, the modulated deformable convolution can modulate the input feature

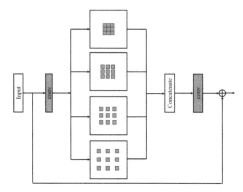

Fig. 3. The architecture of the context block. Instead of downsampling operations, multisize dilated convolutions are implemented to extract different receptive-field features.

amplitudes to further adjust the spatial support regions. Both Δp and Δm are obtained from the previous features.

In each RSAB, we first fuse the extracted features and the reconstructed features from the previous scale as the input. The RSAB is constructed by a modulated deformable convolution followed by a traditional convolution with a short skip connection. Similar to ResBlock, we implement local residual learning to enhance the information flow and improve representation ability of the network. However, unlike ResBlock, we replace the first convolution with modulated deformable convolution and use leaky ReLU as our activation function. Hence, the RSAB can be formulated as

$$F_{RSAB}(x) = F_{cn}(F_{act}(F_{dcn}(x))) + x, \tag{4}$$

where F_{dcn} and F_{cn} denote the modulated deformable convolution and traditional convolution respectively. F_{act} is the activation function (leaky ReLU here). The architecture of RSAB is shown in Fig. 2.

Furthermore, to better estimate the offsets from coarse to fine, we transfer the last-scale offsets Δp^{s-1} and modulation scalars Δm^{s-1} to the current scale s, and then use both $\{\Delta p^{s-1}, \Delta m^{s-1}\}$ and the input features x^s to estimate $\{\Delta p^s, \Delta m^s\}$. Given the small-scale offsets as the initial reference, the related features can be located more accurately on the large scale. The offset transfer can be formulated as follows:

$$\{\Delta p^s, \Delta m^s\} = F_{offset}(x, F_{up}(\{\Delta p^{s-1}, \Delta m^{s-1}\})), \tag{5}$$

where F_{offset} and F_{up} denote the offset transfer and upsampling functions, separately, as shown in Fig. 2. The offset transfer function involves several convolutions, and it extracts features from input and fuses them with the previous offsets to estimate the offsets in the current scale. The upsampling function magnifies both the size and value of the previous offset maps. In our experiment, bilinear interpolation is adopted to upsample the offsets and modulation scalars.

3.2 Context Block

Multiscale information is important for image denoising tasks; therefore, the downsampling operation is often adopted in networks. However, when the spatial resolution is too small, the image structures are destroyed, and information is lost, which is not conducive to reconstructing the features.

To increase the receptive field and capture multiscale information without further reducing the spatial resolution, we introduce a context block into the minimum scale between the encoder and decoder. Context blocks have been successfully used in image segments [6] and deblurring tasks [38]. In contrast to spatial pyramid pooling [13], the context block uses several dilated convolutions with different dilation rates rather than downsampling. It can expand the receptive field without increasing the number of parameters or damaging the structures. Then, the features extracted from the different receptive fields are fused to estimate the output (as shown in Fig. 3). It is beneficial to estimate offsets from a larger receptive field.

In our experiment, we remove the batch normalization layer and only use four dilation rates which are set to 1, 2, 3, and 4. To further simplify the operation and reduce the running time, we first use a 1×1 convolution to compress the feature channels. The compression ratio is set to 4 in our experiments. In the fusion setup, we use a 1×1 convolution to output the fusion features whose channels are equal to the original input features. Similarly, a local skip connection between the input and output features is applied to prevent information blocking.

3.3 Implementation

In the proposed model, we use four scales for the encoder-decoder architecture, and the number of channels for each scale is set to 32, 64, 128, and 256. The kernel size of the first and last convolutional layers is set to 1×1, and the final output is set to 1 or 3 channels depending on the input. Moreover, we use 2×2 filters for up/down-convolutional layers, and all the other convolutional layers have a kernel size of 3×3.

4 Experiments

In this section, we demonstrate the effectiveness of our model on both synthetic datasets and real noisy datasets. We adopt DIV2K [21] which contains 800 images with 2K resolution, and add different levels of noise to synthetic noise datasets. For real noisy images, we use the SIDD [1], RENOIR [3] and Poly [31] datasets. We randomly rotate and flip the images horizontally and vertically for data augmentation. In each training batch, we use 16 patches with size of 128×128 as inputs. We train our model using the ADAM [16] optimizer with $\beta_1 = 0.9$, $\beta_2 = 0.999$, and $\epsilon = 10^{-8}$. The initial learning rate is set to 10^{-4} and then halved after 3×10^5 iterations. Our model is implemented in the PyTorch framework [24] with an Nvidia GeForce RTX 1080Ti. In addition, we employ PSNR and SSIM [29] to evaluate the results.

4.1 Ablation Study

We perform ablation study on the Kodak24 dataset with a noise sigma of 50. The results are shown in Table 1.

Table 1. Ablation study of different components. PSNR values are based on Kodak24 ($\sigma = 50$)

RSAB	$\times$	$\checkmark$	$\times$	$\checkmark$	$\checkmark$	$\checkmark$
Offset transfer	$\times$	$\times$	$\times$	$\times$	$\checkmark$	$\checkmark$
Context block	$\times$	$\times$	$\checkmark$	$\checkmark$	$\times$	$\checkmark$
PSNR	29.05	29.55	29.12	29.60	29.59	29.64

Ablation on RSAB. RSAB is the crucial block in our network. Without it, the network will lose its ability to adapt to image content. When we replace RSAB with an original ResBlock, the performance decreases substantially, which demonstrates its effect.

Ablation on the Context Block. The context block complements the down-sampling operations to capture larger field information. We can observe that the performance improves when the context block is introduced.

Ablation on the Offset Transfer. We remove the offset transfer from coarse to fine and use only the features on the current scale to estimate the offsets for RSAB. This comparison validates the effectiveness of offset transfer.

4.2 Analyses of the Spatial Adaptability

As discussed above, our network introduces the adaptability to spatial textures and edges. The RSABs can extract related features by change the sampling locations based on the image content. We visualize the learned kernel locations of the RSABs in Fig. 4. The visualization results show that in the smooth regions or the homogeneous textured regions, the convolution kernels are approximately uniformly distributed, while in the regions close to the edge, the shapes of the convolution kernels extend along the edge. Most of sampling points fall on the similar texture regions inside the object, which demonstrates that our network has indeed learned spatial adaptability. Moreover, as shown in Fig. 4, the RSAB can extract features from a larger receptive field at the coarse scale, while at the fine scale, the sampled features are located in the neighborhood of the current point. The multiscale structure enables the network to obtain the information of different receptive fields for image reconstruction.

4.3   Comparisons

In this subsection, we compare our algorithm with the state-of-the-art denoising methods. For a fair comparison, all the compared methods employ the default settings provided by the corresponding authors. We first make a comparison on the synthetic noise datasets, since many methods provide only Gaussian noise removal results. Then, we report the denoising results on the real noisy datasets using the state-of-the-art real noise removal methods.

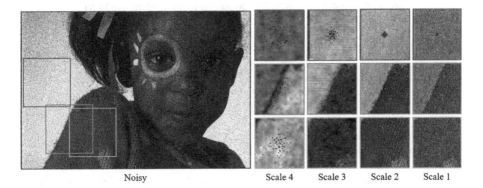

Fig. 4. Visualization of the learned kernels. The scales from 4 to 1 are in order from coarse to fine.

Synthetic Noisy Images. In the comparisons of synthetic noisy images, we use BSD68 and Kodak24 as our test datasets. These datasets include both color and grayscale images for testing. We add AWGN at different noise levels to the clean images. We choose BM3D [9] and CBM3D [8] as representatives of the classical traditional methods as well as some CNN-based methods, including DnCNN [35], MemNet [27], FFDNet [36], RNAN [37], and RIDNet [4], for the comparisons.

Tables 2 shows the average results of PSNR on grayscale images with three different noise levels. Our SADNet achieves the highest values on most of the datasets and tested noise levels. Note that although RNAN can achieve comparable evaluations to our method on partial low noise levels, it requires more parameters and a larger computational overhead. Next, Table 3 reports the quantitative results on color images. We replace the input and output channels from one to three as the other methods. Our SADNet outperforms the state-of-the-art methods on all the datasets with all tested noise levels. In addition, we can observe that our method shows more improvement at higher noise levels, which demonstrates its effectiveness for heavy noise removal.

The visual comparisons are shown in Fig. 5 and Fig. 6. We present some challenging examples from BSD68 and Kodak24. In particular, the birds' feathers and the clothing textures are difficult to separate from heavy noise. The compared methods tend to remove the details along with the noise, resulting in oversmoothing artifacts. Many of the textured areas are heavily smeared in the denoising results. Due to its adaptivity to the image content, our method can restore the vivid textures from noisy images without introducing other artifacts.

Table 2. Average PSNR (dB) results on synthetic **grayscale** noisy images

Dataset	σ	BM3D	DnCNN	MemNet	FFDNet	RNAN	RIDNet	SADNet (ours)
BSD68	30	27.76	28.36	28.43	28.39	28.61	28.54	**28.61**
	50	25.62	26.23	26.35	26.29	26.48	26.40	**26.51**
	70	24.44	24.90	25.09	25.04	25.18	25.12	**25.24**
Kodak24	30	29.13	29.62	29.72	29.70	**30.04**	29.90	30.00
	50	26.99	27.51	27.68	27.63	27.93	27.79	**27.96**
	70	25.73	26.08	26.42	26.34	26.60	26.51	**26.72**

Table 3. Average PSNR (dB) results on synthetic **color** noisy images

Dataset	σ	CBM3D	DnCNN	MemNet	FFDNet	RNAN	RIDNet	SADNet (ours)
BSD68	30	29.73	30.40	28.39	30.31	30.63	30.47	**30.64**
	50	27.38	28.01	26.33	27.96	28.27	28.12	**28.32**
	70	26.00	26.56	25.08	26.53	26.83	26.69	**26.93**
Kodak24	30	30.89	31.39	29.67	31.39	31.86	31.64	**31.86**
	50	28.63	29.16	27.65	29.10	29.58	29.25	**29.64**
	70	27.27	27.64	26.40	27.68	28.16	27.94	**28.28**

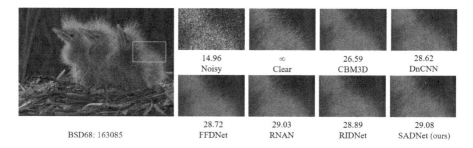

Fig. 5. Synthetic image denoising results on BSD68 with noise level $\sigma = 50$.

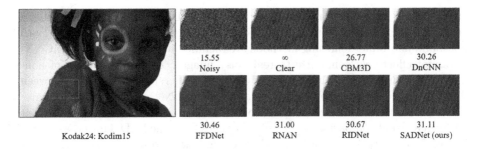

Fig. 6. Synthetic image denoising results on Kodak24 with noise level $\sigma = 50$.

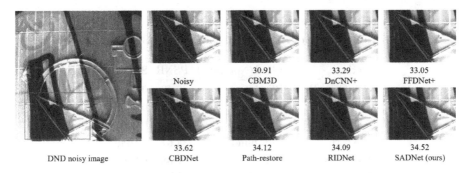

Fig. 7. Real image denoising results from the DnD dataset.

Real Noisy Images. To conduct comparisons on real noisy images, we choose DND [25], SIDD [1] and Nam [22] as test datasets. **DND** contains 50 real noisy images and their corresponding clear images. One thousand patches with a size of 512×512 are extracted from the dataset by the providers for testing and comparison purposes. Since the ground truth images are not publicly available, we can obtain only the PSNR/SSIM results though the online submission system introduced by [25]. The validation dataset of **SIDD** is introduced for our evaluation, which contains 1280 256×256 noisy-clean image pairs. **Nam** includes 15 large image pairs with JPEG compression for 11 scenes. We cropped the images into 512×512 patches and selected 25 patches picked by CBDNet [12] for testing.

We train our model on the SIDD medium dataset and RENOIR for evaluation on the DND and SIDD validation datasets. Then, we finetune our model on the Poly [31] for Nam, which improves the performance on the noisy images with JPEG compression. Furthermore, as comparisons, we choose the state-of-the-art methods whose validity has previously been demonstrated on real noisy images, including CBM3D [8], DnCNN [35], CBDNet [12], PD [39], and RIDNet [4].

DND. The quantitative results are listed in Table 4, which are obtained from the public DnD benchmark website. FFDNet+ is the improved version of FFDNet with a uniform noise level map manually selected by the providers. CDnCNN-B is the original DnCNN model for blind color denoising. DnCNN+ is finetuned on CDnCNN-B with the results of FFDNet+. SADNet (1248) is the modified version of our SADNet with 1, 2, 4, 8 dilation rates in the context block. Both non-blind and blind denoising methods are included for comparisons. CDnCNN-B cannot effectively generalize to real noisy images. The performances of non-blind denoising methods are limited due to the different distributions between AWGN and real-world noise. In contrast, our SADNet outperforms the state-of-the-art methods with respect to both PSNR and SSIM values. We further perform a visual comparison on denoised images from the DnD dataset, as shown in Fig. 7. The other methods corrode the edges with residual noise, while our method can effectively remove the noise from the smooth region and maintain clear edges.

Table 4. Quantitative results on DnD sRGB images

Method	Blind/Non-blind	PSNR	SSIM
CDnCNN-B	Blind	32.43	0.7900
TNRD	Non-blind	33.65	0.8306
BM3D	Non-blind	34.51	0.8507
WNNM	Non-blind	34.67	0.8646
MCWNNM	Non-blind	37.38	0.9294
FFDNet+	Non-blind	37.61	0.9415
DnCNN+	Non-blind	37.90	0.9430
CBDNet	Blind	38.06	0.9421
N3Net	Blind	38.32	0.9384
PD	Blind	38.40	0.9452
Path-Restore	Blind	39.00	0.9542
RIDNet	Blind	39.26	0.9528
SADNet (1248)	Blind	39.37	**0.9544**
SADNet (ours)	Blind	**39.59**	0.9523

Table 5. Quantitative results on SIDD sRGB validation dataset

Method	CBM3D	CDnCNN-B	CBDNet	PD	RIDNET	SADNet (ours)
Blind/Non-blind	Non-blind	Blind	Blind	Blind	Blind	Blind
PSNR	30.88	26.21	30.78	32.94	38.71	**39.46**

SIDD. The images in the SIDD dataset are captured by smartphones, and some noisy images have high noise levels. We employ 1,280 validation images for quantitative comparisons as listed in Table 5. The results demonstrates that our method achieves significant improvements over the other tested methods. For visual comparisons, we choose two challenging examples from the denoised results. The first scene has rich textures, while the second scene has prominent structures. As shown in Fig. 8 and Fig. 9, CDnCNN-B and CBDNet fail at noise removal. CBM3D results in pseudo artifacts, and PD and RIDNet destroy the textures. In contrast, our network recovers textures and structures that are closer to the ground truth.

Nam. The JPEG compression makes the noise more stubborn on the Nam dataset. For a fair comparison, we use the patches chosen by CBDNet [12] for evaluation. Furthermore, CBDNet* [12] is introduced for comparison, which was retrained on JPEG compressed datasets by its providers. We report the average PSNR and SSIM values for Nam in Table 6. With respect to PSNR, Our SADNet achieves 1.88, 1.83 and 1.61 dB gains over RIDNet, PD, and CBDNet*. Similarly, our SSIM values exceed those of all the other methods in the comparison. In the visual comparison shown in Fig. 10, our method again obtains the best result for texture restoration and noise removal.

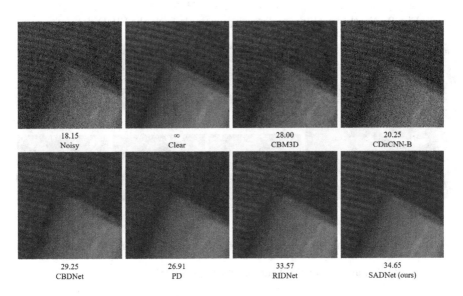

Fig. 8. A real image denoising example from the SIDD dataset.

Table 6. Quantitative results on Nam dataset with JPEG compression

Method	CBM3D	CDnCNN-B	CBDNet*	PD	RIDNET	SADNet (ours)
Blind/Non-blind	Non-blind	Blind	Blind	Blind	Blind	Blind
PSNR	39.84	37.49	41.31	41.09	41.04	**42.92**
SSIM	0.9657	0.9272	0.9784	0.9780	0.9814	**0.9839**

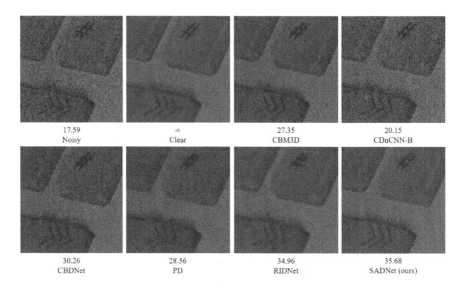

Fig. 9. Another real image denoising example from the SIDD dataset.

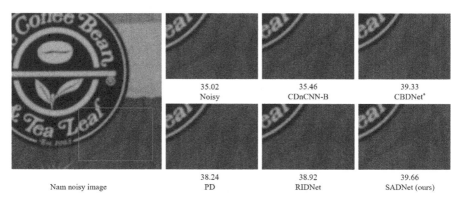

Fig. 10. Real image denoising results from the Nam dataset with JPEG compression.

Table 7. Parameters and time comparisons on 480×320 color images

Method	DnCNN	MemNet	RNAN	RIDNet	SADNET (ours)
Parameters	558k	2,908k	8,960k	1,499k	4,321k
FLOPs	86.1G	449.2G	1163.5G	230.0G	50.1G
Times (ms)	21.3	154.2	1072.2	84.4	26.7

Parameters and Running Times. To compare the running times, we test different methods when denoising 480×320 color images. Note that the running time may depend on the test platform and code; thus, we also provide the number of floating point operations (FLOPs). All the methods are implemented in

PyTorch. As shown in Table 7, although SADNet has high parameter numbers, its FLOPs are minimal, and its running time is short due to the multiple down-sampling operations. Because most of the operations run on smaller-scale feature maps, our model performs faster than many others with fewer parameters.

5 Conclusion

In this paper, we propose a spatial-adaptive denoising network for effective noise removal. The network is built by multiscale residual spatial-adaptive blocks, which sample relevant features for weighting based on the content and textures of images. We further introduce a context block to capture multiscale information and implement offset transfer to more accurately estimate the sampling locations. We find that the introduction of spatially adaptive capability can restore richer details in complex scenes under heavy noise. The proposed SADNet achieves state-of-the-art performances on both synthetic and real noisy images and has a moderate running time.

References

1. Abdelhamed, A., Lin, S., Brown, M.S.: A high-quality denoising dataset for smartphone cameras. In: Proceedings of the IEEE Conference on Computer Vision and Pattern Recognition, pp. 1692–1700 (2018)
2. Aharon, M., Elad, M., Bruckstein, A.: K-SVD: an algorithm for designing overcomplete dictionaries for sparse representation. IEEE Trans. Signal Process. **54**(11), 4311–4322 (2006)
3. Anaya, J., Barbu, A.: RENOIR-a dataset for real low-light image noise reduction. J. Vis. Commun. Image Represent. **51**, 144–154 (2018)
4. Anwar, S., Barnes, N.: Real image denoising with feature attention. arXiv preprint arXiv:1904.07396 (2019)
5. Buades, A., Coll, B., Morel, J.M.: A non-local algorithm for image denoising. In: 2005 IEEE Computer Society Conference on Computer Vision and Pattern Recognition (CVPR 2005), vol. 2, pp. 60–65. IEEE (2005)
6. Chen, L.C., Papandreou, G., Schroff, F., Adam, H.: Rethinking atrous convolution for semantic image segmentation. arXiv preprint arXiv:1706.05587 (2017)
7. Chen, Y., Pock, T.: Trainable nonlinear reaction diffusion: a flexible framework for fast and effective image restoration. IEEE Trans. Pattern Anal. Mach. Intell. **39**(6), 1256–1272 (2016)
8. Dabov, K., Foi, A., Katkovnik, V., Egiazarian, K.: Color image denoising via sparse 3D collaborative filtering with grouping constraint in luminance-chrominance space. In: 2007 IEEE International Conference on Image Processing, vol. 1, pp. I-313. IEEE (2007)
9. Dabov, K., Foi, A., Katkovnik, V., Egiazarian, K.: Image denoising by sparse 3-D transform-domain collaborative filtering. IEEE Trans. Image Process. **16**(8), 2080–2095 (2007)
10. Dai, J., et al.: Deformable convolutional networks. In: Proceedings of the IEEE International Conference on Computer Vision, pp. 764–773 (2017)

11. Gu, S., Zhang, L., Zuo, W., Feng, X.: Weighted nuclear norm minimization with application to image denoising. In: Proceedings of the IEEE Conference on Computer Vision and Pattern Recognition, pp. 2862–2869 (2014)

12. Guo, S., Yan, Z., Zhang, K., Zuo, W., Zhang, L.: Toward convolutional blind denoising of real photographs. In: Proceedings of the IEEE Conference on Computer Vision and Pattern Recognition, pp. 1712–1722 (2019)

13. He, K., Zhang, X., Ren, S., Sun, J.: Spatial pyramid pooling in deep convolutional networks for visual recognition. IEEE Trans. Pattern Anal. Mach. Intell. **37**(9), 1904–1916 (2015)

14. He, K., Zhang, X., Ren, S., Sun, J.: Deep residual learning for image recognition. In: Proceedings of the IEEE Conference on Computer Vision and Pattern Recognition, pp. 770–778 (2016)

15. Jiao, J., Tu, W.C., He, S., Lau, R.W.H.: FormResNet: formatted residual learning for image restoration. In: 2017 IEEE Conference on Computer Vision and Pattern Recognition Workshops (CVPRW) (2017)

16. Kingma, D.P., Ba, J.: Adam: a method for stochastic optimization. arXiv preprint arXiv:1412.6980 (2014)

17. Lefkimmiatis, S.: Non-local color image denoising with convolutional neural networks. In: Proceedings of the IEEE Conference on Computer Vision and Pattern Recognition, pp. 3587–3596 (2017)

18. Liu, D., Wen, B., Fan, Y., Loy, C.C., Huang, T.S.: Non-local recurrent network for image restoration. In: Advances in Neural Information Processing Systems, pp. 1673–1682 (2018)

19. Maas, A.L., Hannun, A.Y., Ng, A.Y.: Rectifier nonlinearities improve neural network acoustic models. In: Proceedings of ICML, vol. 30, p. 3 (2013)

20. Mairal, J., Bach, F.R., Ponce, J., Sapiro, G., Zisserman, A.: Non-local sparse models for image restoration. In: ICCV, vol. 29, pp. 54–62. Citeseer (2009)

21. Martin, D., Fowlkes, C., Tal, D., Malik, J., et al.: A database of human segmented natural images and its application to evaluating segmentation algorithms and measuring ecological statistics. In: ICCV, Vancouver (2001)

22. Nam, S., Hwang, Y., Matsushita, Y., Joo Kim, S.: A holistic approach to cross-channel image noise modeling and its application to image denoising. In: Proceedings of the IEEE Conference on Computer Vision and Pattern Recognition, pp. 1683–1691 (2016)

23. Osher, S., Burger, M., Goldfarb, D., Xu, J., Yin, W.: An iterative regularization method for total variation-based image restoration. Multiscale Model. Simul. **4**(2), 460–489 (2005)

24. Paszke, A., et al.: Pytorch: an imperative style, high-performance deep learning library. In: Advances in Neural Information Processing Systems, pp. 8024–8035 (2019)

25. Plotz, T., Roth, S.: Benchmarking denoising algorithms with real photographs. In: Proceedings of the IEEE Conference on Computer Vision and Pattern Recognition, pp. 1586–1595 (2017)

26. Plötz, T., Roth, S.: Neural nearest neighbors networks. In: Advances in Neural Information Processing Systems (NeurIPS) (2018)

27. Tai, Y., Yang, J., Liu, X., Xu, C.: Memnet: a persistent memory network for image restoration. In: Proceedings of the IEEE International Conference on Computer Vision, pp. 4539–4547 (2017)

28. Wang, X., Girshick, R., Gupta, A., He, K.: Non-local neural networks. In: Proceedings of the IEEE Conference on Computer Vision and Pattern Recognition, pp. 7794–7803 (2018)

29. Wang, Z., Bovik, A.C., Sheikh, H.R., Simoncelli, E.P.: Image quality assessment: from error visibility to structural similarity. IEEE Trans. Image Process. **13**(4), 600–612 (2004)

30. Xu, J., Osher, S.: Iterative regularization and nonlinear inverse scale space applied to wavelet-based denoising. IEEE Trans. Image Process. **16**(2), 534–544 (2007)

31. Xu, J., Li, H., Liang, Z., Zhang, D., Zhang, L.: Real-world noisy image denoising: a new benchmark. arXiv preprint arXiv:1804.02603 (2018)

32. Xu, J., Zhang, L., Zhang, D.: External prior guided internal prior learning for real-world noisy image denoising. IEEE Trans. Image Process. **27**(6), 2996–3010 (2018)

33. Xu, J., Zhang, L., Zhang, D.: A trilateral weighted sparse coding scheme for real-world image denoising. In: Proceedings of the European Conference on Computer Vision (ECCV), pp. 20–36 (2018)

34. Xu, J., Zhang, L., Zhang, D., Feng, X.: Multi-channel weighted nuclear norm minimization for real color image denoising. In: Proceedings of the IEEE International Conference on Computer Vision, pp. 1096–1104 (2017)

35. Zhang, K., Zuo, W., Chen, Y., Meng, D., Zhang, L.: Beyond a gaussian denoiser: residual learning of deep CNN for image denoising. IEEE Trans. Image Process. **26**(7), 3142–3155 (2017)

36. Zhang, K., Zuo, W., Zhang, L.: FFDNet: toward a fast and flexible solution for CNN-based image denoising. IEEE Trans. Image Process. **27**(9), 4608–4622 (2018)

37. Zhang, Y., Li, K., Li, K., Zhong, B., Fu, Y.: Residual non-local attention networks for image restoration. arXiv preprint arXiv:1903.10082 (2019)

38. Zhou, S., Zhang, J., Zuo, W., Xie, H., Pan, J., Ren, J.S.: DAVANet: stereo deblurring with view aggregation. In: Proceedings of the IEEE Conference on Computer Vision and Pattern Recognition, pp. 10996–11005 (2019)

39. Zhou, Y., et al.: When AWGN-based denoiser meets real noises. arXiv preprint arXiv:1904.03485 (2019)

40. Zhu, X., Hu, H., Lin, S., Dai, J.: Deformable ConvNets v2: more deformable, better results. In: Proceedings of the IEEE Conference on Computer Vision and Pattern Recognition, pp. 9308–9316 (2019)

41. Zoran, D., Weiss, Y.: From learning models of natural image patches to whole image restoration. In: 2011 International Conference on Computer Vision, pp. 479–486. IEEE (2011)

Physics-Based Feature Dehazing Networks

Jiangxin Dong[1] and Jinshan Pan[2(✉)]

[1] Max Planck Institute for Informatics, Saarbrucken, Germany
[2] Nanjing University of Science and Technology, Nanjing, China
sdluran@gmail.com

Abstract. We propose a physics-based feature dehazing network for image dehazing. In contrast to most existing end-to-end trainable network-based dehazing methods, we explicitly consider the physics model of the haze process in the network design and remove haze in a deep feature space. We propose an effective feature dehazing unit (FDU), which is applied to the deep feature space to explore useful features for image dehazing based on the physics model. The FDU is embedded into an encoder and decoder architecture with residual learning, so that the proposed network can be trained in an end-to-end fashion and effectively help haze removal. The encoder and decoder modules are adopted for feature extraction and clear image reconstruction, respectively. The residual learning is applied to increase the accuracy and ease the training of deep neural networks. We analyze the effectiveness of the proposed network and demonstrate that it can effectively dehaze images with favorable performance against state-of-the-art methods.

Keywords: Image dehazing · Physics model · Feature dehazing unit · Deep convolutional neural networks

1 Introduction

Haze is a common atmospheric phenomenon. Images captured in hazy environments usually lose the color fidelity and contrast. Restoring clear images from hazy ones has been an active research effort in the computational photography and vision community within the last decade. As low-quality hazy images usually interfere with the subsequent image editing, analysis, and so on, it is of great interest to remove haze and restore high-quality images.

Mathematically, a hazy image I is usually modeled by [9,13]

$$I(\mathbf{x}) = T(\mathbf{x})J(\mathbf{x}) + (1 - T(\mathbf{x}))A, \tag{1}$$

where J, T, and A denote the latent clear image, medium transmission, and global atmospheric light, respectively, and $\mathbf{x}$ denotes the pixel position. This problem is highly ill-posed as only the hazy image I is available.

© Springer Nature Switzerland AG 2020
A. Vedaldi et al. (Eds.): ECCV 2020, LNCS 12375, pp. 188–204, 2020.
https://doi.org/10.1007/978-3-030-58577-8_12

To make this problem well-posed, conventional methods usually develop kinds of priors based on the statistical properties of clear images, transmission, or atmospheric light to constrain the solution space. The commonly used priors include dark channel prior [13], color line prior [10], color attenuation prior [42], sparse gradient prior [7], non-local prior [5], and so on. The methods based on these priors generate promising results. However, those aforementioned priors are manually designed based on specific observations, which do not always model the inherent properties of clear images, transmission, or atmospheric light. In addition, those aforementioned priors usually lead to highly non-convex problems that are difficult to solve.

Motivated by the success of deep learning in high-level vision tasks, the deep neural networks have been developed to overcome the limitation in image dehazing. Several approaches [6,28] develop deep neural networks to estimate the transmission and then follow the conventional method [13] to restore clear images. However, these approaches do not correct the errors of the atmospheric light. To avoid complex estimations of the transmission and atmospheric light, end-to-end trainable deep neural networks have been developed [18,20,22,23,27]. Based on the end-to-end trainable framework, the physics model of the haze process has been utilized to constrain the networks [26,38,39]. Although these methods achieve decent performance, they mainly consider the physics model in the raw image space and few of them explicitly utilize the physics model in the feature space, which does not fully explore the useful feature information for image dehazing.

To overcome these problems, we propose a physics-based feature dehazing network for image dehazing. The critical component is the proposed feature dehazing unit (FDU), which is derived based on the physics model of the haze process (i.e., (1)) in the feature space, so that more feature information that is useful for the clear image reconstruction can be effectively obtained with the constraint of the physics model. The FDU is then embedded into an encoder and decoder network architecture with residual learning, where the encoder module is first adopted to extract features from the hazy image and the decoder is further adopted to process the output of the physics-based feature dehazing block for the clear image reconstruction. The residual learning is used to increase the accuracy of image dehazing and ease the training of deep neural networks. Both quantitative and qualitative evaluation results demonstrate that the proposed approach performs favorably against state-of-the-art methods. The main contributions of this work are summarized as follows:

- We propose an effective feature dehazing unit, which is derived based on the physics model of the haze process in a feature space. We show that it can explicitly utilize the physics model in the feature space and learn useful information that is required in the derived physics model to better remove the haze.
- We develop a physics-based feature dehazing network (PFDN) that embeds the feature dehazing unit into an encoder and decoder network architecture

with residual learning. The encoder module is first adopted for useful feature extraction and the decoder is adopted for the final clear image reconstruction.

– We analyze the effect of the proposed PFDN on image dehazing and demonstrate that the proposed approach achieves state-of-the-art performance on both the dehazing benchmarks and real-world images.

2   Related Work

Recent years have witnessed significant advances in single image dehazing [4]. Existing methods can be roughly categorized into three kinds: adaptive contrast enhancement [11,33], model-based [5,7,9,10,13,42], and data driven-based methods [8,18,20,22,23,26,27,34,37–39,42].

The adaptive contrast enhancement method (e.g., [33]) removes haze by maximizing the local contrast of restored images. These approaches are able to remove haze to some extent, but usually suffer from visual artifacts.

The model-based methods rely on the physics model of the haze process and usually make assumptions on the clear images, transmission, or global atmospheric light. For example, Fattal [9] assumes that the transmission and the surface shading are locally uncorrelated. In [13], He et al. propose a novel dark channel prior to model the properties of the hazy-free images and use this prior to estimate the transmission. Fattal [10] observes that the pixels of the image patches typically exhibit a one-dimensional distribution and proposes a color-line prior to estimate the transmission. Berman et al. [5] find that pixels in a given cluster are often non-local and they use this constraint to solve image dehazing. The priors based on the transmission have been developed by Chen et al. [7]. Although promising results have been achieved, these priors used in image dehazing are designed based on some strong assumptions and do not always hold for some applications.

The data driven-based approaches mainly use some well-known learning methods to learn the most hazy-relevant features for image dehazing. In [34], Tang et al. develop a random forest algorithm to learn priors for the transmission estimation. Zhu et al. [42] formulate the depth estimation using a linear model and learn the color attenuation prior for haze removal. These approaches are able to learn more reliable priors but still heavily depend on hand-crafted priors. Recently, the deep convolutional neural network (CNN) as an effective learning method has been developed to solve image dehazing. In [6,28], they use deep CNNs to estimate the transmission and then use the conventional method [13] to estimate clear images. However, these approaches do not correct the errors of the atmospheric light. To avoid complex estimations of transmission and atmospheric light, several methods [18,20,22,27] develop efficient deep CNNs to solve image dehazing in an end-to-end fashion. Furthermore, to make the deep end-to-end trainable CNNs more compact, the physics model of the haze process has been used [26,38,39]. These deep CNNs-based methods outperform the adaptive contrast enhancement approaches and model-based approaches by large margins. However, few of them discriminatively explore the features of deep CNNs in image dehazing and most of them usually consider the physics model in the raw image space instead of fully exploring the feature information in the feature space. Thus, these methods may generate results with some color distortions and artifacts.

Discriminatively exploring the features of deep CNNs has been demonstrated to be effective in some applications, such as image classification [35], image restoration [41], image super-resolution [40], etc. These methods use the general attention mechanism to discriminatively learn useful features. However, we note that directly applying the attention mechanism in image dehazing does not effectively remove haze. Different from the attention mechanism used in these methods, we develop a physics-based feature dehazing network based on the physics model of the haze process to discriminatively learn useful information in the feature space for image dehazing.

3   Physics-Based Feature Dehazing Network

Most end-to-end trainable deep neural network-based image dehazing methods take the hazy image I as the input and directly output the dehazed result. To better constrain the solution space, some image dehazing methods first estimate the key components (e.g., T and A in (1)) and then compute the final clear image using the physics model (e.g., (1)). These approaches usually utilize the physics model in the raw image space, but rarely consider it in the feature space. To overcome this problem, we develop an effective feature dehazing unit (FDU), which is able to effectively utilize the physics model of the haze process in a feature space for better image dehazing. Then the proposed FDU is embedded into an encoder and decoder architecture with residual learning. Figure 1 shows the network architecture of the proposed physics-based feature dehazing network (PFDN). In the following, we present the details of each component in the PFDN.

3.1   Feature Dehazing Unit

To make full use of the feature information for image dehazing, we develop a feature dehazing unit (FDU). The proposed FDU is motivated by the physics model of the haze process (1). We note that the clear image J can be obtained by

$$J(\mathbf{x}) = I(\mathbf{x})\frac{1}{T(\mathbf{x})} + A(1 - \frac{1}{T(\mathbf{x})}) \qquad (2)$$

according to (1). For simplicity, let k denote a feature extractor, e.g., the filter kernel in a deep CNN. By applying k to (2), we can get

$$k \otimes J = k \otimes (I \odot \frac{1}{T}) + k \otimes A(1 - \frac{1}{T}), \qquad (3)$$

where $\otimes$ denotes the convolution operator and $\odot$ denotes the element-wise product operation. By taking a few algebraic operations using matrix-vector forms, we can get

$$\mathbf{KJ} = \mathbf{KT}_d\mathbf{I} + \mathbf{KA}_t, \qquad (4)$$

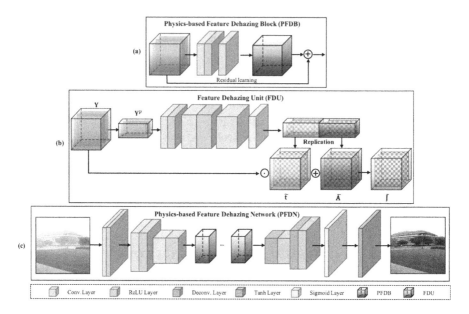

Fig. 1. Proposed network architecture for image dehazing. The whole image dehazing network PFDN in (c) is based on an encoder and decoder architecture with the proposed PFDBs in (a). The proposed PFDB consists of an FDU with a residual learning architecture, where the proposed FDU can make full use of the physics model in the feature space for better image dehazing. Please see text for more details.

where $\mathbf{K}$, $\mathbf{J}$, $\mathbf{I}$, and $\mathbf{A}_t$ denote the matrix-vector forms of k, J, I, and $A(1 - \frac{1}{T})$, respectively; $\mathbf{T}_d$ denotes the diagonal matrix, where the i-th diagonal element corresponds to the i-th element of the vector form of $\frac{1}{T}$. As we can decompose the matrix $\mathbf{K}\mathbf{T}_d$ into the product of two matrices $\mathbf{F}_1\mathbf{F}_2$, (4) can be rewritten as

$$\mathbf{K}\mathbf{J} = \mathbf{F}_1(\mathbf{F}_2\mathbf{I}) + \mathbf{K}\mathbf{A}_t, \tag{5}$$

where $\mathbf{F}_2$ can be regarded as the feature extraction operation. The equation (5) presents the relation between the clear image and the hazy image in a feature space, and is based on (4), which assumes that k is a linear operator. Note that a deep CNN with piece-wise linear activation functions (e.g., ReLU) is inherently locally linear [17,24]. Due to the strong representation ability of deep neural networks, we propose a feature dehazing unit and adopt deep CNNs to approximate the features corresponding to $\mathbf{F}_1$ and $\mathbf{K}\mathbf{A}_t$. Therefore, the discriminatively useful features can be better estimated for the clear image reconstruction.

To that end, the proposed FDU consists of two parts. The first part is used to learn the features that mostly approximate the key components $\mathbf{F}_1$ and $\mathbf{K}\mathbf{A}_t$ associated with the haze formation. The second part is then used to estimate the features $\mathbf{K}\mathbf{J}$ for the clear image reconstruction based on (5).

As the transmission map T is related to the scene depth [13] with the piece-wise constant property and the atmospheric light is usually assumed to be homogeneous, we use the global average pooling (GAP) to remove redundant information in the feature space and remain useful values in the approximated features that correspond to $\mathbf{F}_1$ and $\mathbf{KA}_t$. Specifically, let $\mathbf{Y} = \{y_i\}_{i=1}^{N}$ denote the input of FDU, which has N features with the size of $h \times w$ pixels. We first apply the GAP to $\mathbf{Y}$ and obtain $\mathbf{Y}^p = \{y_i^p\}_{i=1}^{N}$, each element of which is defined as

$$y_i^p = \frac{1}{h \times w} \sum_{q=1}^{h \times w} y_i(q), \tag{6}$$

where q denotes the pixel position and $\times$ denotes the product operation. Then, motivated by the success of the encoder and decoder network architecture in feature exploration, we apply the similar downsampling and upsampling operations [40] to the features in $\mathbf{Y}^p$ by

$$\widetilde{\mathbf{Y}} = \mathcal{R}(\mathcal{C}^N(\mathcal{R}(\mathcal{C}^{\frac{N}{16}}(\mathbf{Y}^p)))), \tag{7}$$

where $\mathcal{C}^N$ denotes the convolution operation with the filter kernel size of 1×1 pixel and N filters; $\mathcal{R}$ denotes the ReLU activation function. With $\widetilde{\mathbf{Y}}$, we further apply the feature upsampling operation $\mathcal{C}^{2N}$ with the Sigmoid function to get the intermediate feature. We respectively use the replication of the first N features ($\widetilde{\mathbf{t}}$) and the remaining N features ($\widetilde{\mathbf{A}}$) to approximate the features corresponding to $\mathbf{F}_1$ and $\mathbf{KA}_t$. Note that by extracting and remaining the most useful information with GAP, the features $\widetilde{\mathbf{t}}$ and $\widetilde{\mathbf{A}}$ are channel-specific. Thus, we generate the output of the FDU by

$$\widetilde{\mathbf{J}} = \mathbf{Y} \odot \widetilde{\mathbf{t}} + \widetilde{\mathbf{A}}. \tag{8}$$

Based on (8), we can discriminatively learn reliable features $\widetilde{\mathbf{J}}$ from $\mathbf{Y}$ for the clear image reconstruction. In Sect. 5, we demonstrate that using $\widetilde{\mathbf{J}}$ instead of $\mathbf{Y}$ is able to help image dehazing. Figure 1(b) shows the network architecture of the FDU.

3.2 Residual Learning

As residual learning [14] has been demonstrated to be effective in lots of vision tasks, we use it in the PFDB to increase the accuracy of image dehazing and ease the training of deep neural networks. Specifically, each PFDB has two convolutional layers with the filter kernel size of 3×3 pixels, where the first one is followed by the ReLU as the activation function and the second one is followed by the FDU. The detailed network architectures of the residual learning and the proposed PFDB are shown in Fig. 1(a).

3.3 PFDN for Image Dehazing

Since the proposed PFDB is performed in the feature space, we embed it into the encoder and decoder network architecture to solve image dehazing. The encoder module is adopted to extract useful features from the hazy image, which contains three scale convolutional blocks. Each convolutional block has one convolutional layer followed by a ReLU layer. The stride value is 1 for the first convolutional layer and 2 for the remaining two convolutional layers. The decoder module is adopted to further process the output of the PFDB and reconstruct the final clear image. It consists of two transposed convolutional blocks and one convolutional block. Each transposed convolutional block has a transposed convolutional layer with stride 2 and a ReLU layer. The stride value of the final convolutional layer is 1. For the parameters in the convolutional and the transposed convolutional layers, we use the same settings as [15]. The network architecture for the proposed PFDN is shown in Fig. 1(c). The detailed parameters of the proposed network are included on our project webpage.

3.4   Implementation Details

We use the ADAM optimizer [16] with the momentum parameters $\beta_1 = 0.5$ and $\beta_2 = 0.999$. The initial learning rate is set as 2×10^{-4} and we follow the decay strategy as [15]. In the training stage, we use the L_1-norm as the loss function to constrain the network output and the ground truth. We implement our network based on the PyTorch using a machine with an NVIDIA GTX 2080Ti GPU. Our un-optimized code takes about 0.09 s to dehaze an image of 512×512 pixels on average.

4   Experimental Results

We evaluate the proposed approach using the publicly available benchmark datasets [19,30,32] and compare it with the state-of-the-art single image dehazing methods.

4.1   Datasets

NYU and Make3D Datasets. For fair comparisons, we first follow the standard protocols adopted by existing methods (e.g., [20,28,29,39]) to generate 2,413 hazy/clear image pairs using the NYU depth v2 dataset [32] and the Make3D dataset [30]. This synthetic dataset contains hazy/clear image pairs of both indoor and outdoor scenes. We randomly choose 2,172 images for training and the remaining 241 images for test, where the training images and test images do not overlap. The image patch size used in the training process is set to be 512×512 pixels.

RESIDE Dataset. The RESIDE dataset [19] is a large-scale benchmark dataset, which contains hazy/clear image pairs in both indoor and outdoor scenarios. The subsets ITS and OTS are used as the training dataset. The subset SOTS is used for test, which contains 500 indoor hazy images and 500 outdoor hazy ones.

Table 1. Quantitative evaluations with the state-of-the-art methods on the synthetic NYU & Make3D test dataset. ↑ and ↓ denote that the better results should achieve higher and lower values of this metric.

Methods	DCP [13]	Nonlocal [41]	MSCNN [28]	DehazeNet [6]	GFN [29]	DCPDN [39]	cGAN [20]	DualCNN [26]	EPDN [27]	GDN [22]	Ours
PSNR↑	18.04	16.69	17.73	20.24	20.79	16.71	26.12	21.30	22.97	26.34	**27.26**
SSIM↑	0.7852	0.7397	0.7725	0.8176	0.8064	0.7676	0.8831	0.8170	0.8265	0.8816	**0.9047**
CIEDE2000↓	10.45	12.80	10.56	7.87	7.65	12.34	4.49	6.85	6.01	4.35	**3.87**

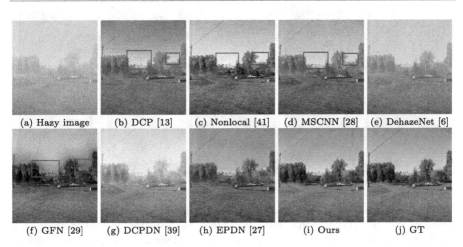

(a) Hazy image   (b) DCP [13]   (c) Nonlocal [41]   (d) MSCNN [28]   (e) DehazeNet [6]

(f) GFN [29]   (g) DCPDN [39]   (h) EPDN [27]   (i) Ours   (j) GT

Fig. 2. Image dehazing results from the synthetic NYU & Make3D test dataset. The parts enclosed in red boxes in (b)–(d) and (f) contain color distortions and artifacts. The result in (h) contains color distortions, while the results in (e) and (g) still have haze residual. The proposed method generates a much clearer image that is visually close to the ground truth image (best viewed on high-resolution display with zoom-in). (Color figure online)

4.2 Comparisons with the State of the Art

To evaluate the performance of the proposed approach, we first compare it against the state-of-the-art methods based on statistical priors [13,41] and deep CNNs [6,20,22,26–29,39] on the synthetic NYU & Make3D test dataset. For fair comparisons with deep learning-based methods, we fine-tune them using the proposed training dataset to achieve the best performance. We use PSNR, SSIM [36], and CIEDE2000 [31] to evaluate the quality of each recovered image.

Table 1 summarizes the quantitative results on the synthetic NYU & Make3D test dataset. The proposed approach performs favorably against state-of-the-art methods, where the average PSNR by our method is at least 0.92 dB higher than those by other image dehazing methods.

Table 2. Quantitative evaluations with state-of-the-art methods on the indoor scenes of SOTS dataset [19].

	DCP [13]	Nonlocal [41]	MSCNN [28]	DehazeNet [6]	AOD-Net [18]	GFN [29]	DCPDN [39]	DualCNN [26]	EPDN [27]	GDN [22]	Ours
PSNR↑	16.61	17.30	19.84	19.82	20.51	24.91	15.85	22.25	25.06	32.16	**32.68**
SSIM↑	0.8546	0.7768	0.8327	0.8209	0.8162	0.9186	0.8175	0.8751	0.9232	0.9836	0.9760

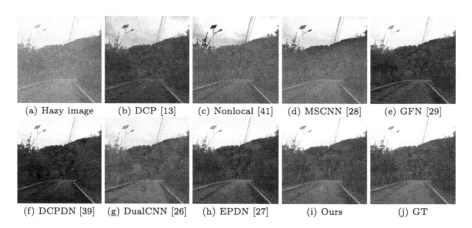

(a) Hazy image　　(b) DCP [13]　　(c) Nonlocal [41]　　(d) MSCNN [28]　　(e) GFN [29]

(f) DCPDN [39]　　(g) DualCNN [26]　　(h) EPDN [27]　　(i) Ours　　(j) GT

Fig. 3. Image dehazing results from the SOTS dataset. The colors of the dehazed images in (b), (e), (f), and (h) look darker than those of the ground truth image. The dehazed images in (c), (d), and (g) contain haze residual. The proposed method generates a much clearer image that is visually close to the ground truth image (best viewed on high-resolution display with zoom-in). (Color figure online)

Figure 2 shows the visual comparison results on the synthetic NYU & Make3D test dataset by the evaluated methods. The results by the statistical priors-based methods [13,41] contain artifacts in the regions of the sky. The methods by [6,28] develop CNNs to estimate the transmission and then use the conventional method [13] to estimate the clear images. However, these approaches do not correct the errors that are caused by the imperfect atmospheric light estimation. In [29], Ren et al. develop a gated neural network for image dehazing. However, as this method uses hand-crafted features to constrain the network, the quality of dehazed results is limited by these hand-crafted features. We note that both DCPDN [39] and EPDN [27] methods develop end-to-end trainable networks for image dehazing. However, the DCPDN method does not remove the haze from the input image, while the EPDN method generates the result with obvious color distortions. In contrast, the proposed approach explicitly considers the physics model of the haze process to discriminatively learns the useful information in the feature space, which accordingly generates high-quality images.

Then, we evaluate the proposed approach against state-of-the-art methods on the SOTS dataset [19]. For fair comparisons, we retrain the proposed method on the training dataset by [19] according to their protocols. Table 2 shows the

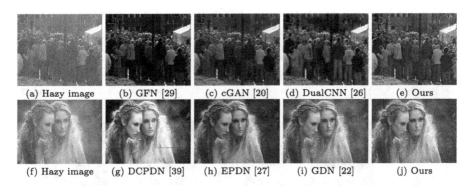

Fig. 4. Image dehazing results on real examples. The colors of the dehazed images in (b), (c), and (h) look darker. The part enclosed in the red box of (d) contains significant artifacts, e.g., the coat. The result in (g) has severe color distortions as shown in the red box. The image in (i) still contains some haze residuals. The proposed approach generates much clearer images (best viewed on high-resolution display with zoom-in). (Color figure online)

evaluation results on the indoor scenes of the SOTS dataset, where the proposed approach performs favorably against state-of-the-art methods in terms of PSNR and SSIM.

Figure 3 shows the dehazed results of the outdoor scene on the SOTS dataset by the evaluated methods. The results by [13,41] have severe color distortions in the sky. The dehazed images generated by [26,28] still have significant haze residuals. The results obtained by [29,39] look too dark. In contrast, the proposed method recovers a clearer image than state-of-the-art methods. More experimental results on the datasets [1–3,19] are included on our project webpage.

Real Examples. We further evaluate the proposed approach using real images in Fig. 4. The state-of-the-art methods [20,27,29] tend to over-estimate the colors of the restored images as shown in Fig. 4(b), (c), and (h). The method [26] generates the result with significant artifacts as shown in Fig. 4(d). The result by the method [22] still contains the haze residual. In contrast, our approach generates much clearer and brighter images than those by the state-of-the-art methods as shown in Fig. 4(e) and (j). More visual comparisons on real-world images are included on our project webpage.

5   Analysis and Discussions

We have shown that using PFDB is able to remove haze and outperforms state-of-the-art methods. To better understand the proposed approach, we perform further analysis and compare with related methods.

Effectiveness of FDU. As the proposed PFDB consists of an FDU and residual learning architecture, one may wonder whether the performance gains merely

Table 3. Effectiveness of FDU on image dehazing on the synthetic NYU & Make3D test dataset. RBs denotes ResBlocks.

Methods	w/o FDU & w/ 9RBs	w/o FDU & w/ 12RBs	w/o FDU & w/ 15RBs	Ours
Avg. PSNRs	26.33	26.49	26.75	**27.26**
Avg. SSIMs	0.8851	0.8897	0.8956	**0.9047**

    (a)   (b)   (c)   (d)   (e)

Fig. 5. Effectiveness of FDU on image dehazing. (a) Hazy image. (b)–(d) denote the results by the baseline methods w/o FDU & w/ 9RBs, w/o FDU & w/ 12RBs, and w/o FDU & w/ 15RBs, respectively. (e) Our result. The methods without using FDU generate the results with obvious haze residual as shown in (b)–(d). In contrast, the proposed approach with the FDU generates a much clearer image in (e).

come from the use of residual learning architecture [14]. To answer this question, we remove the FDU from our network architecture and train this baseline method using the same settings for fair comparisons. We note that the proposed method without using FDU reduces to the one that directly uses the features $\mathbf{Y}$ for the clear image reconstruction in (8) (i.e., "w/o FDU & w/ 9RBs" in Table 3). The comparisons in Table 3 demonstrate that it is more effective to explicitly consider the physics model in the feature space and use the FDU to discriminatively learn useful features from $\mathbf{Y}$, which generates higher-quality images than directly using the features $\mathbf{Y}$.

In addition, as each FDU contains three convolutional layers, one may also wonder whether using more ResBlocks [14] instead of FDU can generate better results. To answer this question, we remove the FDU from our network architecture and adopt more ResBlocks to train the baseline method using the same settings for fair comparisons. Specifically, the baseline models contain 9, 12, and 15 ResBlocks, respectively, which are denoted as "w/o FDU & w/9 RBs", "w/o FDU & w/12 RBs", and "w/o FDU & w/15 RBs" in Table 3. Table 3 shows that using more ResBlocks does not significantly improve the performance. In contrast, the proposed method with FDU performs better than purely stacking ResBlocks due to the discriminatively learned features by the FDU.

The visual comparison results in Fig. 5(b)–(e) further demonstrate the benefit of using the FDU in generating clearer images. We note that the proposed method without using FDU does not effectively remove haze. The generated

(a)   (b)   (c)   (d)   (e)

Fig. 6. Visualizations of the features from the proposed approach. (a) Hazy images. (b) and (c) denote the visualization of the features learned before and after FDU. (d) Our dehazed results. (e) Ground truth images. The principal components of the features are mapped to the principal components of the RGB space for visualization. The features learned after FDU (i.e., $\widetilde{\mathbf{J}}$) have a better color contrast than those before FDU, indicating that the proposed FDU is able to help haze removal (best viewed on high-resolution display with zoom-in). (Color figure online)

results still contain significant haze residual (Fig. 5(b)–(d)). In contrast, the proposed method with FDU generates a much clearer image in Fig. 5(e).

As the proposed FDU performs image dehazing in the feature space, one may also wonder how it generates useful features for haze removal. To better demonstrate the effect of the proposed FDU intuitively, we show some intermediate features from the proposed approach. As the feature space is more complex than the RGB space, we map the principal components of the intermediate features to the principal components of the RGB space according to the visualization method [21]. Figure 6 shows that the features learned after FDU (i.e., $\widetilde{\mathbf{J}}$ in (8)) have a better color contrast than those learned before FDU, suggesting that the proposed FDU is able to remove haze and thus facilitates the clear image restoration. All the above results demonstrate that the proposed FDU is able to help image dehazing.

Effectiveness of Residual Learning in PFDB. As stated in Sect. 3.2, we use residual learning in PFDB to increase the accuracy of image dehazing and ease the training of the deep CNNs. To illustrate the effect of residual learning, we compare with the proposed methods without using residual learning in PFDB. We retrain this baseline method and use the same settings as the proposed approach for fair comparisons. Both Table 4 and Fig. 7 show that using residual learning in PFDB is able to generate high-quality images.

Table 4. Effectiveness of residual learning and GAP on image dehazing on the synthetic NYU & Make3D test dataset.

	w/o residual learning	w/o GAP	Attention-based method [40]	Ours
Avg. PSNRs	23.55	26.56	26.64	**27.26**
Avg. SSIMs	0.7935	0.8973	0.8846	**0.9047**

(a)   (b)   (c)   (d)   (e)

Fig. 7. Effectiveness of the proposed PFDB for discriminatively learning features on image dehazing. (a) Hazy image. (b)–(d) denote the results generated by the proposed method without using residual learning, the proposed method without using GAP, and the feature learning method based on the attention mechanism [40], respectively. (e) Our result. The parts enclosed in red boxes in (b)–(c) contain significant artifacts, while the part enclosed in the red box in (d) still contains some haze residual. In contrast, the proposed approach generates a much clearer image in (e) (best viewed on high-resolution display with zoom-in). (Color figure online)

Effectiveness of GAP in PFDB. As stated in Sect. 3.1, the proposed FDU uses the GAP operation to maintain the most important information for the features that are related to F_1 and KA_t. To demonstrate the effectiveness of the GAP, we compare with the proposed network without using the GAP operation and evaluate this baseline method on the synthetic NYU & Make3D test dataset. Table 4 indicates that using the GAP operation is able to maintain useful information and thus facilitates image dehazing. Although the method without using GAP can also estimate these features, it does not significantly improve the performance. In addition, the comparisons in Fig. 7(c) and (e) demonstrate that using GAP is able to remove artifacts.

Relations with Attention-based Methods. Recently, the attention mechanism has been used to solve image super-resolution [40]. This method develops the channel attention strategy to learn useful features for image super-resolution. In contrast, our approach learns the features using FDU which is based on the physics model of the haze process in a feature space. To further demonstrate the effectiveness of the proposed method, we retain the attention-based method [40] using the proposed training dataset in the same settings for fair comparisons. We evaluate the proposed method against this method on the synthetic NYU & Make3D test dataset. Table 4 shows that directly using the attention mechanism does not always facilitate haze removal. In contrast, explicitly considering the physics model of the haze process in the feature space is able to generate high-quality images. The comparison results in Fig. 7(d) and (e) further demonstrate that the proposed approach is more effective for haze removal than directly using the attention mechanism.

Relations with Deep Physics Model-based Methods. We note that several notable methods [12,18,25,26,38,39] use the physics model of the haze process to constrain the deep neural network for image dehazing. The DualCNN method [26] develops a network based on two branches to estimate the transmission and atmospheric light. In [38], Yang et al. develop a disentangled dehazing network based on the physics model of the haze process to solve image dehazing using unpaired images. Zhang et al. [39] develop a new dense network for image dehazing based on the physics model (1). This method first uses the deep CNNs to estimate the transmission and atmospheric light and then reconstructs clear images based on (1). Similar to [39], Li et al. [18] estimate the clear image based on a re-formulated atmospheric scattering model. Guo et al. [12] use different CNN models to separately estimate the transmission map, atmospheric light, and the latent clear image. Then the final results are generated based on the physics model. As these methods mainly consider the physics model in the raw image space and do not fully explore the physics information in the feature space, the final estimated dehazed images contain haze residual and artifacts as shown in Figs. 2, 3 and 4. In contrast, the proposed approach develops the FDU to explicitly consider the physics model in a feature space, which is able to effectively learn the useful features for image dehazing. Thus, the haze can be well removed and textures of the images are well recovered (see both quantitative and qualitative evaluations in Tables 1 and 2 and Figs. 2, 3 and 4).

Analysis on the Number of PFDBs. The proposed network contains several PFDBs. We further evaluate the effect of the number of PFDBs by setting the number of PFDBs from 7 to 17. Table 5 shows that using more PFDBs does not significantly improve the performance. We empirically use 9 PFDBs as a trade-off between accuracy and efficiency.

Table 5. Ablation study on the number of PFDBs using the synthetic NYU & Make3D test dataset.

Number of PFDBs	7 PFDBs	9 PFDBs	13 PFDBs	15 PFDBs	17 PFDBs
Avg. PSNRs	27.17	27.26	27.32	27.58	27.33
Avg. SSIMs	0.9050	0.9047	0.9068	0.9082	0.9080

Model Size. We evaluate the model size of the proposed approach against state-of-the-art methods. Table 6 shows that the proposed approach has competitive model parameters against state-of-the-art methods.

Table 6. Comparisons of model sizes against the state-of-the-art methods.

Methods	DCPDN [39]	cGAN [20]	GDN [22]	EPDN [27]	Ours
Model size	134M	140M	1M	17M	12M

6 Conclusions

We have presented an effective PFDN for image dehazing. The critical component PFDB consists of an FDU with a residual learning architecture. The FDU is developed to fully explore the useful features for image dehazing based on the physics model of the haze process. The residual learning architecture is applied to the FDU to increase the accuracy and ease the training of deep neural networks. The proposed PFDB is embedded as a backbone into an encoder and decoder network architecture in an end-to-end fashion for image dehazing. We have analyzed the effect of the proposed PFDN on image dehazing. Both quantitative and qualitative results show that the proposed approach performs favorably against state-of-the-art methods.

References

1. Ancuti, C.O., Ancuti, C., Sbert, M., Timofte, R.: Dense-Haze: a benchmark for image dehazing with dense-haze and haze-free images. In: IEEE ICIP, pp. 1014–1018 (2019)
2. Ancuti, C.O., Ancuti, C., Timofte, R., De Vleeschouwer, C.: O-HAZE: a dehazing benchmark with real hazy and haze-free outdoor images. In: IEEE CVPR Workshops, pp. 754–762 (2018)
3. Ancuti, C.O., Ancuti, C., Timofte, R., Vleeschouwer, C.D.: I-HAZE: a dehazing benchmark with real hazy and haze-free indoor images. In: International Conference on Advanced Concepts for Intelligent Vision Systems, pp. 620–631 (2018)
4. Ancuti, C., Ancuti, C.O., Timofte, R.: NTIRE 2018 Challenge on image dehazing: methods and results. In: IEEE CVPR Workshops, pp. 891–901 (2018)
5. Berman, D., Treibitz, T., Avidan, S.: Non-local image dehazing. In: IEEE CVPR, pp. 1674–1682 (2016)
6. Cai, B., Xu, X., Jia, K., Qing, C., Tao, D.: Dehazenet: an end-to-end system for single image haze removal. IEEE TIP **25**(11), 5187–5198 (2016)
7. Chen, C., Do, M.N., Wang, J.: Robust image and video dehazing with visual artifact suppression via gradient residual minimization. In: ECCV, pp. 576–591 (2016)
8. Chen, D., et al.: Gated context aggregation network for image dehazing and deraining. In: IEEE WACV, pp. 1375–1383 (2019)
9. Fattal, R.: Single image dehazing. ACM TOG **27**(3), 72:1–72:9 (2008)
10. Fattal, R.: Dehazing using color-lines. ACM TOG **34**(1), 13:1–13:14 (2014)
11. Galdran, A., Vazquez-Corral, J., Pardo, D., Bertalmío, M.: Enhanced variational image dehazing. SIAM J. Imag. Sci. **8**(3), 1519–1546 (2015)
12. Guo, T., Li, X., Cherukuri, V., Monga, V.: Dense scene information estimation network for dehazing. In: IEEE CVPR Workshops (2019)
13. He, K., Sun, J., Tang, X.: Single image haze removal using dark channel prior. IEEE TPAMI **33**(12), 2341–2353 (2011)

14. He, K., Zhang, X., Ren, S., Sun, J.: Deep residual learning for image recognition. In: IEEE CVPR, pp. 770–778 (2016)
15. Isola, P., Zhu, J.Y., Zhou, T., Efros, A.A.: Image-to-image translation with conditional adversarial networks. In: IEEE CVPR, pp. 5967–5976 (2017)
16. Kingma, D.P., Ba, J.: Adam: a method for stochastic optimization. In: ICLR (2015)
17. Lee, G.H., Alvarez-Melis, D., Jaakkola, T.S.: Towards robust, locally linear deep networks. In: ICLR (2019)
18. Li, B., Peng, X., Wang, Z., Xu, J., Feng, D.: AOD-Net: all-in-one dehazing network. In: IEEE ICCV, pp. 4780–4788 (2017)
19. Li, B., et al.: Benchmarking single-image dehazing and beyond. IEEE TIP **28**(1), 492–505 (2019)
20. Li, R., Pan, J., Li, Z., Tang, J.: Single image dehazing via conditional generative adversarial network. In: IEEE CVPR, pp. 8202–8211 (2018)
21. Liu, C., Yuen, J., Torralba, A.: SIFT flow: dense correspondence across scenes and its applications. IEEE TPAMI **33**(5), 978–994 (2011)
22. Liu, X., Ma, Y., Shi, Z., Chen, J.: GridDehazeNet: attention-based multi-scale network for image dehazing. In: IEEE ICCV (2019)
23. Mei, K., Jiang, A., Li, J., Wang, M.: Progressive feature fusion network for realistic image dehazing. In: ACCV (2018)
24. Montufar, G.F., Pascanu, R., Cho, K., Bengio, Y.: On the number of linear regions of deep neural networks. In: NIPS, pp. 2924–2932 (2014)
25. Pan, J., et al.: Physics-based generative adversarial models for image restoration and beyond. IEEE TPAMI (2020)
26. Pan, J., et al.: Learning dual convolutional neural networks for low-level vision. In: IEEE CVPR, pp. 3070–3079 (2018)
27. Qu, Y., Chen, Y., Huang, J., Xie, Y.: Enhanced pix2pix dehazing network. In: IEEE CVPR (2019)
28. Ren, W., Liu, S., Zhang, H., Pan, J., Cao, X., Yang, M.H.: Single image dehazing via multi-scale convolutional neural networks. In: ECCV, pp. 154–169 (2016)
29. Ren, W., et al.: Gated fusion network for single image dehazing. In: IEEE CVPR, pp. 3253–3261 (2018)
30. Saxena, A., Sun, M., Ng, A.Y.: Make3D: learning 3D scene structure from a single still image. IEEE TPAMI **31**(5), 824–840 (2009)
31. Sharma, G., Wu, W., Dalal, E.N.: The CIEDE2000 color-difference formula: implementation notes, supplementary test data, and mathematical observations. Color Res. Appl. **30**(1), 21–30 (2005)
32. Silberman, N., Hoiem, D., Kohli, P., Fergus, R.: Indoor segmentation and support inference from RGBD images, In: ECCV. pp. 746–760 (2012)
33. Tan, R.T.: Visibility in bad weather from a single image. In: IEEE CVPR (2008)
34. Tang, K., Yang, J., Wang, J.: Investigating haze-relevant features in a learning framework for image dehazing. In: IEEE CVPR, pp. 2995–3002 (2014)
35. Wang, F., et al.: Residual attention network for image classification. In: IEEE CVPR, pp. 6450–6458 (2017)
36. Wang, Z., Bovik, A.C., Sheikh, H.R., Simoncelli, E.P.: Image quality assessment: from error visibility to structural similarity. IEEE TIP **13**(4), 600–612 (2004)
37. Yang, D., Sun, J.: Proximal dehaze-net: a prior learning-based deep network for single image dehazing. In: ECCV, pp. 729–746 (2018)
38. Yang, X., Xu, Z., Luo, J.: Towards perceptual image dehazing by physics-based disentanglement and adversarial training. In: AAAI, pp. 7485–7492 (2018)
39. Zhang, H., Patel, V.M.: Densely connected pyramid dehazing network. In: IEEE CVPR, pp. 3194–3203 (2018)

40. Zhang, Y., Li, K., Li, K., Wang, L., Zhong, B., Fu, Y.: Image super-resolution using very deep residual channel attention networks. In: ECCV, pp. 294–310 (2018)
41. Zhang, Y., Li, K., Li, K., Zhong, B., Fu, Y.: Residual non-local attention networks for image restoration. In: ICLR (2019)
42. Zhu, Q., Mai, J., Shao, L.: A fast single image haze removal algorithm using color attenuation prior. IEEE TIP **24**(11), 3522–3533 (2015)

Learning Surrogates via Deep Embedding

Yash Patel[(✉)], Tomáš Hodaň, and Jiří Matas

Visual Recognition Group, Czech Technical University in Prague,
Czech Republic, Czechia
{patelyas,hodanto2,matas}@fel.cvut.cz

Abstract. This paper proposes a technique for training a neural network by minimizing a surrogate loss that approximates the target evaluation metric, which may be non-differentiable. The surrogate is learned via a deep embedding where the Euclidean distance between the prediction and the ground truth corresponds to the value of the evaluation metric. The effectiveness of the proposed technique is demonstrated in a post-tuning setup, where a trained model is tuned using the learned surrogate. Without a significant computational overhead and any bells and whistles, improvements are demonstrated on challenging and practical tasks of scene-text recognition and detection. In the recognition task, the model is tuned using a surrogate approximating the edit distance metric and achieves up to 39% relative improvement in the total edit distance. In the detection task, the surrogate approximates the intersection over union metric for rotated bounding boxes and yields up to 4.25% relative improvement in the F_1 score.

1 Introduction

Supervised learning of a neural network involves minimizing a differentiable loss function on annotated data. The differentiable nature of the loss function and the network architecture allows the model weights to be updated via backpropagation [53]. The performance on a wide range of computer vision tasks have significantly improved thanks to the progress in deep neural network architectures [21,30,56] and the introduction of large scale supervised datasets [8,35]. As designing architectures often demands detailed domain expertise and creating new datasets is expensive, there has been a substantial effort in automating the process of designing better task-specific architectures [10,54,65] and employing self-supervised methods of learning to reduce the dependence on human-annotated data [7,12,14]. However, little attention has been paid to automate the process of designing the loss functions.

For many practical problems in computer vision, models are trained with simple proxy losses, which may not align with the evaluation metric. The evaluation metric may not always be differentiable, prohibiting its use as a loss function. An example of a non-differentiable metric is the visible surface discrepancy (VSD) [23] used to evaluate 6D object pose estimation methods. Another example is edit distance (ED) defined by counting unit operations (addition,

© Springer Nature Switzerland AG 2020
A. Vedaldi et al. (Eds.): ECCV 2020, LNCS 12375, pp. 205–221, 2020.
https://doi.org/10.1007/978-3-030-58577-8_13

deletion, and substitution) necessary to transform one text string into another and is a common choice for evaluating scene text recognition methods [26,27,43]. Since ED is non-differentiable, the methods use either CTC [17] or per-character cross-entropy [3] as the proxy loss. Yet another popular non-differentiable metric is the intersection over union (IoU) used to compare the predicted and the ground truth bounding boxes when evaluating object detection methods. Although these methods typically resort to using proxy losses such as $smooth$-L_1 [50] or L_2 [49], Rezatofighi $et\ al.$ [51] demonstrate that there is no strong correlation between L_n objectives and IoU. Further, Yu $et\ al.$ [62] show that IoU accounts for a bounding box as a whole whereas regressing using an L_n proxy loss treats each point independently.

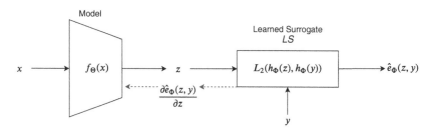

Fig. 1. For the input x with the corresponding ground-truth y, the model being trained outputs $z = f_\Theta(x)$. The learned surrogate provides a differentiable approximation of the evaluation metric: $\hat{e}_\Phi(z,y) = L_2(h_\Phi(z), h_\Phi(y))$, where h_Φ is a learned deep embedding model, and $h_\Phi(z)$ and $h_\Phi(y)$ are embedding representations for the prediction and the ground truth, respectively. Model $f_\Theta(x)$ for the target task ($e.g.$ scene text recognition or detection) is trained with the gradients from the surrogate: $\frac{\partial(\hat{e}_\Phi(z,y))}{\partial z}$.

For popular metrics such as IoU, hand-crafted differentiable approximations have been designed [51,62]. However, hand-crafting a surrogate is not scalable as it requires domain expertise and may involve task-specific assumptions and simplifications. The IoU-loss introduced in [51,62] allows for optimization on the evaluation metric directly but makes a strong assumption about the bounding boxes to be axis-aligned. In numerous practical applications such as aerial image object detection [60], scene text detection [26] and visual object tracking [29], the bounding boxes may be rotated and the methods for such tasks revert to using simple but non-optimal proxy loss functions such as $smooth$-L_1 [2,6,40].

To address the aforementioned issues, this paper proposes to learn a differentiable surrogate that approximates the evaluation metric and use the learned surrogate to optimize the model for the target task. The metric is approximated via a deep embedding, where the Euclidean distance between the prediction and the ground truth corresponds to the value of the metric. The mapping to the embedding space is realized by a neural network, which is learned using only the value of the metric. Gradients of this value with respect to the inputs are not required for learning the surrogate. In fact, the gradients may not even exist, as is the case of the edit distance metric. Throughout this paper, we refer to the proposed method for training with learned surrogates as "LS". Figure 1 provides an overview of the proposed method.

In this paper, the focus on a post-tuning setup, where a model that has converged on a proxy loss is tuned with LS. We consider two different optimization tasks: post-tuning with a learned surrogate for the edit distance (LS-ED) and the IoU of rotated bounding boxes (LS-IoU). To the best of our knowledge, we are the first to optimize directly on these evaluation metrics.

The rest of the paper is structured as follows. Related work is reviewed in Sect. 2, the technique for learning the surrogate and training with it is presented in Sect. 3, experiments are shown in Sect. 4 and the paper is concluded in Sect. 5.

2   Related Work

Training machine learning models by directly minimizing the evaluation metric has been shown effective on various tasks. For example, the state-of-the-art learned image compression [4,33] and super-resolution [9,32] methods directly optimize the perceptual similarity metrics such as MS-SSIM [59] and the peak signal-to-noise ratio (PSNR). Certain compression methods optimize on an approximate of human perceptual similarity, which is learned in a supervised manner using annotated data [44,45]. Image classification methods [21,30,56] are typically trained with the cross-entropy loss that has been shown to align well with the misclassification rate, *i.e.* the evaluation metric, under the assumption of large scale and clean data [5,31].

When designing evaluation metrics for practical computer vision tasks, the primary goal is to fulfil the requirements of potential applications and not to ensure the metrics being amenable to an optimization approach. As a consequence, many evaluation metrics are non-differentiable and cannot be directly minimized by the currently popular gradient-descent optimization approaches. For example, the visible surface discrepancy [23], which is used to evaluate 6D object pose estimation methods, was designed to be invariant under pose ambiguity. This is achieved by calculating the error only over the visible part of the object surface, which requires a visibility test that makes the metric non-differentiable. Another example is the edit distance metric [15,26], which is used to evaluate scene text recognition methods and is calculated via dynamic programming, which makes it infeasible to obtain the gradients.

There have been efforts towards approximating non-differentiable operations in a differentiable manner to enable end-to-end training. Kato *et al.* [28] proposed a neural network to approximate rasterization, allowing for a direct optimization on IoU for 3D reconstruction. Agustsson *et al.* [1] proposed a soft-to-hard vector quantization mechanism. It is based on soft cluster assignments during back-propagation, which allows neural networks to learn tasks involving quantization, *e.g.* the image compression. Our work differs as we propose a general approach to approximate the evaluation metric, instead of approximating task-specific building blocks of neural networks.

Another line of research has focused on hand-crafting differentiable approximates of the evaluation metrics, which either align better with the metrics or enable training on them directly. Prabhavalkar *et al.* [46] proposed a way of optimizing attention based speech recognition models directly on word error rate. As mentioned earlier, [51,62] proposed ways for directly optimizing on intersection-over-union (IoU) as the loss for the case of axis-aligned bounding boxes. Rahman *et al.* [48] proposed a hand-crafted approximation of IoU for semantic segmentation.

Learning task-specific surrogates has been attempted. Nagendra *et al.* [42] demonstrated that learning the approximate of IoU leads to better performance in the case of semantic segmentation. However, the method requires custom operations to estimate true and false positives, and false negatives, which makes the learning approach task-specific. Engilberge *et al.* [11] proposed a learned surrogate for sorting-based tasks such as cross-modal retrieval, multi-label image classification and visual memorability ranking. Their results on sorting-based tasks suggest that learning the loss function could outperform hand-crafted losses.

More closely related to our work is the direct loss method by Hazan *et al.* [20] where a surrogate loss is minimized by embedding the true loss as a correction term. Song *et al.* [57] extended this approach to the training of neural networks. However, it assumes that the loss can be disentangled into per-instance sub-losses, which is not always feasible, *e.g.* the F_1 score [16] involves two non-decomposable functions (recall and precision). An alternative is to directly learn the amount of update values that are applied to the parameters of the prediction model. The framework proposed in [34] includes a controller that uses per-parameter learning curves comprised of the loss values and derivatives of the loss with respect to each parameter. The method suffers from two drawbacks that prohibit its direct application to training on evaluation metrics: a) for large networks, it is computationally infeasible to store the learning curve of every parameter, and b) no gradient information is available for non-differentiable losses.

Our work is similar to the approach by Grabocka *et al.* [16], where the evaluation metric is approximated by a neural network. Their approach differs as the network learning the surrogate takes both the prediction and the ground truth as the input and directly regresses the value of the metric. Since we formulate the task as embedding learning and train the surrogate such that the L_2 in the embedded space corresponds to the metric, our method ensures that the gradients are smaller when the prediction is closer to the ground truth. Furthermore, as illustrated in Sect. 3, we learn the surrogate with an additional gradient penalty term to ensure that the gradients obtained from our learned surrogate are bounded for stable training.

3   Learning Surrogates via Deep Embedding

Say that the supervised task is being learned from samples drawn uniformly from a distribution $(x, y) \sim P_D$. For a given input x and an expected output y, a neural network model outputs $z = f_\Theta(x)$ where Θ are the model parameters learned via backpropagation as:

$$\Theta_{t+1} \leftarrow \Theta_t - \eta \frac{\partial l(z, y)}{\partial \Theta_t} \tag{1}$$

where $l(z, y)$ is a differentiable loss function, t is the training iteration, and η is the learning rate.

The model trained with loss $l(z, y)$ is evaluated using metric $e(z, y)$. When metric $e(z, y)$ is differentiable, it can be directly used as the loss. The technique proposed in this paper addresses the cases when metric $e(z, y)$ is non-differentiable by learning a differentiable surrogate loss denoted as $\hat{e}_\Phi(z, y)$. The learned surrogate is realized by a neural network, which is differentiable and is used to optimize the model. The weight updates are:

$$\Theta_{t+1} \leftarrow \Theta_t - \eta \frac{\partial \hat{e}_\Phi(z, y)}{\partial \Theta_t} \tag{2}$$

3.1   Definition of the Surrogate

The surrogate is defined via a learned deep embedding h_Φ where the Euclidean distance between the prediction z and the ground truth y corresponds to the value of the evaluation metric:

$$\hat{e}_\Phi(z, y) = \|h_\Phi(z) - h_\Phi(y)\|_2 \tag{3}$$

3.2   Learning the Surrogate

Learning the surrogate, *i.e.* approximating the evaluation metric, with a deep neural network is formulated as a supervised learning task requiring three major components: a model architecture, a loss function, and a source of training data.

Architecture. In this paper, the architecture is designed manually, such that it is suitable for the nature of the inputs z and y (details are in Sect. 4). Modern approaches for architecture search, *e.g.* [10,54,65], could yield better results but are computationally expensive.

Training Loss. The surrogate is learned with the following objectives:

1. The learned surrogate corresponds to the value of the evaluation metric:

$$\hat{e}_\Phi(z, y) \approx e(z, y) \tag{4}$$

2. The first order derivative of the learned surrogate with respect to the prediction z is close to 1:

$$\left\| \frac{\partial \hat{e}_\Phi(z, y)}{\partial z} \right\|_2 \approx 1 \tag{5}$$

Both objectives are realized and linearly combined in the training loss:

$$\text{loss}(z, y) = \left\| (\hat{e}_\Phi(z, y) - e(z, y)) \right\|_2^2 + \lambda \left(\left\| \frac{\partial \hat{e}_\Phi(z, y)}{\partial z} \right\|_2 - 1 \right)^2 \tag{6}$$

Bounding the gradients (Eq. 5) has shown to enhance the training stability for Generative Adversarial Networks [18] and has shown to be useful for learning the surrogate. Parameters Φ of the embedding model h_Φ are learned by minimizing the loss (Eq. 6).

Source of Training Data. Source of the training data for learning the surrogate determines the quality of the approximation over the domain. The model $f_\Theta(x) = z$ for the supervised task is trained on samples obtained from a dataset D. Let us assume that R is a random data generator providing examples for the learning of the surrogate, sampled uniformly in the range of the evaluation metric (see Sect. 4 for details). Note that R is independent of $f_\Theta(x)$.

Three possibilities for the data source are considered:

1. *Global approximation:* $(z, y) \sim P_R$.
2. *Local approximation:* $(z, y) \sim P_{f_\Theta(x)}$, where $(x, y) \sim P_D$.
3. *Local-global approximation:* $(z, y) \sim P_{f_\Theta(x) \cup R}$.

The local-global approximation yields a high quality of both the approximation and gradients (Sect. 4.1) and is therefore used in the main experiments.

3.3   Training with the Learned Surrogate

The learned surrogate is used in a post-tuning setup, where model $f_\Theta(x)$ has been pre-trained using a proxy loss. This setup ensures that $f_\Theta(x)$ is not generating random outputs and thus simplifies post-tuning with the surrogate. The parameters of the surrogate Φ are initialized randomly.

Learning of the surrogate $\hat{e}_\Phi$ and post-tuning of the model $f_\Theta(x)$ are conducted alternatively. The surrogate parameters Φ are updated first while the model parameters Θ are fixed. The surrogate is learned by sampling (z, y) jointly from the model and the random generator. Subsequently, the model parameters are trained while the surrogate parameters are fixed. Algorithm 1 demonstrates the overall training procedure.

4   Experiments

The efficacy of LS is demonstrated on two different tasks: post-tuning with a learned surrogate for the edit distance (Sect. 4.2) and for the IoU of rotated bounding boxes (Sect. 4.3). This section provides details of the models for these tasks, design choices for learning the surrogates and empirical evidence showing the efficacy of LS. Unless stated otherwise, the results were obtained using the local-global approximation setup as elaborated in Algorithm 1.

Algorithm 1. Training with LS *(local-global approximation)*

Inputs: Supervised data D, random data generator R, evaluation metric e.
Hyper-parameters: Number of update steps I_a and I_b, learning rates η_a and η_b, number of epochs E.
Objective: Train the model for a given task that is $f_\Theta(x)$ and the surrogate ,*i.e.*, e_Φ.

1: *Initialize* $\Theta \leftarrow$ pre-trained weights, $\Phi \leftarrow$ random weights.
2: **for** epoch $= 1,...,$E **do**
3:     **for** i $= 1,...,I_a$ **do**
4:         sample $(x,y) \sim P_D$, sample $(z_r, y_r) \sim P_R$
5:         inference $z = f_{\Theta_{epoch-1}}(x)$
6:         compute loss $l_{\hat{e}} = loss(z,y) + loss(z_r, y_r)$ (Equation 6)
7:         $\Phi^i \leftarrow \Phi^{i-1} - \eta_a \frac{\partial l_{\hat{e}}}{\partial \Phi^{i-1}}$
8:     **end for**
9:     $\Phi \leftarrow \Phi^{I_a}$
10:    **for** i $= 1,...,I_b$ **do**
11:        sample $(x,y) \sim P_D$
12:        inference $z = f_{\Theta^{i-1}}(x)$
13:        compute loss $l_f = \hat{e}_{\Phi_{epoch}}(z,y)$ (Equation 3)
14:        $\Theta^i \leftarrow \Theta^{i-1} - \eta_b \frac{\partial(l_f)}{\partial \Theta^{i-1}}$
15:    **end for**
16:    $\Theta \leftarrow \Theta^{I_b}$
17: **end for**

4.1 Analysing the Learned Surrogates

The aspects considered for evaluating the surrogates are:

1. The quality of approximation $\hat{e}_\Phi(z,y)$.
2. The quality of gradients $\frac{\partial(\hat{e}_\Phi(z,y))}{\partial z}$.

Both the quality of the approximation and the gradients depend on three components: an architecture, a loss function, and a source of training data (Sect. 3.2). Given an architecture, the choices for the loss function to learn the surrogate and the training data are justified subsequently.

Quality of Approximation. The quality of the approximation is judged by comparing the value of the surrogate with the value of the evaluation metric, calculated on samples obtained from model $f_\Theta(x)$. When learning the surrogate, higher quality of approximation is enforced by the mean squared loss between $e(z,y)$ and $\hat{e}_\Phi(z,y)$ (the first term on the right-hand side of Eq. 6). Figure 2 (left) shows the quality of the approximation measured by the L_1 distance between the learned surrogate and the edit distance. It can be seen that the surrogate approximates the edit distance accurately (the L_1 distance drops swiftly below 0.2, which is negligible for the edit distance).

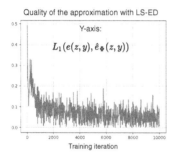
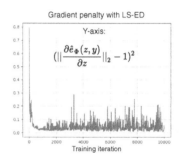

Fig. 2. Left: The error in approximation for the first $10K$ training iterations. The error is obtained by computing the L_1 distance between the true edit distance values and the LS-ED predictions and dividing by the batch size. Note that the edit distance can only take non-negative integer values, thus the error in the range of 0–0.2 is fairly low. **Right**: The gradient penalty term from the optimization of the LS-ED model (Eq. 6).

Quality of Gradients. Judging the quality of gradients is more complicated. When learning the surrogate, the gradient-penalty term attempts to make the gradients bounded, *i.e.* to make the training stable (second term on the right-hand side of the Eq. 6). However, this is not sufficient if the gradients do not optimize $f_\Theta(x)$ on the evaluation metric. We rely on the improvement or the decline in the performance of the model $f_\Theta(x)$ to judge the quality of the gradients. Table 3 shows that the local-global approximation leads to the largest improvements when optimizing on IoU for rotated bounding boxes.

Choice of Training Data. Figure 3 shows the quality of approximation with different choices of training data for learning the surrogate. These empirical observations suggest that using global approximation leads to a low quality of the approximation. This can be accounted to the domain gap between the data obtained from the random generator and the model. Using the local approximation leads to a higher quality of the approximation, however, the gradients obtained from the surrogate are not useful to train $f_\Theta(x)$ (Table. 3), *i.e.* although the quality of the approximation is high, the quality of gradients is not. This can be attributed to surrogate over-fitting on samples obtained from the model and losing generalization capability on samples outside this distribution. Finally, it was observed that using the local-global approximation leads to both properties – high quality of approximation and high quality of gradients.

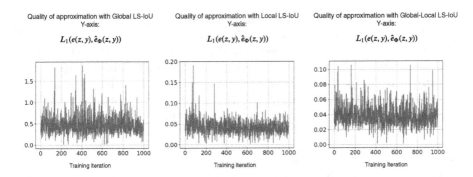

Fig. 3. The error in the approximation of the IoU for rotated bounding boxes is shown for the first $1K$ iterations of the training with LS-IoU. Error is measured by the L_1 distance between IoU and the surrogate. It can be seen that the error is high for the global and low for the local and global-local approximation variants.

4.2   Post-tuning with a Learned Surrogate for ED (LS-ED)

It is experimentally shown that LS can improve scene text recognition models (STR) on edit distance (ED), which is a popularly used metric to evaluate STR methods [26,27,43]. The empirical evidence shows that post-tuning STR models with LS-ED lead to improved performance on various metrics such as accuracy, normalized edit distance, and total edit distance [15].

Scene Text Recognition (STR). Given an input image of a cropped word, the task of STR is to generate the transcription of the word. The state-of-the-art architectures for scene text recognition can be factorized into four modules [3] (in this order): (a) transformation, (b) feature extraction, (c) sequence modelling, and (d) prediction. The feature extraction and prediction are the core modules of any STR model and are always employed. On the other hand, transformation and sequence modelling are not essential but have shown to improve the performance on benchmark datasets. Post-tuning with LS-ED is investigated for two different configurations of STR models.

The transformation module attempts to rectify the curved or tilted text, making the task easier for the subsequent modules of the model. It is learned jointly with the rest of the modules, and a popular choice is thin-plate spline (TPS) [25,36,55]. TPS can be either present or absent in the overall STR model.

The feature extraction module maps the image or its transformed version to a representation that focuses on the attributes relevant for character recognition, while the irrelevant features are suppressed. Popular choices include VGG-16 [56] and ResNet [21]. It is a core module of the STR model and is always present.

The features are the input of the sequence modelling module, which captures the contextual information within a sequence of characters for the next module to predict each character more robustly. BiLSTM [22] is a popular choice.

The output character sequence is predicted from the identified features of the image. The choice of the prediction module depends on the loss function used for training the STR model. Two popular choices of loss functions are CTC [17] (sigmoid output) or attention [55] (per-character softmax output).

Baek *et al.* [3] provides a detailed analysis of STR models and the impact of different modules on the performance. Following [3], LS-ED is investigated with the state-of-the-art performing configuration, which is *TPS-ResNet-BiLSTM-Attn*. To demonstrate the efficacy of LS-ED, results are also shown with *ResNet-BiLSTM-Attn*, *i.e.*, the transformation module is removed. Note that the CTC based prediction has been shown to consistently perform worse compared to the attention counter-part [3], and thus the analysis in this paper has been narrowed down to only the attention-based prediction.

Similar to [3], the STR models are trained on the union of the synthetic data obtained from MJSynth [24] and SynthText [19] resulting in a total of 14.4 million training examples. Furthermore, following the standard setup of [3], there is no fine-tuning performed in a dataset-specific manner before the final testing. Let us say that the STR model is $f_\Theta(x)$, such that $f_\Theta : \mathbb{R}^{100 \times 32 \times 1} \rightarrow \mathbb{R}^{|A| \times L}$. The dimensions of the input cropped word image x is fixed to $100 \times 32 \times 1$ (gray-scale). The output for attention based prediction module is a per-character softmax over the set of characters. Here L is the maximum length of characters in the word and $|A|$ is the number of characters. During inference, argmax is performed at each character location to output the predicted text string. The ground truth y is represented as a per-character one-hot vector.

The STR models are first trained with the proxy loss, *i.e.*, cross-entropy for $300K$ iterations with a mini-batch size of 192. The models are optimized using ADADELTA [63] (same setup as [3]). Once the training is completed these models are tuned with LS-ED on the same set of 14.4 million training examples for another $20K$ iterations. The models trained purely on the synthetic datasets are tested on a collection of real datasets - IIIT-5K [41], SVT [58], ICDAR'03 [38], ICDAR'13 [27], ICDAR'15 [26], SVTP [47] and CUTE [52] datasets.

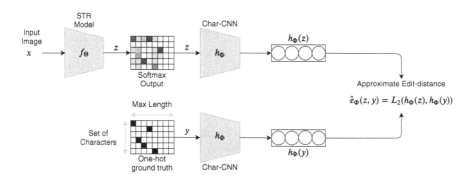

Fig. 4. Training scene text recognition (STR) models with LS-ED. The output of the STR model $z_{|A| \times L}$ and the ground-truth $y_{|A| \times L}$ (L is the maximum length of the word and A is the set of characters) are fed to the Char-CNN embedding model to obtain embedding vectors, $h_\Phi(z)$ and $h_\Phi(y)$ respectively. The approximate edit distance value is obtained by computing $\hat{e}_\Phi(z, y) = L_2(h_\Phi(z), h_\Phi(y))$.

LS-ED Architecture. Char-CNN architecture [64] is used for learning the deep embedding h_Φ. It consists of five $1D$ convolution layers equipped with LeakyReLU activation [61] followed by two fully connected layers. The embedding h_Φ maps the input such that $h_\Phi : \mathbb{R}^{|A| \times L} \rightarrow \mathbb{R}^{1024}$. Note that since h_Φ constitutes of convolution and fully-connected layers, it is differentiable and allows for backpropagation to the STR model. In feed-forward, the two embeddings for the ground-truth y (one-hot) and the model prediction z (softmax) are obtained by performing feed-forward through h_Φ and an approximate of edit distance is computed by measuring the L_2 between the two vectors (Fig. 4).

Post-tuning with LS-ED. A random generator is designed for this task, which generates a pair of words (z_r, y_r) and ensures uniform sampling in the range of the true error. It was observed that the uniform sampling is essential to avert over-fitting of the learned surrogate on a certain range of the true metric. For the edit distance metric $e(z, y) \in \{0, ..., b\}$ (b being the maximum possible value), the generator samples a word randomly from a text corpus and distorts the words by performing random addition, deletion, and substitution operations.

The post-tuning of the STR model $f_\Theta(x)$ with LS-ED follows Algorithm 1. For the case of the edit distance, there is a significant domain gap between the samples obtained from the STR model (z) and the random generator (z_r). This is because the random generator operates directly on the text string, *i.e.*, z_r is one-hot representation. Thus, using the global approximation setting yields a low quality of the approximation. Further, it was observed that training the surrogate purely with the data generated from the STR model, *i.e.*, local approximation, leads to a good approximation but does not lead to an improvement in the performance of the STR model, which indicates a low quality of gradients.

Finally as described in Algorithm 1, the local-global approximation is used. The quality of approximation and the gradient penalty from post-tuning with LS-ED are shown in Fig. 2. Note that the edit distance value is a whole number and the surrogate attempts to approximate it, thus the error in approximation as shown in Fig. 2 is low. The quality of the gradients can be seen by improvement in the performance of the STR models. Thus the local-global approximation guides to a high quality of both the approximation and gradients.

The results for the two configurations of STR models, *i.e.*, *ResNet-BiLSTM-Attn* and *TPS-ResNet-BiLSTM-Attn*, are shown in Table 1 and Table 2, respectively. It can be observed that LS-ED improves the performance of the STR models on all metrics. The most significant gains are observed on total-edit distance (TED) as the surrogate attempts to minimize its approximation.

4.3 Post-tuning with a Learned Surrogate for IoU (LS-IoU)

It is experimentally demonstrated that LS can optimize scene text detection models on intersection-over-union (IoU) for rotated bounding boxes. IoU is a popular metric used to evaluate the object detection [49,50] and scene text detection models [6,15,26,37,40]. Gradients for IoU can be hand-crafted for the case of axis-aligned bounding boxes [51,62], however, it is complex to design the gradients for rotated bounding boxes. The learned surrogate of IoU allows backpropagation for rotated bounding boxes. For the task of rotated scene text detection on ICDAR'15 [26], it is shown that post-tuning the text detection model with LS-IoU leads to improvement on recall, precision, and F_1 score.

Table 1. ResNet-BiLSTM-Attn: The models are evaluated on IIIT-5K [41], SVT [58], ICDAR'03 [38], ICDAR'13 [27], ICDAR'15 [26], SVTP [47] and CUTE [52] datasets. The results are reported using accuracy **Acc.** (higher is better), normalized edit distance **NED** (higher is better) and total edit distance **TED** (lower is better). Relative gains are shown in green and relative declines in red.

Test data	Loss function	↑ Acc.	↑ NED	↓ TED
IIIT-5K	Cross-entropy	84.300	0.954	945
IIIT-5K	LS-ED	86.300	0.953 −0.10%	837
SVT	Cross-entropy	84.699	0.940	229
SVT	LS-ED	86.399	0.947	196
ICDAR'03	Cross-entropy	92.558	0.972	151
ICDAR'03	LS-ED	94.070	0.977	119
ICDAR'13	Cross-entropy	89.754	0.949	260
ICDAR'13	LS-ED	91.133	0.960	157
ICDAR'15	Cross-entropy	71.452	0.889	1135
ICDAR'15	LS-ED	74.655	0.899	1013
SVTP	Cross-entropy	74.109	0.891	424
SVTP	LS-ED	77.519	0.901	381
CUTE	Cross-entropy	68.293	0.838	285
CUTE	LS-ED	71.777	0.868	234

Scene Text Detection. Given a natural scene image, the objective is to obtain precise word-level rotated bounding boxes. The method proposed by Ma *et al.* [40] is used for the task. It extends Faster-RCNN [50] based object detector to incorporate rotations. This is achieved by adding angle priors in anchor boxes to enable rotated region proposals. A sampling strategy using IoU compares these proposals with the ground truth and filter the positive and the negative proposals. Only the filtered proposals are used for the loss computation.

The positive proposals are regressed to fit precisely with the ground truth. Through rotated region-of-interest (RROI) pooling, the features corresponding to the proposals are obtained and used for text/no-text binary classification. The overall loss function for training in [40] is defined as a linear combination of classification loss (negative log-likelihood) and regression loss (*smooth-L*$_1$).

The publicly available implementation of [39,40] is used with the original hyper-parameter settings – the model is trained for $140K$ iterations using the

Table 2. TPS-ResNet-BiLSTM-Attn: The models are evaluated on IIIT-5K [41], SVT [58], ICDAR'03 [38], ICDAR'13 [27], ICDAR'15 [26], SVTP [47] and CUTE [52] datasets. The results are reported using accuracy **Acc.** (higher is better), normalized edit distance **NED** (higher is better) and total edit distance **TED** (lower is better). Relative gains are shown in green and relative declines in red.

Test data	Loss function	↑ Acc.	↑ NED	↓ TED
IIIT-5K	Cross-entropy	87.500	0.961	722
IIIT-5K	LS-ED	87.933 +0.49%	0.963 +0.20%	645 +10.66%
SVT	Cross-entropy	87.172	0.952	180
SVT	LS-ED	86.708 −0.53	0.954 +0.21%	163 +9.44%
ICDAR'03	Cross-entropy	94.302	0.979	110
ICDAR'03	LS-ED	94.535 +0.24%	0.981 +0.20%	99 +10.00%
ICDAR'13	Cross-entropy	92.020	0.966	137
ICDAR'13	LS-ED	92.299 +0.30%	0.979 +1.34%	108 +21.16%
ICDAR'15	Cross-entropy	78.520	0.915	868
ICDAR'15	LS-ED	78.410 −0.14%	0.915 ±0.00%	837 +3.57%
SVTP	Cross-entropy	78.605	0.912	346
SVTP	LS-ED	79.225 +0.78%	0.913 +0.10%	333 +3.75%
CUTE	Cross-entropy	73.171	0.871	224
CUTE	LS-ED	74.216 +1.42%	0.875 +0.45%	219 +2.23%

Table 3. RRPN-ResNet-50 [39,40]: Evaluations on Incidental Scene Text ICDAR'15 [26]. Relative gains are shown in green and relative declines in red.

Loss function	↑ Recall	↑ Precision	↑ F_1 score
Smooth-L_1	71.21%	84.71%	77.37%
LS-IoU (global)	66.97% −5.95%	84.71% ±0.00%	74.81% −3.30%
LS-IoU (local)	70.92% −0.40%	86.60% +2.23%	77.98% +0.78%
LS-IoU (local-global)	76.79% +7.83%	84.93% +0.25%	80.66% +4.25%

SGD optimizer and batch-size of 1. The model is trained on a union of ICDAR'15 [26] and ICDAR-MLT [43] datasets, providing 6295 training images.

LS-IoU Architecture. The embedding model for LS-IoU consists of five fully-connected layers with ReLU activation [13]. A rotated bounding box is represented with six parameters, two for the coordinates of the centre of the box, two for the height and the width and two for *cosine* and *sine* of the rotation angle. The centre coordinates and the dimensions of the box are normalized with image dimensions to make the representation invariant to the image resolution.

The embedding model maps the representation of a positive box proposal and the matching ground-truth into a vector as $h_\Phi : \mathbb{R}^6 \to \mathbb{R}^{16}$. The approximation of the IoU between two bounding boxes is computed by the L_2 distance between the two vector representations.

Post-tuning with LS-IoU. The random generator for LS-IoU samples rotated bounding boxes from the set of training labels and modifies the boxes by changing the centre locations, dimensions, and rotation angle within certain bounds to create a distorted variant. Since uniform sampling over the range of IoU is difficult, we store roughly 3 million such examples along with the IoU values and sample from this collection.

Note that since the overall loss for training [40] is a combination of a regression loss and a classification loss, LS-IoU only replaces the regression component (*smooth-L_1*) with the learned surrogate for IoU. For post-tuning with LS-IoU, the results are shown for all three setups, that is, global approximation, local approximation and global-local approximation (Algorithm 1). For each of these, the model trained with proxy losses is post-tuned with LS-IoU for $20K$ iterations. The quality of the approximations for the first $1K$ iterations of the training is shown in Fig. 3. Since the range of IoU is in $[0, 1]$, it can be seen that the error is high for the global approximation. For both local and global-local, the quality of the approximation is significantly better (roughly 10 times lower error).

As mentioned earlier, the quality of gradients is judged by the improvement or deterioration of the model ($f_\Theta(x)$) post-tuned with LS-IoU. The results for scene text detection on the ICDAR'15 [26] dataset are shown in Table 3. It is observed that post-tuning the detection model with LS-IoU (global) leads to deterioration. Post-tuning with LS-IoU (local) improves the precision but makes recall worse. Finally, LS-IoU (local-global) from Algorithm 1 improves both the precision and recall, boosting the F_1 score by relative 4.25%.

5   Conclusions

A technique is proposed for training neural networks by minimizing learned surrogates that approximate the target evaluation metric. The effectiveness of the proposed technique has been demonstrated in a post-tuning setup, where a trained model is tuned on the learned surrogate. Improvements have been achieved on the challenging tasks of scene-text recognition and detection. By post-tuning, the model with LS-ED, relative improvements of up to 39% on the total edit distance has been achieved. On detection, post-tuning with LS-IoU has shown to provide a relative gain of 4.25% on the F_1 score.

Acknowledgement. The authors thank R. Manmatha, Dmytro Mishkin, Michal Bušta, Klára Janoušková, Viresh Ranjan and Abhijeet Kumar for the feedback. This research was supported by Research Center for Informatics (project CZ.02.1.01/0.0/0.0/16019/0000765 funded by OP VVV) and CTU student grant (SGS OHK3-019/20).

References

1. Agustsson, E., et al.: Soft-to-hard vector quantization for end-to-end learning compressible representations. In: NeurIPS (2017)
2. Azimi, S.M., Vig, E., Bahmanyar, R., Körner, M., Reinartz, P.: Towards multiclass object detection in unconstrained remote sensing imagery. In: Jawahar, C.V., Li, H., Mori, G., Schindler, K. (eds.) ACCV 2018. LNCS, vol. 11363, pp. 150–165. Springer, Cham (2019). https://doi.org/10.1007/978-3-030-20893-6_10
3. Baek, J., et al.: What is wrong with scene text recognition model comparisons? dataset and model analysis. In: ICCV (2019)
4. Ballé, J., Minnen, D., Singh, S., Hwang, S.J., Johnston, N.: Variational image compression with a scale hyperprior. In: ICLR (2018)
5. Berrada, L., Zisserman, A., Kumar, M.P.: Smooth loss functions for deep top-k classification. In: ICLR (2018)
6. Bušta, M., Patel, Y., Matas, J.: E2E-MLT – an unconstrained end-to-end method for multi-language scene text. In: Carneiro, G., You, S. (eds.) ACCV 2018. LNCS, vol. 11367, pp. 127–143. Springer, Cham (2019). https://doi.org/10.1007/978-3-030-21074-8_11
7. Caron, M., Bojanowski, P., Joulin, A., Douze, M.: Deep clustering for unsupervised learning of visual features. In: ECCV (2018)
8. Deng, J., Dong, W., Socher, R., Li, L.J., Li, K., Fei-Fei, L.: Imagenet: a large-scale hierarchical image database. In: CVPR (2009)
9. Dong, C., Loy, C.C., He, K., Tang, X.: Image super-resolution using deep convolutional networks. IEEE Trans. Pattern Anal. Mach. Intell. $38(2)$, 295–307 (2015)
10. Elsken, T., Metzen, J.H., Hutter, F.: Neural architecture search: a survey. arXiv preprint arXiv:1808.05377 (2018)
11. Engilberge, M., Chevallier, L., Pérez, P., Cord, M.: Sodeep: a sorting deep net to learn ranking loss surrogates. In: CVPR (2019)
12. Gidaris, S., Singh, P., Komodakis, N.: Unsupervised representation learning by predicting image rotations. In: ICLR (2018)
13. Glorot, X., Bordes, A., Bengio, Y.: Deep sparse rectifier neural networks. In: AISTATS (2011)
14. Gomez, L., Patel, Y., Rusiñol, M., Karatzas, D., Jawahar, C.: Self-supervised learning of visual features through embedding images into text topic spaces. In: CVPR (2017)
15. Gomez, R., et al.: ICDAR2017 robust reading challenge on coco-text. In: ICDAR (2017)
16. Grabocka, J., Scholz, R., Schmidt-Thieme, L.: Learning surrogate losses. arXiv preprint arXiv:1905.10108 (2019)
17. Graves, A., Fernández, S., Gomez, F., Schmidhuber, J.: Connectionist temporal classification: labelling unsegmented sequence data with recurrent neural networks. In: ICML (2006)
18. Gulrajani, I., Ahmed, F., Arjovsky, M., Dumoulin, V., Courville, A.C.: Improved training of Wasserstein GANs. In: NeurIPS (2017)
19. Gupta, A., Vedaldi, A., Zisserman, A.: Synthetic data for text localisation in natural images. In: CVPR (2016)
20. Hazan, T., Keshet, J., McAllester, D.A.: Direct loss minimization for structured prediction. In: NeurIPS (2010)
21. He, K., Zhang, X., Ren, S., Sun, J.: Deep residual learning for image recognition. In: CVPR (2016)

22. Hochreiter, S., Schmidhuber, J.: Long short-term memory. Neural Comput. **9**(8), 1735–1780 (1997)
23. Hodan, T., et al.: Bop: benchmark for 6D object pose estimation. In: ECCV (2018)
24. Jaderberg, M., Simonyan, K., Vedaldi, A., Zisserman, A.: Synthetic data and artificial neural networks for natural scene text recognition. CoRR (2014)
25. Jaderberg, M., Simonyan, K., Zisserman, A., Kavukcuoglu, K.: Spatial transformer networks. In: NeurIPS (2015)
26. Karatzas, D., et al.: ICDAR 2015 competition on robust reading. In: ICDAR (2015)
27. Karatzas, D., et al.: ICDAR 2013 robust reading competition. In: ICDAR (2013)
28. Kato, H., Ushiku, Y., Harada, T.: Neural 3D mesh renderer. In: CVPR (2018)
29. Kristan, M., et al.: The seventh visual object tracking vot2019 challenge results. In: ICCV Workshops (2019)
30. Krizhevsky, A., Sutskever, I., Hinton, G.E.: Imagenet classification with deep convolutional neural networks. In: NeurIPS (2012)
31. Lapin, M., Hein, M., Schiele, B.: Loss functions for top-k error: analysis and insights. In: CVPR (2016)
32. Ledig, C., et al.: Photo-realistic single image super-resolution using a generative adversarial network. In: CVPR (2017)
33. Lee, J., Cho, S., Beack, S.K.: Context-adaptive entropy model for end-to-end optimized image compression. In: ICLR (2019)
34. Li, K., Malik, J.: Learning to optimize neural nets. arXiv preprint arXiv:1703.00441 (2017)
35. Lin, T.-Y., et al.: Microsoft COCO: common objects in context. In: Fleet, D., Pajdla, T., Schiele, B., Tuytelaars, T. (eds.) ECCV 2014. LNCS, vol. 8693, pp. 740–755. Springer, Cham (2014). https://doi.org/10.1007/978-3-319-10602-1_48
36. Liu, W., Chen, C., Wong, K.K., Su, Z., Han, J.: Star-net: a spatial attention residue network for scene text recognition. In: BMVC (2016)
37. Liu, X., Liang, D., Yan, S., Chen, D., Qiao, Y., Yan, J.: FOTS: fast oriented text spotting with a unified network. In: CVPR (2018)
38. Lucas, S.M., Panaretos, A., Sosa, L., Tang, A., Wong, S., Young, R.: ICDAR 2003 robust reading competitions. In: ICDAR (2003)
39. Ma, J.: RRPN in pytorch. https://github.com/mjq11302010044/RRPNpytorch (2019)
40. Ma, J., et al.: Arbitrary-oriented scene text detection via rotation proposals. IEEE Trans. Multimedia **20**(11), 3111–3122 (2018)
41. Mishra, A., Alahari, K., Jawahar, C.: Scene text recognition using higher order language priors. In: BMVC (2012)
42. Nagendar, G., Singh, D., Balasubramanian, V.N., Jawahar, C.: Neuro-IoU: learning a surrogate loss for semantic segmentation. In: BMVC (2018)
43. Nayef, N., et al.: ICDAR 2019 robust reading challenge on multi-lingual scene text detection and recognition–RRC-MLT-2019. arXiv preprint arXiv:1907.00945 (2019)
44. Patel, Y., Appalaraju, S., Manmatha, R.: Deep perceptual compression. arXiv preprint arXiv:1907.08310 (2019)
45. Patel, Y., Appalaraju, S., Manmatha, R.: Hierarchical auto-regressive model for image compression incorporating object saliency and a deep perceptual loss. arXiv preprint arXiv:2002.04988 (2020)
46. Prabhavalkar, R., et al.: Minimum word error rate training for attention-based sequence-to-sequence models. In: ICASSP (2018)
47. Quy Phan, T., Shivakumara, P., Tian, S., Lim Tan, C.: Recognizing text with perspective distortion in natural scenes. In: ICCV (2013)

48. Rahman, M.A., Wang, Y.: Optimizing intersection-over-union in deep neural networks for image segmentation. In: Bebis, G. (ed.) ISVC 2016. LNCS, vol. 10072, pp. 234–244. Springer, Cham (2016). https://doi.org/10.1007/978-3-319-50835-1_22

49. Redmon, J., Divvala, S., Girshick, R., Farhadi, A.: You only look once: unified, real-time object detection. In: CVPR (2016)

50. Ren, S., He, K., Girshick, R., Sun, J.: Faster R-CNN: Towards real-time object detection with region proposal networks. In: NeurIPS (2015)

51. Rezatofighi, H., Tsoi, N., Gwak, J., Sadeghian, A., Reid, I., Savarese, S.: Generalized intersection over union: a metric and a loss for bounding box regression. In: CVPR (2019)

52. Risnumawan, A., Shivakumara, P., Chan, C.S., Tan, C.L.: A robust arbitrary text detection system for natural scene images. Expert Syst. Appl. **41**(18), 8027–8048 (2014)

53. Rumelhart, D.E., Hinton, G.E., Williams, R.J.: Learning representations by back-propagating errors. Nature **323**, 533–536 (1986)

54. Ryoo, M.S., Piergiovanni, A., Tan, M., Angelova, A.: Assemblenet: searching for multi-stream neural connectivity in video architectures. In: NeurIPS (2019)

55. Shi, B., Wang, X., Lyu, P., Yao, C., Bai, X.: Robust scene text recognition with automatic rectification. In: CVPR (2016)

56. Simonyan, K., Zisserman, A.: Very deep convolutional networks for large-scale image recognition. arXiv preprint arXiv:1409.1556 (2014)

57. Song, Y., Schwing, A., Urtasun, R., et al.: Training deep neural networks via direct loss minimization. In: ICML (2016)

58. Wang, K., Babenko, B., Belongie, S.: End-to-end scene text recognition. In: ICCV (2011)

59. Wang, Z., Simoncelli, E.P., Bovik, A.C.: Multiscale structural similarity for image quality assessment. In: ACSSC (2003)

60. Xia, G.S., et al.: DOTA: a large-scale dataset for object detection in aerial images. In: CVPR (2018)

61. Xu, B., Wang, N., Chen, T., Li, M.: Empirical evaluation of rectified activations in convolutional network. CoRR (2015)

62. Yu, J., Jiang, Y., Wang, Z., Cao, Z., Huang, T.: Unitbox: an advanced object detection network. In: ACM MM (2016)

63. Zeiler, M.D.: ADADELTA: an adaptive learning rate method. CoRR (2012)

64. Zhang, X., Zhao, J.J., LeCun, Y.: Character-level convolutional networks for text classification. In: NeurIPS (2015)

65. Zoph, B., Le, Q.V.: Neural architecture search with reinforcement learning. arXiv preprint arXiv:1611.01578 (2016)

An Asymmetric Modeling for Action Assessment

Jibin Gao[1,4], Wei-Shi Zheng[1,2,5(✉)], Jia-Hui Pan[1], Chengying Gao[1(✉)],
Yaowei Wang[2], Wei Zeng[3], and Jianhuang Lai[1]

[1] School of Data and Computer Science, Sun Yat-sen University, Guangzhou, China
{gaojb5,panjh7}@mail2.sysu.edu.cn
{zhwshi,mcsgcy,stsljh}@mail.sysu.edu.cn
[2] Peng Cheng Laboratory, Shenzhen 518005, China
wangyw@pcl.ac.cn
[3] School of Electronics Engineering and Computer Science, Peking University,
Beijing, China
weizeng@pku.edu.cn
[4] Pazhou Lab, Guangzhou, China
[5] Key Laboratory of Machine Intelligence and Advanced Computing,
Ministry of Education, Guangzhou, China

Abstract. Action assessment is a task of assessing the performance
of an action. It is widely applicable to many real-world scenarios such
as medical treatment and sporting events. However, existing methods
for action assessment are mostly limited to individual actions, especially lacking modeling of the asymmetric relations among agents (e.g.,
between persons and objects); and this limitation undermines their ability to assess actions containingasymmetrically interactive motion patterns, since there always exists subordination between agents in many
interactive actions. In this work, we model the asymmetric interactions among agents for action assessment. In particular, we propose an
asymmetric interaction module (AIM), to explicitly model asymmetric interactions between intelligent agents within an action, where we
group these agents into a primary one (e.g., human) and secondary ones
(e.g., objects). We perform experiments on *JIGSAWS* dataset containing surgical actions, and additionally collect a new dataset, *TASD-2*,
for interactive sporting actions. The experimental results on two interactive action datasets show the effectiveness of our model, and our
method achieves state-of-the-art performance. The extended experiment
on *AQA-7* dataset also demonstrates the generalization capability of our
framework to conventional action assessment.

1 Introduction

Action assessment [1,4,9,13,18] has attracted much attention in recent years. It
is widely applicable to many practical scenarios. For instance, action assessment

Electronic supplementary material The online version of this chapter (https://
doi.org/10.1007/978-3-030-58577-8_14) contains supplementary material, which is
available to authorized users.

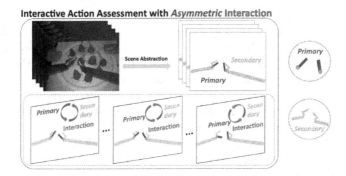

Fig. 1. Our asymmetric interaction module is designed to assess action performance. For egocentric surgical videos, we regard motions of the master tool-tips as the *primary* (in red), and those of the slave tool-tips and handles, which are relatively inactive, as the *secondary* (in blue). Best viewed in colour. (Color figure online)

models can be applied in sports events to assist the referee in scoring, as well as to assist athletes in training [1,16,18,20]. Athletes can make reasonable corrections to their motions according to the feedback from the assessment model to achieve better training effects. In modern medical treatment, rehabilitation treatment has received increasing attention. Action assessment can be applied to the rehabilitation training of patients [13,22,28]. By monitoring and assessing the daily rehabilitation training of patients, doctors can give follow-up rehabilitation treatment suggestions to patients according to the assessment report, aiming to achieve efficient treatment [7,14,31,32].

However, existing action assessment methods [15,17,20] are mostly designed for individual actions, such as diving and vaulting. In real-world applications, there are many non-individual actions, which are defined by interaction, especially there are subordination between agents in an interaction. For example, in the view of egocentric surgical videos, only motions of two tool-tips are captured. Accordingly, the actions involving interactions between human (featured as motion of tool-tips) and the "two tool-tips and handles" should be explored explicitly for the assessment. More importantly, such an interaction is asymmetric. The roles in asymmetrically interactive actions mentioned above can semantically be categorized into the primary agent and the secondary ones. While some work such as [15] can be applied to handle the assessment on interactive action, they treat all agents equally, and thus they cannot provide particular modeling on the subordination between agents (e.g. between human and objects).

In this work, we propose a new framework for addressing asymmetrically interactive action assessment with an asymmetric interaction module (AIM) that provides a general and novel proposal for action interaction among multiple people or parts. In this module, it is assumed that there is a primary agent that is dominant relative to the others it interacts with; correspondingly, the others are viewed as the secondary agents in interactive actions. This assumption makes sense, since the multiple parts involved in the interactive actions are always naturally semantically categorized as two parts, an important part (e.g., human) and secondary parts (e.g., objects). In AIM, we exploit the difference between the *primary* and the *secondary* in the same latent space, and utilize the *primary* equipped with the difference to learn the interaction in the temporal domain, since the *primary* is dominant relative to *secondary* in representing an action. With this module, our framework can explicitly learn the latent criterion of interactive action assessment. Afterwards, we construct an attention fusion module inspired by the attention mechanism [24] to pay different amounts of attention to the whole-scene feature and AIM feature (Fig. 1).

Moreover, apart from assessing interactive actions in strong asymmetric relation among each part, our method can also be applied to interactive actions whose agents are in weak asymmetric or equal relation, such as synchronous sports. We therefore generalize our model by a multi-task learning for operating general interactive action assessment.

To the best of our knowledge, *AQA-7* [18] and MTL-AQA [16] are the only two available datasets that contain the events involving two players; however, these events are from side view, and thus it is unsuitable to investigate the interaction between players as they overlap seriously most of the time. Therefore, we have additionally collected a new dataset, named Two-person Action Synchronized Diving dataset (*TASD-2*) for evaluating interactive action assessment.

In summary, our contributions are three-fold: (1) A novel module, called the asymmetric interaction module (AIM), is constructed to reasonably extract the interactive relation for asymmetric interaction; (2) a general framework for interactive action assessment is proposed that can be easily generalized to different kinds of action assessment tasks; (3) a new dataset, *TASD-2*, is collected in our work containing two-person interactive actions captured from the front view. We have reported experiments to validate the effectiveness of the proposed method. Project homepage: https://www.isee-ai.cn/~gaojibin/ProjectAIM.html.

2   Related Work

Action Quality Assessment. Action assessment is the evaluation of how well an action is performed. For the tasks of action assessment, the existing works mainly modelled the problem in three manners: 1) casting it as a classification task to classify the action performance as expert or novice [32,33]; 2) casting it as a regression problem that fits the scores of multiple action performances [17,18,20,26,30]; 3) casting it as a pair-wise ranking task [1,4,5,29]. Our work follows the second branch. However, few works have assessed the action quality by explicitly exploring the interaction in actions, and especially lack of modeling on asymmetric interaction for the assessment. To learn the relation among joints of performers' skeleton, Pan et al. [15] computed action quality based on joint relation modeling by GNN [21]. Compared to [15], our asymmetric interaction module is a non-symmetric modeling (i.e. we treat primary and secondary non-equally), while existing method [15] treats them equally. Nevertheless, they can be applied to model the actions in joint-based interaction, but they ignore the subordination modeling in the asymmetric interaction.

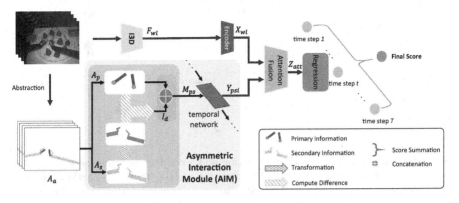

Fig. 2. An overview of our proposal. We uniformly divide an input video into T time steps, and present the process of the asymmetric interaction module (AIM) in a clear manner at time step t. The kinetic information of mobile objects is extracted, including the *primary* and the *secondary* ones. We perform asymmetric interaction between the *primary* and the *secondary* and obtain the AIM feature. Afterwards, we perform attentive contextual interaction between the whole-scene feature, which is extracted via I3D [2], and the AIM feature with an attention fusion. Finally, a regression module is utilized to learn regressing the action quality.

Interactive Models. Recently, an increasing number of interactive models [10, 15,17,25] have been proposed. Wang et al. [25] constructed a channel-wise interaction learning method to evaluate the interaction of each part of an image to preserve information with a binary feature map through prior knowledge graphs. Li and Cai [10] paid different attention to the interaction of each individual part extracted from the images. In action assessment, many related works [17,20] fed the key-points feature into an LSTM [8] framework to explore the interaction in the temporal domain, while other works [15,27] have exploited the interactive relations among the human skeleton through GNNs. However, these methods either yield poor interpretability of the interaction process or are limited by the connection of nodes, resulting in poor generalization to various assessment tasks.

3 Approach

In this section, we will introduce our model for asymmetrically interactive action assessment in detail. The overall structure of our model is shown in Fig. 2. In this framework, the asymmetric interaction module is particularly designed to explore the asymmetric interaction between primary agent and secondary agents. An attentive textual interaction with an attention fusion module is developed to further fuse the AIM feature and whole-scene feature. Finally, a regression module is used to learn regressing the action quality.

3.1 Asymmetric Interaction Module

It is common that the interactions among multiple people or multiple agents, in particular the asymmetric interaction among humans and objects, play

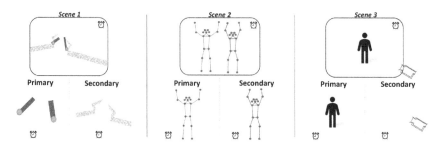

Fig. 3. Examples of the primary information and secondary information partitioning. The clock icon indicates the motion of that part. Specially, there exist actions in which two performers are in weak asymmetric relation or equal. In these cases, any one of the performers can be assigned as the *primary* and the other as the *secondary*.

important roles. In order to reduce noise interference, we extract subtle but informative feature at an abstraction level, which we denote as A_a; that is only indispensable kinetic information for describing an action is considered, such as human pose. For example, on surgery tasks in egocentric views, we assign A_a with the kinetic information of the tool-tips which contains the object information (e.g., tool orientations) and speed information; while for most types of action performance, the entire human body should be considered, so A_a is the pose information provided by some pose estimator.

Before performing interaction, we divide A_a into two parts according to their semantics, the primary information, denoted as A_p, and the secondary information, denoted as A_s. An example diagram is shown in Fig. 3. For egocentric surgical videos, where only motions of two tool-tips are captured, it is more intuitive since each tool-tip consists of a master part and a slave one. For semantic consistence, we regard motions of the master tool-tips as the *primary*, and those of the slave tool-tips and handles, which are relatively inactive, as the *secondary*.

To explicitly explore the asymmetric relations between primary information and secondary information, we design the asymmetric interaction module (AIM), as shown in Fig. 2. With different semantics, it is natural that the primary information and secondary information come from different domains. Thus, to explore the potential relation and asymmetric interaction between the *primary* and the *secondary*, we first pass secondary information through a transformation module to map it into a latent space, the same as that of the primary. When the *primary* and *secondary* are from the same domain, the transformation will tend to learn an identity function [3]. Afterwards, we determine the difference between the *primary* and the *secondary* after the transformation, where the difference operation is an effective operation to explore the relation between visual instances [15], and the process can be formed as

$$I_d^{(t)} = \mathcal{D}(A_p^{(t)}, \mathcal{T}(A_s^{(t)})), \tag{1}$$

where $A_*^{(t)}$ denotes a certain feature A_* in time step t in Fig. 2, $I_d^{(t)} \in \mathbb{R}^N$ and $\mathcal{T}(\cdot)$ is a function to conduct the transformation operation, and $\mathcal{D}(\cdot)$ is a function to determine the difference between the *primary* and the *secondary*. Here, N denotes the dimension of the $I_d^{(t)}$ and $A_p^{(t)}$.

According to the discussion above, the primary information is dominant relative to the secondary in the capability of representation for interactive actions. To utilize the superiority of the primary information, we then concatenate the difference feature and primary information, called the primary-secondary information and denoted as M_{ps}. We present it as

$$M_{ps}^{(t)} = A_p^{(t)} \oplus I_d^{(t)}, \qquad (2)$$

where $M_{ps}^{(t)} \in \mathbb{R}^{2N}$ and $\oplus$ represents the concatenating operator.

The process mentioned above can be regarded as the interaction in the spatial domain. Moreover, since the interactions occur over time, temporal relations for asymmetrically interactive action assessment are essential. Then, we use a temporal network to learn the temporal interaction and obtain the complete AIM feature, which can be expressed as

$$Y_{psi}^{(t)} = \mathcal{P}(M_{ps}^{(t)}), \qquad (3)$$

where $Y_{psi}^{(t)} \in \mathbb{R}^d$ and $\mathcal{P}(\cdot)$ is a temporal network, for which we use LSTM in this work, and d is the dimension of the hidden layer in the LSTM model.

3.2   Attentive Contextual Interaction

To assist the learned AIM feature, we further employ I3D [2] to extract the whole-scene feature of videos, denoted as F_{wl}. To some extent, the whole-scene feature contains extra information complement to our AIM feature, even though noise exists. Now, we obtain two-stream outputs, the whole-scene one F_{wl} and the AIM one Y_{psi}. Before fusion of these outputs, we pass F_{wl} through an encoder to map F_{wl} into the latent space the same as the AIM feature, Y_{psi}. Then, $X_{wl}^{(t)}$ is obtained, where $X_{wl}^{(t)} \in \mathbb{R}^d$ and d is the dimension of the encoder feature.

In our fusion modeling, we perform attentive contextual interaction between the whole-scene feature and the AIM feature; that is the whole-scene feature F_{wl} is utilized to learn a key map as attention for fusion of the whole-scene feature and our AIM feature because it contains the whole-scene context. Inspired by self-attention [24], we regard $(X_{wl}^{(t)} \oplus Y_{psi}^{(t)})$ as the queries and values of attention mechanism. To be detailed, we form the fusion process as follows:

$$Z_{att}^{(t)} = W^{(t)} \circ (X_{wl}^{(t)} \oplus Y_{psi}^{(t)})', \qquad (4)$$

$$W^{(t)} = softmax((X_{wl}^{(t)} \oplus Y_{psi}^{(t)})' \circ O_{key}^{(t)}), \quad O_{key}^{(t)} = \mathcal{FC}_{key}(F_{wl}^{(t)}), \qquad (5)$$

where $\circ$ represents the matrix multiplication, $\oplus$ represents the concatenating operator, and $softmax(\cdot)$ is the softmax function, and $\mathcal{FC}_{key}(\cdot)$ is a fully connected layer to learn the key mapping. Here, $X_{wl}^{(t)}, Y_{psi}^{(t)}, Z_{att}^{(t)}, O_{key}^{(t)} \in \mathbb{R}^d$, and A' denotes the transpose of matrix A.

3.3   Scoring for Action Assessment

In the final step, our method should give a final score for the action performance through the regression module shown in Fig. 2. The overall assessment result will be presented in a score given by

$$S = \sum_{}^{T} \mathcal{R}(Z_{att}^{(t)}), \tag{6}$$

where S denotes the predicted score for the action performance, $Z_{att}^{(t)}$ is the output of attentive contextual interaction, T is the number of time steps in the video, and $\mathcal{R}(\cdot)$ represents the regression module implemented with two FCs.

In the training stage, we use the Mean-Squared Error (MSE) as loss function for model optimization, which is defined as $\delta = \frac{1}{C} \sum_{i}^{C} (y_i - \hat{y}_i)^2$, where y and $\hat{y}$ represent the ground truth and the predicted value respectively, and C denotes the number of samples.

4   Extension to General Interactive Action Assessment: A Multi-task Training

The asymmetric interaction module can be generalized to general interactive action assessment, even though there is no explicit primary and secondary roles between performers. The *second row* in Fig. 3 does not show a strong asymmetric relation between two performers[1], and then we choose any one of them as the *primary* and the other as the *secondary*.

To be detailed, we generalize our model by a multi-task training. The multiple tasks can naturally align to the two-stream features in reasonable semantics; the whole-scene feature can be utilized for learning action assessment on overall performance, and the AIM feature can be designed for learning action assessment on interactive actions. For instance, for synchronized diving, the execution score and synchronization score are given by referees during scoring for the entire action performance. We could assess the execution of action using the whole-scene feature, which several existing methods [17,18] have conducted, while feature extracted by AIM is capable to be utilized for learning the synchronization of action reasonably since AIM mainly explores the interaction between two players. Thus, we can use the whole-scene feature X_{wl} to learn scoring for the execution and Y_{psi} for synchronization of the action. Their assessment results are given by

$$S_{ex} = \sum_{}^{T} \mathcal{R}_A(X_{wl}^{(t)}), \; S_{sn} = \sum_{}^{T} \mathcal{R}_B(Y_{psi}^{(t)}), \tag{7}$$

where S_{ex} and S_{sn} denote the predicted execution score and synchronization score, respectively. $\mathcal{R}_*(\cdot)$ represents the regression module implemented with two fully connected layers.

---

[1] For interactive actions involving more than two performers, the important people detection [11,12,23] can be utilized to divide performers into the *primary* (the most important one) and the *secondary* (the rest).

Fig. 4. Samples of the *TASD-2 dataset.*

The loss function in this setting could be formulated as

$$\mathcal{L} = \mathcal{L}_{fn} + \theta * \mathcal{L}_{ex} + (1 - \theta) * \mathcal{L}_{sn}, \tag{8}$$

where $\mathcal{L}_{fn}$, $\mathcal{L}_{ex}$ and $\mathcal{L}_{sn}$ denote the loss functions of regression for final scores, execution scores and synchronization scores, respectively. θ denotes a trade-off weight. Similarly, the loss function is the Mean-Squared Error(MSE).

The overall loss function shown in Eq. (8) is meaningful, since in synchronized diving, a great performance must be excellent in both synchronization and execution. Therefore, apart from the final score, the execution score and synchronization score are also utilized to perform a multi-task training.

4.1 TASD-2 Dataset

For assessing general interactive action, we also collect a new dataset. While *AQA-7* [18] contains two events involving two performers, namely the synchronized 3-m springboard and 10-m platform, these events were captured from the side view, and thus it is hard to investigate the interaction between two performers as they overlap seriously for most of the time. Therefore, we collected a new diving dataset whose videos provide a better view to capture the interaction between two performers on synchronized diving videos. The construction details of our TASD-2 dataset can be found in the supplementary materials (Table 1).

5   Experiments

We mainly conducted experiments on assessment of interactive action on *JIG-SAWS* and *TASD-2* (Fig. 4). In addition, conventional action assessment for single person can be regarded as a special extension of our method, and an evaluation of it on *AQA-7* [18] was also conducted.

Table 1. Details of the *TASD-2 dataset*

Sport	SyncDiving-3m	SyncDiving-10m
#Frames of a sample	102	102
#Samples	119	184
#Augmented samples	238	368
#Training set	188	293
#Testing set	50	75

Knot Tying   Needle Passing   Suturing

Fig. 5. Frames of samples in *JIGSAWS*.

- **Dataset introduction.** We conduct experiments on *JIGSAWS* [6] and *TASD-2*. *JIGSAWS* contains egocentric videos of three surgical tasks, including suturing, needle passing and knot tying. There are 206 videos in this dataset, of which 78 are for suturing, 56 for needle passing, and 72 for knot tying. Samples are shown in Fig. 5. The videos are captured in stereo recordings with left views and right views by using two cameras. All videos will be used in our experiments. For each video in *JIGSAWS*, 3D kinetics information of the master tool manipulators and patient-side manipulators is provided. The details of *TASD-2* can be found in Sect. 4.1.

5.1   Implementation Details

- **Model training setting.**[2] Our model is implemented in PyTorch. Without specific explanation, our model uses Adam Optimizer with a weight decay rate of 0.5. In the training process, the batch size is 64. We use cyclic learning rates of {1e-4, 1e-5, 1e-6} changing according to the {20, 50, 100}th epoch in every 100 epochs. For each task, we train our model for 3000 epochs. T is set to 10. The encoder was implemented with a fully connected layer of shape 400×512 with ReLU activation, and the LSTM is a single layer with a 512-dimensional output. In AIM, we used fully connected layers with input and output in the same dimension as the learnable transformation operation. The difference operation is vector subtraction. In the regression module, two FC layers were utilized. The first has a shape of 512×128 with ReLU activation, and the second has a shape of 128×1 without an activation function to avoid dead ReLU during score regression. The dropout parameter is set to 0.2 and θ is 0.4.

- **Evaluation Metric.** For comparison with previous works [15,17,18,20], we use Spearman's rank correlation as the evaluation metric of our model. It is defined as $\rho = \frac{\sum_i (p_i - \bar{p})(q_i - \bar{q})}{\sqrt{\sum_i (p_i - \bar{p})^2 \sum_i (q_i - \bar{q})^2}}$, where p and q represent the ranking of two sequences and $-1 \leq \rho \leq 1$. The higher the Spearman's rank correlation is, the more positive the ranking relation between two sequences is. It will be used to evaluate the ranking relation between the predicted and ground-truth assessment results of our model. In order to better reflect the performance of our methods, we run the model 10 times and report the average as the final model performance. Moreover, for multiple actions in a dataset, we compute

---

[2] Details of data preprocessing can be found in the supplementary materials.

Table 2. The results (%) of our proposal compared with the state-of-the-art methods and our baseline on *JIGSAWS*.

	Suturing	Needle Passing	Knot Tying	Avg. Corr.
ST-GCN [27]	31	39	58	43
TSN [4]	34	23	72	46
JR-GCN [15]	36	51	75	57
Baseline	5	9	11	8
Baseline+Kinetic	17	37	73	46
Ours	**63**	**65**	**82**	**71**

the average Spearman's rank correlation across actions from individual action correlations by using Fisher's z-value, as in [18].

5.2 Comparison

Experiments on Interactive Actions. We first evaluate our model on *JIG-SAWS*, and the results are shown in Table 2, comparing with the state-of-the-art methods and our baseline. To the best of our knowledge, the methods proposed in [4,15,27] had achieved state-of-the-art performance for skill action assessment on *JIGSAWS*, and we used 4-fold cross validation on *JIGSAWS* by following [15]. In comparison, the results show that our model outperforms the previous state-of-the-art methods and achieves the best results, with an improvement of 14% on average. According the structure of our model, it is common to find that the great performance of our method partially benefits from the well-performed I3D [2] and partially profits from the asymmetric interaction. Thus, we remove the AIM part in Fig. 2, and evaluate our baseline by only using the I3D feature. We also concatenate I3D feature and kinetics feature as a stronger baseline. The results in Table 2 (the last three rows) indicate that the asymmetric interaction is much important in our model. The effectiveness of AIM is confirmed. Moreover, ablation study in Sect. 5.3 demonstrates that the roles of *primary* and *secondary* could not be exchanged for their asymmetric relation on *JIGSAWS*.

We also compared our method with the best non-deep learning approach reported in [30] using leave-one-user-out(LOUO) in Table 3. As shown, both JR-GCN [15] and ours have their own strength. However, since the LOUO setting is demanding for the model's gerenration ability, our model is better and less specialized than [15], in which each joint is modelled in a specialized manner.

Table 3. Evaluation (%) on *JIGSAWS* with LOUO.

	Suturing	Needle Passing	Knot Tying	Avg. Corr.
DTC+DFT +ApEn [30]	37	25	60	41
JR-GCN [15]	35	**67**	19	40
Ours	**45**	34	**61**	**47**

Table 4. Results (%) of our model on *TASD-2*.

	SyncDiving-3m	SyncDiving-10m	Avg. Corr.
RANDOM	−3	3	0
C3D-LSTM [17]	−14	1	−7
I3D [2]-SVR-L	77	73	75
I3D [2]-SVR-P	84	83	83
I3D [2]-SVR-RBF	71	77	74
JR-GCN [15]	89	81	86
Baseline	84	79	82
Baseline+Pose	88	80	84
Ours (Single-task)	89	**85**	87
Ours (Multi-task)	**92**	**85**	**89**

To confirm the generalization of our framework to actions in weak relations between the *primary* and the *secondary*, experiments on *TASD-2* are performed, and the results are shown in Table 4. Since *TASD-2* is brand new, we utilize a naive model (RANDOM) that predicts scores for actions performance randomly in the range of [0, 100]. The results illustrate that the distribution of samples in *TASD-2* is relatively reasonable. We also evaluate C3D-LSTM [17] on *TASD-2*, but it did not work based on the experimental setting in [17,18]. Then, we use I3D [2] and SVR with different kernels, including linear polynomial and RBF kernels, on *TASD-2*. The results show that I3D-SVR models gain great performance, which reflects the strong ability of I3D to some extent. With the multi-task training in our model, our proposal achieved state-of-the-art performance on *TASD-2*, with a more than 3% improvement on average.

5.3   Ablation Study

Table 5 shows the results of an ablation study on our model. To explore the contributions of each main module in our model, we conduct experiments by removing one of the components from our full model, including the attention

Table 5. Ablation study (%) for exploring the effectiveness of each main module of our model on *JIGSAWS*.

	Suturing	Needle Passing	Knot Tying	Avg. Corr.
Full model	**63**	**65**	**82**	**71**
w/o AIM	7	41	64	40 (−31)
w/o attention fusion module	61	55	80	67 (−4)
Exchange *primary* and *secondary*	55	62	80	67 (−4)
w/o transformation module	61	62	79	68 (−3)
w/o difference module	60	61	80	68 (−3)
Whole-scene (Baseline)	5	9	11	8 (−63)

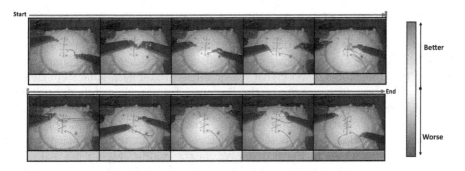

Fig. 6. The action assessment results of our model on a suturing case. The assessment results of our model indicate good (in green) and bad (in red) action performance for each time step. Best viewed in colour. (Color figure online)

fusion module and AIM. When replacing the attention fusion module with a fusion in each half, the model performance decreases by 4% on average. This result implies that paying different amounts of attention to whole-scene feature and AIM feature exactly makes a positive difference. Removing the transformation or difference module respectively, a 3% reduction was observed, indicating that these modelings are necessary. Moreover, when simply removing AIM part, the the performance decreases by 31%. These results indicate the significance of the asymmetric interaction and the effectiveness of AIM structure.

We also exchanged the *primary* and the *secondary* when performing model training and evaluating. The resulting performance reduction of 4% implies that the *primary* and the *secondary* really play their semantic roles with asymmetric interaction in the model evaluation. Moreover, from the last two rows of Table 4, we find that our proposal with multi-task training increases by more than 2% in model performance on average, compared to that with single-task setting. Thus, the results indicate that the multi-task training is effective.

Moreover, we exchanged the *primary* and the *secondary* in our modeling when evaluating on *TASD-2*. The results are shown in Table 6. There was little difference as compared to the performance without exchange. Thus, it indicates that our proposal is adapted to interactive actions in weak asymmetric relation in semantics, such as synchronized diving.

Table 6. Results (%) of exchanging the *primary* and the *secondary* on *TASD-2*.

	SyncDiving-3m	SyncDiving-10m
Before exchanging *primary* and *secondary*	**91.50**	**85.13**
After exchanging *primary* and *secondary*	**91.75**	**85.10**

5.4   Visualization of the Assessment Process

In order to view the process of assessment, we output the predicted sub-score defined in Eq. (6). Figure 6 shows an example about scoring in each time step[3]. We find that our model could give a reasonable score for each time step. Before accomplishing the first passing of the line used for suturing, it is difficult for most of us to control the surgical line expertly with tool-tips. Thus, it is not suitable to judge clearly a good or bad performance at this stage. Accordingly, we could observe that the proposed model gave relatively neutral judgement in the first few time steps in Fig. 6. However, in the middle stage of the suturing case, we found that two tool-tips performed relatively abnormally, causing the surgical line to be staggered in the air; thus, bad judgements were obtained during this time. Correspondingly, when approaching to finishing the suturing task, our model scored with positive judgements for great performance in this process. Therefore, the visualization also confirms that our framework is effective and interpretable.

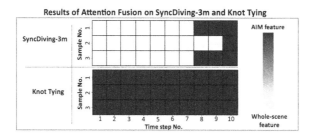

Fig. 7. Visualization of the attention fusion. We output the results of attention fusion on different actions, including synchronized 3-m springboard and knot tying. "Sample No." represents number of three randomly selected samples, and "Time step No." represents number of ten time steps for each video sample. The results indicate that our attention fusion could pay different amounts of attention on different time steps.

In addition, we also visualized the attention fusion through observing the computed results of Eq. (4) in Fig. 7. For the synchronized 3-m springboard, the attention fusion module could pay different amounts of attention on different time steps in a sample. The AIM feature is more important after time step 8 for SyncDiving-3m, because the interaction between two actors when they were approaching entry is more importance for synchronized diving assessment. It was obvious that our attention fusion also did make a difference by comparing different actions. It indicates that our attentive contextual interaction with an attention fusion is effective.

---

[3] Videos can be found in the supplementary materials.

5.5 Extended Experiment on Single-Person Actions

The secondary information is relatively difficult to determine for single-person actions due to semantically only one motion in videos. For generalization, we define a condition that if the secondary information is ambiguous, we can use the motion of the camera capturing the action performance for replacement, as shown in the third row in Fig. 3. We additionally evaluated our framework on *AQA-7* [18] under such an assumption; this dataset is collected from summer and winter Olympics and contains 1106 videos in total composed by six actions. As discussed in Sect. 4.1, *AQA-7* [18] contains two-person actions, but only captured from the side view. The performers are not visually separable. Thus visually there is only one agent in the videos, and we regard it as the *primary* without other choices. Then, we extract the motion feature of the camera as the *secondary*, by computing the optical flow (using the TV-L1 algorithm [19]) at the region near the edge of images. In this task, we fix the weight decay rate of Adam Optimizer in our model to 0.8. For consistency, the performance results that we report in Table 7 are obtained and the experimental setting follows [15]; the results demonstrate that our method is competitive compared with current state-of-the-art methods, with the best performance on sync. 3m action assessment. Our proposal outperforms most of the state-of-the-art methods except JR-GCN [15], and its performance score is only 0.6% less than that of JR-GCN [15] on average. Therefore, the extended experiment demonstrates that our framework is capable to generalize effectively to common action assessment tasks.

Table 7. Results (%) of our model applied to *AQA-7*. To illustrate the competitive results, the average of the rank among existing methods is used.

	diving	Gymvault	skiing	snowboard	sync. 3m	sync.10m	Avg. Corr.	Avg. Rank.
Pose+DCT [20]	53.00(5)	–	–	–	–	–	–	5
ST-GCN [27]	32.86(6)	57.70(4)	16.81(5)	12.34(5)	66.00(4)	64.83(5)	44.33(5)	4.9
C3D-LSTM [17]	60.47(4)	56.36(5)	45.93(4)	50.29(2)	79.12(3)	69.27(4)	61.65(4)	3.7
C3D-SVR [17]	79.02(1)	68.24(3)	52.09(3)	40.06(4)	59.37(5)	91.20(2)	69.37(3)	3
JR-GCN [15]	76.30(2)	73.58(1)	60.06(1)	54.05(1)	90.13(2)	92.54(1)	78.49(1)	1.3
Ours	74.19(3)	72.96(2)	58.90(2)	49.60(3)	92.98(1)	90.43(3)	77.89(2)	2.3

6 Conclusion

In this work, we proposed a novel asymmetric interaction model for asymmetrically interactive action assessment. In our model, we categorize the roles in an asymmetrically interactive action as a primary agent and secondary ones. With the asymmetric interaction, we can model the interactive actions in strong asymmetric relation. We evaluated our model on *JIGSAWS* [6] and our method achieved the state-of-the-art performance. Moreover, experimental results on

TASD-2, a new dataset (to be released) collected in our work, also demonstrated our method could be generalized to general interactive actions in weak asymmetric relation. The extra experiments on *AQA-7* [18] have also indicated that our model can be adapted to perform conventional action assessment. For future development, our method can also be extended to actions involving more than two people, with the help of the important people detectors [11,12,23]. It will be explored in the future work along with constructing relevant datasets.

Acknowledgement. This work was supported partially by the National Key Research and Development Program of China (2018YFB1004903), NSFC(U1911401, U1811461), Guangdong Province Science and Technology Innovation Leading Talents (2016TX03X157), Guangdong NSF Project (No. 2018B030312002), Guangzhou Research Project (201902010037), and Research Projects of Zhejiang Lab (No. 2019KD0AB03).

References

1. Bertasius, G., Soo Park, H., Yu, S.X., Shi, J.: Am i a baller? basketball performance assessment from first-person videos. In: Proceedings of the IEEE International Conference on Computer Vision, pp. 2177–2185 (2017)
2. Carreira, J., Zisserman, A.: Quo vadis, action recognition? a new model and the kinetics dataset. In: Proceedings of the IEEE Conference on Computer Vision and Pattern Recognition, pp. 6299–6308 (2017)
3. Chen, J., Wang, Y., Qin, J., Liu, L., Shao, L.: Fast person re-identification via cross-camera semantic binary transformation. In: The IEEE Conference on Computer Vision and Pattern Recognition (CVPR), July 2017
4. Doughty, H., Damen, D., Mayol-Cuevas, W.: Whoś better, whoś best: skill determination in video using deep ranking. In: Proceedings of the IEEE Conference on Computer Vision and Pattern Recognition (2018)
5. Doughty, H., Mayol-Cuevas, W., Damen, D.: The pros and cons: Rank-aware temporal attention for skill determination in long videos, June 2019
6. Gao, Y., et al.: Jhu-isi gesture and skill assessment working set (jigsaws): a surgical activity dataset for human motion modeling. In: MICCAI Workshop: M2CAI, vol. 3, p. 3 (2014)
7. Gattupalli, S., Ebert, D., Papakostas, M., Makedon, F., Athitsos, V.: Cognilearn: a deep learning-based interface for cognitive behavior assessment. In: Proceedings of the 22nd International Conference on Intelligent User Interfaces, pp. 577–587. ACM (2017)
8. Gers, F.A., Schmidhuber, J., Cummins, F.: Learning to forget: Continual prediction with LSTM. In: IET Conference Proceedings, vol. 5, pp. 850–855, January 1999
9. Ilg, W., Mezger, J., Giese, M.: Estimation of skill levels in sports based on hierarchical spatio-temporal correspondences. In: Michaelis, B., Krell, G. (eds.) DAGM 2003. LNCS, vol. 2781, pp. 523–531. Springer, Heidelberg (2003). https://doi.org/10.1007/978-3-540-45243-0_67
10. Li, H., Cai, Y., Zheng, W.S.: Deep dual relation modeling for egocentric interaction recognition. In: The IEEE Conference on Computer Vision and Pattern Recognition (CVPR), June 2019
11. Li, W.H., Hong, F.T., Zheng, W.S.: Learning to learn relation for important people detection in still images. In: Computer Vision and Pattern Recognition (2019)

12. Li, W.H., Li, B., Zheng, W.S.: Personrank: detecting important people in images. In: International Conference on Automatic Face & Gesture Recognition (FG 2018) (2018)
13. Malpani, A., Vedula, S.S., Chen, C.C.G., Hager, G.D.: Pairwise comparison-based objective score for automated skill assessment of segments in a surgical task. In: Stoyanov, D., Collins, D.L., Sakuma, I., Abolmaesumi, P., Jannin, P. (eds.) IPCAI 2014. LNCS, vol. 8498, pp. 138–147. Springer, Cham (2014). https://doi.org/10.1007/978-3-319-07521-1_15
14. Paiement, A., Tao, L., Hannuna, S., Camplani, M., Damen, D., Mirmehdi, M.: Online quality assessment of human movement from skeleton data. In: British Machine Vision Conference, pp. 153–166. BMVA Press (2014)
15. Pan, J.H., Gao, J., Zheng, W.S.: Action assessment by joint relation graphs. In: The IEEE International Conference on Computer Vision (ICCV), October 2019
16. Parmar, P., Morris, B.T.: What and how well you performed? a multitask learning approach to action quality assessment. In: The IEEE Conference on Computer Vision and Pattern Recognition (CVPR), June 2019
17. Parmar, P., Tran Morris, B.: Learning to score olympic events. In: Proceedings of the IEEE Conference on Computer Vision and Pattern Recognition Workshops, pp. 20–28 (2017)
18. Parmar, P., Tran Morris, B.: Action quality assessment across multiple actions. In: 2019 IEEE Winter Conference on Applications of Computer Vision (WACV), pp. 1468–1476, January 2019. https://doi.org/10.1109/WACV.2019.00161
19. Pérez, J.S., Meinhardt-Llopis, E., Facciolo, G.: Tv-l1 optical flow estimation. Image Processing On Line, pp. 137–150 (2013)
20. Pirsiavash, H., Vondrick, C., Torralba, A.: Assessing the quality of actions. In: Fleet, D., Pajdla, T., Schiele, B., Tuytelaars, T. (eds.) ECCV 2014. LNCS, vol. 8694, pp. 556–571. Springer, Cham (2014). https://doi.org/10.1007/978-3-319-10599-4_36
21. Scarselli, F., Gori, M., Tsoi, A.C., Hagenbuchner, M., Monfardini, G.: The graph neural network model. IEEE Trans. Neural Netw. **20**(1), 61–80 (2009)
22. Sharma, Y., et al.: Video based assessment of osats using sequential motion textures. Georgia Institute of Technology (2014)
23. Solomon Mathialagan, C., Gallagher, A.C., Batra, D.: VIP: finding important people in images. In: Computer Vision and Pattern Recognition (2015)
24. Vaswani, A., et al.: Attention is all you need. In: Advances in Neural Information Processing Systems 30, pp. 5998–6008. Curran Associates, Inc. (2017). http://papers.nips.cc/paper/7181-attention-is-all-you-need.pdf
25. Wang, Z., Lu, J., Tao, C., Zhou, J., Tian, Q.: Learning channel-wise interactions for binary convolutional neural networks. In: The IEEE Conference on Computer Vision and Pattern Recognition (CVPR), June 2019
26. Xu, C., Fu, Y., Zhang, B., Chen, Z., Jiang, Y.G., Xue, X.: Learning to score the figure skating sports videos. arXiv preprint arXiv:1802.02774 (2018)
27. Yan, S., Xiong, Y., Lin, D.: Spatial temporal graph convolutional networks for skeleton-based action recognition. In: Thirty-Second AAAI Conference on Artificial Intelligence (2018)
28. Zhang, Q., Li, B.: Video-based motion expertise analysis in simulation-based surgical training using hierarchical dirichlet process hidden markov model. In: Proceedings of the 2011 international ACM workshop on Medical multimedia analysis and retrieval, pp. 19–24. ACM (2011)

29. Zhang, Q., Li, B.: Relative hidden markov models for video-based evaluation of motion skills in surgical training. IEEE transactions on pattern analysis and machine intelligence **37**(6), 1206–1218 (2015)
30. Zia, A., Essa, I.: Automated surgical skill assessment in RMIS training. Int J CARS **13**, 731–739 (2018)
31. Zia, A., Sharma, Y., Bettadapura, V., Sarin, E.L., Clements, M.A., Essa, I.: Automated assessment of surgical skills using frequency analysis. In: Navab, N., Hornegger, J., Wells, W.M., Frangi, A.F. (eds.) MICCAI 2015. LNCS, vol. 9349, pp. 430–438. Springer, Cham (2015). https://doi.org/10.1007/978-3-319-24553-9_53
32. Zia, A., Sharma, Y., Bettadapura, V., Sarin, E.L., Essa, I.: Video and accelerometer-based motion analysis for automated surgical skills assessment. Int. J. Comput. Assisted Radiol. Surgery **13**(3), 443–455 (2018)
33. Zia, A., et al.: Automated video-based assessment of surgical skills for training and evaluation in medical schools. Int. J. Comput. Assisted Radiol. Surgery **11**(9), 1623–1636 (2016)

High-Quality Single-Model Deep Video Compression with Frame-Conv3D and Multi-frame Differential Modulation

Wenyu Sun⬤, Chen Tang, Weigui Li, Zhuqing Yuan, Huazhong Yang⬤,
and Yongpan Liu$^{(\boxtimes)}$⬤

Tsinghua University, Beijing, China
ypliu@tsinghua.edu.cn

Abstract. Deep learning (DL) methods have revolutionized the paradigm of computer vision tasks and DL-based video compression is becoming a hot topic. This paper proposes a deep video compression method to simultaneously encode multiple frames with Frame-Conv3D and differential modulation. We first adopt Frame-Conv3D instead of traditional Channel-Conv3D for efficient multi-frame fusion. When generating the binary representation, the multi-frame differential modulation is utilized to alleviate the effect of quantization noise. By analyzing the forward and backward computing flow of the modulator, we identify that this technique can make full use of past frames' information to remove the redundancy between multiple frames, thus achieves better performance. A dropout scheme combined with the differential modulator is proposed to enable bit rate optimization within a single model. Experimental results show that the proposed approach outperforms the H.264 and H.265 codecs in the region of low bit rate. Compared with recent DL-based methods, our model also achieves competitive performance.

1 Introduction

Since video data have contributed to more than 80% of internet traffic [8], video is playing an increasingly important role in human life. Therefore, an efficient video compression system is in highly demand. In the past decades, standard video compression algorithms require large amounts of hand-crafted modules. While they have been well engineered and thoroughly tuned for each local module, they cannot be end-to-end optimized for emerging video applications such as object detection, VR applications, video understanding, and so on. Recently, deep learning (DL) methods achieve great success in various computer vision tasks. DL-based image and video compression approaches have achieved comparable or even better performance than the traditional codecs [4,6,14,21,22,26]. Inspired by recent advance in DL-based image and video compression works, we

Electronic supplementary material The online version of this chapter (https://doi.org/10.1007/978-3-030-58577-8_15) contains supplementary material, which is available to authorized users.

propose a fully end-to-end deep video compression model. The keynotes of this paper are summarized as follows.

Frame 3D Convolution for Efficient Multi-frame Fusion. 3D convolution (Conv3D) has been proved as an efficient module for video processing [7,10,13]. Traditional Conv3D slides a window along three dimensions: depth, height and width. Then the partial results are added up along channels. For video tasks, the depth dimension is commonly along multiple frames and for clarity we denote this traditional way as Channel-Conv3D. Although Channel-Conv3D can extract spatial and temporal information of multiple frames at the same time, it needs deeper network to slide the window through all input frames, thus causes high computing cost. To resolve this, we introduce Frame-Conv3D which sets the channel as depth and adds up the partial result along frames. This operation can learn the temporal features of all input frames within one convolution layer.

Multi-frame Differential Modulation to Alleviate the Effect of Quantization Noise in Training. Despite various block structures, most works follow an auto-encoder framework with an extra quantizer after the middle bottle layer [14,17,21,22,26]. The quantizer is necessary to achieve the binary representation and generate the bit-streams. However, it complicates gradient based learning since it is inherently non-differentiable. To make the quantizer trainable, one way is to use soft assignment to approximate quantization [4] and another is to model it with additive uniform noise [21]. Although these methods can enable propagating the gradients, the binarization is in pixel-level and the similarity of frames is not adopted, thus introduce much quantization noise. In this paper, we propose the multi-frame differential modulation that binarize the residual between multiple frames. This method can effectively remove the redundant information between frames and thus minimize the effects of quantization noise during training.

Bit Rate Optimization Within Single Model. To enable bit rate optimization, we propose a dropout scheme to optimize different bit-rate levels within a single model. Our approach is dropping some of the binary representations for transmitting bit. This is an important setting to save computing cost and transmitting bandwidth when variable video qualities are required.

We compare the proposed model with state-of-art video compression codecs (H.264, H.265) and recent DL-based methods [14,26] on two standard datasets of uncompressed video: UVG [2] and HEVC Standard Test Sequences [19]. Our video codec outperforms the standard H.264, H.265 and DL-based method [26] in compression quality measured by peak signal-to-noise ratio (PSNR) and multiscale structural similarity (MS-SSIM) [24]. It also on par with the deep codecs of [14].

The rest of this paper is organized as follows. Section 2 reviews the related works in DL-based image and video compression. Section 3 describes the proposed framework and illustrates the detailed information in implementation. Section 4 gives the experimental results to show the efficiency of the proposed method, followed with a conclusion in Sect. 5.

2 Related Work

2.1 DL-Based Image Compression

DL-based image compressing methods have attracted extensive attention in recent years. Auto-encoders [3–6,12,15,16,20] and recurrent neural networks (RNNs) [21,22] are two popular architectures in image compression. Common loss functions for optimizing the network are the mean-squared error (MSE) [4,5,20] and MS-SSIM [6,22]. To improve the subjective visual quality, generative adversarial networks (GANs) are adopted in [3,16]. In addition, other techniques to improve the image compression performance includes generalized divisive normalization (GDN) [5], multi-scale image decomposition [16], and importance map [12]. These learned approaches provide well guidance for video compression.

2.2 DL-Based Video Compression

Compared with image compression, video needs efficient methods to remove the inter-picture redundancy. For this, spatio-temporal auto-encoder is an effective structure, as it extend the convolutional auto-encoder formulation with the ability to extract features of spatial and temporal information at the same time. Chen et al. [23] divide video frames into 32×32 blocks and use an anto-encoder to compress the block. They perform motion estimation and compensation with traditional methods and the encoded representations are directly quantized and coded by the Huffman method. This approach is not totally end-to-end and cannot be competitive with H.264. Wu et al. [26] propose image interpolation and Conv-LSTM to compress frames iteratively. To reconstruct high-quality video, the Conv-LSTM based codec has to work for several recurrent loops, which results in long running times for both encoding and decoding. Lu et al. [14] propose a real end-to-end deep video coding scheme. They employ an extra motion compensation network to calculate the optical flow and compensate for motion between current and previous frames. This scheme achieves better compression efficiency than H.264, and can be competitive with H.265 when evaluated with MS-SSIM. However, multiple models have to be trained in [14] for different levels of bit rate.

2.3 Quantizer for Deep Learning

To generate the bit-stream for image or video compression, a trainable quantizer is necessary in DL framework. Binarizing feature maps in neural networks has been studied in several works for network quantization [9] and image compression [4,21,22]. To propagate gradients through the non-differentiable quantizer, methods can be divided into two categories: stochastic regularization [9,22] and soft assignment [4]. The former replaces quantization by adding noise and the later adopts a soft function to approximate the $sign(x)$. Although these quantizers can be plugged in the DL framework for backward optimization, the methods are pixel independent. They binarize each pixel without considering the dependency between nearing pixels. As a result, the mentioned quantizer cannot benefit from frame similarity to compress the features into binary. Instead, this paper propose a quantizer based on differential modulation, which can effectively eliminating redundant information of multiple frames for binary compression.

3   Proposed Method

3.1   Overview of the Proposed Method

A video is composed by a sequence of N frames: $F^N = \{f_1, f_2, ..., f_N\}$, where each frame $f_i \in R^{C \times H \times W}$ has $H \times W$ pixels and C channels (for RGB frames, C is 3). If we stack these N frames in the second dimension together, the input video sequence can also be described as a 4D tensor: $F^N \in R^{C \times N \times H \times W}$ with spatial and temporal information. Figure 1 gives a high-level description of the proposed video compressing framework. We encode and decode multiple frames together to improve the compressing efficiency. As an auto-encoder, the network structure is composed by an encoder E_ϕ, a differential quantizer Q and a decoder D_θ, where ϕ and θ are the parameters of neural network. Assuming N frames $\hat{F}^N_{t-1}$ have been decoded at last time $t-1$, to encode the N frames F^N_t at current time t, we first apply a prediction network P_σ on $\hat{F}^N_{t-1}$ and then encode the residual $r_t = F^N_t - P_\sigma(\hat{F}^N_{t-1})$, where σ is the parameter of the prediction network. The decoder will reconstruct the residual $\hat{r}_t$ as side information. Algorithm flow for time step t is shown as follows.

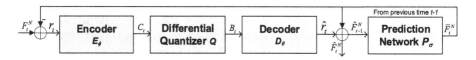

Fig. 1. Proposed end-to-end video compression network composed by the prediction network P_σ, the encoder E_ϕ, the quantizer with differential modulation, and the decoder D_θ.

Input: previous decoded N frames $\hat{F}^N_{t-1} \in R^{3 \times N \times H \times W}$ and current N frames $F^N_t \in R^{3 \times N \times H \times W}$ to encode.

Step 1: prediction network. Based on previous decoded frames $\hat{F}^N_{t-1}$, we use a 3D resnet model P_σ in Fig. 2(d) to predict the current frames $\bar{F}^N_t = P_\sigma(\hat{F}^N_{t-1})$. 3D resnet block (ResB3D) in Fig. 2(a) has been proved effective for many video tasks such as classification and super-resolution [7,10,13].

Step 2: encoder. Based on the predicted frames $\bar{F}^N_t$, we encode the residual $r_t = F^N_t - \bar{F}^N_t$. The encoder E_ϕ is composed by four parts, which is shown in Fig. 2(b). First, we up-sample the channel number from 3 to 64 by Channel-Conv3D. Then a stack of ResB3D with max-pooling layers in height and width are introduced to expand the receptive field and scale down the frame size

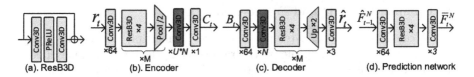

Fig. 2. Network architectures of sub-modules. The yellow modules are built with Channel-Conv3D and the blue are with Frame-Conv3D. The number after × is either the amount of convolution filters or the replica of blocks. (a). The 3D resnet block (ResB3D). (b). The encoder with a Frame-Conv3D to up-sample frames. (c). The decoder with a Frame-Conv3D to down-sample frames. (d). The prediction network using ResB3D (Color figure online).

for compression. After that, we up-sample the number of frames with ratio U by Frame-Conv3D for latter multi-frame differential modulation. The difference between Channel-Conv3D and Frame-Conv3D is discussed in Sect. 3.2. Finally, we down-sample the channel number to 1 with Channel-Conv3D for compression. Thus the encoded tensor C_t before quantizer can be denoted in Eq. 1, and it should have a shape $C_t \in R^{1 \times (U*N) \times H/2^M \times W/2^M}$, where M is the number of max-pooling with 2 kernel size.

$$C_t = E_\phi(r_t) = E_\phi(F_t^N - \bar{F}_t^N) = E_\phi(F_t^N - P_\sigma(\hat{F}_{t-1}^N)) \tag{1}$$

Step 3: quantization with multi-frame differential modulation. A quantizer is necessary to generate the bit-stream $B_t \in \{-1, 1\}^{s(C_t)}$ from float tensor C_t. Trainable quantizer has been well studied in previous works for model quantization [9] and image compression [4,21]. However, these works mainly focus on how to propagate the gradients with this non-differentiable operation, but fail to make full use of the signal and noise characteristic. Instead, we introduce a differential quantizer to alleviate the effect of quantization noise. The key here is to quantize the differential information between up-sampled frames. Details will be given in Sect. 3.3. After this step, the output bit-stream B_t have the same shape of C_t, i.e. $B_t \in R^{1 \times (U*N) \times H/2^M \times W/2^M}$, but each pixel of B_t is quantized into -1 or 1. To measure the number of bits for compressing, we use bits per pixel (BPP) to represent the required bits for each pixel in the current frames. It can be calculated that the maximum BPP is $\frac{U}{4^M}$ in our model.

Step 4: decoder. The decoder D_θ in Fig. 2(c) is nearly the inverse process of the encoder. First, we up-sample the channel from 1 to 64 with Channel-Conv3D and down-sample the frames from $U*N$ to the original N with Frame-Conv3D. Then we decode the side information using 3D resnet and scale up the features to the original size with bilinear interpolation. After that we down-sample the channel number to 3 to reconstruct the side information $\hat{r}_t$. Finally, the decoded N frames $\hat{F}_t^N$ at time step t can be described as Eq. 2.

$$\hat{F}_t^N = \bar{F}_t^N + \hat{r}_t = P_\sigma(\hat{F}_{t-1}^N) + \hat{r}_t = P_\sigma(\hat{F}_{t-1}^N) + D_\theta(B_t). \tag{2}$$

Output: the quantized bit-stream B_t and the reconstructed N frames $\hat{F}_t^N$.

3.2   Channel-Conv3D and Frame-Conv3D

Conv3D can learn the spatio-temporal features of multiple frames effectively. Traditional Conv3D sums up the partial results along the input channel dimension. The output can be denoted in top of Eq. 3, where $*$ is the convolution operation, and C_{out} is the number of 3D filters that determines the output channel. We denote this operation as Channel-Conv3D. The receptive field of one Channel-Conv3D layer is limited by the small kernel size which is usually 3. Multiple layers can extend the receptive field but introduce high computing cost. Therefore, we proposed the Frame-Conv3D that sums up the partial results along the input frame dimension. It is described in bottom of Eq. 3, where N_{out} is the number of output frames. It can be noticed that all features of input frames are fused to the output by summing operation regardless of the kernel size. We replace some traditional Channel-Conv3D with Frame-Conv3D in encoder and decoder to further fuse the features of multiple frames. Another potential benefit is that the number of output frames can be arbitrarily assigned, which is useful for the mult-frame differential modulation (discussed in Sect. 3.3).

$$\text{Channel-Conv3D: out}(C_{out_j}) = \text{bias}(C_{out_j}) + \sum_{i=0}^{C_{in}-1} \text{weight}(C_{out_j}) * \text{input}$$

$$\text{Frame-Conv3D: out}(N_{out_j}) = \text{bias}(N_{out_j}) + \sum_{i=0}^{N_{in}-1} \text{weight}(N_{out_j}) * \text{input} \tag{3}$$

3.3   Differential Quantizer Q

In order to generate a binary feature maps, previous works adopt a pixel-level quantizer. They quantize each pixel of the activation without considering the dependency between frames. Different from any previous methods, we propose a differential quantizer. It is based on the multi-frame differential modulation and can effectively eliminate the redundant information for binary compression. This section first gives the detailed implementation of the differential quantizer in DL framework, and then explains the effectiveness of this method in theory.

Implementation of differential quantizer in DL framework. The differential quantizer is composed by a differential modulator and a trainable quantizer Q_b. After we get the up-sampled frames $C_t = \{c_t[1], c_t[2], ..., c_t[U*N]\}, c_t \in R^{1 \times H/2^M \times W/2^M}$ in Sect. 3.1, we first apply differential modulation to each frame c_t and generated the modulated result y_t. Then a trainable quantizer is applied on y_t to generate the final binary output b_t.

1) **Trainable quantizer Q_b.** We employ the binary technique proposed in [21] to realize a trainable quantizer Q_b for DL framework. We first apply a tanh activation on the input to normalize it into the range of $[-1, 1]$. Then the binary function $B(x)$ for $x \in [-1, 1]$ is defined as $B(x) = x + \epsilon$, $B(x) \in \{-1, 1\}$, ϵ has $(1+x)/2$ probability to be $1 - x$ and $(1-x)/2$ probability to be $-1 - x$. The ϵ is the quantization noise. Therefore, the full function of the

quantizer Q_b according to the modulated signal y_t is $Q_b(y_t) = B(\tanh(y_t))$. For the backward pass of gradients, the derivative of the expectation is taken [21]. Since $\mathbb{E}[B(x)] = x$, the gradients pass through B is unchanged. Once the network are trained, $B(x)$ is replaced by the sign function to get a fixed representation for a particular input.

2) **Forward computing of the differential quantizer Q.** Based on the trainable quantizer Q_b, the forward computing flow of the differential quantizer Q is presented in Algorithm 1. The input C_t has been up-sampled with Frame-Conv3D and each c_t represents a frame in C_t. Q_b is the defined trainable quantizer, $Y_t = \{y_t[1], ..., y_t[n]\}$ are the modulated frames before quantizer and $B_t = \{b_t[1], ..., b_t[n]\}$ are the quantized outputs. Line 3 in Algorithm 1 is the definition of the differential modulation. It should be noticed that the modulation is not simply the difference between $c_t[i]$ and $c_t[i-1]$. The current $y_t[i]$ is modulated not only with $c_t[i]$, but also the $y_t[i-1]$ and $b_t[i-1]$ of last frame. We theoretically identify that the modulator can effectively deal with the quantization noise, which is explained in next part of this section. Since the computing flow only includes a trainable quantizer and operations of addition and subtraction, it is compatible with the DL framework and can be optimized by back propagation algorithm.

Algorithm 1. The forward computing flow of differential quantizer Q

Input: Up-sampled frames $C_t = \{c_t[1], c_t[2], ..., c_t[n]\}$
Output: Quantized output $B_t = \{b_t[1], b_t[2], ..., b_t[n]\}$
1: **Initialization** $y_t[0] \Leftarrow 0$, $b_t[0] \Leftarrow 0$
2: **for** i from 1 to n **do**
3:     $y_t[i] \Leftarrow y_t[i-1] + c_t[i] - b_t[i-1]$ (Differential Modulator)
4:     $b_t[i] \Leftarrow Q_b(y_t[i])$ (Trainable Quantizer)
5: **end for**

---

3) **Back propagation of the differential quantizer Q.** To see how the weights update during back propagation with differential modulation, we calculate the gradients of Algorithm 1. In traditional quantizer, only Q_b has been used, i.e., $b'_t = B(\tanh(c_t))$, where $B(x)$ is the binarization function defined above and has derivative 1. The gradients propagated from c_t to b'_t is $\partial b'_t / \partial c_t = 1 - \tanh^2(c_t)$, which is pixel independent. For the differential quantizer Q, we can rewrite the updating function of line 3 and 4 in Algorithm 1 as

$$b_t[m] = Q_b(y_t[m]) = Q(c_t[m], c_t[m-1], ..., c_t[1]) \tag{4}$$

It can be noticed that $b_t[m]$ is updated not only from $c_t[m]$, but also all previous frames. So the gradients of Q should also propagate from latter frames to previous. The gradients from $b_t[m]$ to $c_t[m-k]$ can be calculated as Eq. 5 (detailed proof is given in the supplementary material).

$$\frac{\partial b_t[m]}{\partial c_t[m-k]} = \begin{cases} 1 - \tanh^2(y_t[m]), & k = 0 \\ (1 - \tanh^2(y_t[m])) \times \prod_{i=1}^{k} \tanh^2(y_t[m-i]), & 1 \leq k \leq m-1. \end{cases}$$

(5)

The first gradient at $k = 0$ is consistent with tradition methods that updating from $b_t[m]$ to $c_t[m]$. It's interesting to analyze the second gradients from $b_t[m]$ to $c_t[m-k]$. According to this equation, the $b_t[m]$ propagates gradients to all previous frames $c_t[m-k]$ with weight $w_k = \prod_{i=1}^{k} \tanh^2(y[m-i])$. This indicates that the network not only optimizes in pixel level, but also makes use of information in previous frames to remove the redundancy between frames. Therefore, this method is more likely to encode a binary distribution with less loss of information for compressing. Since $|\tanh^2(x)| < 1$, an important property of w_k is that $w_0 = 1 > w_1 > w_2 > ... > w_{m-1}$. This distribution of w is intuitively correct since two frames with long distance (larger k) should be less relevant and have smaller weight. This analysis also indicates the importance of tanh activation for network training.

Explanation for the effectiveness of differential quantizer. Based on the computing flow, we can theoretically analyze the effectiveness of the proposed differential quantizer. Notice that the trainable quantizer Q_b in line 4 of Algorithm 1 can be modeled by Eq. 6, where $B(x)$ is the defined binary function, $y_t[i]$ is the modulated frames before quantization, and $e[i]$ is the additive noise. To simplify the formula, we adopt a linear approximation to get the last equality with an equivalent noise $e'[i]$. We identify that in a well-trained network, $y_t[i]$ often have values in $[-1, 1]$. Since $\tanh(y_t[i]) - y_t[i]$ is close to zero when $y_t[i]$ is in $[-1, 1]$, this approximation is acceptable.

$$Q_b(y_t[i]) = B(\tanh(y_t[i])) = \tanh(y_t[i]) + e[i]$$
$$= y_t[i] + (\tanh(y_t[i]) - y_t[i] + e[i]) = y_t[i] + e'[i]$$

(6)

Combining Eq. 6 with the updating function of the differential modulator in Algorithm 1 (line 3), we can subsequently infer the following equation,

$$b_t[i] = c_t[i] + e'[i] - e'[i-1]$$

(7)

Now it's obviously to see how the differential modulation deal with the quantization noise from Eq. 7. It differential modulates the quantization noise between $e'[i]$ and $e'[i-1]$ to generate the quantized output b_t. So the modulator can effectively alleviate the effect of quantization. More specifically, if we adopt a U-tap moving average filter on b_t to recover r^{th} frame $\hat{c}_t[r]$, i.e.,

$$\hat{c}_t[r] = \frac{1}{U} \sum_{i=r+1}^{r+U} b_t[i] = \frac{1}{U} \sum_{i=r+1}^{r+U} c_t[i] + \frac{1}{U}(e'[r+U] - e'[r])$$

(8)

The first item of Eq. 8 is the average of input c_t, and the second is accumulated noise. The noise comes to zero with large U. U is actually corresponding to the up-sample ratio by the Frame-Conv3D mentioned in Sect. 3.1. In implementation, we replace the average filter with trainable convolution to improve the performance. The analysis shows that combing Frame-Conv3D and differential quantizer together can effectively reduce the quantization noise. In fact, this up-sampling and modulation principle is similar to the theory of 1st order $\Delta\Sigma$ modulation [18] in design of analog-to-digital converter (ADC) circuits. Although signal in $\Delta\Sigma$ modulation is usually one-dimensional while the frame is 3-dimensional in video, the method of frequency-domain analysis in $\Delta\Sigma$ can be adopted to visually understand the benefit of differential modulation. Figure 3(a) shows the computing flow of $\Delta\Sigma$ modulation at current time step n in discrete time domain, which is similar to our proposed differential quantizer. In frequency domain, we can compute the signal transfer function (STF) and noise transfer function (NTF). The power spectral density (PSD) of STF and NTF is plotted in Fig. 3(b). The NTF of the modulator is a high-pass filter function which shapes noise e from low frequency band to high. Owing to this noise shaping characteristic, the in-band noise after quantizaion is suppressed and the signal-to-noise ratio (SNR) is improved greatly by adopting $\Delta\Sigma$ modulation, which is shown in Fig. 3(c). Although the same theory can be adopted to analyze the noise effect, we should mention that the frameworks of these two algorithms are totally different. $\Delta\Sigma$ modulation is a forward flow while the proposed multi-frame differential modulation needs to back propagate gradients. The techniques in $\Delta\Sigma$ modulation cannot be simply copied into DL frameworks. For example, higher order $\Delta\Sigma$ modulation can achieve better performance of noise shaping in ADC. However, we identify that using higher order differential modulation cannot achieve improvement, which is discussed in Sect. 4.

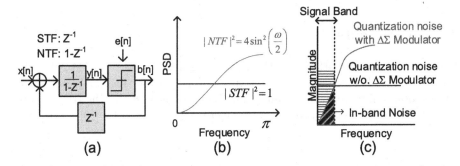

Fig. 3. Theory of the 1st order $\Delta\Sigma$ modulation. (a). Computing flow in Z domain. (b). The power spectral density (PSD) of NTF (red) and STF (black). (c). Noise shaping of $\Delta\Sigma$ (Color figure online).

3.4   Single Model Supporting Multiple Bit Rate

The rate-distortion optimization is significant for video compression. Inspired from the forward and backward flows of differential modulation, we propose a dropout scheme for bit rate optimization. Noticing that the mth quantized frame $b[m]$ receives information from all previous frames $\{c_t[1], ..., c_t[m]\}$, once some later frames $\{c_t[k], c_t[k+1], ..., c_t[m]\}$ are dropped, the former $\{c_t[1], ..., c_t[k]\}$ can still help to update the $b_t[m]$. As a result, we can directly dropout some latter up-sampled frames to decrease bit cost for rate optimization. From experimental results in Sect. 4.2 we find that the proposed differential quantizer can still recover high-quality video frames while traditional methods are failed in case of dropping frame bits. Moreover, this scheme is similar to the dropout training and is compatible with DL framework. Involving this scheme in training can further improve the network performance at different levels of bit rate.

3.5   Training Strategy

To reduce the distortion between original frames F_t^N and reconstructed $\hat{F}_t^N$, we adopt the following loss function:

$$\mathcal{L} = d(F_t^N, \hat{F}_t^N) + \lambda d(F_t^N, \bar{F}_t^N) \tag{9}$$

where $d(F_t^N, \hat{F}_t^N)$ is the reconstructed distortion and $d(F_t^N, \bar{F}_t^N)$ is to control the distortion of prediction network. We find that the second loss can help to achieve some improvement in PSNR. In implementation, we use l_1 loss to measure the distortion and set λ as 0.05. Since previous reconstructed frames $\hat{F}_{t-1}^N$ are required for training, we adopt the same online updating strategy proposed in [14]. The mini-batch size is set as 8 and the training patch size of input frames is 128×128. We use the Adam optimizer [11] by setting the initial learning rate (lr) as 2e-4, β_1 as 0.9, and β_2 as 0.99. The lr is divided by 2 at every 50k iterations.

4   Experiments

We train the models with ZuriVID dataset, which is adopted in AIM 2019 challenge [1] for video extreme super-resolution. The dataset has 50 1080p videos with a total number of 69604 frames. The UVG dataset [2] and the HEVC Standard Test Sequences (Class B, Class C, Class D, and Class E) [19] are used to evaluate our model.

4.1   Network Parameters

To search for suitable parameters of coding frame number N and the down-scale times M with max-pooling, control experiments are implemented. For fairly comparison, the BPPs of all models are set the same as 0.25. A subset of UVG dataset including 224 frames of 7 videos is used for validation during training. We analyze four cases and the training results are shown in Fig. 4(a). We find that fewer frame number of 4 can converge faster than 8, and both achieve a similar performance at last. Down-scaling the feature to $1/4 \times 1/4$ with $M = 2$ has a better performance than down-scaling to $1/8 \times 1/8$ with $M = 3$. We also identify that tanh activation is important for back propagation. Based on the experiments, we set N as 4 and M as 2 in our final model.

The paper is focused on the proposed differential quantizer in 1st order. To confirm whether higher order modulation takes effect, we compare 1st and 2nd order differential modulation. The 2nd order modulator utilizes former two frames to modulate the current, and the detailed implementation is provided in the supplementary material. The results of evaluated PSNR and MS-SSIM are presented in Fig. 4 (b). It shows that introducing higher order modulator takes little effects, and may even cause a decreasing in metrics. Theoretically, although 2nd order modulator uses extra information at $n-1$ and $n-2$ to update the current $c_t[n]$ while 1st order only use $n-1$, the updating paths of gradients are similar, i.e., propagating from $b_t[n]$ to $c_t[n]$ and all previous $c_t[n-1], ..., c_t[1]$. So they should get similar results ultimately with the back propagation algorithm. However, high order modulation introduces higher computing costs and makes the rule of updating gradients extremely complex. From this perspective, 1st order $\Delta\Sigma$ quantizer is superior.

Besides, we also remove the prediction network or the second prediction loss $d(F_t^N, \bar{F}_t^N)$ in Eq. 9 for further comparison. We notice that either removing the prediction network or loss can substantially decreasing the performance, as shown in Fig. 4 (b). This suggests that motion estimation and compensation from previous decoded frames to currents is important for video compression. The current prediction network is a multi-layer 3D CNN. To improve the performance, more complex motion estimation model can be considered, such as models with optical flow and block motion vectors.

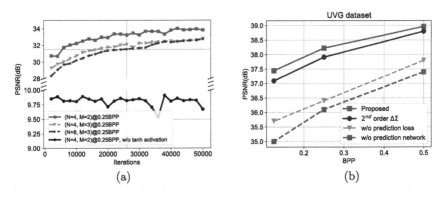

(a)   (b)

Fig. 4. (a). Evaluated PSNR during training with different parameters: frame number N and down-scale times M. (b). Comparison between models of the 1st order differential quantizer (proposed), the 2nd order differential quantizer, and removing the prediction loss.

4.2   Advantages of Differential Quantizer with Visual Results

To illustrate the advantages of differential quantizer in visual, we train 3 models for ablation study, i.e., up-sampling and down-sampling in encoder and decoder by:

(A). only using Channel-Conv3D with traditional quantizer Q_b;
(B). using Frame-Conv3D with traditional quantizer Q_b;
(C). using Frame-Conv3D with differential quantizer Q.

For a better visual comparison, we include more frames as input and set N as 8 to show the efficient frame fusion of differential modulation. In order to better visualize the middle quantized results in case of dropping frames, the prediction network is removed when comparing between model A/B/C. All hyperparameters and model blocks are set with no difference for these 3 models except for the settings of quantizer. Model A replaces all the Frame-Conv3D with $\times 64$ Channel-Conv3D and change the last Conv3D to $\times U$ in Fig. 2(b) to guarantee the same BPP level with model B and C. We set the number of max-pooling M as 2 and the up-sampling ratio U as 16 and 4, which results in 1 and 0.25 maximum BPP, respectively. All models are trained for up to 50×10^3 iterations.

Figure 5 shows the evaluated PSNR of the subset in training process. It can be observed that model B with Frame-Conv3D trains better than model A. Model C with differential quantizer can improve the metric significantly within 50k training iterations, especially at low BPP, e.g., 0.25. The curves show that the proposed method can effectively train a network for binary representation, which has been analyzed in Sect. 3.3.

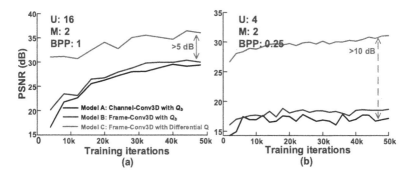

Fig. 5. Evaluated PSNR of model A,B, and C during the training phase. (a). One quarter down-scaled with 1 BPP. (b). One eighth down-scaled with 0.25 BPP.

To identify the effectiveness of differential quantizer in case of dropping bits, we evaluate the well-trained three models only with former half frames at case of $U = 16$. Since U is 16 and N is 8, there are 128 up-sampled frames in total. We reserve the former 64 frames and set the latter 64 frames to zero, which is shown in Fig. 6(a). Model A cannot fuse the frame information properly and the latter recovered frames will be noisy gradually. The average PSNR decreases up to 10 dB. Model B can reconstruct all frames because of the efficient feature fusion along the frame dimension with Frame-Conv3D. However, the reconstructed frames still lose much detailed information, resulting in a low video quality according to PSNR (around 8 dB loss). For model C with differential quantizer, all frames are decoded in good quality and the decreasing in PSNR is less than 5dB. It should be mentioned that all the above models are not particularly optimized in training phase for the case of dropping bits. It shows that differential modulation can make full use of all past frame information for recovering. Adopting dropout scheme in training phase can certainly further improve the performance.

The entropy of generated bit streams for each model is also calculated and compared in Fig. 6(b). When coding with different bit width, model C achieves the highest entropy and there is nearly no room for compression, which means it successfully remove the redundant information between frames and no extra entropy coding is needed. However, margins of compression still exist for the cases without differential modulation. When the coding bit width is 16, model A and B can further achieve nearly 30% compressing gain by entropy coding.

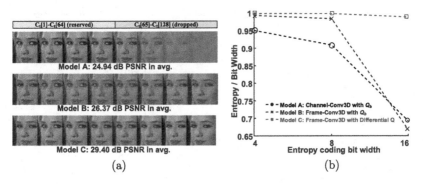

Fig. 6. Experimental comparison between Model A, B, and C. (a). Reconstructed frames of three models in case of dropping bits. (b). Entropy of bit-streams generated by three models.

4.3   Comparison to Previous Works

We train the final model in two steps. First, a model based on differential quantizer is obtained by 100k training iterations without dropout scheme. Then the pre-trained model is re-trained combined with dropout scheme for bit rate optimization. Although we can drop arbitrary number of frames, in implementation we only define 3 fixed patterns that drop {50%, 75%, 87.5%} frames of all to speed up training. As a result, the trained single model can encode videos into {0.5, 0.25, 0.125} BPPs respectively. We compare our method with H.264, H.265 and two DL-based methods of [14, 26] in terms of PSNR and MS-SSIM metrics.

We follow the setting in [14] and use the FFmpeg tool with very fast mode to generate the compressed frames of H.264 and H.265.

We find that our model achieves high performance on UVG and HEVC Class D and Class E dataset, which is shown in Fig. 7. Compared with H.264 and H.265 codecs on the three dataset, the model shows superiority especially in low BPP region. It also outperforms the DL-based method of [26] and is competitive with [14] on UVG dataset. It should be noticed that our method jointly trains a single model to support different BPP levels, while the work by Lu *et al.* [14] needs to train multiple models for each BPP level.

However, our model shows some weaknesses on HEVC Class B and Class C dataset (results are provided in supplementary materials) because of the fast moving objects (BasketballDrill, BasketballDrive, RaceHorses, etc.) in such videos. It can be improved by adopting optical flow or motion vectors in the prediction network. Despite this, we pay more attentions to the efficient training method for the quantizer and achieve high performance on vast majority of videos. Other techniques for the motion estimation can be easily plugged into this framework to further improve the performance, which is our future work.

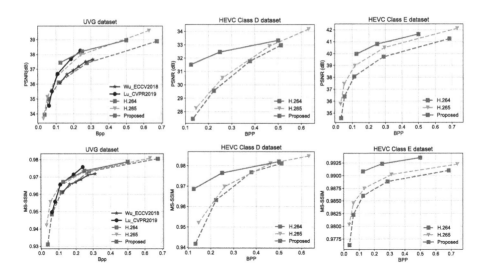

Fig. 7. Comparison between our model and H.264 [25], H.265 [19], the DL-based video method in [26] and [14]. The data for DL-based codecs are from paper report, and the H.264 and H.265 results are from FFmpeg tool.

5　Conclusions

This paper proposes an end-to-end model for deep video compression. We introduce the Frame-Conv3D and multi-frame differential modulation in codecs to reconstruct high-quality video. Different from traditional quantizer with pixel-level binarization, the proposed method of differential modulation can quantize the difference and propagate gradients between frames to remove the redundant information as much as possible. We identify the differential quantizer can alleviate the effect of quantization noise by theoretical analysis and experimental results. Inspired from the gradient propagating path of differential quantizer, we propose a dropout scheme to optimize bit rate within single model. Experimental results show that the proposed model outperforms standard H.264 and H.265 in low BPP region and is competitive with the recent deep codecs. Based on the proposed model, other techniques of optical flow or motion vector estimation can be easily plugged into this framework to further improve the performance.

Acknowledgment. This work is supported in part by National Key R&D Program (2019YFB2204204), NSFC Grant (No. 61934005,61674094), Beijing National Research Center for Information Science, Technology, and Beijing Innovation Center for Future Chip.

References

1. Advances in image manipulation workshop and challenges on image and video manipulation. http://www.vision.ee.ethz.ch/aim19/. Accessed 10 Nov 2019
2. Ultra video group test sequences. http://ultravideo.cs.tut.fi. Accessed 7 Nov 2019
3. Agustsson, E., Tschannen, M., Mentzer, F., Timofte, R., Van Gool, L.: Generative adversarial networks for extreme learned image compression. In: 2019 IEEE/CVF International Conference on Computer Vision (ICCV), pp. 221–231, October 2019. https://doi.org/10.1109/ICCV.2019.00031
4. Agustsson, E., et al.: Soft-to-hard vector quantization for end-to-end learning compressible representations. In: Advances in Neural Information Processing Systems, pp. 1141–1151 (2017)
5. Ballé, J., Laparra, V., Simoncelli, E.P.: End-to-end optimized image compression. arXiv preprint arXiv:1611.01704 (2016)
6. Ballé, J., Minnen, D., Singh, S., Hwang, S.J., Johnston, N.: Variational image compression with a scale hyperprior (2018)
7. Caballero, J., Ledig, C., Aitken, A., Acosta, A., Shi, W.: Real-time video super-resolution with spatio-temporal networks and motion compensation (2016)
8. Cisco, V.: Cisco visual networking index: Forecast and trends, 2017–2022. White Paper 1 (2018)
9. Courbariaux, M., Hubara, I., Soudry, D., Elyaniv, R., Bengio, Y.: Binarized neural networks: training deep neural networks with weights and activations constrained to +1 or −1. arXiv: Learning (2016)
10. Diba, A., Pazandeh, A.M., Van Gool, L.: Efficient two-stream motion and appearance 3D CNNS for video classification. arXiv preprint arXiv:1608.08851 (2016)
11. Kingma, D.P., Ba, J.: Adam: A method for stochastic optimization. arXiv: Learning (2014)
12. Li, M., Zuo, W., Gu, S., Zhao, D., Zhang, D.: Learning convolutional networks for content-weighted image compression. In: 2018 IEEE/CVF Conference on Computer Vision and Pattern Recognition. pp. 3214–3223 (June 2018). https://doi.org/10.1109/CVPR.2018.00339

13. Li, S., He, F., Du, B., Zhang, L., Xu, Y., Tao, D.: Fast spatio-temporal residual network for video super-resolution. arXiv preprint arXiv:1904.02870 (2019)
14. Lu, G., Ouyang, W., Xu, D., Zhang, X., Cai, C., Gao, Z.: DVC: an end-to-end deep video compression framework. In: 2019 IEEE/CVF Conference on Computer Vision and Pattern Recognition (CVPR). pp. 10998–11007, June 2019. https:// doi.org/10.1109/CVPR.2019.01126
15. Minnen, D., Ballé, J., Toderici, G.: Joint autoregressive and hierarchical priors for learned image compression (2018)
16. Rippel, O., Bourdev, L.: Real-time adaptive image compression. arXiv: Machine Learning (2017)
17. Rippel, O., Nair, S., Lew, C., Branson, S., Anderson, A.G., Bourdev, L.: Learned video compression. In: Proceedings of the IEEE International Conference on Computer Vision, pp. 3454–3463 (2019)
18. Schreier, R., Pavan, S., Temes, G.C.: Understanding delta-sigma data converters— the delta-sigma toolbox https://doi.org/10.1002/9781119258308, pp. 499–537 (2016)
19. Sullivan, G.J., Ohm, J.R., Han, W.J., Wiegand, T.: Overview of the high efficiency video coding (HEVC) standard. IEEE Trans. Circuits Syst. Video Technol. **22**(12), 1649–1668 (2012)
20. Theis, L., Shi, W., Cunningham, A., Huszár, F.: Lossy image compression with compressive autoencoders (2017)
21. Toderici, G., et al.: Variable rate image compression with recurrent neural networks. Computer Science (2015)
22. Toderici, G., et al.: Full resolution image compression with recurrent neural networks. In: Proceedings of the IEEE Conference on Computer Vision and Pattern Recognition, pp. 5306–5314 (2017)
23. Tong, C., Liu, H., Qiu, S., Tao, Y., Zhan, M.: Deepcoder: a deep neural network based video compression. In: 2017 IEEE Visual Communications and Image Processing (VCIP) (2017)
24. Wang, Z., Simoncelli, E.P., Bovik, A.C.: Multiscale structural similarity for image quality assessment. In: 2003. Conference Record of the Thirty-Seventh Asilomar Conference on Signals, Systems and Computers (2003)
25. Wiegand, T., Sullivan, G.J., Bjontegaard, G., Luthra, A.: Overview of the h. 264/AVC video coding standard. IEEE Trans. Circuits Syst. Video Technol. **13**(7), 560–576 (2003)
26. Wu, C.Y., Singhal, N., Krahenbuhl, P.: Video compression through image interpolation. In: Proceedings of the European Conference on Computer Vision (ECCV), pp. 416–431 (2018)

Instance-Aware Embedding for Point Cloud Instance Segmentation

Tong He, Yifan Liu, Chunhua Shen$^{(\boxtimes)}$, Xinlong Wang, and Changming Sun

The University of Adelaide, Adelaide, Australia
{tong.he,yifan.liu04,chunhua.shen,xinlong.wang}@adelaide.edu.au,
changming.sun@data61.csiro.au

Abstract. Although recent works have made significant progress in encoding meaningful context information for instance segmentation in 2D images, the works for 3D point cloud counterpart lag far behind. Conventional methods use radius search or other similar methods for aggregating local information. However, these methods are unaware of the instance context and fail to realize the boundary and geometric information of an instance, which are critical to separate adjacent objects. In this work, we study the influence of instance-aware knowledge by proposing an Instance-Aware Module (IAM). The proposed IAM learns discriminative instance embedding features in two-fold: (1) Instance contextual regions, covering the spatial extension of an instance, are implicitly learned and propagated in the decoding process. (2) Instance-dependent geometric knowledge is included in the embedding space, which is informative and critical to discriminate adjacent instances. Moreover, the proposed IAM is free from complicated and time-consuming operations, showing superiority in both accuracy and efficiency over the previous methods. To validate the effectiveness of our proposed method, comprehensive experiments have been conducted on three popular benchmarks for instance segmentation: ScannetV2, S3DIS, and PartNet and achieve state-of-the-art performance. The flexibility of our method allows it to handle both indoor scenes and CAD objects.

Keywords: 3D point cloud · Instance segmentation · Instance-aware

1 Introduction

The task of instance segmentation has recently gained popularity. As an extension to semantic segmentation, this task needs to separate pixels/points that have identical categories into individual groups. In the 2D image domain, many approaches [4,5,10,12,18] have been proposed and achieve promising results. With the growth of the availability of 3D sensors, more and more researches have focused on 3D scene understanding, which is a fundamental necessity for robotic vision, autonomous driving, and virtual reality. Although instance segmentation in the 3D domain has started to draw attention and has been discussed

© Springer Nature Switzerland AG 2020
A. Vedaldi et al. (Eds.): ECCV 2020, LNCS 12375, pp. 255–270, 2020.
https://doi.org/10.1007/978-3-030-58577-8_16

Input Point Cloud   **With Instance-Aware**   **Without Instance-Aware**
Knowledge   **Knowledge**

Fig. 1. Comparison of the instance segmentation results with and without the proposed Instance-Aware Module (IAM). The proposed IAM successfully encodes instance-aware information and geometric knowledge, which are critical for separating adjacent instances. Note that different instances can be presented in different colours. (Color figures online)

in [21,29,30,33,34], it still lags behind its 2D image counterpart and far from being solved.

Similar to the tasks of dense prediction in 2D images [2,16,35], context is also important in 3D domain. For 3D point clouds, PointNet++ [24] is the first work that captures local structure information and has been successfully utilized in the task of semantic segmentation. It maintains an encoder-decoder architecture, which includes several set-abstraction layers and feature-propagation layers for down-sampling and up-sampling, respectively. Algorithms such as radius search and k nearest neighbours (K-NN) search are utilized for aggregating local context knowledge. Building on this powerful network, many methods [21,29,30] have been proposed to tackle the task of instance segmentation on point clouds. To encode meaningful context information, ASIS [30] is proposed to associate two tasks together so they can cooperate with each other. JSIS3D [21] applied multi-value Conditional Random Field (CRF) that formulates a joint optimization for semantic segmentation and instance segmentation in a unified framework. However, these methods fail to explicitly encode the *instance contextual knowledge* and *geometric information*, which are extremely critical for separating adjacent instances and handling complex situations. For example, two neighbouring chairs can be easily confused and grouped as one united instance if boundaries and geometric information are not encoded in the embedding space (e.g., the second row in Fig. 1). In this paper, we address the problem by proposing an Instance-Aware Module (IAM) to learn the instance level context by locating representative regions for each input point. Moreover, geometric knowledge is explicitly encoded in the embedding space, which is an informative indicator to identify the points belonging to the same instance. The whole framework can be trained in an end-to-end manner to tackle instance segmentation and semantic segmentation simultaneously with little computation resource overhead.

Specifically, as shown in Fig. 2, our method maintains an encoder-decoder architecture. Different from previous methods that only maintain an instance grouping branch and a semantic segmentation branch, we come up with a novel light-weight instance-aware module, which localizes representative points within the same instance for each input point. The information from these representative points is then aggregated into the decoding process of the instance branch, generating instance-aware contexts for learning discriminative point-level embeddings. Moreover, the normalized geometric centroids of these representative points (predicted by every input point feature), are directly added to the embedding space, which provides critical geometric knowledge for identifying and reducing the ambiguity of adjacent instances.

The training of the instance-aware module is regularized jointly by the bounding box and instance segmentation supervision, such that the meaningful semantic regions can be tightly bonded by the spatial extension of the instance and guided towards representative regions of the instance.

Compared with the conventional representation of an instance by using vertexes to represent a bounding box, learning semantically meaningful regions helps to remove unrelated background and noise information. As it is applied in the bottleneck layer, very few additional computations are introduced. Compared with ASIS [30], which needs to search neighbours of every input point exhaustively, our approach shows superiority in both efficiency and effectiveness.

To validate the effectiveness of our proposed method, extensive experiments have been conducted on three popular benchmarks. The flexibility of our method allows it to be applied in not only indoor scenes but objects with fine-grained part labels. State-of-the-art performances are achieved on these datasets. To summarize, our main contributions are listed as follows.

- We propose a novel Instance-Aware Module, which successfully encodes instance-dependent context information for point cloud instance segmentation.
- Our method explicitly encodes instance-related geometric information, which is informative and helpful to produce discriminative embedding features.
- The proposed framework can be trained in an end-to-end manner and shows superiority over previous methods on both efficiency and effectiveness. With the proposed method, state-of-the-art results are achieved on different tasks.

2 Related Works

Instance segmentation on point clouds has just started to be discussed recently. In this section, we briefly review some existing approaches that are related to this field.

2.1 Deep Learning on Point Clouds

Deep learning-based methods for 3D feature extraction can be roughly categorized into three classes: voxel-based, multi-view-based, and point-based. Voxel-based methods [9,19,25,32] utilize 3D convolution neural networks for feature extraction on voxelized spatial grids, which can be easily influenced by the density of the points. Meanwhile, it is highly constrained by the huge memory occupation and lower running speed because a large proportion of computation is wasted on vacant voxels. Many approaches have been proposed to address the problem [9,25]. Octree [25] tries to modify the convolution operation by generating average hidden states in empty space. SparseConv [9] is proposed to process spatially sparse data more efficiently by encoding with a Hash Table to avoid unnecessary memory usage in vacant space. The second category is multi-view-based methods [13,23,26], which first project 3D shapes or point clouds into 2D images and utilize conventional 2D CNN for feature extraction. Hou *et al.* proposed 3D-SIS [13] by leveraging both RGB 2D input and 3D geometrical information. 2D features are then back-projected into 3D grids. Unlike the above methods, directly extracting features on point clouds is more efficient and straightforward. PointNet [22] is the pioneering work that directly learns a spatial encoding of each point. A symmetrical function is utilized to process disordered point sets. To effectively encode local context information to obtain representative features, many approaches [14,15,24,27,28] have been proposed. Qi *et al.* proposed PointNet++ [24] which applied PointNet recursively on a nested partitioning of the input point clouds. Thomas *et al.* came up with KPConv [27] by designing a continuous weight space through interpolating with several kernel points. In our experiments, we utilize PointNet++ as the backbone to verify the effectiveness of our method.

2.2   Instance Segmentation on Point Cloud

Although the task of instance segmentation on 2D images has made huge progress since Mask-RCNN [10] was proposed, its 3D point cloud counterpart lags far behind. SGPN [29] is the first deep-learning-based method developed in this field. It tried to generate point cloud groups by predicting three objectives: the similarity matrix, the confidence map, and the semantic prediction map. Due to the pair-wise term, the method occupies a large amount of GPU memories and suffers from slow running speed and small batch size for training. On the other hand, generating instance groups from three matrices requires many hyper-parameters, making it less stable for different scenarios. Wang *et al.* proposed ASIS [30] to address the problem by removing the pair-wise prediction and introducing a discriminative loss for instance embedding. The loss pulls the embeddings of the same instance towards the cluster center and pushes the cluster centers away from each other. However, the method fails to utilize the geometrical information and is unaware of the spatial distribution of the instances. GSPN [34], proposed by Yi *et al.*, generates shape proposals using a generative model for instance segmentation. Due to its emphasis on geometric understanding for object proposal, it achieved promising performance on both indoor dataset and part instances dataset. Due to the large requirement of GPU memory and a two-step training procedure, it is ineffective with limited computation resources. MPNet [11] proposed a memory-based module to deal with the imbalance of the point cloud data. In this work, we propose an Instance-Aware Module (IAM) to encode instance context knowledge and geometric information. The state-of-the-art performance on three large open benchmarks shows superiority over previous methods in both effectiveness and efficiency.

3   Method

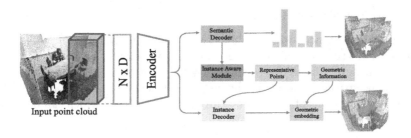

Fig. 2. The whole framework of our proposed one-stage method, which is a simple and clear encoder-decoder architecture. The input point clouds first go through a shared encoder network, and two parallel decoders are followed: one for semantic segmentation, one for instance grouping. A novel instance aware module (IAM) is proposed to generate representative points for instance segmentation. We use the coordinates of representative points to select argument features for instance segmentation module and the geometric information of the coordinates to extend the instance embedding. The whole framework is end-to-end trainable.

In this section, we describe our proposed Instance-Aware Module (IAM), which can encode both instance-aware context and instance-related geometric information. Details of the approach are presented below.

3.1   Network Framework

As shown in Fig. 2, we apply an encoder-decoder architecture. The encoder is shared by two tasks and takes point sets $P \in \mathbb{R}^{N \times D}$ as input, where N denotes the total number of the points and D refers to the input feature dimension. The input features can consist of colour and position information, e.g., X, Y, Z, R, G, and B. The decoder contains two parallel branches: one for semantic segmentation, one for instance embedding. The semantic segmentation branch generates per-point classification results $S \in \mathbb{R}^{N \times D_c}$, where D_c is the category number. Focal loss [17] L_{fl} is applied to address the category imbalance during the training process. Besides, the instance branch outputs per-point embedding features $E \in \mathbb{R}^{N \times D_e}$ for learning a distance metric, where D_e is the embedding dimension. The embeddings belonging to the same instance should end up close together, and the embeddings belonging to the different instances should end up far apart. During the inference, a clustering algorithm is applied to obtain the final grouping results. A novel IAM for producing instance aware knowledge is achieved by detecting the spatial extension of an instance. Through IAM, representative points locating on the corresponding instance provide instance-aware knowledge, which contains two parts: (1) instance-related contextual information via detection a set of regions that are tightly covering the spatial extension of an instance. (2) instance geometric knowledge that is critical for separating adjacent objects.

3.2 Instance-Aware Module

We propose an instance-aware module (IAM) mainly for selecting representative points that capture spatial instance context. For point p_i with position x_i, y_i and z_i, point-level offsets are predicted by the contextual detection branch to represent the spatial extension of the instance, denoted as $\{\Delta x_i^k, \Delta y_i^k, \Delta z_i^k\}_{k=1}^K$. Representative regions of the instance predicted by p_i is $\mathcal{R}_i$, which can be simply represented as:

$$\mathcal{R}_i = \{(x_i + \Delta x_i^k, y_i + \Delta y_i^k, z_i + \Delta z_i^k)\}_{k=1}^K, \tag{1}$$

where K is the number of representative points and i represent the i-th point. The axis-aligned bounding box predicted by every point can be formulated as $\mathcal{B}_i$ through a min-max function F: $\mathcal{B}_i = F(\mathcal{R}_i)$

Learning these representative regions is jointly driven by both the spatial bounding boxes and the instance grouping labels, such that $\mathcal{R}_i$ can tightly compass the instance. To achieve this, three losses are provided: L_{bnd}, L_{cen} and L_{ins} (the last two will be discussed in the next section). L_{bnd} is to maximize the overlaps of the bounding boxes between the prediction and the ground truth. 3D IoU loss is utilized in our paper:

$$L_{bnd} = \frac{1}{N} \sum_{i=1}^{N} 1 - IoU(GT_i, \mathcal{B}_i), \tag{2}$$

where N is the total number of points, $\mathcal{B}_i$ is the predicted bounding box of the i-th point, and GT_i is the 3D axis-aligned bounding box ground truth of the i-th point. To have a better understanding of the detection branch, we visualize $\mathcal{R}_i$ in Fig. 3. Green points are selected p_i, and red points are the predicted $\mathcal{R}_i$. We choose the number of representative points as 18, which empirically works well in our experiments. Employing more points will have limited improvements. Therefore, in terms of efficiency, we choose $K = 18$. Instance related regions are located and successfully cover the spatial extension. In the next section, we provide details of how to incorporate these instance contextual information.

3.3 Instance Branch

Conventionally, the inputs of the instance decoder are down-sampled bottleneck points $P_b \subseteq P$, and the corresponding features are denoted as F_b. These features

(a)   (b)   (c)

Fig. 3. Visualization of detected representative points. The green point is randomly selected, and red points are the corresponding meaningful regions output by the IAM. Due to the encoded instance context information, our method can separate adjacent objects. (Figure best viewed in color) (Color figures online)

are gradually propagated to the full set of points through several up-sampling layers. To encode the instance context during the propagation process, we utilize the meaningful semantic regions of $\mathcal{R}_b$ for the bottleneck points.

Encode Instance-Aware Context. Representations of F_b are augmented by aggregating information from $\mathcal{R}_b$ that covers the instance spatial extent. As these detected points are not necessarily located on the input points, the features of $\mathcal{R}_b$ are interpolated by using K-NN. The interpolated features are then added to the original F_b, generating features containing both local representation and instance context. Compared with ASIS [30], which has to search neighbours for every input point, our method, on the other hand, is more efficient. As K-NN is applied in the bottleneck layer, the searching space in P_b is much smaller than that in P, introducing very limited computation overhead. The combined features are gradually upsampled during the decoding process, propagating the instance-aware context through all points.

Encode Geometric Information. Geometric information is critical for identifying two close objects. To learn a discriminative embedding feature, we directly concatenate the normalized centroids of coordinates to the embedding space. Considering the centroid $C(\mathcal{B}_i)$ predicted by point p_i, where $C(\cdot)$ is the function for computing geometric centroids of a given bounding box, the final per-point embedding feature can be represented as $\hat{E}_i = \text{Concat}(E_i, C(\mathcal{B}_i))$, where E_i is the embedding feature produced from the instance branch. Besides, to force the geometric information to be consistent for the points that have identical instance label, we pull the predicted geometric centroids from the same instance towards the cluster center by:

$$L_{cen} = \frac{1}{M} \sum_{m=1}^{M} \frac{1}{N_m} \sum_{i=1}^{N_m} [\|C(\mathcal{B}_i) - \mu_m\| - \sigma_v]_+^2, \tag{3}$$

where M is the total number of instances, and N_m is the point number for m-th instance. μ_m refers to the average predicted geometric centroids of m-th instance. $[x]_+$ is defined as $[x]_+ = \max(0, x)$ and σ_v is the loose margin. The L_{cen} is designed for forcing the additional geometric information to have less variation and to be informative for separating adjacent objects.

The informative per-point embedding $\{\hat{E}\}_{n=1}^N$ is applied for learning a distance metric that could pull intra-instance embedding toward the cluster center and push instances centers away from each other. The loss function is formulated as:

$$
L_{ins} = \underbrace{\frac{1}{M(M-1)} \sum_{a=1}^{M} \sum_{\substack{b=1 \\ b \neq a}}^{M} [2\sigma_d - \|\mu_a - \mu_b\|]_+^2}_{inter-instance}
$$
$$
+ \underbrace{\frac{1}{M} \sum_{i=1}^{M} \frac{1}{N_m} \sum_{m=1}^{N_m} [\|\mu_m - \hat{E}_m\| - \sigma_v]_+^2,}_{intra-instance} \tag{4}
$$

where M is the total instance number, N_m is the point number of the m-th instance. σ_d and σ_v are relaxation margins. During the training process, the first term pushes instance clusters away from each other and the second term pulls the embedding towards the cluster center. During the inference process, a fast mean-shift algorithm is applied for clustering different instances in the embedding spaces.

To summarize, our method is end-to-end trainable and supervised by four losses. The loss weights for the four losses are all set to 1 in all our experiments.

$$
L = L_{fl} + L_{bnd} + L_{cen} + L_{ins}, \tag{5}
$$

4   Experiments

In this section, we evaluate the effectiveness of our proposed method. Both qualitative and quantitative experiments are conducted and reported.

4.1   Datasets

We introduce three popular datasets that have instance annotations: Stanford 3D Indoor Semantic Dataset (S3DIS) [1], ScanNetV2 [3], and PartNet [20]. S3DIS is collected in 6 large-scale indoor areas, covering 272 rooms. The whole dataset contains more than 215 million points and is consisted of 13 common semantic categories. ScanNetV2 [3] is an RGB-D video dataset. It contains more than 1500 scans, which is split into 1201, 300, and 100 scans for training, validation, and testing, respectively. The dataset contains 40 classes in total, and 13 categories are evaluated. Different from the above two datasets, PartNet [20] is a consistent large-scale dataset with fine-grained object annotations. It consists of more than 570k part instances covering 24 object categories. Each object contains 10000 points. Similar to GSPN [34], we select five categories that have the largest number of training examples.

4.2 Evaluation Metrics

On the S3DIS dataset, we conduct 6-fold cross-validation. Similar to SGPN [29] and ASIS [30], the performance on Area-5 is also reported. On ScanNetV2 [3], we report our results on the validation set, which contains more instances and has more stable results. On the PartNet [20] dataset, five selected categories are Chair, Storage, Table, Lamp, and Vase. Both coarse and fine-grained results are included. Different levels of different categories are trained separately and independently. The evaluation metrics for semantic segmentation are the overall pixel-wise accuracy ($mAcc$), category-wise mean accuracy ($oAcc$) and average intersection-over-union ($mIoU$). The instance segmentation is evaluated by the average instance-wise coverage ($mCov$), mean weighted instance-wise coverage ($mWCov$), mean instance precision ($mPrec$) and recall ($mRec$) with IoU threshold of 0.5. The weights for $mWCov$ is calculated by $w_i = \frac{|N_i|}{\sum_k |N_k|}$, where i is the i-th instance and N_k is the point number of k-th ground truth instance.

4.3 Implementation Details

For the S3DIS [1] and ScanNetV2 [3], each scan contains millions of points, making it hard to process all data at one time. In our experiments, we split each scene into $1m \times 1m$ overlapped blocks with 0.5 m stride. Then, 4,096 points are randomly sampled across each block. Similar to SGPN [29], every point is represented by a 9-D feature (X, Y, Z, R, G, B, and normalized positions in blocks N_X, N_Y, N_Z). PartNet [20], on the other hand, is proposed for shape analysis which contains 10,000 points for each instance. We randomly select 8,000 for training and 10,000 for testing.

Although our method is not restricted to any specific network, all experiments are conducted with vanilla PointNet++ [24] as the backbone (without multi-scale grouping) and leave the other choices for future study. One single GTX1080Ti GPU card is used for training with the batch size set to 16. The initial learning rate is set to 0.01 (0.001 for S3DIS) and divided by 2 in every 300k iterations. We use Adam optimizer with momentum set to 0.9, and the whole network is trained for 100 epochs. The hyper-parameters for discriminative loss are identical with original setting in [30]: $\sigma_v = 0.5$, $\sigma_d = 1.5$. Besides, for testing the whole scene on S3DIS and ScanNetV2, a method named BlockMerging [29] is used for grouping blocks according to the segmentation information of the overlapped areas.

4.4 Ablation Studies

We first build a strong baseline that contains two decoder branches: one is the semantic segmentation, and the other is the instance embedding branch. Two losses are used for supervising the two branches: the cross-entropy loss for the segmentation task and the discriminative loss for instance grouping. The discriminative loss forces points belonging to the same instance to lie close together in the embedding space and keep a large margin for points belonging to different instances. The loss weights are set to 1.0. We conduct our experiments on the ScanNetV2 validation set.

Table 1. Ablation study on ScanNetV2 dataset. Both AP_{50} and AP_{25} are reported on the validation set. **FL** refers to focal loss. **InsContext** refers to instance-aware context. $\mathbf{L_{cen}}$ refers to centroid constrain loss in Eq. 3. **GE** refers to geometric embedding.

Method	FL	InsContext	L_{cen}	GE	AP_{50}	AP_{25}
Baseline					22.0	45.2
	✓				24.0	45.5
	✓	✓			27.6	48.2
	✓	✓	✓		28.9	48.9
Ours	✓	✓	✓	✓	**31.5**	**50.4**

Focal Loss. Focal loss [17] is first proposed in the object detection task to address the problem of data imbalance between positive and negative samples. Due to the imbalance of categories introduced in the point cloud, we apply focal loss in the segmentation branch with default parameters identical to [17]. The results are shown in Table 1, and the focal loss can improve the results by 2.0 for AP_{50}, from 22.0 to 24.0.

Instance Aware Module. We study the influence of the proposed instance-aware module, which first finds out representative points of the instance, and then features from these sampled points are aggregated. Encoding the spatial extension knowledge helps to separate and distinguish close instances. As shown in Table 1, the instance aware decoder boosts the performance by a large margin, improving AP_{50} from 24.0 to 27.6 and AP_{25} from 45.5 to 48.2. Besides, simply enlarging the dimension of the embedding space can not bring further improvement in performance (presented in ASIS [30]). The proposed geometric embedding provides informative knowledge, which brings about 2.6% improvement in AP_{50}, demonstrating the effectiveness of our proposed method. Qualitative results are shown in Fig. 4. Our method shows robustness to the intensive scenes, which require more discriminative features to separate different instances.

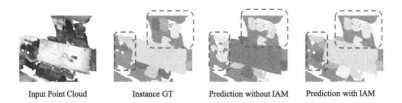

Input Point Cloud   Instance GT   Prediction without IAM   Prediction with IAM

Fig. 4. Comparison of the results with and without the Instance-Aware Module. Due to the successfully encoded instance context and geometric information, our method generates discriminative results, especially for nearby objects.

Table 2. Instance segmentation results on the S3DIS dataset. Both Area-5 and 6-fold performance are reported. **mCov:** mean instance-wise IoU coverage. **mWCov:** mean size-weighted IoU coverage. **mPrec:** mean precision with IoU threshold 0.5. **mRec:** mean recall with IoU threshold of 0.5. All our results are achieved on a vanilla PointNet++ [24] backbone without multi-scale grouping for fair comparison.

Method	Year	mCov	mWCov	mPrec	mRec
Test on Area 5					
SGPN [29]	2018	32.7	35.5	36.0	28.7
ASIS [30]	2019	44.6	47.8	55.3	42.4
3D-BoNet [33]	2019	-	-	57.5	40.2
JSNet [36]	2020	48.7	51.5	**62.1**	46.9
Ours	-	**49.9**	**53.2**	61.3	**48.5**
Test on 6-fold					
SGPN [29]	2018	37.9	40.8	31.2	38.2
MT-PNet [21]	2019	-	-	24.9	-
MV-CRF [21]	2019	-	-	36.3	-
ASIS [30]	2019	51.2	55.1	63.6	47.5
3D-BoNet [33]	2019	-	-	65.6	47.6
PartNet [20]	2019	-	-	56.4	43.4
Ours	-	**54.5**	**58.0**	**67.2**	**51.8**

Table 3. Comparison per-class performance of our proposed method with state-of-the-arts on the S3DIS semantic segmentation task, tested on all areas (6-fold). Our result utilize the vanilla PointNet++ [24] without multi-scale group. Even with a simple baseline, the proposed method surpassed the complex graph-based methods. **mA:** mean pixel-wise accuracy. **mI:** mean category-wise IoU.

	mA	mI	cei.	flo.	wall	beam	col.	win.	door	tab.	cha.	sofa	boo.	boa.	clu.
[22]	78.5	47.6	88.0	88.7	69.3	42.4	23.1	47.5	51.6	54.1	42.0	9.6	38.2	29.4	35.2
[7]	79.2	47.8	88.6	**95.8**	67.3	36.9	24.9	48.6	52.3	51.9	45.1	10.6	36.8	24.7	37.5
[7]	81.1	49.7	90.3	92.1	67.9	44.7	24.2	52.3	51.2	58.1	47.4	6.9	39.0	30.0	41.9
[24]	-	53.2	90.2	91.7	73.1	42.7	21.2	49.7	42.3	62.7	59.0	19.6	45.8	48.2	45.6
[8]	-	58.3	92.1	90.4	**78.5**	37.8	35.7	51.2	65.4	64.0	**61.6**	25.6	51.6	49.9	53.7
[31]	84.1	56.1	-	-	-	-	-	-	-	-	-	-	-	-	-
[14]	85.9	60.0	93.1	95.3	78.2	33.9	**37.4**	56.1	68.2	64.9	61.0	34.6	51.5	51.1	**54.4**
Ours	**86.5**	**60.2**	**94.0**	94.1	76.6	**53.4**	33.6	54.2	62.7	**70.2**	60.2	**36.6**	53.4	54.3	53.5

Centroid Constrain Loss. The centroid constraint loss L_{cen} is designed for maintaining consistency for points belonging to the same instance. The loss function serves as a regularizer to constrain the embedding features from the same instance to have a small variance. Moreover, it also helps stabilize the centroids when concatenated to the embedding space. As can be inferred from

Table 1, the utilization of L_{cen} improves the AP_{50} from 27.6 to 28.9. By further combing the geometric embeddings with the per-point features, we achieve an improvement on the AP_{50} from 28.9 to 31.5.

Table 4. Instance segmentation results on ScanNetV2 benchmark (validation set). The metric of mAP@0.25 is reported. All methods except [8] are based on PointNet or PointNet++. (Categories of Table, Toilet, and Window are not presented in the table.)

	mAP	bat	bed	she	cab	cha	cou	cur	des	doo	oth	pic	ref	shc	sin	sof
[10]	26.1	33.3	0.2	0.0	5.3	0.2	0.2	2.1	0.0	4.5	2.4	23.8	6.5	0.0	1.4	10.7
[6]	-	66.7	56.6	7.6	3.5	39.4	2.7	3.5	9.8	9.9	3.0	2.5	9.8	37.5	12.6	60.4
[29]	35.1	20.8	39.0	16.9	6.5	27.5	2.9	6.9	0.0	8.7	4.3	1.4	2.7	0.0	11.2	35.1
[30]	47.4	57.3	52.1	1.4	18.5	46.1	19.2	20.3	13.3	13.8	18.8	6.6	17.6	33.1	8.8	32.1
Ours	50.4	63.0	60.9	0.2	22.9	67.2	10.2	18.6	10.5	15.5	22.7	9.5	16.5	55.2	13.6	34.3

4.5   Comparison with State-of-the-Art Methods

In this section, we make a comprehensive comparison with other state-of-the-art methods on three popular benchmarks. Our method can not only be applied to indoor scenes but also achieved promising results on the hierarchical 3D part dataset. The results on S3DIS [1], ScanNetV2 [3], and PartNet [20] show the superiority of our method on both efficiency and effectiveness.

Training and Testing Efficiency. As the first method to solve instance segmentation on the point cloud, SGPN [29] needs to predict a pair-wise similarity matrix, which requires a lot of memory. Each sample requires about 2.7G for training. GSPN [34] needs two training stages, and each sample has to take about 6G memory for training due to the generative network. ASIS [30] addresses the problem by removing the memory consuming parts and learning a discriminative embedding. However, due to the massive usage of K-NN for every point, training ASIS requires a memory of more than 700M for every sample and the inference time for the network requires 60 ms for each block. As we only utilize K-NN in the bottleneck layer, training IAM needs only about 400M for each sample and reduces the running time to 42 ms for each block, showing the superiority in both the effectiveness and efficiency of our method.

Quantitative Results on S3DIS. Instance segmentation performance on Area-5 and k-fold cross validation results are reported in Table 2. We compare our method with other start-of-the-art results. Equipped with instance-aware knowledge, 2.4%, and 7.7% improvement are achieved with metric $mPrec$ and $mRec$ for instance segmentation. Although employing a simple backbone, our

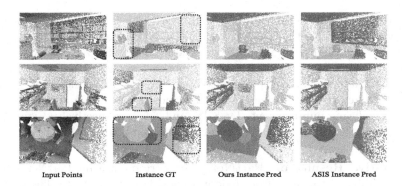

Input Points   Instance GT   Ours Instance Pred   ASIS Instance Pred

Fig. 5. Visualization of the instance segmentation results on the S3DIS indoor scenes. From left to right are: input point cloud, the ground truth of instance segmentation, the results of our proposed method, and the results of ASIS [30]. As shown in the figure, our methods have discriminative embedding features for distinguishing adjacent objects. We should note that: different instances are presented with different colors, and the same instances in different methods are not necessarily sharing the same color. (Color figure online)

method surpasses previous methods, which need more complex operations and more memories for training. Moreover, we also report the performance on the semantic segmentation task in Table 3. The results are evaluated with 6-fold cross-validation. Our method is built upon vanilla PointNet++ [24] and achieves better results compared with methods that applied multi-view [7] or even graph CNN [14,31]. Qualitative instance grouping results are shown in Fig. 5. We compare the performance of our method with ASIS [30], showing the effectiveness of the encoded instance-aware knowledge.

Table 5. Instance segmentation results on PartNet. We report part-category mAP (%) under IoU threshold 0.5. There are three different levels for evaluation: coarse-grained level, middle-grained level, and fine-grained level. We select five categories with the most data amount for training and evaluation.

	Level1					Level 2					Level 3				
	Cha	Sto	Tab	Lam	Vas	Cha	Sto	Tab	Lam	Vas	Cha	Sto	Tab	Lam	Vas
[29]	72.4	32.9	49.2	32.7	46.6	25.4	30.5	18.9	21.7	-	19.4	21.5	14.6	14.4	36.5
[20]	74.4	**45.2**	54.2	**37.2**	49.8	35.5	35.0	31.0	26.9	-	29.0	27.5	23.9	18.7	52.0
[34]	-	-	-	-	-	-	-	-	-	-	26.8	26.7	21.9	18.3	-
[34]	77.1	43.2	55.0	34.1	48.5	36.0	35.5	31.3	24.8	-	26.8	26.7	21.9	18.3	51.9
Ours	**79.5**	44.2	**56.1**	36.1	**49.9**	**38.6**	**37.1**	**33.0**	**26.9**	-	**31.2**	**28.9**	**25.5**	**19.4**	**53.1**

Quantitative Results on ScanNetV2. The quantitative performance on ScanNetV2 is presented in Table 4. It is evaluated on the validation set. Both $mAP@0.25$ and $mAP@0.5$ are reported. The results of [30] and [34] are reproduced via the open source code. For fair comparison, methods based on PointNet [22] or PointNet++ [24] are reported. Compared with state-of-the-art ASIS [30], our method achieves promising results and boosts $mAP@0.25$ and $mAP@0.5$ with a significant improvement, by 8.4% and 6.5%, respectively. Figure 6(a) shows qualitative results of instance segmentation on ScanNetV2.

Quantitative Results on PartNet. The performance on PartNet [20] is shown in Table 5. Different from indoor scenes, PartNet provides fine-grained and hierarchical object parts annotations. Level-1 contains the coarsest annotations and level-3 contains the finest annotations. Similar to GSPN [34], we report the performance of the five categories that have the largest number of training samples: Chair, Storage, Table, Lamp, and Vase. $mAP@0.5$ is reported. Each category of different levels is trained separately. Our method achieved state-of-the-art results on most categories and levels, substantially improving the performance. Figure 6(b) shows qualitative results of instance segmentation on PartNet. Different categories and fine-grained levels are provided.

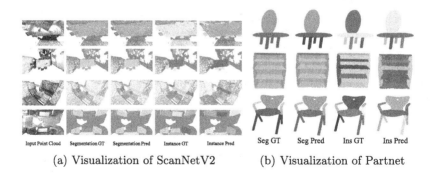

(a) Visualization of ScanNetV2   (b) Visualization of Partnet

Fig. 6. Visualization of the instance segmentation results on (a) ScanNetV2 and (b) Partnet. Our method successfully discriminates adjacent objects that are difficult to separate. Noting: different instances are presented with different colors, and the same instance in different methods are not necessarily sharing the same color. (Color figure online)

5   Conclusion

In this paper, we present a novel method for solving point cloud instance segmentation and semantic segmentation simultaneously. An instance-aware module (IAM) is proposed to encode both instance-aware context and geometric information. Extensive experimental results show that our method has achieved state-of-the-art performance on several benchmarks and shown superiority in both effectiveness and efficiency.

References

1. Armeni, I., et al.: 3D semantic parsing of large-scale indoor spaces. In: Proceedings of the IEEE Conference on Computer Vision and Pattern Recognition (2016)
2. Chen, L.C., Zhu, Y., Papandreou, G., Schroff, F., Adam, H.: Encoder-decoder with atrous separable convolution for semantic image segmentation. In: Proceedings of the European Conference on Computer Vision (2018)
3. Dai, A., Chang, A.X., Savva, M., Halber, M., Funkhouser, T., Nießner, M.: Scan-Net: richly-annotated 3D reconstructions of indoor scenes. In: Proceedings of the IEEE Conference on Computer Vision and Pattern Recognition (2017)
4. Dai, J., He, K., Sun, J.: Instance-aware semantic segmentation via multi-task network cascades. In: Proceedings of the European Conference on Computer Vision (2016)
5. Dai, J., et al.: Deformable convolutional networks. In: Proceedings of the IEEE International Conference on Computer Vision (2017)
6. Elich, C., Engelmann, F., Kontogianni, T., Leibe, B.: 3D-BEVIS: Bird's-Eye-View Instance Segmentation. arXiv preprint arXiv:1904.02199 (2019)
7. Engelmann, F., Kontogianni, T., Hermans, A., Leibe, B.: Exploring spatial context for 3D semantic segmentation of point clouds. In: Proceedings of the IEEE International Conference on Computer Vision Workshops (2017)
8. Engelmann, F., Kontogianni, T., Schult, J., Leibe, B.: Know What Your Neighbors Do: 3D Semantic Segmentation of Point Clouds. arXiv:1810.01151 (2018)
9. Graham, B., Engelcke, M., van der Maaten, L.: 3D semantic segmentation with submanifold sparse convolutional networks. In: Proceedings of the IEEE Conference on Computer Vision and Pattern Recognition (2018)
10. He, K., Gkioxari, G., Dollár, P., Girshick, R.: Mask R-CNN. In: Proceedings of the IEEE International Conference on Computer Vision (2017)
11. He, T., Gong, D., Tian, Z., Shen, C.: Learning and Memorizing Representative Prototypes for 3D Point Cloud Semantic and Instance Segmentation. arXiv preprint arXiv:2001.01349 (2020)
12. He, T., Shen, C., Tian, Z., Gong, D., Sun, C., Yan, Y.: Knowledge adaptation for efficient semantic segmentation. In: Proceedings of the IEEE Conference on Computer Vision and Pattern Recognition (2019)
13. Hou, J., Dai, A., Nießner, M.: 3D-SIS: 3D semantic instance segmentation of RGB-D scans. In: Proceedings of the IEEE Conference on Computer Vision and Pattern Recognition (2019)
14. Li, G., Müller, M., Thabet, A., Ghanem, B.: DeepGCNs: can GCNs go as deep as CNNs? In: Proceedings of the IEEE International Conference on Computer Vision (2019)
15. Li, Y., Bu, R., Sun, M., Wu, W., Di, X., Chen, B.: PointCNN: convolution on X-transformed points. In: Advances in Neural Information Processing Systems (2018)
16. Lin, T.Y., Dollár, P., Girshick, R., He, K., Hariharan, B., Belongie, S.: Feature pyramid networks for object detection. In: Proceedings of the IEEE Conference on Computer Vision and Pattern Recognition (2017)
17. Lin, T.Y., Goyal, P., Girshick, R., He, K., Dollár, P.: Focal loss for dense object detection. In: Proceedings of the IEEE International Conference on Computer Vision (2017)
18. Liu, Y., Chen, K., Liu, C., Qin, Z., Luo, Z., Wang, J.: Structured knowledge distillation for semantic segmentation. In: Proceedings of the IEEE Conference on Computer Vision and Pattern Recognition (2019)

19. Maturana, D., Scherer, S.: VoxNet: a 3D convolutional neural network for real-time object recognition. In: Proceedings of the IEEE International Conference on Intelligent Robots and Systems (2015)

20. Mo, K., et al.: PartNet: a large-scale benchmark for fine-grained and hierarchical part-level 3D object understanding. In: Proceedings of the IEEE Conference on Computer Vision and Pattern Recognition (2019)

21. Pham, Q.H., Nguyen, D.T., Hua, B.S., Roig, G., Yeung, S.K.: JSIS3D: joint semantic-instance segmentation of 3D point clouds with multi-task pointwise networks and multi-value conditional random fields. In: Proceedings of the IEEE Conference on Computer Vision and Pattern Recognition (2019)

22. Qi, C.R., Su, H., Mo, K., Guibas, L.J.: PointNet: deep learning on point sets for 3D classification and segmentation. In: Proceedings of the IEEE Conference on Computer Vision and Pattern Recognition (2017)

23. Qi, C.R., Su, H., Nießner, M., Dai, A., Yan, M., Guibas, L.J.: Volumetric and multi-view CNNs for object classification on 3D data. In: Proceedings of the IEEE Conference on Computer Vision and Pattern Recognition (2016)

24. Qi, C.R., Yi, L., Su, H., Guibas, L.J.: PointNet++: deep hierarchical feature learning on point sets in a metric space. In: Proceedings of Advances in Neural Information Processing Systems (2017)

25. Riegler, G., Ulusoy, A.O., Geiger, A.: OctNet: Learning Deep 3D Representations at High Resolutions. arXiv preprint arXiv:1611.05009 (2016)

26. Su, H., Maji, S., Kalogerakis, E., Learned-Miller, E.: Multi-view convolutional neural networks for 3D shape recognition. In: Proceedings of the IEEE International Conference on Computer Vision (2015)

27. Thomas, H., Qi, C.R., Deschaud, J.E., Marcotegui, B., Goulette, F., Guibas, L.J.: KPConv: flexible and deformable convolution for point clouds. In: Proceedings of the IEEE International Conference on Computer Vision (2019)

28. Wang, L., Huang, Y., Hou, Y., Zhang, S., Shan, J.: Graph attention convolution for point cloud semantic segmentation. In: Proceedings of the IEEE Conference on Computer Vision and Pattern Recognition (2019)

29. Wang, W., Yu, R., Huang, Q., Neumann, U.: SGPN: similarity group proposal network for 3D point cloud instance segmentation. In: Proceedings of the IEEE Conference on Computer Vision and Pattern Recognition (2018)

30. Wang, X., Liu, S., Shen, X., Shen, C., Jia, J.: Associatively segmenting instances and semantics in point clouds. In: Proceedings of the IEEE Conference on Computer Vision and Pattern Recognition (2019)

31. Wang, Y., Sun, Y., Liu, Z., Sarma, S.E., Bronstein, M.M., Solomon, J.M.: Dynamic graph CNN for learning on point clouds. ACM Trans. Graph. **38**, 1–12 (2019)

32. Wu, Z., et al.: 3D ShapeNets: a deep representation for volumetric shapes. In: Proceedings of the IEEE Conference on Computer Vision and Pattern Recognition (2015)

33. Yang, B., et al.: Learning object bounding boxes for 3D instance segmentation on point clouds. In: Proceedings of the Advances in Neural Information Processing Systems (2019)

34. Yi, L., Zhao, W., Wang, H., Sung, M., Guibas, L.J.: GSPN: generative shape proposal network for 3D instance segmentation in point cloud. In: Proceedings of the IEEE Conference on Computer Vision and Pattern Recognition (2018)

35. Zhang, H., et al.: Context encoding for semantic segmentation. In: Proceedings of the IEEE Conference on Computer Vision and Pattern Recognition (2018)

36. Zhao, L., Tao, W.: JSNet: joint instance and semantic segmentation of 3D point clouds. In: Proceedings of AAAI Conference on Artificial Intelligence (2019)

Self-Paced Deep Regression Forests with Consideration on Underrepresented Examples

Lili Pan[1]([⊠]), Shijie Ai[1], Yazhou Ren[1], and Zenglin Xu[1,2,3]

[1] University of Electronic Science and Technology of China, Chengdu, China
{lilipan,yazhou.ren}@uestc.edu.cn,asj1995@163.com,zenglin@gmail.com
[2] Harbin Institute of Technology, Shenzhen, China
[3] Center for Artificial Intelligence, Peng Cheng Lab, Shenzhen, China

Abstract. Deep discriminative models (*e.g.* deep regression forests, deep neural decision forests) have achieved remarkable success recently to solve problems such as facial age estimation and head pose estimation. Most existing methods pursue robust and unbiased solutions either through learning discriminative features, or reweighting samples. We argue what is more desirable is learning gradually to discriminate like our human beings, and hence we resort to self-paced learning (SPL). Then, a natural question arises: *can self-paced regime lead deep discriminative models to achieve more robust and less biased solutions?* To this end, this paper proposes a new deep discriminative model—self-paced deep regression forests with consideration on underrepresented examples (SPUDRFs). It tackles the fundamental ranking and selecting problem in SPL from a new perspective: fairness. This paradigm is fundamental and could be easily combined with a variety of deep discriminative models (DDMs). Extensive experiments on two computer vision tasks, i.e., facial age estimation and head pose estimation, demonstrate the efficacy of SPUDRFs, where state-of-the-art performances are achieved.

Keywords: Underrepresented examples · Self-paced learning · Entropy · Deep regression forests

1 Introduction

Deep discriminative models (*e.g.* deep regression forests, deep neural decision forests) have recently been applied to many computer vision problems with remarkable success. They compute the input to output mapping for regression or classification by virtue of deep neural networks [3,4,14,23,44,45]. In general, DDMs probably perform better when large amounts of effective training data (less noisy and balanced) is available. However, such ideal data is hard to collect, especially when large amounts of labels are required.

Computer vision literatures are filled with scenarios in which we are required to learn DDMs, not only robust to confusing and noisy examples, but also capable

© Springer Nature Switzerland AG 2020
A. Vedaldi et al. (Eds.): ECCV 2020, LNCS 12375, pp. 271–287, 2020.
https://doi.org/10.1007/978-3-030-58577-8_17

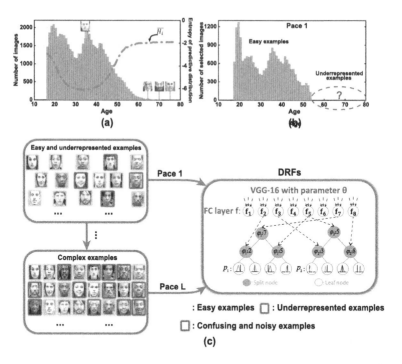

Fig. 1. The motivation of considering underrepresented examples in DRFs. **(a):** The histogram shows the number of face images at different ages, and the average entropy curve represents the predictive uncertainty. We observe the high entropy values correspond to *underrepresented samples*. **(b):** The histogram of the selected face images at pace 1 in SPL. **(c):** The proposed new self-paced learning paradigm: easy and underrepresented samples first.

to conquer imbalanced data problem [5,22,24,37,48]. One typical approach is to learn discriminative features through rather deep neural networks, and feed them into a *cost-sensitive* discriminative function, often with regularization [28]. The other typical approach reweights training samples according to their cost values [5,22] or gradient directions [37] (*i.e.* meta learning). These strategies are unlike our human beings, who lean things gradually—start with easy concepts and build up to complex ones, and can exclude extremely hard ones. More importantly, we have a sense of *uncertainty* for some samples (*e.g.* seldom seen) and progressively improve our capability to recognize them. Thus, the main challenge towards realistic discrimination lies how to mimic our human discrimination system might work.

This line of thinking makes us resort to self-paced learning—a gradual learning regime inspired by the manner of humans [26]. In fact, there are rare studies on the problem of self-paced DDMs. Then, a natural question arises: *can the self-paced regime lead DDMs to achieve more robust and less biased solutions?*

Motivated by this, we propose a new self-paced learning paradigm for DDMs, which tackles the fundamental ranking and selecting problem in SPL from a new perspective: fairness. To the best of our knowledge, this is the first work considering *ranking fairness* in SPL. Specifically, we focus on deep regression forests (DRFs), a typical discriminative method, and propose self-paced deep regression forests with consideration on underrepresented examples (SPUDRFs). First, by virtue of SPL, our model distinguishes confusing and noisy examples from regular ones, and emphasizes more on "good" examples to obtain robust solutions. Second, our method considers underrepresented examples, which may incur neglect in SPL since visual data is often imbalanced, rendering less biased solutions. Third, we build up a new self-paced learning paradigm: ranking samples on the basis of both likelihood and entropy (predictive uncertainty), as shown in Fig. 1, which could be easily combined with a variety of DDMs.

For verification, we apply the SPUDRFs framework on two computer vision problems: (i) facial age estimation, and (ii) head pose estimation. Extensive experimental results demonstrate the efficacy of our proposed new self-paced paradigm for DDMs. Moreover, on both aforementioned problems, SPUDRFs almost achieve the state-of-the-art performances.

2   Related Work

This section reviews the deep discriminative methods for facial age estimation and head pose estimation, and SPL methods.

Facial Age Estimation. DDM based facial age estimation methods, for example [4,8,29,34,44], employ DNNs to precisely model the mapping from image to age. Ordinal-based approaches [4,34] resort to a set of sequential binary queries—each query refers to a comparison with a predefined age, to exploit the inter-relationship (ordinal information) among age labels. Improved deep label distribution learning (DLDL-v2) [8] explores the underlying age distribution patterns to effectively accommodates age ambiguity. Besides, deep regression forests (DRFs) [44] connect random forests to deep neural networks and achieve promising results. BridgeNet [29] uses local regressors to partition the data space and gating networks to provide continuity-aware weights. The final age estimation result is the mixture of the weighted regression results. Overall, these DDM based approaches have enhanced age estimation performance largely; however, they plausibly ignore one problem: the interference arising from confusing and noisy examples—facial images with PIE (*i.e.* pose, illumination and expression) variation, occlusion, misalignment and so forth.

Head Pose Estimation. For head pose estimation, Riegler [41] *et al.* utilized convolutional neural networks (CNNs) to learn patch features of facial images and achieved better performance. In [16], Huang *et al.* adopted multi-layer perceptron (MLP) networks for head pose estimation and proposed multi-modal deep regression networks to fuse RGB and depth information. In [46], Wang *et al.* proposed a deep coarse-to-fine network for head pose estimation. In [43], Ruiz *et al.* used a large synthetically expanded head pose dataset to train rather deep multi-loss CNNs for head pose estimation and gained satisfied accuracy. In [25], Kuhnke *et al.* proposed domain adaptation for head pose estimation, assuming shared and continuous label spaces. Despite seeing much success, these methods seldom consider the potential problems caused by imbalanced and noisy training data, which may exactly exist in visual problems.

Self-Paced Learning. The SPL is a gradual learning paradigm, which builds on the intuition that, rather than considering all training samples simultaneously, the algorithm should be presented with the training data from easy to difficult, which facilitates learning [26,33]. Variants of SPL methods have been proposed recently with varying degrees of success. For example, in [21], Zhao *et al.* generalized the conventional binary (hard) weighting scheme for SPL to a more effective real valued (soft) weighting manner. In [32], Ma *et al.* proposed self-paced co-training which applies self-paced learning to multi-view or multi-modality problems. In [13], Han *et al.* made some efforts on mixture of regressions with SPL strategy, to avoid poorly conditioned linear sub-regressors. In [38,39], Ren *et al.* introduced soft weighting schemes of SPL to reduce the negative influence of outliers and noisy samples. In fact, the majority of these mentioned methods can be cast as the combination of SPL and shallow classifiers, where SVM and logistic regressors are usually involved. In computer vision, due to the remarkable performance of DNNs, some authors have realized SPL may guide DDMs to achieve more robust solutions recently. In [27], Li *et al.* sought to enhance the learning robustness of CNNs with SPL, and proposed SP-CNNs. However, [27] omits one important problem in the discriminative model: the imbalance of training data. In contrast to SP-CNNs, our SPUDRFs model has three advantages: (i) it emphasizes ranking fairness (*i.e.* considering underrepresented examples) in SPL, and hence tends to achieve less biased solutions; (ii) its learning regime is fundamental and can be easily combined with other DDMs, especially the ones with predictive uncertainty; (iii) it creatively explores how SPL can integrate with DMMs with a probabilistic interpretation.

Our work is inspired by the existing works [20,47] which take the class diversity in the sample selection of SPL into consideration. Jiang *et al.* [20] encouraged the class diversity in sample selection at the early paces of self-paced training. Yang *et al.* [47] defined a metric, named complexity of image category, to measure sample number and recognition difficult jointly, and adopted this measure for sample selection in SPL. In fact, the aforementioned two methods realize the lack of class diversity in SPL's sample selection may achieve biased solutions since visual data is often imbalanced. But what causes lack of class diversity is exactly the ranking unfairness as underrepresented examples may often have large loss (particular in DDMs). Not only that, [20,47] are only suitable for classification, but not regression (with continuous and high dimensional output). In this paper, we will go further along this direction, aiming to tackle the fundamental problem in SPL: ranking unfairness.

3   Preliminaries

In this section, we review the basic concepts of deep regression forests (DRFs) [44].

Deep Regression Tree. DRFs usually consist of a number of deep regression trees. A deep regression tree, given input-output pairs $\{\mathbf{x}_i, y_i\}_{n=1}^{N}$, where $\mathbf{x}_i \in \mathbb{R}^{D_x}$ and $y_i \in \mathbb{R}$, models the mapping from input to output through DNNs coupled with a regression tree. A regression tree $\mathcal{T}$ consists of split nodes $\mathcal{N}$ and leaf nodes $\mathcal{L}$ [44]. More specifically, each split node $n \in \mathcal{N}$ possesses a split to determine whether input $\mathbf{x}_i$ goes to the left or right subtree; each leaf node $\ell \in \mathcal{L}$ corresponds to a Gaussian distribution $p_\ell(y_i)$ with mean μ_l and variance σ_l^2.

Split Node. Split node has a split function, $s_n(\mathbf{x}_i; \boldsymbol{\Theta}) : \mathbf{x}_i \to [0, 1]$, which is parameterized by $\boldsymbol{\Theta}$—the parameters of DNNs. Conventionally, the split function is formulated as $s_n(\mathbf{x}_i; \boldsymbol{\Theta}) = \sigma\left(\mathbf{f}_{\varphi(n)}(\mathbf{x}_i; \boldsymbol{\Theta})\right)$, where $\sigma(\cdot)$ is the sigmoid function, $\varphi(\cdot)$ is an index function to specify the $\varphi(n)$-th element of $\mathbf{f}(\mathbf{x}_i; \boldsymbol{\Theta})$ in correspondence with a split node n, and $\mathbf{f}(\mathbf{x}_i; \boldsymbol{\Theta})$ denotes the learned deep features. An example to illustrate the sketch chart of the DRFs is shown in Fig. 1, where φ_1 and φ_2 are two index functions for two trees. The probability that $\mathbf{x}_i$ falls into the leaf node ℓ is given by:

$$\omega_\ell(\mathbf{x}_i | \boldsymbol{\Theta}) = \prod_{n \in \mathcal{N}} s_n(\mathbf{x}_i; \boldsymbol{\Theta})^{[\ell \in \mathcal{L}_{n_l}]} \left(1 - s_n(\mathbf{x}_i; \boldsymbol{\Theta})\right)^{[\ell \in \mathcal{L}_{n_r}]}, \tag{1}$$

where $[\mathcal{H}]$ denotes an indicator function conditioned on the argument $\mathcal{H}$. In addition, $\mathcal{L}_{n_l}$ and $\mathcal{L}_{n_r}$ correspond to the sets of leaf nodes owned by the subtrees $\mathcal{T}_{n_l}$ and $\mathcal{T}_{n_r}$ rooted at the left and right children n_l and n_r of node n, respectively.

Leaf Node. For tree $\mathcal{T}$, given $\mathbf{x}_i$, each leaf node $\ell \in \mathcal{L}$ defines a predictive distribution over age y_i, denoted by $p_\ell(y_i)$. To be specific, $p_\ell(y_i)$ is assumed to be a Gaussian distribution: $\mathcal{N}\left(y_i | \mu_l, \sigma_l^2\right)$. Thus, considering all leaf nodes, the final distribution of y_i conditioned on $\mathbf{x}_i$ is averaged by the probability of reaching each leaf:

$$p_\mathcal{T}(y_i | \mathbf{x}_i; \boldsymbol{\Theta}, \boldsymbol{\pi}) = \sum_{\ell \in \mathcal{L}} \omega_\ell(\mathbf{x}_i | \boldsymbol{\Theta}) p_\ell(y_i), \tag{2}$$

where $\boldsymbol{\Theta}$ and $\boldsymbol{\pi}$ represent the parameters of DNNs and the distribution parameters $\{\mu_l, \sigma_l^2\}$, respectively. It can be viewed as a mixture distribution, where $\omega_\ell(\mathbf{x}_i | \boldsymbol{\Theta})$ denotes mixing coefficients and $p_\ell(y_i)$ denotes the Gaussian distributions associated with the ℓth leaf node. Note that $\boldsymbol{\pi}$ varies along with tree $\mathcal{T}_k$, and thus we rewrite it as $\boldsymbol{\pi}_k$ below.

Forests of Regression Trees. Since a forest comprises a set of deep regression trees $\mathcal{F} = \{\mathcal{T}_1, ..., \mathcal{T}_k\}$, the predictive output distribution, given $\mathbf{x}_i$, is obtained by averaging over all trees:

$$p_\mathcal{F}(y_i | \mathbf{x}_i, \boldsymbol{\Theta}, \boldsymbol{\Pi}) = \frac{1}{K} \sum_{k=1}^{K} p_{\mathcal{T}_k}(y_i | \mathbf{x}_i, \boldsymbol{\Theta}, \boldsymbol{\pi}_k), \tag{3}$$

where K is the number of trees and $\boldsymbol{\Pi} = \{\boldsymbol{\pi}_1, ..., \boldsymbol{\pi}_K\}$. $p_{\mathcal{F}}(y_i|\mathbf{x}_i, \boldsymbol{\Theta}, \boldsymbol{\Pi})$ can be viewed as the likelihood that the ith sample has output y_i.

4   Self-Paced DRFs with Consideration on Underrepresented Examples

The problems in training DDMs for visual tasks arise from: (i) the noisy and confusing examples, and (ii) the imbalance of training data. Intuitively inspired by the gradual learning manner of humans, we resort to self-paced learning and explore whether the DDMs, by virtue of SPL, tend to achieve more robust solutions. Perhaps not easily, in existing SPL, we observe ranking unfairness, as shown in Fig. 1. Motivated by this observation, we propose SPUDRFs, which starts learning with easy yet underrepresented examples, and build up to complex ones. Such a paradigm avoids overlooking the "minority" of training samples, leading to less biased solutions.

4.1   Underrepresented Examples

Underrepresented examples mean "minority", as which the examples with similar or the same labels are scarce. Unsurprisingly, we observe that they may incur unfairness treatment in the early paces of SPL (see Fig. 1(b)), due to imbalanced data distribution. The underrepresented level could be measured by predictive uncertainty. Given the sample $\mathbf{x}_i$, its predictive uncertainty is formulated as the entropy of its predictive output distribution $p_{\mathcal{F}}(y_i|\mathbf{x}_i, \boldsymbol{\Theta}, \boldsymbol{\Pi})$:

$$H\left[p_{\mathcal{F}}(y_i|\mathbf{x}_i, \boldsymbol{\Theta}, \boldsymbol{\Pi})\right] = \frac{1}{K}\sum_{k=1}^{K} H\left[p_{\mathcal{T}_k}(y_i|\mathbf{x}_i, \boldsymbol{\Theta}, \boldsymbol{\pi}_k)\right], \tag{4}$$

where $H\left[\cdot\right]$ denotes entropy, and the entropy corresponds to the kth tree is:

$$H\left[p_{\mathcal{T}_k}(y_i|\mathbf{x}_i, \boldsymbol{\Theta}, \boldsymbol{\pi}_k)\right] = -\int p_{\mathcal{T}_k}(y_i|\mathbf{x}_i, \boldsymbol{\Theta}, \boldsymbol{\pi}_k) \ln p_{\mathcal{T}_k}(y_i|\mathbf{x}_i, \boldsymbol{\Theta}, \boldsymbol{\pi}_k) \, dy_i, \tag{5}$$

The large the entropy is, the more uncertain the prediction should be, *i.e.* , the more underrepresented the sample is. Considering underrepresented samples can be interpreted as adequately utilizing the "information" inherent in such examples in SPL training.

As previously discussed, $p_{\mathcal{T}_k}(y_i|\mathbf{x}_i; \boldsymbol{\Theta}, \boldsymbol{\pi}_k)$ is a mixture distribution, taking the form $\sum_{\ell\in\mathcal{L}} \omega_\ell(\mathbf{x}_i|\boldsymbol{\Theta}) p_\ell(y_i)$, where $\omega_\ell(\mathbf{x}_i|\boldsymbol{\Theta})$ denotes mixing coefficients and $p_\ell(y_i)$ denotes the Gaussian distribution associated with the ℓth leaf node. In Eq. (5), the integral of mixture of Gaussians is non-trivial. Monte Carlo sampling provides a way to calculate it, but incurs large computational cost [17]. Here, we use the lower bound of this integral to approximate its true value:

$$H\left[p_{\mathcal{T}_k}(y_i|\mathbf{x}_i, \boldsymbol{\Theta}, \boldsymbol{\pi}_k)\right] \approx \frac{1}{2}\sum_{\ell\in\mathcal{L}} \omega_\ell(\mathbf{x}_i|\boldsymbol{\Theta})\left[\ln\left(2\pi\sigma_\ell^2\right) + 1\right]. \tag{6}$$

The underrepresented examples are often scarce, and have not been treated fairly, resulting in large prediction uncertainty (*i.e.* entropy).

4.2   Objective Function

Rather than considering all the samples simultaneously, our proposed SPUDRFs are presented with the training data in a meaningful order, that is, easy and underrepresented examples first. Specifically, we define a latent variable v_i that indicates whether the ith sample is selected ($v_i = 1$) or not ($v_i = 0$) depending on how easy and underrepresented it is for training. Our objective is to jointly maximize the log likelihood with respect to DRFs' parameters Θ and Π, and learn the latent selecting variables $\mathbf{v} = (v_1, ..., v_N)^T$. We prefer to select the underrepresented examples, which probably have higher predictive uncertainty (*i.e.* entropy), particularly in the early paces. It builds on the intuition that the underrepresented examples may incur neglect since they are the "minority" in training data. Therefore, we maximize a self-paced term regularized likelihood function, meanwhile considering predictive uncertainty,

$$\max_{\Theta, \Pi, \mathbf{v}} \sum_{i=1}^{N} v_i \left\{ \log p_\mathcal{F} (y_i | \mathbf{x}_i, \Theta, \Pi) + \gamma H_i \right\} + \lambda \sum_{i=1}^{N} v_i, \tag{7}$$

where λ is a parameter controlling the learning pace, $\lambda > 0$, γ is the parameter imposing on entropy, and H_i denotes the predictive uncertainty of the ith sample, as previously discussed in Sect. 4.1. When γ decays to 0, the objective function is equivalent to the log likelihood function with respect to DRFs' parameters Θ and Π. Equation (7) indicates each sample is weighted by v_i, and whether $\log p_\mathcal{F} (y_i | \mathbf{x}_i, \Theta, \Pi) + \gamma H_i > -\lambda$ determines the ith sample is selected or not. That is, the sample with high likelihood value or high predictive uncertainty may be selected. The optimal v_i^* is:

$$v_i^* = \begin{cases} 1 \text{ if } \log p_{\mathcal{F}i} + \gamma H_i > -\lambda \\ 0 \text{ otherwise} \end{cases}, \tag{8}$$

where $p_\mathcal{F} (y_i | \mathbf{x}_i, \Theta, \Pi)$ is written as $p_{\mathcal{F}i}$ for simplicity.

One might argue the noisy and hard examples tend to have high predictive uncertainty also, rendering being selected in the early paces. In fact, from Eq. (8), we observe whether one sample is selected is determined by both its predictive uncertainty and the log likelihood of being predicted correctly. The noisy and hard examples probably have relatively large loss *i.e.* low log likelihood, avoiding being selected at the very start.

Iteratively increasing λ and decreasing γ, samples are dynamically involved in the training of DRFs, starting with easy and underrepresented examples and ending up with all samples. Note every time we retrain DRFs, that is, maximizing Eq. (7), our model is initialized to the result of the last iteration. As such, our model is initialized progressively by the result of the previous pace—adaptively calibrated by "good" examples. This also means we place more emphasis on easy and underrepresented examples rather than confusing and noisy ones. Thus, SPUDRFs are prone to have more robust and less biased solutions since we adequately consider the underrepresented examples.

Mixture Weighting. In the previous section, we adopt a hard weighting scheme to assign data points to paces, in which one sample is either selected ($v_i = 1$) or not ($v_i = 0$). Such a weighting scheme appears to be less accurate as it omits the importance of samples. Hence, we adopt a mixture weighting scheme [19], where the selected samples are weighted by its importance, ling in the range $0 \leq v_i \leq 1$. The objective function with mixture weighting is defined as:

$$\max_{\Theta,\Pi,\mathbf{v}} \sum_{i=1}^{N} v_i \left\{ \log p_{\mathcal{F}}\left(y_i|\mathbf{x}_i, \Theta, \Pi\right) + \gamma H_i \right\} + \zeta \sum_{i=1}^{N} \log\left(v_i + \zeta/\lambda\right), \qquad (9)$$

where ζ is a parameter controlling the learning pace. We set $\zeta = \left(\frac{1}{\lambda'} - \frac{1}{\lambda}\right)^{-1}$, and $\lambda > \lambda' > 0$ to construct a reasonable soft weighting formulation. The self-paced regularizer in Eq. (9) is convex with respect to $v \in [0, 1]$. Then, setting the partial gradient of Eq. (9) with respect to v_i as zero will lead to the following:

$$\log p_{\mathcal{F}}\left(y_i|\mathbf{x}_i, \Theta, \Pi\right) + \gamma H_i + \frac{\zeta}{v_i + \zeta/\lambda} = 0. \qquad (10)$$

Then, the optimal solution of v_i is given by:

$$v_i^* = \begin{cases} 1 & \text{if } \log p_{\mathcal{F}i} + \gamma H_i \geq -\lambda' \\ 0 & \text{if } \log p_{\mathcal{F}i} + \gamma H_i \leq -\lambda \\ \frac{-\zeta}{\log p_{\mathcal{F}i} + \gamma H_i} - \zeta/\lambda & \text{otherwise} \end{cases} \qquad (11)$$

If either the log likelihood or the predictive uncertainty is too large, v_i^* equals to 1. In addition, if the likelihood and the predictive uncertainty are both too small, v_i^* equals to 0. Except the above two situations, the soft weighting calculation (i.e., the last line of Eq. (11)) is adopted.

Curriculum Reconstruction. The underrepresented examples play an important role in our SPUDRFs algorithm. As previously mentioned, the proposed new self-paced regime coupled with a mixture weighting scheme emphasizes more on underrepresented examples, rendering better solutions. Since the intrinsic reason that causes predictive uncertainty is plausibly the imbalanced training data, we further re-balance data distribution via a curriculum reconstruction strategy. More specifically, we distinguish the underrepresented examples (whose H_i is lager than β) from regular ones at each pace, and augment them into the training data.

4.3   Optimization

We propose a two-step alternative search strategy (ASS) algorithm to solve SPUDRFs: (i) update $\mathbf{v}$ for sample selection with fixed Θ and Π, and (ii) update Θ and Π with current fixed sample weights $\mathbf{v}$.

Optimizing Θ and Π. The parameters $\{\Theta, \Pi\}$ and weights $\mathbf{v}$ are optimized alternatively. With fixed $\mathbf{v}$, our DRFs is learned by alternatively updating Θ

and $\boldsymbol{\Pi}$. In [44], the parameters $\boldsymbol{\Theta}$ for split nodes (*i.e.* parameters for VGG) are updated through gradient descent since the loss is differentiable with respect to $\boldsymbol{\Theta}$. While the parameters $\boldsymbol{\Pi}$ for leaf nodes are updated by virtue of variational bounding [44] when fixing $\boldsymbol{\Theta}$.

Optimizing v. As previously discussed, v_i is a binary variable or real variable ranged in $[0, 1]$. It indicates how to weight the ith sample during training. The parameter λ could be initialized to obtain 50% samples to train the model, and is then progressively increased to involve 10% more data in each pace. The parameter γ could be initialized empirically and is progressively decayed to zero. The training stops when all the samples are selected, at $\gamma = 0$. Along with increasing λ and decreasing γ, DRFs are trained to be more "mature". This learning process is like how our human beings learn one thing from easy and uncertain to complex.

5   Experimental Results

5.1   Tasks and Benchmark Datasets

Age Estimation. The Morph II [40] dataset contains 55,134 unique face images of 13618 individuals with unbalanced gender and ethnicity distributions, and is the most popular publicly available real age dataset. The FG-NET [35] dataset includes 1,002 color or gray images of 82 people with each subject almost accompanied by more than 10 photos at different ages. Since all images were taken in a totally uncontrolled environment, there exists a large deviation on lighting, pose and expression (*i.e.* PIE) of faces inside the dataset.

Head Pose Estimation. The BIWI dataset [7] contains 20 subjects, of which 10 are male and 6 are female, besides, 4 males have been chosen twice with wearing glasses or not. It includes 15678 images collected by a Kinect sensor device for different persons and head poses with pitch, yaw and roll angles mainly ranging within $\pm 60°$, $\pm 75°$ and $\pm 50°$.

5.2   Experimental Setup

Dataset Setting. The settings of different datasets are given below.

- **Morph II.** Following the recent relevant work [44], the images in Morph II were divided into two sets: 80% for training and the rest 20% for testing. The random division was repeated 5 times and the reported performance was averaged over these 5 times. The VGG-Face [36] networks were chosen as the pre-trained model.
- **FG-NET.** The leave-one-person-out scheme [44] was adopted, where the images of one person were selected for testing and the remains for training. The VGG-16 networks were pre-trained on the IMDB-WIKI [42] dataset.

- **BIWI.** Similarly, 80% of the whole data was randomly chosen for training and the rest 20% for testing, and this operation was repeated 5 times. Moreover, the VGG-FACE networks were the pre-trained model.

Evaluation Metrics. The first evaluation metric is the mean absolute error (MAE), which is defined as the average absolute error between the ground truth and the predicted output: $\sum_{i=1}^{N} |\hat{y}_i - y_i| / N$, $\hat{y}_i$ represents the estimated output of the ith sample, and N is the total number of testing images. The other evaluation metric is cumulative score (CS), which denotes the percentage of images sorted in the range of $[y_i - L, y_i + L]$: $CS(L) = \sum_{i=1}^{N} [|\hat{y}_i - y_i| \le L] / N \cdot 100\%$, where $[\cdot]$ denotes an indicator function and L is the error range.

Preprocessing and Data Augmentation. On the Morph II and FG-NET datasets, MTCNN [49] was used for joint face detection and alignment. Furthermore, following [44], we augmented training images in three ways: (i) random cropping (5 times); (ii) adding Gaussian white noise with variance of 0.0001 (2 times); (iii) random horizontal flipping (2 times). The whole number of samples was increased by 20 times after augmentation. On the BIWI dataset, we utilized the depth images for training and did not augment training images.

Parameters Setting. The VGG-16 [45] was employed as the fundamental backbone networks of SPUDRFs. The hyper-parameters of VGG-16 were: training batch size (32 on Morph II and BIWI, 8 on FG-NET), drop out ratio (0.5), max iterations of each pace (80k on Morph II, 20k on FG-NET, and 40k on BIWI), stochastic gradient descent (SGD), initial learning rate (0.2 on Morph II, 0.1 on BIWI, 0.02 on FG-NET) by reducing the learning rate ($\times 0.5$) per 10k iterations. The hyper-parameters of SPUDRFs were: tree number (5), tree depth (6), output unit number of feature learning (128), iterations to update leaf node predictions (20), number of mini-batches used to update leaf node predictions (50). In the first pace, 50% samples which are easy or underrepresented were selected for training. Here, λ was set to guarantee the first 50% samples with large $\log p_{\mathcal{F}i} + \gamma H_i$ values involved. λ' was set to ensure 10% of selected samples with soft weighting. γ was initialized to be 15 on the Morph II and BIWI datasets, and 5 on the FG-NET dataset. β was set to select 1180 and 2000 samples as the ones needed to be augmented twice at each pace on the Morph II and BIWI datasets. The number of paces was empirically set to be 10, 3 and 6 on the Morph II, FG-NET, and BIWI datasets, and except the first pace, an equal proportion of the rest data was gradually involved at each pace.

5.3   Validity of Our Proposed Method

Self-paced Learning Strategy. The validity of self-paced strategy in training DDMs is mainly demonstrated by the following experiments on the MorphII dataset. We first used all training images in the Morph II datasets to train DRFs so as to rank samples at the beginning pace. Retraining proceeded with progressively increasing λ such that every 1/9 of the rest data was gradually involved at each pace, where γ was decreased to the half of its previous value every time. In the last pace, the value γ was constrainedly set to be 0. The visualization of this process can be found in Fig. 2.

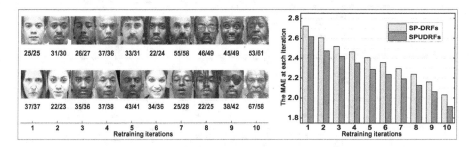

Fig. 2. The gradual learning process of SP-DRFs and SPUDRFs. **Left:** The typical worst cases at each iteration become more confusing and noisy along with iteratively increasing λ and decreasing γ. The two numbers below each image are the real age (left) and predicted age (right). **Right:** The MAEs of SP-DRFs and SPUDRFs at each pace descend gradually. The SPUDRFs show its superiority of taking predictive uncertainty into consideration, when compared with SP-DRFs.

Figure 2 illustrates the representative face images in each learning pace of SPUDRFs, along with increasing λ and decreasing γ. The two numbers below each image are the real age (left) and predicted age (right). We observe that the training images in the latter paces are obviously more confusing and noisy than the ones in the early paces. Since our model is initialized by the results of the previous retraining pace, meaning adaptively calibrated by "good" examples. As a result, it has improved performance than DRFs, where the MAE is improved from 2.17 to 1.91, and the CS is promoted from 92.79% to 93.31% (see Fig. 4(a)).

Figure 2 also shows the comparison between SP-DRFs and SPUDRFs on the Morph II datasets. The yellow bar denotes the MAE of SP-DRFs, while the orange bar denotes for SPUDRFs. We find the MAE of SPUDRFs is lower than SP-DRFs at each pace, particularly the last pace (1.91 against 2.02). As we discussed previously, as in Fig. 1, SPUDRFs are prone to reach less biased solutions due to the wider covering range of leaf nodes, owing to considering underrepresented examples. This experiment could be regarded as an ablation study of considering ranking fairness in SPL.

Considering Underrepresented Examples. On the BIWI dataset, the necessity of considering ranking fairness in SPUDRFs is further demonstrated. In SP-DRFs, DRFs was first trained on the basis of all data, and the samples were ranked and selected for the first pace according to this result. Subsequently, every 10% of the rest samples were progressively involved for retraining. λ was progressively increased while γ was progressively decreased until zero. In SP-DRFs, the same self-paced strategy was adopted as in SPUDRFs, but without considering ranking fairness (*i.e.* underrepresented examples).

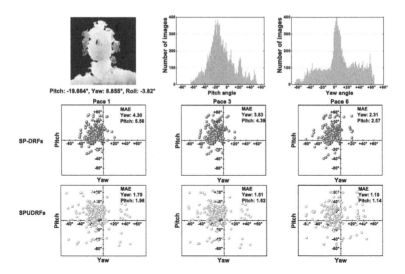

Fig. 3. The leaf node distribution of SP-DRFs and SPUDRFs in gradual learning process. Three paces, *i.e.* pace 1, 3, and 6, are randomly chosen for visualization. For SP-DRFs, the Gaussian means of leaf nodes (the red points in the second row) are concentrated in a small range, incurring seriously biased solutions. For SPUDRFs, the Gaussian means of leaf nodes (the orange points in the third row) distribute widely, leading to much better MAE performance. (Color figure online)

Figure 3 visualizes the leaf node distributions of SP-DRFs and SPUDRFs in the progressive learning process. The Gaussian means μ_l associated with the 160 leaf nodes, where each 32 leaf nodes are defined for 5 trees, are plotted in each sub-figures. Three paces, *i.e.* pace 1, 3, and 6, are randomly chosen for visualization. Only pitch and yaw angles are shown for clarity. Besides, the distribution of angle labels (*i.e.* pitch and yaw) are also shown, where the imbalance problem of data distribution is obvious.

In Fig. 3, the comparison results between SP-DRFs and SPUDRFs demonstrate the efficacy of considering ranking fairness in SPL. For SP-DRFs, the Gaussian means of leaf nodes (red points in the second row) are concentrated in a small range, incurring seriously biased solutions. That means the underrepresented examples have been neglected in SPL training. The poor MAEs are the evidence for this, which are even inferior to DRFs (see Fig. 6(a)). SPUDRFs rank samples by log likelihood coupled with entropy, and are prone to achieve less biased solutions, as shown in the third rows of Fig. 3. Such an experiment could be also regard as an ablation study of the proposed ranking algorithm.

5.4   Comparison with State-of-the-Art Methods

We compared our SPUDRFs with other state-of-the-art methods on the Morph II, FG-NET and BIWI datasets.

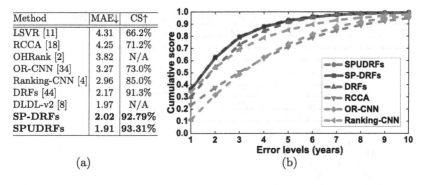

Method	MAE↓	CS↑
LSVR [11]	4.31	66.2%
RCCA [18]	4.25	71.2%
OHRank [2]	3.82	N/A
OR-CNN [34]	3.27	73.0%
Ranking-CNN [4]	2.96	85.0%
DRFs [44]	2.17	91.3%
DLDL-v2 [8]	1.97	N/A
SP-DRFs	**2.02**	**92.79%**
SPUDRFs	**1.91**	**93.31%**

(a)   (b)

Fig. 4. The comparison results on the Morph II dataset. (a) The MAE comparison with the state-of-the-art methods, (b) the CS curves of the comparison methods.

Results on Morph II. Figure 4(a) compares SPUDRFs with other baseline methods: LSVR [11], RCCA [18], OHRank [2], OR-CNN [34], Ranking-CNN [4], DRFs [44], and DLDL-v2 [8]. Firstly, owing to the effective feature learning ability of DNNs, the SPUDRFs method is much superior to the shallow model based approaches, such as LSVR [11] and OHRank [2]. Secondly, during to the valid self-paced regime, our SPUDRFs outperform other DDMs, and lead to more robust and less biased solutions. Thirdly, SPUDRFs outperform SP-DRFs on both MAE and CS, and achieve state-of-the-art performance. Figure 4(b) shows the CS comparison on this dataset. We observe that the CS of SPUDRFs reaches 93.31% at error level $L = 5$, which is significantly better than DRFs and obtained 2.01% increment.

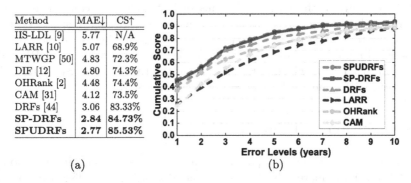

Method	MAE↓	CS↑
IIS-LDL [9]	5.77	N/A
LARR [10]	5.07	68.9%
MTWGP [50]	4.83	72.3%
DIF [12]	4.80	74.3%
OHRank [2]	4.48	74.4%
CAM [31]	4.12	73.5%
DRFs [44]	3.06	83.33%
SP-DRFs	**2.84**	**84.73%**
SPUDRFs	**2.77**	**85.53%**

(a)   (b)

Fig. 5. The comparison results on the FGNET dataset. (a) The MAE comparison with the state-of-the-art methods, (b) the CS curves of the comparison methods.

Results on FG-NET. Figure 5(a) shows the comparison results of SPUDRFs with the state-of-the-art approaches on FG-NET dataset. As can be seen, SPU-DRFs reach an MAE of 2.77 years, which reduces the MAE of DRFs by 0.29

years. Besides, the CS comparison is shown in Fig. 5(b), SPUDRFs consistently outperform other recent proposed methods at different error levels, proving that our method is effective in enhancing the robustness of facial age estimation.

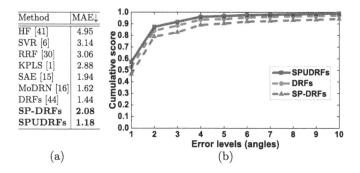

Method	MAE↓
HF [41]	4.95
SVR [6]	3.14
RRF [30]	3.06
KPLS [1]	2.88
SAE [15]	1.94
MoDRN [16]	1.62
DRFs [44]	1.44
SP-DRFs	**2.08**
SPUDRFs	**1.18**

(a)   (b)

Fig. 6. The comparison results on the BIWI dataset. (a) The MAE comparison with the state-of-the-art methods, (b) the CS curves the comparison methods.

Results on BIWI. Figure 6(a) shows the comparison results of our method with several state-of-the-art approaches. The experimental results reveal the proposed SPUDRFs method achieves the best performance with an MAE of 1.18, which is state-of-the-art. *Besides, we observe one important phenomenon: the MAE of SP-DRFs is even much worse than DRFs. This further demonstrates the obvious drawback of the ranking and selecting algorithm in original SPL— incurring seriously biased solutions.* In the first pace of the original SPL, as illustrated in Fig. 3, the Gaussian means of leaf nodes are concentrated in a small range, leading biased solutions. Incorporating underrepresented examples in the early pace of SPUDRFs renders to more reasonable distributions of the leaf nodes. Figure 6(b) plots only three CS curves for brevity, *i.e.* , DRFs, SP-DRFs and SPUDRFs, which is the average of the three angles. SPUDRFs also outperform DRFs and SP-DRFs at different error levels.

6   Conclusion and Future Work

This paper explored how self-paced regime leads deep discriminative models (DDMs) to achieve more robust and less biased solutions on different computer vision tasks (*e.g.* facial age estimation and head pose estimation). Specifically, a novel self-paced paradigm, which considers ranking fairness, was proposed. The new ranking scheme jointly considers loss and predictive uncertainty. Such a paradigm was combined with deep regression forests (DRFs), and led to a new model, namely self-paced deep regression forests with consideration on underrepresented examples (SPUDRFs). Extensive experiments on two well-known computer vision tasks demonstrated the efficacy of the proposed paradigm.

We are currently applying self-paced DDMs for other computer vision tasks, *e.g.* viewpoint estimation, indoor scene classification, where the ability to handle ranking unfairness is fundamental to the success. Thus, investigating the causes of algorithm unfairness in DDMs is a worthy direction. Obviously, except data imbalance, there exist some other causing factors. In addition to this, exploring how to combine the new self-paced paradigm with other DDMs, including deep regressors and classifiers, will also be our future work.

Acknowledgement. The authors gratefully acknowledge the support of China Postdoctoral Science Foundation No. 2017M623007.

References

1. Al Haj, M., Gonzalez, J., Davis, L.S.: On partial least squares in head pose estimation: how to simultaneously deal with misalignment. In: CVPR, pp. 2602–2609 (2012)
2. Chan, K., Chen, C., Hung, Y.: Ordinal hyperplanes ranker with cost sensitivities for age estimation. In: CVPR, pp. 585–592 (2011)
3. Chen, S., Zhang, C., Dong, M.: Deep age estimation: from classification to ranking. IEEE Trans. Multimed. **20**(8), 2209–2222 (2018)
4. Chen, S., Zhang, C., Dong, M., Le, J., Rao, M.: Using ranking-CNN for age estimation. In: CVPR, pp. 742–751 (2017)
5. Cui, Y., Jia, M., Lin, T.Y., Song, Y., Belongie, S.: Class-balanced loss based on effective number of samples. In: CVPR, pp. 9268–9277 (2019)
6. Drucker, H., Burges, C.J., Kaufman, L., Smola, A.J., Vapnik, V.: Support vector regression machines. In: NeurIPS, pp. 155–161 (1997)
7. Fanelli, G., Dantone, M., Gall, J., Fossati, A., Van Gool, L.: Random forests for real time 3D face analysis. Int. J. Comput. Vis. **101**(3), 437–458 (2013). https://doi.org/10.1007/s11263-012-0549-0
8. Gao, B., Zhou, H., Wu, J., Geng, X.: Age estimation using expectation of label distribution learning. In: IJCAI, pp. 712–718 (2018)
9. Geng, X., Yin, C., Zhou, Z.H.: Facial age estimation by learning from label distributions. IEEE Trans. Pattern Anal. Mach. Intell. **35**(10), 2401–2412 (2013)
10. Guo, G., Fu, Y., Dyer, C., Huang, T.: Image-based human age estimation by manifold learning and locally adjusted robust regression. IEEE Trans. Image Process. **17**(7), 1178–1188 (2008)
11. Guo, G., Guowang, M., Fu, Y., Huang, T.S.: Human age estimation using bio-inspired features. In: CVPR, pp. 112–119 (2009)
12. Han, H., Otto, C., Liu, X., Jain, A.K.: Demographic estimation from face images: human vs. machine performance. IEEE Trans. Pattern Anal. Mach. Intell. **37**(6), 1148–1161 (2015)
13. Han, L., et al.: Self-paced mixture of regressions. In: IJCAI, pp. 1816–1822 (2017)
14. He, K., Zhang, X., Ren, S., Sun, J.: Deep residual learning for image recognition. In: CVPR, pp. 770–778 (2016)
15. Hinton, G.E., Salakhutdinov, R.R.: Reducing the dimensionality of data with neural networks. Science **313**(5786), 504–507 (2006)
16. Huang, Y., Pan, L., Zheng, Y., Xie, M.: Mixture of deep regression networks for head pose estimation. In: ICIP, pp. 4093–4097 (2018)

17. Huber, M.F., Bailey, T., Hugh, D.W., Hanebeck, U.D.: On entropy approximation for gaussian mixture random vectors. In: IEEE International Conference on Multi-Sensor Fusion and Integration for Intelligent Systems, pp. 181–188 (2008)
18. Huerta, I., Fernández, C., Prati, A.: Facial age estimation through the fusion of texture and local appearance descriptors. In: Agapito, L., Bronstein, M.M., Rother, C. (eds.) ECCV 2014. LNCS, vol. 8926, pp. 667–681. Springer, Cham (2015). https://doi.org/10.1007/978-3-319-16181-5_51
19. Jiang, L., Meng, D., Mitamura, T., Hauptmann, A.G.: Easy samples first: self-paced reranking for zero-example multimedia search. In: ACM MM, pp. 547–556 (2014)
20. Jiang, L., Meng, D., Yu, S.I., Lan, Z., Shan, S., Hauptmann, A.: Self-paced learning with diversity. In: NeurIPS, pp. 2078–2086 (2014)
21. Jiang, L., Meng, D., Zhao, Q., Shan, S., Hauptmann, A.G.: Self-paced curriculum learning. In: AAAI, pp. 2694–2700 (2015)
22. Khan, S., Hayat, M., Zamir, S.W., Shen, J., Shao, L.: Striking the right balance with uncertainty. In: CVPR, pp. 103–112 (2019)
23. Kontschieder, P., Fiterau, M., Criminisi, A., Bulo, S.R.: Deep neural decision forests. In: ICCV (2015)
24. Kortylewski, A., Egger, B., Schneider, A., Gerig, T., Morel-Forster, A., Vetter, T.: Analyzing and reducing the damage of dataset bias to face recognition with synthetic data. In: CVPR Workshops (2019)
25. Kuhnke, F., Ostermann, J.: Deep head pose estimation using synthetic images and partial adversarial domain adaption for continuous label spaces. In: ICCV, pp. 10164–10173 (2019)
26. Kumar, M.P., Packer, B., Koller, D.: Self-paced learning for latent variable models. In: NeurIPS, pp. 1189–1197 (2010)
27. Li, H., Gong, M.: Self-paced convolutional neural networks. In: IJCAI, pp. 2110–2116 (2017)
28. Li, K., Xing, J., Su, C., Hu, W., Zhang, Y., Maybank, S.: Deep cost-sensitive and order-preserving feature learning for cross-population age estimation. In: CVPR (2018)
29. Li, W., Lu, J., Feng, J., Xu, C., Zhou, J., Tian, Q.: BridgeNet: a continuity-aware probabilistic network for age estimation. In: CVPR, pp. 1145–1154 (2019)
30. Liaw, A., Wiener, M., et al.: Classification and regression by random forest. R News 2(3), 18–22 (2002)
31. Luu, K., Seshadri, K., Savvides, M., Bui, T.D., Suen, C.Y.: Contourlet appearance model for facial age estimation. In: ICB, pp. 1–8 (2013)
32. Ma, F., Meng, D., Xie, Q., Li, Z., Dong, X.: Self-paced co-training. In: ICML, pp. 2275–2284 (2017)
33. Meng, D., Zhao, Q., Jiang, L.: A theoretical understanding of self-paced learning. Inf. Sci. 414, 319–328 (2017)
34. Niu, Z., Zhou, M., Wang, L., Gao, X., Hua, G.: Ordinal regression with multiple output CNN for age estimation. In: CVPR, pp. 4920–4928 (2016)
35. Panis, G., Lanitis, A., Tsapatsoulis, N., Cootes, T.F.: Overview of research on facial ageing using the FG-NET ageing database. IET Biometr. 5(2), 37–46 (2016)
36. Parkhi, O.M., Vedaldi, A., Zisserman, A.: Deep face recognition (2015)
37. Ren, M., Zeng, W., Yang, B., Urtasun, R.: Learning to reweight examples for robust deep learning. arXiv:1803.09050 (2018)
38. Ren, Y., Huang, S., Zhao, P., Han, M., Xu, Z.: Self-paced and auto-weighted multi-view clustering. Neurocomputing 383, 248–256 (2020)

39. Ren, Y., Zhao, P., Sheng, Y., Yao, D., Xu, Z.: Robust softmax regression for multi-class classification with self-paced learning. In: IJCAI, pp. 2641–2647 (2017)
40. Ricanek, K., Tesafaye, T.: MORPH: a longitudinal image database of normal adult age-progression. In: FG, pp. 341–345 (2006)
41. Riegler, G., Ferstl, D., Rüther, M., Bischof, H.: Hough networks for head pose estimation and facial feature localization. Int. J. Comput. Vis. **101**(3), 437–458 (2013)
42. Rothe, R., Timofte, R., Van Gool, L.: Deep expectation of real and apparent age from a single image without facial landmarks. Int. J. Comput. Vis. **126**(2–4), 144–157 (2018). https://doi.org/10.1007/s11263-016-0940-3
43. Ruiz, N., Chong, E., Rehg, J.M.: Fine-grained head pose estimation without keypoints. In: CVPR Workshops, pp. 2074–2083 (2018)
44. Shen, W., Guo, Y., Wang, Y., Zhao, K., Wang, B., Yuille, A.: Deep regression forests for age estimation. In: CVPR, pp. 2304–2313 (2018)
45. Simonyan, K., Zisserman, A.: Very deep convolutional networks for large-scale image recognition. arXiv:1409.1556 (2014)
46. Wang, Y., Liang, W., Shen, J., Jia, Y., Yu, L.F.: A deep coarse-to-fine network for head pose estimation from synthetic data. Pattern Recogn. **94**, 196–206 (2019)
47. Yang, J., et al.: Self-paced balance learning for clinical skin disease recognition. IEEE Trans. Neural Netw. Learn. Syst. **38**, 2832–2846 (2019)
48. Zeng, X., Ding, C., Wen, Y., Tao, D.: Soft-ranking label encoding for robust facial age estimation. arXiv:1906.03625 (2019)
49. Zhang, K., Zhang, Z., Li, Z., Qiao, Y.: Joint face detection and alignment using multi-task cascaded convolutional networks. IEEE Sig. Process. Lett. **23**(10), 1499–1503 (2016)
50. Zhang, Y., Yeung, D.Y.: Multi-task warped gaussian process for personalized age estimation. In: CVPR, pp. 2622–2629 (2010)

Manifold Projection for Adversarial Defense on Face Recognition

Jianli Zhou[1,2], Chao Liang[1,2(✉)], and Jun Chen[1,2]

[1] National Engineering Research Center for Multimedia Software,
School of Computer Science, Wuhan University, Wuhan, China
`cliang@whu.edu.cn`
[2] Key Laboratory of Multimedia and Network Communication Engineering,
Beijing, Hubei Province, China

Abstract. Although deep convolutional neural network based face recognition system has achieved remarkable success, it is susceptible to adversarial images: carefully constructed imperceptible perturbations can easily mislead deep neural networks. A recent study has shown that in addition to regular off-manifold adversarial images, there are also adversarial images on the manifold. In this paper, we propose Adversarial Variational AutoEncoder (A-VAE), a novel framework to tackle both types of attacks. We hypothesize that both off-manifold and on-manifold attacks move the image away from the high probability region of image manifold. We utilize variational autoencoder (VAE) to estimate the lower bound of the log-likelihood of image and explore to project the input images back into the high probability regions of image manifold again. At inference time, our model synthesizes multiple similar realizations of a given image by random sampling, then the nearest neighbor of the given image is selected as the final input of the face recognition model. As a preprocessing operation, our method is attack-agnostic and can adapt to a wide range of resolutions. The experimental results on LFW demonstrate that our method achieves state-of-the-art defense success rate against conventional off-manifold attacks such as FGSM, PGD, and C&W under both grey-box and white-box settings, and even on-manifold attack.

Keywords: Face recognition · Adversarial defense

1 Introduction

As one of the most popular real-world applications of computer vision tasks, such as multimedia analysis and surveillance [48], face recognition has been

Electronic supplementary material The online version of this chapter (https://doi.org/10.1007/978-3-030-58577-8_18) contains supplementary material, which is available to authorized users.

© Springer Nature Switzerland AG 2020
A. Vedaldi et al. (Eds.): ECCV 2020, LNCS 12375, pp. 288–305, 2020.
https://doi.org/10.1007/978-3-030-58577-8_18

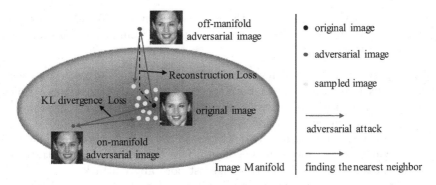

Fig. 1. The main idea of our model to defend adversarial images. Estimated by VAE, off-manifold and on-manifold adversarial images reduce the lower bound of log-likelihood by causing reconstruction loss and KL divergence loss, respectively (we explain it in Sect. 3.1). Our model samples images in high probability region and replaces the adversarial images with their nearest neighbors.

improved by a large margin with the aid of DNNs [41,42,44], and some models have even exceeded humans [28,36]. Despite the prominent role, DNNs suffer from adversarial examples [13,43], which constructed by maliciously adding some customized small perturbations to original input. Many recent works have shown that adversarial examples can be easily found with a simple gradient method [8,13,25,31]. In response, many defense strategies are proposed as a hedge against the adversarial examples. Several recent works are devoted to imposing transformations on input images due to the advantage of being model-agnostic. This type of method attempts to remove or alleviate hidden perturbations from input images. For example, Defense-GAN [35] employs generative adversarial networks (GANs) [12] to project the input images onto the range of the generator. In [40], Sun et al. design a sparse transformation layer to project the input images to a low dimensional quasi-natural space. However, these defense methods are limited to defend adversarial images which leave the manifold, but cannot handle on-manifold adversarial images.

In this work, we devise a novel *Adversarial Variational AutoEncoder* (A-VAE) to tackle both types of attacks. We assume that both types of adversarial images are moved away from the data distribution of high probability [38], and we focus on pulling them back onto their original regions, e.g. Fig. 1. By taking advantage of VAE [24], we estimate the lower bound of the log-likelihood of a given image, and project it onto the high probability region of image manifold using a unified process. Specifically, given a downsampled image as input, the proposed model randomly sample various outputs on the high probability region of image manifold which are approximately similar to the input, then the nearest neighbor is selected to be classified by the face recognition model instead of the original image. Moreover, we notice there is a contradiction: the traditional reconstruction loss in training phase will lead to false retention of perturbation at inference time. To solve this problem, we utilize the well-known adversarial training as GANs to substitute the reconstruction loss. We experimentally demonstrate this property helps the model strip noise from images. In summary, our contributions are:

- We propose a generative defense method against diverse adversarial attacks on complex face dataset under both grey-box and white-box settings, which leverages the capacity of VAE to project input images onto the high probability region of image manifold.
- Except for the regular off-manifold attacks, we also consider novel on-manifold attack which is even more challenging on face dataset. We demonstrate the effectiveness of our defense method against both above attacks.
- We introduce a simple training mechanism that makes the goal of training and testing more consistent, which enhances the robustness of our model.

2 Related Works

2.1 Adversarial Attacks

The existence of the adversarial examples is first shown in [43]. Regular adversarial attacks fool a well trained classifier through adding a small perturbation δ to a real image X. ϵ_p limits the l_p norm of the perturbations. We introduce several state-of-the-art attacks.

Fast Gradient Sign Method (FGSM). FGSM [43] generates adversarial examples by minimizing the probability of true class y:

$$X^{adv} = X + \epsilon_p \cdot sign(\nabla_x J(X, y)), \tag{1}$$

where $J(X, y)$ is cross-entropy loss. $sign(\cdot)$ is the sign function. FGSM performs one-step update towards the direction of gradient ascent.

Projected Gradient Descent (PGD). Madry et al. [30] improve FGSM by applying it multiple times with a small step size α. The adversarial examples can be formally expressed as:

$$X_0^{adv} = X, X_{n+1}^{adv} = Clip_{\epsilon_p}^X(X_n^{adv} + \epsilon_p \cdot sign(\nabla_x J(X_n^{adv}, y))), \tag{2}$$

where $Clip_{\epsilon_p}^X(\cdot)$ clips updated images to constrain it within the ϵ-ball of X. They also set the initial perturbation to a random point within the allowed ϵ-ball and restart the search multiple times to avoid falling into a local minimum.

Carlini & Wagner (C&W). C&W [3] is a strong optimization-based attack method. They find the smallest perturbations by optimizing the L_p norm of perturbation δ iteratively:

$$\arg\min_{\delta} \quad \|\delta\|_p + c \cdot \mathcal{L}(X + \delta), \tag{3}$$

where the loss function $\mathcal{L}$ is chosen to make the examples to be misclassified and c is a hyperparameter.

2.2 Adversarial Attacks on Face Recognition

Face recognition relies on distinguishing subtle differences, which often concentrated on small landmark locations, so it is more vulnerable to adversarial attacks. Different from the attacks described above, many attacks on face recognition are more concerned with the semantic structure of the face. Sharif et al. [37] employ a pair of eyeglass frames to fool face recognition systems. In [4], the adversarial face images are created by manipulating landmark locations. With the help of GANs, AdvFaces [5] learns to generating minimal perturbations in the salient facial regions. While regular attacks usually produce adversarial images leave the manifold, semantic-based attacks can generate high-fidelity images on manifold, which is more challenging.

2.3 Adversarial Defense

Adversarial defense methods can be roughly categorized into two groups:

Target Model Enhancement: Adversarial training [13] is the most popular defense method by injecting adversarial examples into training datasets. This approach works well for adversarial trained attacks but remains vulnerable to black-box attacks. He et al. [17] propose a trainable parametric noise injection technique to improve model regularization. Xu et al. [14] squeeze the images via color bit depth squeezing and spatial smoothing. In [46], Xie et al. propose to use feature denoising to increase adversarial robustness.

Input-Transformation: As a pre-processing strategy, input-transformation is model-agnostic and can complement other defenses. HGD [27] trains a image denoiser using a loss function defined by the difference in the top layers of the target model. Liu et al. [29] propose a feature distillation method to effectively defend against adversarial examples. Recently, a variety of work assume that the adversarial images leaves the manifold and aim to project it back. Among them, PixelDefend [38] and Defense-GAN [35] leverage generative networks to transform adversarial images into clean images. The disadvantage to these approaches is that they are limited by the expressiveness of generative networks, and cannot be applied to large-scale datasets. Abhimanyu et al. [10] approximate the image manifold using a web-scale image datasets. The main idea here is to localize nearest neighbors of adversarial image in the image datasets and classify them instead of the adversarial image.

2.4 Generative Adversarial Networks(GANs)

GANs was first introduced by Goodfellow et al. [12], it has important applications in various fields, such as image generation [21,33] and image-to-image translation [19,32,47]. The GANs framework has an excellent capacity to fit data distribution because of its min-max two-player game mechanism. Isola et al. [19] have shown that conditional generative adversarial networks are competent at image-to-image translation tasks. Variational Autoencoder (VAE) [24] consists of an encoder that represents the input as a distribution over the latent space and a decoder that reconstructs the input from the latent code. Donahue et al. [7] employ bidirectional GANs to learn an inverse mapping from data to latent representation. Analogously, ALI [11] proposes the same framework to learn mutually inference. In our work, to learn image distributions, we use a VAE-based architecture with adversarial training. Different from the prior work on image-to-image translation tasks, we replace the explicit reconstruction loss with GANs to mitigate the effects of small disturbances in the input.

3 Method

3.1 Motivation

As mentioned in Sect. 2, we conclude two types of attacks: off-manifold attacks and on-manifold attacks. On-manifold attacks can create adversarial images which are perceptually close to the original images, e.g., advFaces [4], A^3GN [26] and GFLM [3]. Therefore, we cannot make the common assumptions that all the adversarial images are out of the image manifold, but we assume that they are moved away from data distribution of high probability [38], at least.

These observations inspire us to project the data back onto the region of high probability. We use VAEs estimate the log-likelihood $\log p(x)$ of data x with a posterior $p(z|x)$ and a conditional distribution $q(z|x)$:

$$\log p(x) \geq -D_{KL}(q(z|x)||p(z)) + \mathbb{E}_{q(z|x)}[log(p(x|z))], \tag{4}$$

where $p(z)$ is a prior distribution and D_{KL} is the Kullback-Leibler divergence. By learning a probabilistic encoder $E(x)$ to represent $q(z|x)$ and an probabilistic decoder $Dec(z)$ to represent $p(x|z)$, we can find a tight estimate of the lower bound on the likelihood of images by optimizing z.

The key to solving the problem is that, the two terms in Eq. 4 correspond to the two types of attacks: For attacks leave the manifold, the adversarial images are outside of the distribution and it causes reconstruction loss inevitably (the second term). For on-manifold attacks, the adversarial images deviate from the high probability regions of the image distribution, it will increase the KL divergence of the posterior from the prior (the first term). Both types of adversarial images reduce the estimation of the likelihood, so the model can distinguish them from legal images and try to project them back.

Another intractable problem is how to accurately restore high fidelity faces. For general datasets, e.g. ImageNet [34], F-MNIST [45] and MNIST [26], the defense methods may only need to extract the coarse-grained information to satisfy the recognition conditions, such as shape and contour, while face recognition pays more attention to the distinction in local facial details. Therefore, face reconstruction quality is the premise of defense performance. Since the adversarial disturbance is imperceptible, we make a hypothesis that an image consists of low-level information and high-level information, and adversarial attacks ruin or manipulate with high-level information, but the low-level information is retained.

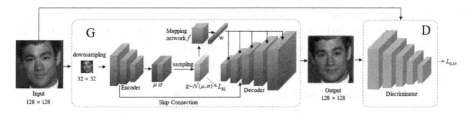

Fig. 2. The architecture of our network. Before passing the input layer, the input image is first downsampled to 32×32. The mapping network f transforms the latent code z to a new latent code w, which controls the variation of output images. The loss function consists only KL divergence loss and adversarial loss, does not include pixel-wise loss.

Instead of hallucinating the whole image, we constrain the expression space of the model to the high-level part of the images, so the network search in a smaller image space and less burden of generation.

In the next section, by constructing a specific network architecture and training strategy, we try to subtly achieve the above theories to move the adversarial face images back to the high probability regions. The proposed model consists of a generator G and a discriminator D. The overall framework is shown in Fig. 2.

3.2 Objective

To optimize the lower bound on the likelihood of datapoint x, the objective of a VAE can be expressed as

$$\mathcal{L}_{VAE} = D_{KL}(q(z|x)||p(z)) - \mathbb{E}_{q(z|x)}[log(p(x|z))]. \tag{5}$$

The first term tries to move the approximate posterior $q(z|x)$ closer to the prior $p(z)$, while the second term acts as an reconstruction error. Normally, the reconstruction loss will be Euclidean distance.

However, in our case, we *do not* adapt any explicit reconstruction loss. The reason is that, the pixel-based loss prompt the generator to faithfully preserve all the information from the input. During training, the training images are clean so there are no problems. However, in the testing phase, the input may be an adversarial image, and the generator will also tend to retain the perturbations, which is unreasonable. In this case, we use the adversarial loss to act as an implicit reconstruction loss:

$$\mathcal{L}_{GAN} = \mathbb{E}_x[logD(x)] + \mathbb{E}_x[log(1 - D(G(x)))], \tag{6}$$

where G is composed of the encoder E and the decoder Dec, and D is an discriminator. The generator G learns to synthesize images which cannot be distinguished by D, while D tries to classify the generated images and real images. Intuitively, during the training process, the generator will treat the input as a significant reference and try to extract useful components from it. Therefore, we expect the generator to generate images similar to the input in some ways. If this goal can be realized, the implicit loss will be better than the traditional loss. In this way, the reconstruction loss can be learned by the network during the training process, rather than being heuristically stipulated. With a large number of experiments, we implement this using a specific network architecture, which will covered in Sect. 3.3.

Hence, the final objective function for A-VAE is

$$\arg \min_{G} \max_{D} \mathcal{L}_{GAN} + KL(q(z|x)||p(z)), \tag{7}$$

where $p(z)$ is assumed to be $\mathcal{N}(\mathbf{0}, \mathbf{I})$. Ablation study of loss function is shown in Appendix A.

3.3 Architecture of Generator

We employ variational autoencoder networks as the basic framework of the generator. Since the adversarial perturbations only changes the high-level information in the images, we then design the network around this core.

As mentioned in Sect. 3.2, We expect the network to extract semantic information from the input. However, we found that when the input size is consistent with the output, the network just copies the input as it is, so that the network does not have to understand the image content. To solve this problem, we perform a downsampling operation on the input images. This strategy motivates the generator to focus the attention on low-level information and create some vivid details to complement the input faces. The prerequisite of this is that the generator must comprehend the semantic information of the input. Downsampling also mitigates the impact of the adversarial perturbations on the generator.

Given the latent code z generated by the encoder E, we aim to make it control the high-level variations of the synthesized images. A traditional decoder receives z at the input layer and transmits the required information of the upper layers by consuming the capacity of the network, which greatly limits the expressiveness power. Inspired by the recent success of style-based generator [22], we map z to an intermediate latent space $\mathcal{W}$ using a mapping network f, and inject it into each convolution layer of the decoder Dec through AdaIN operation [22]. The mapping network f is composed of 4 fully connected layers. Furthermore, we add a skip connection between two layers in the network to help the network transmit information more easily and reduce the burden on the bottleneck layer. Architecture details of A-VAE can be found in Appendix D.

3.4 Inference

At inference time, given an image x, the downsampled version is fed into the encoder E, then the encoder randomly sample M latent code z to generate different output images with diverse details, where z is clipped to meet a prerequisite that $KL(\mathcal{N}(z, \sigma^2 \mathbf{I})||\mathcal{N}(\mathbf{0}, \mathbf{I})) \leq \tau$, and τ is a threshold. Finally, we calculate Euclidean distance in pixel-level between these images and the original input image, and take the smallest one as the input of the face recognition model.

3.5 Discussion

Relevance to Existing Methods. As a pre-processing defense using generative models, our work is most similar to Defense-GAN [35]. It also finds a closest clean output which exists in image manifold to a given image. However, there are several fundamental differences between our method and Defense-GAN. First, we utilize a whole encoding-decoding network to generate the output, which has an input image as a valuable prior. Yet, Defense-GAN can only rely on random latent code to generate the image. Therefore, no matter which image is given, the potential generation space is always the entire image space, it needs better expressive power to recover the same quality image as ours, which is very tricky on face datasets. Second, at inference time, we find the nearest neighbor of input in a large number of randomly sampled output images, but Defense-GAN approximates the input by optimizing the latent code. Under on-manifold attacks, optimizing the latent code will make the generator reconstruct the image which is the same as the input, thus making the defense meaningless.

In [9,50], nearest neighbor search is also used to defend against adversarial images, equipped with a web-scale image dataset. The dataset needs to collect large-scale images and register in advance with images having same identities as test images, both of which are unrealistic in practical applications, especially for face recognition. Furthermore, both of them show their defense methods are more easily compromised by white-box attacks, we believe that one main reason is because they use the extracted features for nearest neighbor retrieval, but the feature extractor itself is vulnerable to adversarial attacks. On the contrary, we perform the nearest neighbor retrieval directly at the pixel-level.

A-VAE resembles ALI [11] and BIGAN [7], which have three components in training process: an encoder, a decoder and a discriminator. They ask the discriminator to distinguish generated data from real images, and between latent code z from the posterior distribution $q(z|x)$ and prior distribution $p(z)$. In our case, the KL-divergence term is employed to supervise the approximate posterior, and we replace the explicit reconstruction loss with GANs to learn a similarity metric. The main difference is that they expect the model to learn meaningful hidden features, whereas we are looking for a robust reconstruction process. It is worth mentioning that although ALI's inference network also samples stochastic latent code, unlike our method, it does not get different reconstructions of a same input.

4 Experiments

4.1 Experimental Settings

Training. We train our models on the publicly available CASIA-WebFace dataset [49] consists of 494,414 face images belonging to 10,575 different individuals. We use aligned CASIA-WebFace and crop to 128 × 128 image size. The hyperparameters λ of the loss function is empirically set to 1. We use ADAM optimizers [23] with $\beta_1=0$ and $\beta_2 = 0.99$. We train A-VAE for 140,000 steps

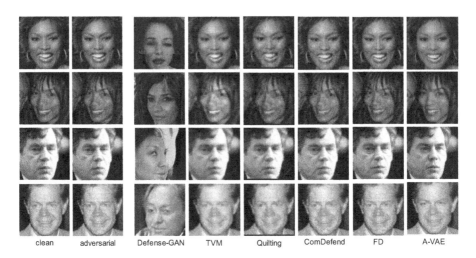

| clean | adversarial | Defense-GAN | TVM | Quilting | ComDefend | FD | A-VAE |

Fig. 3. Qualitative comparison results of image reconstruction results on LFW. The adversarial images is generated by FGSM grey-box attacks with $\epsilon = 8$. FD refers to Feature Distillation [29].

with the batch size of 16 on a single NVIDIA GTX 2080Ti. The learning rate is set to 0.001 during the training process.

Testing. We test our models on LFW [18]. LFW is a standard face verification testing dataset which includes 13,233 web-collected face images from 5,749 identities. We evaluate the verification accuracy on the 6,000 face pairs. Among them, 3,000 pairs are from the same identity and another 3,000 pairs represent the different identities. sampling times M is 1000, and threshold τ is set to 0.03.

Adversarial Attacks. We evaluate the performance of A-VAE against the state-of-the-art attacks including FGSM [43], PGD [30] and C&W [3]. C&W is constrained by L_2 norm with an allowed maximum value ϵ, and others are constrained by L_∞ norm. We implement the C&W attack with learning rate 10.0. We consider grey-box attacks and white-box attacks. All attack methods performs obfuscation attacks on same identity pairs, and impersonation attacks on different identities pairs. We perform attacks for images in the probe set.

Evaluation Metrics. We evaluate our method on a ResNet-50 [16] trained from VGG-Face2 [2]. More results on ArcFace [6] are given in Appendix E. The target model verifies whether the two images belong to one person by cosine distance between them. We measure the percentage of pairs which are successfully verified. The cosine distance threshold is set at 1% FAR.

4.2 Qualitative Results

We present our image reconstruction results against FGSM black-box attacks with $\epsilon = 8$ on LFW and compare to other input-transformation defense methods

Fig. 4. stochastic generated results on LFW. Each line presents input image(leftmost) and different output of latent code.

in Fig. 3, including Defense-GAN [35], TVM [15], Quilting [15], ComDefend [20] and Feature Distillation [29]. As can be seen, limited by expressive power, the images reconstructed by Defense-GAN [35] have a huge gap with the original images, so that the identity information is completely lost. TV minimization [15] almost completely eliminates adversarial perturbations, but makes the images blurry at the same time, which may lead to the loss of some crucial details. Compared to other methods, A-VAE is the only one that generates high-fidelity face images and captures the true identity information under large perturbations. The qualitative results show robustness of A-VAE to adversarial attacks on face images. Since it is just constrained by implicit reconstruction loss, our model can determine the important components of the image and focus on reconstructing it. This property relies on the model to understand the image from the high-level semantic features, instead of the pixels.

Figure 4 shows stochastic generated results of the same input on LFW, using different latent code. We can find that the variation in latent code has a significant impact on the appearance of the eyebrows, eyes, nose and mouth. These factors combine to express a wide variety of different identities, which allows our model to mine appropriate image that match the original identity well. More qualitative results are shown in Appendix F.

4.3   Gray-Box Attacks

In this section, we evaluate the performance of A-VAE against the gray-box attacks. In this setting, the attacker knows the details of the classifier but does not have access to the details of the defense strategy. The accuracy comparison results on the LFW dataset is shown in Table 1. Adversarial training [13], feture denoising [46] and HGD [27] obtain the FGSM adversarial images with

Table 1. Verification accuracies of different defense methods on the LFW dataset, under FGSM, PGD, C&W grey-box attacks.

LFW (Same identity pairs/Different identities pairs/Average)					
Defense	Clean	FGSM $\epsilon = 4$	FGSM $\epsilon = 8$	PGD $\epsilon = 8$	C& W
No Defense	0.992/0.992 /0.992	0.487/0.417 /0.452	0.190/0.300 /0.245	0.000/0.007 /0.003	0.000/0.017 /0.008
Adversarial Training [13]	0.981/0.993 /0.987	0.513/0.787 /0.650	0.177/0.737 /0.457	0.023/0.190 /0.107	0.000/0.417 /0.208
Feature Denoising [46]	0.950/0.953 /0.952	0.647/0.717 /0.682	0.213/0.730 /0.472	0.073/0.260 /0.167	0.020/0.570 /0.295
TVM [15]	**0.990**/0.991 /0.991	0.737/0.683 /0.710	0.343/0.357 /0.350	0.307/0.353 /0.330	0.007/0.020 /0.013
Quilting [15]	0.980/0.993 /0.987	0.813/0.890 /0.852	0.593/0.680 /0.637	0.667/0.783 /0.725	0.230/0.037 /0.134
ComDefend [20]	0.989/0.990 /0.990	0.523/0.637 /0.630	0.281/0.389 /0.335	0.022/0.148 /0.085	0.000/0.017 /0.008
Feature Distillation [29]	**0.990**/0.993 /**0.992**	0.667/0.580 /0.623	0.383/0.380 /0.382	0.143/0.190 /0.167	0.003/0.027 /0.015
HGD [27]	0.943/**1.000** /0.972	0.650/0.860 /0.755	0.387/0.790 /0.588	0.150/0.647 /0.398	0.040/0.597 /0.318
A-VAE	0.927/**1.000** /0.963	**0.830/0.957** /**0.893**	**0.637/0.863** /**0.753**	**0.697/0.960** /**0.828**	**0.423/0.797** /**0.610**

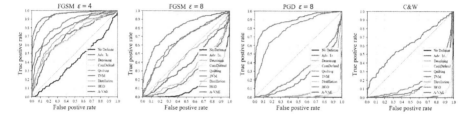

Fig. 5. ROC curves of different defense methods under FGSM, PGD and C&W attacks (Setting: gray-box, LFW).

$\epsilon = 8$. The evaluations involve the state-of-the-art attack methods including FGSM, PGD, C&W. It has been shown that A-VAE significantly improves the accuracy against off-manifold attacks and also outperforms other defense methods prominently. For example, although adversarial training defense with FGSM images achieves an average accuracy of 0.457 against FGSM attacks with $\epsilon = 8$, our method achieves an accuracy of 0.753 without any knowledge of adversarial images. Moreover, adversarial training has a poor generalization across different attacks, however, our method does not over-fit a specific attack. Figure 5 shows the Receiver Operating Characteristic (ROC) curves for different defense methods. We further show that our method also achieves better performance at different resolutions (Appendix B).

Table 2. Verification accuracies of different defense methods on the LFW dataset, under FGSM, PGD, C&W white-box attacks.

LFW (Same identity pairs/Different identities pairs/Average)				
Defense	Clean	FGSM $\epsilon = 4$	FGSM $\epsilon = 8$	PGD $\epsilon = 8$
No Defense	0.992/0.992/0.992	0.487/0.417/0.452	0.190/0.300/0.245	0.000/0.007/0.003
adversarial Training [13]	0.981/0.993/0.987	0.363/0.650/0.507	0.177/0.610/0.393	0.000/0.010/0.005
Feature Denoising [46]	0.950/0.953/0.952	0.401/0.440/0.423	0.170/0.450/0.310	0.000/0.020/0.010
ComDefend [20]	**0.989**/0.990/**0.990**	0.467/0.543/0.507	0.303/0.513/0.408	0.187/0.277/0.232
HGD [27]	0.943/**1.000**/0.972	0.243/0.677/0.460	0.103/ 0.627/0.365	0.323/0.713/0.518
A-VAE	0.927/**1.000**/0.963	**0.720/0.743/0.732**	**0.468/0.637/0.552**	**0.557/0.594/0.573**

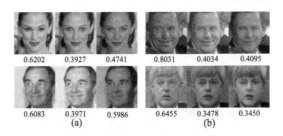

0.6202 0.3927 0.4741   0.8031 0.4034 0.4095

0.6083 0.3971 0.5986   0.6455 0.3478 0.3450
(a)   (b)

Fig. 6. Reconstruction results on the LFW dataset, under the on-manifold attack. Each pair presents clean face (left), adversarial face (middle) and reconstruction face (right). Cosine similarity score is calculated by comparing to gallery image.

4.4 White-Box Attacks

We present experimental results on white-box attacks using FGSM, PGD and C&W. Since our method is non-differentiable, to perform white-box attacks, we apply Backward Pass Differentiable Approximation (BPDA) [1] to estimate the gradient of the output of the classifier with respect to the input. We compare our defense with adversarial training [13], feature denoising [46], ComDefend [20] and HGD [27]. ComDefend causes gradient masking by adding random Gaussian noise in the image compression process, hence we employ Expectation over Transformation (EOT) [1] to correctly compute the gradient. Table 2 shows the accuracy comparison results on LFW. By comparison with Table 1, We can see that A-VAE does not suffer seriously when the attacker knows the defense strategy. Through this phenomenon, we conclude that our method does not entirely rely on gradient masking to improve robustness.

4.5 On-Manifold Attacks

On-manifold attacks generate high-quality adversarial images which are perceptually similar to the original images, e.g., GFLM [4] and advFaces [5]. However, in practice, we found when they deceive the target model successfully, the real label

Table 3. Verification accuracies of different defense methods on the LFW dataset, under on-manifold attack.

LFW (Same identity pairs/Different identities pairs/Average)	
Defense	On-manifold attack
No Defense	0.080/0.451/0.321
Adversarial Training [13]	0.237/0.867/0.537
Feature Denoising [46]	0.273/**0.867**/0.570
TVM [15]	0.337/0.623/0.480
Quilting [15]	0.393/0.664/0.529
Feature Distillation [29]	0.327/0.593/0.460
ComDefend [20]	0.407/0.553/0.480
HGD [27]	0.301/0.543/0.422
A-VAE	**0.644**/0.687/**0.667**

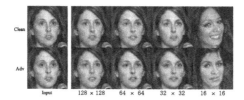

Fig. 7. Reconstruction results of different models.

of the adversarial images is also changed. In other words, neither of them achieve legitimate adversarial face images that only causing imperceptible perturbation. In our case, similar to [39], we directly compute the latent code z generated by the encoder to realize onmanifold attack, instead of original images. The latent code z is optimized n times with a step size ϵ:

$$z_0^{adv} = z, z_n^{adv} = z_{n-1}^{adv} + \epsilon \cdot \bigtriangledown_z J(F(Dec(z_{n-1}^{adv})), y), \qquad (8)$$

where J is the cosine similarity loss. The on-manifold adversarial image is obtained by $Dec(z_n^{adv})$. We use $n = 5$ and $\epsilon = 20$. Figure 6 shows the reconstruction results on LFW. As can be seen in Fig. 6 (b), this attack still inevitably produces some images that are notably different from the original images, and our method cannot handle this situation. The verification performance is shown in Table 3.

4.6   Tradeoff Between Quality and Robustness

In fact, A-VAE achieves the tradeoff between quality and robustness with a very simple strategy: We resize the scale of the input images. The more image information given, the model can supplement details in a smaller search space, the

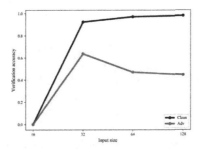

Fig. 8. Verification accuracies of different models on same identity pairs, using clean and adversarial images. (Setting: gray-box, FGSM with $\epsilon = 8$).

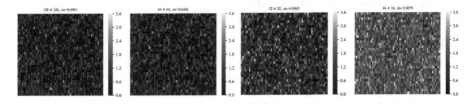

Fig. 9. Standard deviation of feature over 128 different realizations of the same input image, using different models. The feature of 2048 dimensions is shown as 32×64.

final resconstructed image will be more similar, but the adversarial perturbations will tend to be preserved. Conversely, the less image information given, the adversarial perturbations remains less, which makes the model more robust but requires a better expressiveness. To explore the effects of different scaling weights, we construct multiple generators with input size of 128×128, 64×64, 32×32 and 16×16, and correspondingly increse or delete the number of downsampling layers in the encoder block.

The performance of different models on clean images and adversarial images is shown in Fig. 7. Interestingly, when the input size drops to 16×16, the model seems to ignore the input and behaves like a standard GAN. The performance of different models on clean images and adversarial images is shown in Fig. 8. As can be seen, when the input size increases, the gap between the two accuracies enlarge, which means that the robustness declines. The accuracy of clean images continues to rise, which means that the quality increases. Figure 9 further illustrates standard deviation of extracted feature over 128 different realizations of the same image. We find it interesting that the expression space of the model is negatively related to the input size, although we do not explicitly constrain the model to generate images that are consistent with the input. This proves that our implicit reconstruction loss is effective. We set the input size as 32×32 so that the model retains enough capacity to reconstruct the original image well.

5 Conclusion

In this paper, we have proposed an adversarial defense method which projects the image into the high probability regions of image manifold. By constructing a special architecture and training mechanism, we enhance the robustness against both off-manifold and on-manifold attacks. The evaluation results on LFW show the superiority of our method.

Acknowledgements. This work is supported by National Nature Science Foundation of China (No. U1611461, U1903214, 61876135, 61862015), National Key R&D Program of China (No. 2017YFC0803700), National Nature Science Foundation of Hubei Province (2019CFB472) and Hubei Province Technological Innovation Major Project (2018AAA062, 2018CFA024, 2017AAA123).

References

1. Athalye, A., Carlini, N., Wagner, D.A.: Obfuscated gradients give a false sense of security: Circumventing defenses to adversarial examples. In: Dy, J.G., Krause, A. (eds.) Proceedings of the 35th International Conference on Machine Learning, ICML 2018, Stockholmsmässan, Stockholm, Sweden, July 10–15, 2018. Proceedings of Machine Learning Research, vol. 80, pp. 274–283. PMLR (2018). http://proceedings.mlr.press/v80/athalye18a.html
2. Cao, Q., Shen, L., Xie, W., Parkhi, O.M., Zisserman, A.: Vggface2: a dataset for recognising faces across pose and age. In: 2018 13th IEEE International Conference on Automatic Face & Gesture Recognition (FG 2018), pp. 67–74. IEEE (2018)
3. Carlini, N., Wagner, D.A.: Towards evaluating the robustness of neural networks. In: 2017 IEEE Symposium on Security and Privacy, SP 2017, San Jose, CA, USA, May 22–26, 2017, pp. 39–57. IEEE Computer Society (2017). https://doi.org/10.1109/SP.2017.49, https://doi.org/10.1109/SP.2017.49
4. Dabouei, A., Soleymani, S., Dawson, J., Nasrabadi, N.: Fast geometrically-perturbed adversarial faces. In: 2019 IEEE Winter Conference on Applications of Computer Vision (WACV), pp. 1979–1988. IEEE (2019)
5. Deb, D., Zhang, J., Jain, A.K.: Advfaces: Adversarial face synthesis. arXiv preprint arXiv:1908.05008 (2019)
6. Deng, J., Guo, J., Xue, N., Zafeiriou, S.: Arcface: Additive angular margin loss for deep face recognition. In: Proceedings of the IEEE Conference on Computer Vision and Pattern Recognition, pp. 4690–4699 (2019)
7. Donahue, J., Krähenbühl, P., Darrell, T.: Adversarial feature learning. In: 5th International Conference on Learning Representations, ICLR 2017, Toulon, France, April 24–26, 2017, Conference Track Proceedings. OpenReview.net (2017)
8. Dong, Y., Liao, F., Pang, T., Su, H., Zhu, J., Hu, X., Li, J.: Boosting adversarial attacks with momentum. In: Proceedings of the IEEE Conference on Computer Vision and Pattern Recognition, pp. 9185–9193 (2018)
9. Dubey, A., van der Maaten, L., Yalniz, Z., Li, Y., Mahajan, D.: Defense against adversarial images using web-scale nearest-neighbor search. In: IEEE Conference on Computer Vision and Pattern Recognition, CVPR 2019, Long Beach, CA, USA, June 16–20, 2019, pp. 8767–8776. Computer Vision Foundation/IEEE (2019)

10. Dubey, A., Maaten, L.v.d., Yalniz, Z., Li, Y., Mahajan, D.: Defense against adversarial images using web-scale nearest-neighbor search. In: Proceedings of the IEEE Conference on Computer Vision and Pattern Recognition, pp. 8767–8776 (2019)
11. Dumoulin, V., et al.: Adversarially learned inference. In: 5th International Conference on Learning Representations, ICLR 2017, Toulon, France, April 24–26, 2017, Conference Track Proceedings. OpenReview.net (2017)
12. Goodfellow, I., Pouget-Abadie, J., Mirza, M., Xu, B., Warde-Farley, D., Ozair, S., Courville, A., Bengio, Y.: Generative adversarial nets. In: Advances in neural information processing systems. pp. 2672–2680 (2014)
13. Goodfellow, I.J., Shlens, J., Szegedy, C.: Explaining and harnessing adversarial examples. arXiv preprint arXiv:1412.6572 (2014)
14. Gu, S., Yi, P., Zhu, T., Yao, Y., Wang, W.: Detecting adversarial examples in deep neural networks using normalizing filters. UMBC Student Collection (2019)
15. Guo, C., Rana, M., Cissé, M., van der Maaten, L.: Countering adversarial images using input transformations. In: 6th International Conference on Learning Representations, ICLR 2018, Vancouver, BC, Canada, April 30 - May 3, 2018, Conference Track Proceedings. OpenReview.net (2018), https://openreview.net/forum?id=SyJ7ClWCb
16. He, K., Zhang, X., Ren, S., Sun, J.: Deep residual learning for image recognition. In: Proceedings of the IEEE Conference on Computer Vision and Pattern Recognition, pp. 770–778 (2016)
17. He, Z., Rakin, A.S., Fan, D.: Parametric noise injection: trainable randomness to improve deep neural network robustness against adversarial attack. In: Proceedings of the IEEE Conference on Computer Vision and Pattern Recognition, pp. 588–597 (2019)
18. Huang, G.B., Mattar, M., Berg, T., Learned-Miller, E.: Labeled faces in the wild: a database for studying face recognition in unconstrained environments (2008)
19. Isola, P., Zhu, J.Y., Zhou, T., Efros, A.A.: Image-to-image translation with conditional adversarial networks. In: Proceedings of the IEEE Conference on Computer Vision and Pattern Recognition, pp. 1125–1134 (2017)
20. Jia, X., Wei, X., Cao, X., Foroosh, H.: Comdefend: an efficient image compression model to defend adversarial examples. In: IEEE Conference on Computer Vision and Pattern Recognition, CVPR 2019, Long Beach, CA, USA, June 16–20, 2019. pp. 6084–6092. Computer Vision Foundation/IEEE (2019)
21. Karras, T., Aila, T., Laine, S., Lehtinen, J.: Progressive growing of GANs for improved quality, stability, and variation. In: 6th International Conference on Learning Representations, ICLR 2018, Vancouver, BC, Canada, 30 April–3 May 2018, Conference Track Proceedings. OpenReview.net (2018)
22. Karras, T., Laine, S., Aila, T.: A style-based generator architecture for generative adversarial networks. In: Proceedings of the IEEE Conference on Computer Vision and Pattern Recognition, pp. 4401–4410 (2019)
23. Kingma, D.P., Ba, J.: Adam: a method for stochastic optimization. In: Bengio, Y., LeCun, Y. (eds.) 3rd International Conference on Learning Representations, ICLR 2015, San Diego, CA, USA, May 7–9, 2015, Conference Track Proceedings (2015). http://arxiv.org/abs/1412.6980
24. Kingma, D.P., Welling, M.: Auto-encoding variational Bayes. In: Bengio, Y., LeCun, Y. (eds.) 2nd International Conference on Learning Representations, ICLR 2014, Banff, AB, Canada, April 14–16, 2014, Conference Track Proceedings (2014). http://arxiv.org/abs/1312.6114
25. Kurakin, A., Goodfellow, I., Bengio, S.: Adversarial examples in the physical world. arXiv preprint arXiv:1607.02533 (2016)

26. LeCun, Y., Bottou, L., Bengio, Y., Haffner, P.: Gradient-based learning applied to document recognition. Proc. IEEE **86**(11), 2278–2324 (1998)

27. Liao, F., Liang, M., Dong, Y., Pang, T., Hu, X., Zhu, J.: Defense against adversarial attacks using high-level representation guided denoiser. In: 2018 IEEE Conference on Computer Vision and Pattern Recognition, CVPR 2018, Salt Lake City, UT, USA, 18–22 June 2018, pp. 1778–1787. IEEE Computer Society (2018)

28. Liu, J., Deng, Y., Bai, T., Wei, Z., Huang, C.: Targeting ultimate accuracy: face recognition via deep embedding. arXiv preprint arXiv:1506.07310 (2015)

29. Liu, Z., et al.: Feature distillation: DNN-oriented JPEG compression against adversarial examples. In: IEEE Conference on Computer Vision and Pattern Recognition, CVPR 2019, Long Beach, CA, USA, 16–20 June 2019, pp. 860–868. Computer Vision Foundation/IEEE (2019)

30. Madry, A., Makelov, A., Schmidt, L., Tsipras, D., Vladu, A.: Towards deep learning models resistant to adversarial attacks. In: 6th International Conference on Learning Representations, ICLR 2018, Vancouver, BC, Canada, 30 April–3 May 2018, Conference Track Proceedings. OpenReview.net (2018). https://openreview.net/forum?id=rJzIBfZAb

31. Moosavi-Dezfooli, S.M., Fawzi, A., Frossard, P.: Deepfool: a simple and accurate method to fool deep neural networks. In: Proceedings of the IEEE Conference on Computer Vision and Pattern Recognition, pp. 2574–2582 (2016)

32. Park, T., Liu, M., Wang, T., Zhu, J.: Semantic image synthesis with spatially-adaptive normalization. In: IEEE Conference on Computer Vision and Pattern Recognition, CVPR 2019, Long Beach, CA, USA, 16–20 June 2019, pp. 2337–2346. Computer Vision Foundation/IEEE (2019)

33. Radford, A., Metz, L., Chintala, S.: Unsupervised representation learning with deep convolutional generative adversarial networks. In: Bengio, Y., LeCun, Y. (eds.) 4th International Conference on Learning Representations, ICLR 2016, San Juan, Puerto Rico, May 2–4, 2016, Conference Track Proceedings (2016)

34. Russakovsky, O., et al.: Imagenet large scale visual recognition challenge. Int. J. Comput. Vision **115**(3), 211–252 (2015)

35. Samangouei, P., Kabkab, M., Chellappa, R.: Defense-GAN: protecting classifiers against adversarial attacks using generative models. In: 6th International Conference on Learning Representations, ICLR 2018, Vancouver, BC, Canada, April 30 - May 3, 2018, Conference Track Proceedings. OpenReview.net (2018) https://openreview.net/forum?id=BkJ3ibb0-

36. Schroff, F., Kalenichenko, D., Philbin, J.: FaceNet: a unified embedding for face recognition and clustering. In: Proceedings of the IEEE Conference on Computer Vision and Pattern Recognition, pp. 815–823 (2015)

37. Sharif, M., Bhagavatula, S., Bauer, L., Reiter, M.K.: Accessorize to a crime: real and stealthy attacks on state-of-the-art face recognition. In: Proceedings of the 2016 ACM SIGSAC Conference on Computer and Communications Security, pp. 1528–1540 (2016)

38. Song, Y., Kim, T., Nowozin, S., Ermon, S., Kushman, N.: Pixeldefend: leveraging generative models to understand and defend against adversarial examples. In: 6th International Conference on Learning Representations, ICLR 2018, Vancouver, BC, Canada, 30 April–3 May 2018, Conference Track Proceedings. OpenReview.net (2018), https://openreview.net/forum?id=rJUYGxbCW

39. Stutz, D., Hein, M., Schiele, B.: Disentangling adversarial robustness and generalization. In: Proceedings of the IEEE Conference on Computer Vision and Pattern Recognition, pp. 6976–6987 (2019)

40. Sun, B., Tsai, N.H., Liu, F., Yu, R., Su, H.: Adversarial defense by stratified convolutional sparse coding. In: Proceedings of the IEEE Conference on Computer Vision and Pattern Recognition, pp. 11447–11456 (2019)

41. Sun, Y., Chen, Y., Wang, X., Tang, X.: Deep learning face representation by joint identification-verification. In: Advances in Neural Information Processing Systems, pp. 1988–1996 (2014)

42. Sun, Y., Liang, D., Wang, X., Tang, X.: Deepid3: face recognition with very deep neural networks. arXiv preprint arXiv:1502.00873 (2015)

43. Szegedy, C., et al.: Intriguing properties of neural networks. In: Bengio, Y., LeCun, Y. (eds.) 2nd International Conference on Learning Representations, ICLR 2014, Banff, AB, Canada, 14–16 April 2014, Conference Track Proceedings (2014). http://arxiv.org/abs/1312.6199

44. Taigman, Y., Yang, M., Ranzato, M., Wolf, L.: DeepFace: closing the gap to human-level performance in face verification. In: Proceedings of the IEEE Conference on Computer Vision and Pattern Recognition, pp. 1701–1708 (2014)

45. Xiao, H., Rasul, K., Vollgraf, R.: Fashion-MNIST: a novel image dataset for benchmarking machine learning algorithms. arXiv preprint arXiv:1708.07747 (2017)

46. Xie, C., Wu, Y., Maaten, L.v.d., Yuille, A.L., He, K.: Feature denoising for improving adversarial robustness. In: Proceedings of the IEEE Conference on Computer Vision and Pattern Recognition, pp. 501–509 (2019)

47. Xue, Y., Xu, T., Zhang, H., Long, L.R., Huang, X.: Segan: adversarial network with multi-scale L 1 loss for medical image segmentation. Neuroinformatics **16**(3–4), 383–392 (2018)

48. Ye, M., et al.: Person reidentification via ranking aggregation of similarity pulling and dissimilarity pushing. IEEE Trans. Multimedia **18**(12), 2553–2566 (2016)

49. Yi, D., Lei, Z., Liao, S., Li, S.Z.: Learning face representation from scratch. CoRR abs/1411.7923 (2014). http://arxiv.org/abs/1411.7923

50. Zhao, J., Cho, K.: Retrieval-augmented convolutional neural networks against adversarial examples. In: IEEE Conference on Computer Vision and Pattern Recognition, CVPR 2019, Long Beach, CA, USA, 16–20 June 2019, pp. 11563–11571. Computer Vision Foundation/IEEE (2019)

Weakly Supervised Learning with Side Information for Noisy Labeled Images

Lele Cheng$^{(\boxtimes)}$, Xiangzeng Zhou$^{(\boxtimes)}$, Liming Zhao$^{(\boxtimes)}$, Dangwei Li$^{(\boxtimes)}$, Hong Shang$^{(\boxtimes)}$, Yun Zheng$^{(\boxtimes)}$, Pan Pan$^{(\boxtimes)}$, and Yinghui Xu$^{(\boxtimes)}$

Machine Intelligence Technology Lab, Damo Academy, Alibaba Group,
Hangzhou, China
{yinan.cll,xiangzeng.zxz,lingchen.zlm,dangwei.ldw,shanghong.sh,
zhengyun.zy,panpan.pp}@alibaba-inc.com, renji.xyh@taobao.com

Abstract. In many real-world datasets, like WebVision, the performance of DNN based classifier is often limited by the noisy labeled data. To tackle this problem, some image related side information, such as captions and tags, often reveal underlying relationships across images. In this paper, we present an efficient weakly-supervised learning by using a Side Information Network (SINet), which aims to effectively carry out a large scale classification with severely noisy labels. The proposed SINet consists of a visual prototype module and a noise weighting module. The visual prototype module is designed to generate a compact representation for each category by introducing the side information. The noise weighting module aims to estimate the correctness of each noisy image and produce a confidence score for image ranking during the training procedure. The propsed SINet can largely alleviate the negative impact of noisy image labels, and is beneficial to train a high performance CNN based classifier. Besides, we released a fine-grained product dataset called AliProducts, which contains more than 2.5 million noisy web images crawled from the internet by using queries generated from 50,000 fine-grained semantic classes. Extensive experiments on several popular benchmarks (i.e. Webvision, ImageNet and Clothing-1M) and our proposed AliProducts achieve state-of-the-art performance. The SINet has won the first place in the 5000 category classification task on WebVision Challenge 2019, and outperforms other competitors by a large margin.

Keywords: Weakly supervised learning · Noisy labels · Side information · Large scale web images

1 Introduction

In recent years, the computer vision community has witnessed the significant success of Deep Neural Networks (DNNs) on several benchmark datasets of image classification, such as ImageNet [1] and MS-COCO [22]. However, obtaining large-scale data with clean and reliable labels is expensive and time-consuming.

© Springer Nature Switzerland AG 2020
A. Vedaldi et al. (Eds.): ECCV 2020, LNCS 12375, pp. 306–321, 2020.
https://doi.org/10.1007/978-3-030-58577-8_19

When noisy labels are introduced in training data, it is widely known that the performance of a deep model can be significantly degraded [2,3,23,36], which prevents deep models from being quickly employed in real-world noisy scenarios.

A common solution is to collect a large amount of image related side information (e.g. surrounding texts, tags and descriptions) from the internet, and directly take them as the ground-truth for model training. Though this solution is more efficient than manual annotation, the obtained labels usually contain noise due to the heterogeneous sources. Therefore, improving the robustness of deep learning models against noisy labels has become a critical issue.

To estimate the noise in labels, some works propose new layers [26,27] or loss functions [18,28–30] to correct the noisy label during training. However, these works rely on a strict assumption that there is a single transition probability between the noisy labels and the ground-truth labels. Owning to this assumption, these methods may show good performance on hand-crafted noisy datasets but are inefficient on real noisy datasets such as Clothing1M [36]. In some situations, it is possible to annotate a small fraction of training samples as additional supervision. By using additional supervision, works like [11,31,32] could improve the robustness of deep networks against label noises. But still, the requirement on clean samples make them less flexible to apply in large scale real-world scenarios.

Many data cleaning algorithms [33–35] are developed to discard those samples with wrong label ahead of the training procedure. The major difficulty of these algorithms is how to distinguish informative hard samples from harmful mislabeled ones. CleanNet [11] achieves state-of-the-art performance on the real-world noisy dataset Clothing1M [36]. CleanNet generates a single representative sample (class prototype) for each class and uses it to estimate the correctness of sample labels. With the observation that samples have wide-spread distribution in noisy classes, SMP [20] takes multiple prototypes to represent a noisy class instead of single prototype in CleanNet. In both CleanNet and SMP, extra clean supervision is required to train models.

In most of previous works, image related side information or annotations (e.g. titles and tags) from web are commonly regarded as noisy labels. These works may not fully take advantage of the side information. Based on our observations, these image related side information reveal underlying similarity among images and classes, which has great potential to help tackle label noises. By analyzing the label structure and text descriptions, we explore an weakly-supervised learning strategy to deal with noisy samples. For example, the label "apple" may refer to a fruit or an Apple mobile phone. When acquiring images from web using the label "apple", images of apple fruits and Apple mobile phones will be wrongly put under a same class. Fortunately, titles or text descriptions about the images could imply the misplacement. In this paper, we propose an efficient weakly-supervised learning strategy to evaluate the correctness of each image sample in each class by exploiting the label structure and label descriptions. Moreover, we release a large scale fine-grained product dataset to facilitate further research on visual recognition. To our knowledge, the proposed product dataset contains the largest number of product categories by now.

Fig. 1. Images of WebVision 2019 dataset [37] from the categories of phalarope, horseman, candied apple, tulipa, gesneriana. The dataset was collected from the Internet by textual queries generated from 5, 000 semantic concepts in WordNet. Obviously, each category includes a lot of noisy images as shown above.

Our contributions in this paper are summarized as follows:

1) A weakly supervised learning with side information network (SINet) is proposed for noisy labeled image classification. SINet infers the relationship between images and labels without any human annotation, and enable us to train high-performance and robust CNN models against large scale label noises.
2) A noisy and fine-grained product dataset called AliProducts is released, which contains more than 2.5 million web images crawled from the Internet by using queries generated from the 50, 000 fine-grained semantic classes. In addition, side information (e.g., hierarchical relationships between classes) are also provided for the convenience of extending research.
3) Extensive experiments are conducted on a number of benchmarks, including WebVision, ImageNet, Clothing1M and AliProducts, in which the proposed SINet obtains the state-of-the-art performance. Our SINET also won the first place on the WebVision Challenge 2019, and outperforms the other competitors by a large margin.

2   Related Work

Recent studies have shown that the performances of DNNs degraded substantially when training on data with noisy labels [2,3]. To alleviate this problem, a number of approaches have been introduced and can be generally summarized as below.

Some methods design robust loss functions against label noises [4–9]. Zhang et al. [5] found that the mean absolute error (MAE) is inherently more robust to label noises than the commonly-used categorical cross entropy (CCE) in many circumstances. However, MAE performs poorly with DNNs and challenging datasets due to slow convergence. Generalized Cross Entropy (GCE) loss [9] applies a Box-Cox transformation to probabilities (power law function of probability with exponent q) and can be seen as a generalization of MAE and CCE, thus can be easily applied with existing DNN architecture and yield good performance in certain noisy datasets.

Re-weighting training samples aims to evaluate the correctness of each sample on a given label, and has been widely studied in [10–16,21]. In [12], meta learning paradigm is used to determine the sample weighting factors. [13] takes open-set noisy labels into consideration and train a Siamese network to detect noisy labels. In each iteration, sample weighting factors will be re-estimated, and the classifier will be updated at the same time. [14] also presents a method to separate clean samples from noisy samples in an iterative fashion. The biggest challenge encountering these data cleaning algorithms is how to distinguish informative hard samples from harmful mislabeled ones. To prevent discarding valuable hard samples, noisy samples is weighted according to their noisiness level which is estimated by pLOF [15]. In CleanNet [11], an additional network is designed to decide whether a sample is mislabelled or not. CleanNet aims to produce weights of samples during the training procedure. CurriculumNet [16] designs a learning curriculum by measuring the complexity of data and ranking samples in an unsupervised manner. However, most of these approaches either requires extra clean samples as additional information or adopts a complicated training procedure, making them less suitable for being widely applied in many real-world scenarios.

Self-learning pseudo-labels has been studied in many scenarios to deal with noisy labels. Reed et al. [17] propose to jointly train model with both noisy labels and pseudo-labels. However, [17] over-simplifies the assumption of the noisy label distribution, which leads to sub-optimal results. In the joint optimization process of [18], original noisy labels are completely replaced by pseudo-labels. This often discards some valuable information in the original noisy labels. Li et al. [19] proposes to simulate actual training by generating synthetic noisy labels, and train the model such that after one gradient update using each set of synthetic noisy labels, thus the model does not overfit to the specific noise.

Our method is similar to the work of [20], in which each class is represented by a learnable prototype. For each sample, a similarity is calculated between the sample and the corresponding prototype to correct its label. A final classifier is trained by using both the corrected label and the noisy label. However, [20] only takes visual information into consideration to construct class prototypes. Our approach integrate visual with side information to generate more reliable prototype for each class.

3   Approach

We focus on learning a robust image classifier from large-scale noisy images with side information. Let $\mathbb{D} = \{(x_1, y_1), ..., (x_N, y_N)\}$ be a noisily labeled dataset of N images. $y_i \in \{1, 2, ..., C\}$ is the noisy label corresponding to the image x_i, and C is the number of classes in the dataset.

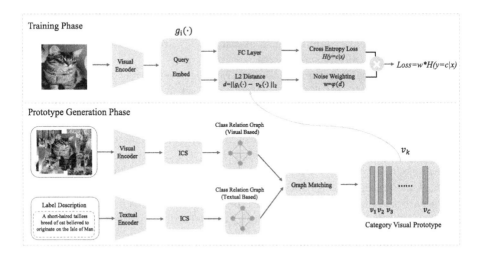

Fig. 2. Illustration of the framework of SINet on the noisy dataset, which includes three sub-modules, including *Class Relation Graph*, *Visual Prototype Generation*, and *Noise Weighting*. First, we construct visual-based and textual-based class relation graph with representation of image and textual using Inter Class Similarity (ICS), respectively. Second, we generate a visual prototype for each class using the compliance between the visual and textual class relations. Finally, *Noise Weighting* is used to weight all noisily labeled images with class prototypes before the training procedure.

Training with noisily labeled images, deep neural networks may over-fit these noisy labels and perform poorly. To alleviate this problem, we introduce a conceptually simple but effective side information network (SINet) for training against noisy labels. Based on the knowledge of "different classes look different" [38], we train the network under the constraint that the produced visual similarity between classes should have potential relevance with their natural semantic similarity. The semantic similarity is derived from a class relation graph constructed with the image related side information such as image titles, long text descriptions and image tags.

For each class, a prototype v_k, $k \in \{1, 2, ..C\}$ is generated from reliable training samples whose visual similarity graph aligns well with the constructed Class Relation Graph. Subsequently, we can decide whether an training sample is mislabeled or not by comparing its visual representation with v_k which is considered as a clean and reliable representation of k-th class. During the training phrase, an image is recognized as a noisy sample or not in the light of the distance of the image feature and the class prototype. Instead of directly discarding noisy images by a predefined threshold, we assign each image a correctness weight in a noise weighting module.

Training a deep model $\mathcal{G}$ parameterized by θ on the dataset $\mathbb{D}$, the overall optimization objective is formulated as

$$\theta^* = argmin \sum_{i=1}^{N} w_i * L(y_i, \mathcal{G}(\theta, x_i)) \tag{1}$$

where L is a conventional cross entropy loss, and w_i is the image weight generated by the noise weighting module.

In the following sections, we elaborate the proposed SINet that using image related side information to facilitate a classification task on noisy images. The SINet comprises three modules, i.e. class relation graph generation, visual prototype generation and noise weighting. As shown in Fig. 2, an overview of the SINet architecture is illustrated. In Sect. 3.1, two kinds of category relation graphs are constructed using label embeddings and WordNet information, respectively. In Sect. 3.2, the KL divergence is used to estimated the compliance between the two kinds of graphs, so as to generate a visual prototype for each class. Given the class prototypes, a noise weighting module is presented to weight all noisily labeled images before the training procedure in Sect. 3.3.

3.1   Class Relation Graph

In some classification scenarios, for each class we can obtain both the long-text description of label and the hierarchical structure of class relationships using WordNet [40]. Both the label descriptions and WordNet structure reveal rich semantic information across classes. In this section, we attempt to exploit the inter-class semantic knowledge by constructing two kinds of class relation graphs. In the graphs, each node represents a class and the edges between nodes are built using two different similarity metrics.

Firstly, a straightforward way to build a class relation (marked as $\mathbb{G}_w$) is using the tree structure of WordNet. In the graph $\mathbb{G}_w$, an edge of two class nodes is created using the distance of the shortest path in the WordNet tree. Here we represent the WordNet-based class relation graph $\mathbb{G}_w$ as a matrix $S_w \in R^{C \times C}$.

Secondly, we learn a label embedding for each class node with the text description of label from WordNet. For instance, a label description *Manx cat: A short-haired tailless breed of cat believed to originate on the Isle of Man* provides rich semantic knowledge of the corresponding class. Moreover, these text descriptions reveal underlying relationship across classes from the perspective of natural language. To obtain a semantic representation of each class, we use a BERT [39] language model to learn a sequence of word embeddings from text description of each class label. We then feed them into a bidirectional LSTM module to achieve a class label level embedding (called label embedding). Please note that the number of label text descriptions available for training is too small, so we use a pretrained BERT and freeze it in the training procedure, and only finetune the LSTM module. Meanwhile, the number of trainable parameters is significantly reduced.

We then build a graph $\mathbb{G}_l$ based on the label embeddings, in which the set of nodes is $\mathcal{V} = \{\mathbf{v}_1, \mathbf{v}_2, ..., \mathbf{v}_C\}$, and $\mathbf{v}_i \in R^d$ represents the label embedding of the i-th class. We then calculate the cosine similarity between all pairs of label embeddings to build edges of the graph $\mathbb{G}_l$. For convenience, the graph $\mathbb{G}_l$ is formulated as a inter-class similarity (ICS) matrix $S_l \in R^{C \times C}$ as below.

$$S_l^{ij} = \frac{\mathbf{v}_i^T \mathbf{v}_j}{||\mathbf{v}_i||_2 ||\mathbf{v}_j||_2} \tag{2}$$

Then S_l^{ij} is regarded as a kind of similarity between two class embeddings $\mathbf{v}_i$ and $\mathbf{v}_j$. Larger S_l^{ij} indicates higher similarity between the classes i and j.

Eventually, we blend the two class relation graphs $\mathbb{G}_w$ and $\mathbb{G}_l$ generated using two kinds of semantic knowledge, and obtain a hybrid graph $\mathbb{G}_t$, formulated as below.

$$S_t = S_l + S_w \tag{3}$$

3.2   Visual Prototype Generation

This section introduce an effective visual prototype generation module for training robust CNNs with noisy images. The key idea of visual prototype module is to generate a clean visual prototype v_k, $k \in \{1, 2, ..C\}$ for each class. The visual prototype v_k can be interpreted as a reliable and effective representation of k-th class, and can be used to identify the reliability of all training data.

In order to generate visual prototype v_k, we need to obtain some reliable images from k-th class, and evaluate their contributions to v_k. Since noisy images is ubiquitous within each class, it is an intractable problem to directly collect reliable images. In this paper, we resort to the class relation graph $\mathbb{G}_t$ constructed in Sect. 3.1 to help this collection. Intuitively, the inter-class relation in visual representation space for reliable images should be closely related to that in class relation graph. For example, the k nearest classes of *siamese cat* in class relation graph are *persian cat*, *tiger cat*, *manx cat*, etc. If the k nearest classes of an image in visual representation space are also the same, then this image is probably reliable, and should contribute to the generation of visual prototype in a high confidence.

To be specific, we consider an image sample x_i and its current labelled class c. The semantic similarity vector of class c can be obtained from the class relation graph $\mathbb{G}_t$, and is denoted as a vector s_t^i of length C. To compute the visual similarity vector between x_i and all C classes, it is required to generate an initial prototype for each class first. We first extract visual features from all images using the CNN model in the proposed SINet. Then for each class, top-k ranked image features according to their classification confidence score are averaged to generate initial class prototype. Then the visual similarity vector of x_i is computed as the cosine similarity score of the CNN feature g_i with all initial class prototypes, which is denoted a vector s_v^i of length C. Finally the consistence score p_i of image sample $x_{i'}$ is estimated based on KL divergence between s_t^i and s_v^i.

$$p_i = \frac{1}{\left(KL\left(\psi(s_t^i), \psi(s_v^i)\right) + \epsilon\right)^\gamma} \tag{4}$$

where ψ is a normalize function, e.g. $L2$ norm or softmax, γ is used to control the "contrast" of two similarity vectors, and ϵ is a small positive constant to prevent the denominator going to zero.

Eventually, we generate a visual prototype v_c for class c by using the weighted sum of the image features, as formulated in Eq. (5):

$$v_c = \frac{\sum_{i=1}^N g_i p_i}{\sum_{i=1}^N p_i} \tag{5}$$

where the g_i is the visual CNN feature of image x_i in the base model.

As the training proceeds, the visual prototype for each class will be updated iteratively, then more reliable samples could contribute to train the CNN model better.

3.3 Noise Weighting

In this section, we use the class prototypes generated above to weight noisily labelled images before training. Considering an image x_i and its current labelled class c, we estimate an importance weight $w_{i,c}$ by calculating the Euclidean distance of the visual feature g_i of image x_i and the prototype v_c. As formulated in Eq. (6), the importance weight $w_{i,c}$ is computed with two hyper-parameters α and β to control the shift and contrast of different visual features.

$$w_{i,c} = max\{0, [\alpha - ||v_c - g_i||_2]^\beta\} \tag{6}$$

We finally use a weighted cross entropy loss for model training as shown in Equ.(7).

$$\text{Loss}_{ce} = \sum_{i=1}^N \sum_{c=1}^C w_{i,c} \cdot log(p_{i,c}) \tag{7}$$

where $p_{i,c}$ is the softmax output of image x_i on class c.

3.4 Implementation Details and WebVision Challenge

Implementation Details. The scale of WebVision data is significantly larger than that of ImageNet, it is important to considering the computational cost when extensive experiments are conducted in evaluation and comparisons. In our experiments, we employ the resnext-101 as our standard architecture. The resnext-101 model is trained by adopting the proposed SINet. The network weights are optimized with mini-batch stochastic gradient decent (SGD), where the batch size is set to 2,500. The learning rate starts from 0.1, and decreases by a factor of 10 at the epochs of 30, 60, 80, 90. The whole training process stop at 100 epochs. To reduce the risk of over-fitting, we use common data augmentation technologies which include random cropping, mirror flip and autoaugment. We also add a dropout operation with a ratio of 0.25 after the global pooling layer.

Topk Label Smoothing. Since there exists massive noise images in WebVision, if we directly utilize the one-hot target of ground truth to train CNN, it is inevitable to over-fit the noisy labels. To alleviate this problem, we proposed Adaptive Label Smoothing to assist the model training. Specifically, we select a small subset of high confidence images to train an initial model, and then we use the model to predict probability distribution of rest images. We use the topk predictions and ground truth to construct a smoothing label, and use this smoothing label to train the model. The Adaptive Label Smoothing enhance the tolerance of noisy labels, leading to about 0.2% performance improvements on top-5 accuracy in WebVision challenge.

Adaptive Spatial Resolution. There exists a lot of fine-grained categories in WebVision, which are hard to distinguish. Many studies have show that high-resolution images can improve the performance of fine-grained recognition. Inspired by this, we first train an initial model with fixed image resolution of 224×224, and then finetune the model with large input resolutions, e.g. 256×256 and 312×312. Specifically, the adaptive average pooling is used before the classifier layer to keep the feature dimension unchanged. The large input resolutions enhance the tolerance of noisy labels, leading to about 0.5% performance improvements on top-5 accuracy in WebVision challenge.

4   Experiments

In this section, we mainly evaluate our SINet on four popular benchmarks for noisy-labeled visual recognition, i.e., WebVision, ImageNet, Clothing1M and AliProducts. Particularly, we investigate the learning capability on large-scale web images without any human annotation.

4.1   Datasets

WebVision 1.0 [37] is an object-centric dataset, which is larger than ImageNet for object recognition and classification. The images are crawled from both Flickr and Google images search, by using queries generated from the $1,000$ semantic concepts of the WordNet. Meta information along with those web images (e.g., title, description, tags, etc.) are also crawled. The dataset contains 1,000 object categories, including 2.4 millions images in total, but without any human annotation. 50K images with human annotation are used as validation set, and another 50K images with human annotation for testing. The evaluation measure is based on top-5 accuracy, where each algorithm provides a list of at most 5 object categories to match the ground truth.

WebVision 2.0 is similar with WebVision 1.0 [37]. It also contains images crawled from the Flickr website and Google Images search. The number of visual concepts was extended from 1,000 to 5,000, and the total number of training images reaches 16 million. It includes massive noisy labels, as shown in Fig. 1. There are 290K images

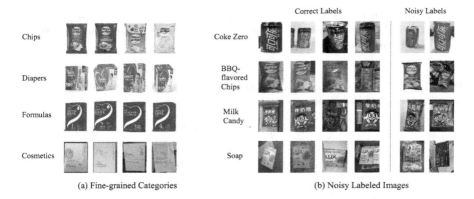

Fig. 3. Image samples of AliProducts, (a) fine-grained categories: from top to bottom are respectively chips in different flavors , diaper with different sizes, formula with different stages, cosmetic with different functions, and each column is a fine-grained category; (b) noisy labeled images: images from coke zero, bbq-flavored chips, milk candy, soap. As can been see, each category includes massive noisy images.

with human annotation are used as validation set, and another 290K images with human annotation for testing. The evaluation measure is the same as WebVision 1.0.

ImageNet. [1] is an image classification dataset, which contains 1000 classes. The original dataset has been splitted into 1.28 million training images, 50k validation images. In this paper, we randomly select 40% training images for each category and assign them with class label uniformly sampled from the rest categories. The generate new dataset thus have lot of noises which could be used to evaluate the effectiveness of popular algorithms on noisy image classification.

Clothing1M. [36] is a large-scale fashion dataset, which includes 14 clothes categories. It contains 1 million noise label images and 74,000 manually annotated images. We call the annotated images as clean set, which is divided into training data,validation data and testing data, with numbers of 50,000,14,000, and 10,000 images, respectively. There are some images overlap between the clean set and the noisy set. The dataset was designed for learning robust models from noisy data without human supervision.

AliProducts[1] is a large-scale noisy and fine-grained product dataset, which includes 50,000 categories. The images are crawled from image search engine and other web sources by using 50,000 product SKU names. The dataset covers foods, snacks, drinks, cosmetics and other daily products and the categories are in SKU (Stock Keeping Unit) level and specific to flavor, capacity, function or even the batch of the production. Therefore, some of the categories might have great difficulty in visual distinguishing due to the fine-grained attribute. AliProducts contains 2.5 million training images without any human annotation, consequently contain massive noisy labels, as show in Fig. 3. Totally 148K manually annotated images are used as validation set, and another 250K manually annotated images for testing. In addition, we released side information (e.g., hierarchical relationships between classes) concerning these image data, which could be exploited to learn better representations and models. The main difference

---

[1] https://tianchi.aliyun.com/competition/entrance/231780/information.

between AliProducts and other noisy datasets (e.g., Clothing 1M and WebVision) is that AliProducts contains massive fine-grained and real-world noisy images, which is relatively difficult for robust DNN methods to improve.

Table 1. Top1/Top5 accuracy of three different models with ResNext-101 architecture on validation set of WebVision.

Method	Model-A	Model-B	Model-C
Top1	51.05%	47.81%	55.57%
Top5	74.94%	72.08%	78.34%

4.2   Experiments on WebVision 2.0

In this subsection, we conduct extensive experiments on WebVision 2.0 dataset to evaluate and demonstrate the effectiveness of proposed SINet. All experiments are implemented using ResNext-101 backbone if there is no special instructions.

Training Strategy and Comparison. We conduct three training strategies with a standard ResNext-101 architecture, resulting in three models, which are described as follow.

Model-A. The model was trained by directly using all the training data.

Model-B. The model was trained by using the high-confidence images without reweighting in training loss.

Model-C. The model was trained with proposed training strategy, where the confidence score is multiplied on the loss of corresponding image for reweighting.

The top1/top5 results of three models on the validation set of WebVision are reported in Table 1. The result shows Model-A with all training data significantly outperforms the Model-B with subset of clean data, with improvements of 3.24%/2.86% for top1/top5 accuracy. This is due to that it is hard to distinguish all the clean labeled samples from those images with heavy noises. In addition, Model-C with our proposed method significantly outperforms Model-A, with improvements of 4.52%/3.40% on top1/top5 accuracy. It is obviously that our proposed method could better explore those clean labels from those noise samples. These improvements are significant on WebVision Challenge with such a large scale noisy dataset, which demonstrate the effectiveness of our method.

Class Relation Graph. We investigate different ways for constructing Class Relation Graph. (I) Category Name: We use category name, i.e., cat, lion, apple, with BERT model to extract the word embeddings, then use the similarity comparison of word embeddings to construct the Class Relation Graph. (II) Category Description: We use category descriptions in WordNet, i.e., *Siamese cat: A slender short-haired blue-eyed breed of cat having a pale coat with dark ears paws face and tail tip*, with BERT and LSTM to extract the textual embeddings, then use the similarity comparison of textual embeddings to construct the Class Relation Graph. (III) Hierachical WordNet: We directly use the prior knowledge of Hierachical WordNet based on the shortest path

Table 2. Class Relation Graph construction with different strategies. (I) Category Name (CN) (II) Category Description (CD) (III) Hierachical WordNet (HW)

Strategy	Top1	Top5
CN	54.18%	76.84%
CD	55.63%	78.15%
HW	55.71%	78.42%
CN+HW	55.79%	78.46%
CD+HW	**55.98%**	**78.62%**

Table 3. Visual Prototype Generation with different strategies. (I) with Reweighting: Topk matched images with matching score as weighting coefficient (II) without Reweighting: Topk matched images averaged

Strategy	Top1	Top5
Constant	55.68%	78.45%
Weighting	**55.98%**	**78.62%**

Table 4. The effect of Shift factor α on the performance of WebVision validation set.

α	Top1	Top5
0.8	53.25%	76.24%
1.0	55.43%	78.31%
1.2	**55.98%**	**78.62%**
1.4	55.76%	78.48%

Table 5. The effect of Contrast factor β on the performance of WebVision validation set.

β	Top1	Top5
1.0	55.68%	78.45%
1.5	**55.98%**	**78.62%**
2.0	55.82%	78.56%
2.5	55.35%	78.24%

between two category to establish the Class Relation Graph. Experimental results of using these three types of class relation graph are shown in Table 2. Obviously, by introducing these side information, the performance could improve a lot than original Model-A, which shows the effectiveness of proposed class relation graphs. Also, these three types of class relation graph are complementary, and combining of them could also boost the performance.

Visual Prototype. We investigate different ways for visual prototype generation. (I) Constant: We does not use any weight operations for the images in Visual Prototype candidates. In this case, the visual prototype representation in Eq. (5) is reduced as the mean of candidates representations. (II) Weighting: We use the method described in Eq. (5) as the weighting operation. Since p_i is the importance score, we use the soft weighted representation as the visual prototype. Experiments of using two types of strategies are shown in Table 3. It shows that when paying more attention on top-ranked images, the generated visual prototypes are better then simple average all the feature representations.

Noise Weighting. In this section, we conduct ablation analysis on the hyper-parameters for our proposed noisy weighting method, and discuss how they affect the recognition performance. (I) Shift factor α: α controls the amount of noisy data actually participating in the model training, the weight of images whose Euclid distance larger than α is 0, that is equivalent to deleting them from the training set, and only use the images with Euclid distance smaller than α to train the model. (II) Contrast factor β: We introduce the contrast parameter β to sharpen the differences of the scores. Experiments of different α and β are shown in Table 4 and 5, respectively. Typically, $\alpha = 1.2$ could keep most of cleaning samples. $\beta = 1.5$ is a proper value to map the score to sampling weight and handle the noise data.

Final Results on the WebVision Challenge. We further evaluate the performance of our proposed SINet with various networks architectures, including ResNext101, SE-ResNext101, SE-Net154. Results are reported in Table 6. As can be found, SE-Net154 substantially performs ResNext101 and SE-ResNext101 on WebVision validation set, with top1/top5 improvements of 1.31%/1.46% and 1.30%/1.27%, while SE-ResNext101 and ResNext101 has similar performance with a marginal performance gain obtained. Our final results were obtained with ensemble of five models. We had the best performance at a Top 5 accuracy of 82.54% on the WebVision challenge 2019.

Table 6. Performance of SINet with various networks on WebVision validation set.

Method	ResNext101	SE-ResNext101	SENet154
Top1	55.56%	55.57%	56.87%
Top5	78.15%	78.34%	79.61%

4.3   Comparisons with the State-of-the-Art Methods

To further explore the effectiveness of our proposed SINet, we conduct extensive comparisons with recent state-of-the-art approaches developed specifically for learning from noisy labels, such as CleanNet, MetaCleaner and MentorNet. For fairness, our comparisons are based on the same CNN backbone, i.e., ResNet50.

WebVision1.0 and ImageNet. We evaluate our SINet on ImageNet, by adding 40% noise ratio with uniform flip. The top-1 accuracy is 66.47/69.12 for ResNet50 without/with SINet. It further shows the power of SINet for large-scale noisy image recognition. By following [16], we use the training set of WebVision1.0 to train the models, and test on the validation sets of WebVision1.0 and ImageNet. Both of them has same 1000 categories. Full results are presented in Table 7. SINet improves the performance of our baseline significantly, and our results compare favorably against recent CurriculumNet, CleanNet and MentorNet with consistent improvements.

Table 7. Comparisons on Webvision1.0 and ImageNet. The models are trained on WebVision1.0 training set and tested on WebVision1.0 and ImageNet validation sets.

Method	WebVision1.0	ImageNet
	Top1/Top5	Top1/Top5
Baseline [11]	67.8(85.8)	58.9(79.8)
CleanNet [11]	70.3(87.8)	63.4(84.6)
MentorNet [10]	70.8(88.0)	62.5(83.0)
CurriculumNet [16]	72.1(89.2)	64.8(84.9)
Our Baseline	69.9(87.4)	63.2(83.8)
SINet	**73.8(90.6)**	**66.8(85.9)**

Clothing1M. For Clothing1M, we consider the state of the art results in [21], which use both noisy and clean set to train the model. Following [21], we conduct two experiments. First we use the 25k clean set to construct the visual prototype and apply noise weighting to one million noisy data, and then use the images with confidence score to train the model. Second, we conduct the same experiment, but with all the clean training set (50k). As shown in Table 8, our SINet outperforms CleanNet, MetaCleaner and DeepSelf, which demonstrates its effectiveness.

Table 8. Experimental results on Clothing1M. Clean set is used in CleanNet [11] to obtain the validation set. To keep same data setting, we use the 25k clean images to construct the visual prototype and use 1M noisy training set with confidence scores to train our SINet, and then fine-tune it on 25k clean images. Furthermore, we achieve the state-of-the-art performance on the setting of Noise1M+Clean(50k), which illustrate the robustness of our SINet on noisy label recognition.

Noise1M+Clean(25k)	Method	CleanNet [11]	MetaCleaner [21]	DeepSelf [20]	Ours
	Accuray	74.69	76.00	76.44	**77.26**
Noise1M+Clean(50k)	Method	CleanNet [11]	MetaCleaner [21]	DeepSelf [20]	Ours
	Accuray	79.9	80.78	81.16	**81.32**

AliProducts. The existed benchmarks with noisy labels are relatively small in the scale of categories or images. To further explore the effectiveness of SINet, we conduct experiments on our released AliProducts, which is a large-scale product dataset with noisy labels and the hierarchical category relations is also provided. We use the hierarchical category relations to construct the class relation graph, and then combine it with images to construct the visual prototype. Finally, we use the images with confidence scores to train the model. As shown in Table 9, our SINet outperforms all other approaches, which illustrates that SINet is more robust to noisy labels.

Table 9. Comparison with the state-of-the-art on AliProducts dataset.

Method	Baseline	CurriculumNet [16]	CleanNet [11]	MetaCleaner [21]	Ours
Accuray	85.35%	85.69%	86.13%	85.92%	**86.29%**

5   Conclusions

In this paper, we presented a novel method, which can learn to generate a visual prototype for each category, for training deep CNNs with large-scale real-world noisy labels. It mainly consists of two submodules. The first module, Visual Prototype can generate a clean representation from the noisy images for every category by integrate the noisy images with side information. The second module, namely Noise Weighting, can estimate the confidence scores of all the noisy images and rank images with confidence scores by analyzing their deep features and Visual Prototype. Via SINet, we can train a high-performance CNN model, where the negative impact of noisy labels can be reduced substantially. We conduct extensive experiments on WebVision, ImageNet, Clothing1M, as well as collected AliProducts, where it achieves state-of-the-art performance on all benchmarks. Future work could aim to train an end-to-end DNNs with Side Information to handle the noisy label recognition.

References

1. Deng, J., Dong, W., Socher, R., Li, J., Li, K., Li, F.: Imagenet: a large-scale hierarchical image database. In: CVPR (2009)
2. Nettleton, D., Orriols, P., Fornells, A.: A study of the effect of different types of noise on the precision of supervised learning techniques. Artif. Intell. Rev. **33**(4), 275–306 (2010)
3. Pechenizkiy, M., Tsymbal, A., Puuronen, S., Pechenizkiy, O.: Class noise and supervised learning in medical domains: the effect of feature extraction. In: IEEE Symposium on Computer-Based Medical Systems (CBMS), pp. 708–713 (2006)
4. Brooks, J.: Support vector machines with the ramp loss and the hard margin loss. Oper. Res. **59**(2), 467–479 (2011)
5. Ghosh, A., Kumar, H., Sastry, P.: Robust loss functions under label noise for deep neural networks. In: AAAI (2017)
6. Ghosh, A., Manwani, N., Sastry, P.: Making risk minimization tolerant to label noise. Neurocomputing **160**, 93–107 (2015)
7. Shirazi, H., Vasconcelos, N.: On the design of loss functions for classification: theory, robustness to outliers, and savageboost. In: NeurIPS (2009)
8. Rooyen, B., Menon, A., CWilliamson, R.: Learning with symmetric label noise: the importance of being unhinged. In: NeurIPS (2015)
9. Zhang, Z., Sabuncu, M.: Generalized cross entropy loss for training deep neural networks with noisy labels. In: NeurIPS (2018)
10. Jiang, L., Zhou, Z., Leung, T., Li, T., Li, F.: Mentornet: regularizing very deep neural networks on corrupted labels. arXiv preprint arXiv:1712.05055 (2017)
11. Lee, K., He, X., Zhang, L., Yang, L.: Cleannet: transfer learning for scalable image classifier training with label noise. arXiv preprint arXiv:1711.07131 (2017)
12. Ren, M., Zeng, W., Yang, B., Urtasun, R.: Learning to reweight examples for robust deep learning. arXiv preprint arXiv:1803.09050 (2018)
13. Wang, Y., et al.: Iterative learning with open-set noisy labels. In: CVPR (2018)
14. Xue, C., Dou, Q., Shi, X., Chen, H., Heng, P.: Robust learning at noisy labeled medical images: applied to skin lesion classification. arxiv.org (2019)
15. Kriegel, H., Kroger, P., Schubert, E., Zimek, A.: Loop: local outlier probabilities. In: CIKM (2009)
16. Guo, S., et al.: CurriculumNet: weakly supervised learning from large-scale web images. In: Ferrari, V., Hebert, M., Sminchisescu, C., Weiss, Y. (eds.) ECCV 2018. LNCS, vol. 11214, pp. 139–154. Springer, Cham (2018). https://doi.org/10.1007/978-3-030-01249-6_9
17. Reed, S., Lee, H., Anguelov, D., Szegedy, C., Erhan, D., Rabinovich, A.: Training deep neural networks on noisy labels with bootstrapping. arXiv preprint arXiv:1412.6596 (2014)
18. Tanaka, D., Ikami, D., Yamasaki, T., Aizawa, K.: Joint optimization framework for learning with noisy labels. In: CVPR (2018)
19. Li, J., Wong, Y., Zhao, Q., Kankanhalli, M.: Learning to learn from noisy labeled data. In: CVPR (2019)

20. Han, J., Luo, P., Wang, X.: Deep self-learning from noisy labels. arXiv preprint arXiv:1908.02160 (2019)
21. Zhang, W., Wang, Y., Qiao, Y.: MetaCleaner: learning to hallucinate clean representations for noisy-labeled visual recognition. In: CVPR (2019)
22. Lin, T.-Y., et al.: Microsoft COCO: common objects in context. In: Fleet, D., Pajdla, T., Schiele, B., Tuytelaars, T. (eds.) ECCV 2014. LNCS, vol. 8693, pp. 740–755. Springer, Cham (2014). https://doi.org/10.1007/978-3-319-10602-1_48
23. Zhu, X., Wu, X.: Class noise vs attribute noise: a quantitative study. Artif. Intell. Rev. **22**(3), 177–210 (2004)
24. Simonyan K., Zisserman A.: Very deep convolutional networks for large-scale image recognition. arXiv:1409.1556 (2014)
25. Szegedy, C., et al.: Going deeper with convolutions. arXiv:1409.4842 (2014)
26. Sukhbaatar, S., Bruna, J., Paluri, M., Bourdev, L., Fergus, R.: Training convolutional networks with noisy labels. arXiv preprint arXiv:1406.2080 (2014)
27. Goldberger, J., Reuven, E.: Training deep neural-networks using a noise adaptation layer. In: ICLR (2017)
28. Patrini, G., Rozza, A., Menon, A., Nock, R., Qu, L.: Making deep neural networks robust to label noise: a loss correction approach. In: CVPR (2017)
29. Hendrycks, D., Mazeika, M., Wilson, D., Gimpel, K.: Using trusted data to train deep networks on labels corrupted by severe noise. In: NeurIPS (2018)
30. Zhang Z., Sabuncu, M.: Generalized cross entropy loss for training deep neural networks with noisy labels. In: NeurIPS (2018)
31. Li, Y., Yang, J., Song, Y., Cao, L., Luo, J., Li, L.: Learning from noisy labels with distillation. In: CVPR (2017)
32. Veit, A., Alldrin, N., Chechik, G., Krasin, I., Gupta, A., Belongie, S.: Learning from noisy large-scale datasets with minimal supervision. In: CVPR (2017)
33. Brodley, C., Friedl, M.: Identifying mislabeled training data. arXiv:1106.0219 (2011)
34. Miranda, A., Garcia, L., Carvalho A., Lorena, A.: Use of classification algorithms in noise detection and elimination. In: HAIS (2009)
35. Barandela, R., Gasca, E.: Decontamination of training samples for supervised pattern recognition methods. In: ICAPR (2000)
36. Xiao, T., Xia, T., Yang, Y., Huang, C., Wang, X.: Learning from massive noisy labeled data for image classification. In: CVPR (2015)
37. Li, W., Wang, L., Li, W., Agustsson, E., Gool, L.: Webvision database: visual learning and understanding from web data. CoRR abs/1708.02862 (2017)
38. Alexander, B., Denzler, J.: Not just a matter of semantics: the relationship between visual and semantic similarity. In: German Conference on Pattern Recognition (2019)
39. Devlin, J., Chang, M., Lee, K., Toutanova, K.: BERT: pre-training of deep bidirectional transformers for language understanding. In: ACL (2019)
40. Miller, G.A.: WordNet: a lexical database for English. Commun. ACM **38**(11), 39–41 (1995)

Not only Look, But Also Listen: Learning Multimodal Violence Detection Under Weak Supervision

Peng Wu, Jing Liu$^{(\boxtimes)}$, Yujia Shi, Yujia Sun, Fangtao Shao, Zhaoyang Wu, and Zhiwei Yang

School of Artificial Intelligence, Xidian University, Xi'an, China
xdwupeng@gmail.com, neouma@163.com, shiyujiaaaa@163.com,
yjsun@stu.xidian.edu.cn, shaofangtao96@163.com, 15191737495@163.com,
zwyang97@163.com

Abstract. Violence detection has been studied in computer vision for years. However, previous work are either superficial, e.g., classification of short-clips, and the single scenario, or undersupplied, e.g., the single modality, and hand-crafted features based multimodality. To address this problem, in this work we first release a large-scale and multi-scene dataset named XD-Violence with a total duration of 217 h, containing 4754 untrimmed videos with audio signals and weak labels. Then we propose a neural network containing three parallel branches to capture different relations among video snippets and integrate features, where holistic branch captures long-range dependencies using similarity prior, localized branch captures local positional relation using proximity prior, and score branch dynamically captures the closeness of predicted score. Besides, our method also includes an approximator to meet the needs of online detection. Our method outperforms other state-of-the-art methods on our released dataset and other existing benchmark. Moreover, extensive experimental results also show the positive effect of multimodal (audio-visual) input and modeling relationships. The code and dataset will be released in https://roc-ng.github.io/XD-Violence/.

Keywords: Violence detection · Multimodality · Weak supervision · Relation networks

1 Introduction

Everyone hopes for peaceful life, and it is our duty to safeguard peace and oppose violence. Violence detection [4,13,15,23,24,29,34,35] in videos has been studied in computer vision community for years. However, due to limited application and challenging nature, this specific task received far less attention than other

Electronic supplementary material The online version of this chapter (https://doi.org/10.1007/978-3-030-58577-8_20) contains supplementary material, which is available to authorized users.

© Springer Nature Switzerland AG 2020
A. Vedaldi et al. (Eds.): ECCV 2020, LNCS 12375, pp. 322–339, 2020.
https://doi.org/10.1007/978-3-030-58577-8_20

popular tasks, e.g., video classification [41], action recognition [11], and temporal action detection [12], in the past decades. Along with the advance in video technology in recent years, the application of violence detection is becoming more and more extensive. For example, violence detection is not only used for real-world scenarios, e.g., intelligent surveillance, but also used for Internet, e.g., video content review (VCR). Violence detection aims to timely locate the start and the end of violent events with minimum human resource cost.

The earliest task of violence detection [15,24] can be considered as video classification. Within this context, most methods assume well-trimmed videos, where violent events last for nearly the entire video. However, such solutions restrict their scope to short clips and cannot be generalized to locate violent events in untrimmed videos, therefore they render a limited use in practice. A small step towards addressing violence detection is to develop algorithms to focus on untrimmed videos. Such as the violent scene detection (VSD) task on MediaEval [4], and the Fighting detector [29]. However, assigning frame-level annotations to videos is a time-consuming procedure which is adverse to building large-scale datasets.

Recently, several research [35,50] focus on weakly supervised violence detection, where only video-level labels are available in the training set. Compared with annotating frame-level labels, assigning video-level labels is labor-saving. Thus, forming large-scale datasets of untrimmed videos and training a data-driven and practical system is no longer a difficult challenge. In this paper, we aim to study weakly supervised violence detection.

Furthermore, we leverage multimodal cue to address violence detection, namely, incorporating both visual and audio information. Multimodal input is beneficial for violence detection as compared to unimodal input. In most cases, visual cue can precisely discriminate and locate events. At times, visual signals are ineffective and audio signals can separate visually ambiguous events. For example, it is difficult for visual signals to figure out what happen in the violently shaking video accompanied by the sounds of explosion, rather, audio signals are the prime discriminators in this case. Therefore, audiovisual fusion can make full use of complementary information and become an extensive tendency in computer vision and speech recognition communities [1,2,10,19,25,31]. We are certainly not the first to attempt to detect violence by multimodal signals, there were precedents for multimodal violence detection before, such as, [7,22,28,48]. However, the above methods have several drawbacks, e.g., relying on small-scale datasets, using subjective hand-crafted features, and the single scene, that indicates a pragmatic system with high generalization is still in the cradle. Unlike these, we are intended to design a reliable neural network based algorithm on large-scale data.

To support research toward leveraging multimodal information (vision and audio) to detect violent events in weakly supervised perspective, we first release a large-scale video violence dataset consisting of 4754 untrimmed videos. Unlike previous datasets [4,15,24,35], our dataset has audio signals and is collected from both movies and in-the-wild scenarios. With the dataset in hands, we then view weakly supervised violence detection as a multiple instance learning (MIL) task; that is, a video is cast as a bag, which consists of several instances (snippets), and instance-level annotations are learned via bag-level labels. Based on this, we attempt to learn more powerful representations to remedy weak labels. Therefore, we propose a holistic and localized network (HL-Net) that explicitly exploits relations of snippets and learns powerful representations based on these relations, where holistic branch captures long-range dependencies by similarity prior of snippets, and localized branch models short-range interactions within a local neighborhood. In addition, we introduce a holistic and localized cue (HLC) approximator for online violence detection since HL-Net need the whole video to compute relations of snippets. The HLC approximator only processes a local neighborhood and learns precise prediction guided by HL-Net. Even better, HLC approximator brings a dynamic score branch paralleling to holistic branch and localized branch, which computes the response at a position by a weighted sum of all features, and weights depend on predicted scores.

To summarize, contributions of this paper are threefold,

We release a audio-visual violence dataset termed XD-Violence, which consists of 4754 untrimmed videos and covers six common types of violence. To our knowledge, XD-Violence is by far the largest scale violence dataset, with a total of 217 h. Unlike previous datasets, the videos of XD-Violence are captured from multi scenarios, e.g. movies and YouTube.

We introduce a HL-Net to simultaneously capture long-range relations and local distance relations, of which these two relations are based on similarity prior and proximity prior, respectively. In addition, we also propose an HLC approximator for online detection. Based on this, we use a score branch to dynamically obtain an additional holistic relation.

We conducted extensive experiments to verify the effectiveness of our proposed method, and our method shows clear advantage over existing baselines on two benchmarks, i.e., XD-Violence (Ours), and UCF-Crime. Furthermore, experimental results also demonstrate the superiority of multimodal information as compared with the unimodality.

2   Related Work

Violence Detection. In the last years, many researchers proposed different methods for violence detection. For instance, Bermejo *et al.* [24] released two well-known fighting datasets. Gao *et al.* [15] proposed violent flow descriptors to detect violence in crowded videos. Mohammadi *et al.* [23] proposed a new behavior heuristic based approach to classify violent and non-violent videos. Most of prior work utilized hand-crafted features to detect violence on small-scale datasets. Common features include, scale-invariant feature transform (SIFT), spatial-temporal interest point (STIP), histogram of oriented gradient (HOG), histograms of oriented optical flow (HOF), motion intensity, and so on.

With the rise of deep convolutional neural networks (CNNs), many work have looked into designing effective deep convolutional neural networks for violence detection. For example, Sudhakaran and Lanz [34] used a convolutional long short-term memory (LSTM) network for the purpose of recognizing violent videos. Similarly, Hanson *et al.* [13] built a bidirectional convolutional LSTM architecture for violence detection in videos. Peixoto *et al.* [27] used two deep neural network frameworks to learn the spatial-temporal information under different scenarios, then aggregated them by training a shallow neural network to describe violence. Recently, an interesting research [33] was proposed, which used a scatter net hybrid deep learning network for violence detection in drone surveillance videos.

There are several attempts to detect violence with multimodality or audio [7–9,22,28,48]. To our knowledge, vast majority of methods use hand-crafted features to extract audio information, e.g., spectrogram, energy entropy, audio energy, chroma, Mel-scale frequency cepstral coefficients (MFCC), zero-crossing rate (ZCR), pitch, etc. Hand-crafted features are easy to extract but are low-level and not robust. Unlike them, our method uses a CNN based model to extract high-level features.

Relation Networks. Several work apply the graph neural networks (GCNs) [20,38] over the graph to model relations among different nodes and learn powerful representations for computer vision. For instance, GCN are used for temporal action localization [49], video classification [36,40], anomaly detection [50], skeleton-based action recognition [32,44], point cloud semantic segmentation [21], image captioning [45], and so on. Besides GCN, temporal relation networks [51], designed to learn and reason about temporal dependencies between video frames, are proposed to address video classification. Recently, self-attention networks [5,18,39,46] have been successfully applied in vision problems. An attention operation can affect an individual element by aggregating information from a set of elements, where the aggregation weights are automatically learned.

3 XD-Violence Dataset

3.1 Selecting Violence Categories

The World Health Organization (WHO) defines violence as "the intentional use of physical force or power, threatened or actual, against oneself, another person, or against a group or community, which either results in or has a high likelihood of resulting in injury, death, psychological harm, maldevelopment, or deprivation." Because of the multiple facets of violence, no common and generic enough definition for violent events was ever proposed, even when restricting ourselves to physical violence. However, establishing a clear definition of violence is a key issue because human annotators can rely on a ground truth reference to reduce ambiguity. To mitigate the problem, we consider six physically violent classes, namely, *Abuse, Car Accident, Explosion, Fighting, Riot,* and *Shooting*. We take these violence into account due to clear definition, frequent occurrence, widespread use [4,35], and adverse impact to safety.

Fig. 1. Sample videos from the XD-Violence dataset.

3.2 Collection and Annotation

Video Collection. Previous violence datasets are collected from either movies or in-the-wild scenes, almost no dataset is collected from both. Unlike them, our dataset is collected from both movies and YouTube (in-the-wild scenes). There is a total of 91 movies, of which violent movies are used to collect both violent and non-violent events, and non-violent movies are only used to collect non-violent events. We also collect in-the-wild videos by YouTube. We search and download a mass of video candidates using text search queries. In order to prevent violence detection systems from discriminating violence based on the background of scenarios rather than occurrences, we specifically collect large amounts of non-violent videos whose background is consistent with that of violent videos. A dataset with video-level labels is completed after elaborate efforts of several months. Several example videos from each category are shown in Fig. 1. More details are given in the supplementary material.

Video Annotation. Our dataset has a total of 4754 videos, which consists of 2405 violent videos and 2349 non-violent videos. We split it into two parts: the training set containing 3954 videos and the test set including 800 videos, where the test set consists of 500 violent videos and 300 non-violent videos. To evaluate the performance of violence detection methods, we need to make frame-level (temporal) annotations for test videos. To be specific, for each violent video of the test set, we mark the start and ending frames of violent events. As [35], we also assign the same videos to multiple annotators to label the temporal extent of each violence and average annotations of different annotators to make final temporal annotations more precise. Both training and test sets contain all 6 kinds of violence at various temporal locations in the videos.

3.3 Dataset Statistics

Multi-scenario includes but not limited to the following sources: movies, cartoons, sports, games, music, fitness, news, live scenes, captured by CCTV cameras, captured by hand-held cameras, captured by car driving recorders, etc.

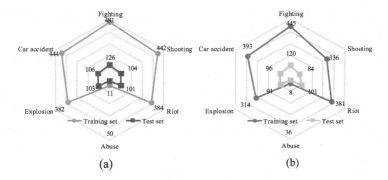

Fig. 2. Dataset Statistics. (a) Distribution of the number of videos belonging to each category according to multi-label. (b) Distribution of the number of videos belonging to each category according to the first label.

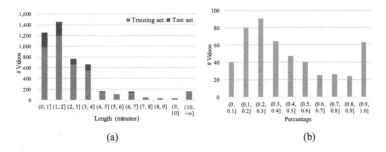

Fig. 3. Dataset Statistics. (a) Distribution of videos according to length (minutes). (b) Distribution of violent videos according to percentage of violence (in each video) in test set.

(some of them may be overlapped.) We also assign multi violent labels ($1 \leq$ #labels ≤ 3) to each violent video owing to the co-occurrence of violent events. The order of labels of each video is based on the importance of different violent events occurring in the video. The distribution of the violent videos in terms of the number of labels is shown in Fig. 2. In addition, our dataset consists of untrimmed videos, therefore, we show the distribution of videos in terms of length in Fig. 3(a). We also present the percentage of violence in each test video in Fig. 3(b).

3.4   Dataset Comparisons

In order to highlight the traits of our dataset, we compare our dataset with other widely-used datasets for violence detection. These datasets can be split into three types: small scale, medium scale, and large scale, of which Hockey [24], Movie [24], and Violent-Flows [15] are small-scale, VSD [4] and CCTV-Fights [29] are medium-scale, and the remaining UCF-Crime [35] is large-scale. Table 1 compares several characteristics of these datasets. Our dataset is by far the

Table 1. Comparisons of different violence datasets.*means quite a few videos are silent or only contain background music.

Dataset	#Videos	Length	Source of scenarios	#Violence types	Audio
Hockey [24]	1000	27 min	Ice hockey	1	No
Movie [24]	200	6 min	Movies and sports	1	No
Violent-Flows [15]	246	15 min	Streets, school, and sports	1	Yes*
CCTV-Fights [29]	1000	18 h	CCTV and mobile cameras	1	Yes*
VSD [4]	25	35 h	Movies	8	Yes
UCF-Crime [35]	1900	128 h	CCTV camera	9	No
XD-Violence (Ours)	4754	217 h	Movies, sports, games, hand-held cameras, CCTV, car cameras, etc.	6	Yes

largest dataset, which is more than 300 times than the total of small datasets, 4 times than the total of medium datasets, and almost 2 times than the UCF-Crime dataset. Besides, variations in scenes of previous datasets are also limited, by contrast, our dataset embraces a wide variety of scenarios. In addition, Hockey, Movie, Violent-Flows and CCTV-Fights are only used for fighting detection. Intriguingly, though Violent-Flows and CCTV-Fights contain audio signals, in fact, there are quite a few videos that are silent or only contain background music, which is inoperative or even harmful for training detection methods taking multimodality as input. The UCF-Crime dataset is used to detect violence in surveillance videos, but lacks audio.

Overall, our dataset has three good traits: 1) large scale, which is beneficial for training generalizable methods for violence detection; 2) diversity of scenarios, so that violence detection methods actively respond to complicated and diverse environments and are more robust; 3) containing audio-visual signals, making algorithms leverage multimodal information and more confidence.

4   Methodology

4.1   Multimodal Fusion

Our proposed method is summarized in Fig. 4. Consider that we have a training set of videos, we denote an untrimmed video and the corresponding label as V and y, where $y \in \{0,1\}, y = 1$ denotes V covers violent events. With a video V in hands, we use feature extractors F^V and F^A to extract visual and audio feature matrix X^V and X^A using the sliding window mechanism, where $X^V \in \mathbb{R}^{T' \times d^V}, X^A \in \mathbb{R}^{T' \times d^A}$, T' is the length of feature matrix, x_i^V and x_i^A are visual and audio features of the ith snippet, respectively.

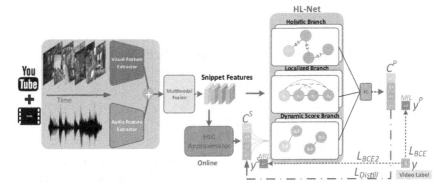

Fig. 4. The pipeline of our proposed method.

Quite a few fusion manners have been proposed for multimodal input, we there opt to the simple yet effective concatenation fusion. More precisely, X^V and X^A are first concatenated in the channels, then the concatenation passes through two stacked fully connected layers (FC) with 512 and 128 nodes respectively, where each FC layer is followed by ReLU and dropout. Finally, the output is taken as the fusion features, which is denoted by X^F.

4.2   Holistic and Localized Networks

Revisit Relations. We first recap the long-range dependencies of neural networks, which can be acquired by two prevalent types of networks, namely, GCNs and non-local networks (NL-Net, self-attention networks).

A general graph convolution operation can be formulated as follows [21],

$$X_{l+1} = Update\left(Aggregate(X_l, W_l^{agg}), W_l^{update}\right) \tag{1}$$

which contains two essential operations, aggregation and update, and corresponding learnable weights, where the aggregation operation is used to compile information from the global vertices (long-range dependencies), while update functions perform a non-linear transform to compute new representations.

An instantiated non-local operation can be formulated as follows [39,46],

$$X_{l+1} = \left[softmax(X_l^T W_\theta^T W_\phi X_l)(W_g X_l)\right] W_\psi \tag{2}$$

Although they have different original intentions (GCN is mainly used to address problems with non-Euclidean data, capturing long-range dependencies is an avocation), they are similar on capturing long-range dependencies. Because the term within the outer brackets in Eq. (2) can be viewed as an aggregation operation based on feature similarity, which is followed by the update operation.

Holistic Branch Implementation. Inspired by the GCN for video understanding [40,49,50], we define the holistic relation matrix by feature similarity prior, as shown below:

$$A_{ij}^H = g\left(f(x_i, x_j)\right) \tag{3}$$

where $A^H \in T' \times T'$, A_{ij}^H measures the feature similarity between the ith and j^{th} features. g is the normalization function, and function f computes the similarity of a pair of features, we define f as follows,

$$f(x_i, x_j) = \frac{x_i^T x_j}{\|x_i\|_2 \cdot \|x_j\|_2} \tag{4}$$

we also define other versions of f, and present them on the supplementary material. f bounds the similarity to the range of $(0, 1]$ for the sake of the thresholding that filters weak relations and strengthens correlations of more similar pairs. The thresholding operation can be defined as follows,

$$f(x_i, x_j) = \begin{cases} f(x_i, x_j) & f(x_i, x_j) > \tau \\ 0 & f(x_i, x_j) \leq \tau \end{cases} \tag{5}$$

where τ is the threshold. After that, we adopt the softmax as the normalization function g to make sure the sum of each row of A is 1, as shown in Eq. (6):

$$A_{ij}^H = \frac{exp(A_{ij}^H)}{\sum_{k=1}^{T'} exp(A_{ik}^H)} \tag{6}$$

We emphasize that X used in Eq. (3) is the concatenation of raw features (X^A and X^V) to capture the original feature prior.

In order to capture long-range dependencies, we follow the GCN paradigm and design the holistic layer as,

$$X_{l+1}^H = Dropout\left(ReLU(A^H X_l^H W_l^H)\right) \tag{7}$$

which allows us to compute the response of a position defined by the similarity prior based on the global filed rather than its neighbors.

Localized Branch Implementation. Holistic branch captures long-range dependencies directly by computing interactions between any two positions, regardless of their positional distance. However, positional distance has positive effects on temporal events detection [50], and to retain it, we devise the local relation matrix based on proximity prior as,

$$A_{ij}^L = exp\left(\frac{-|i-j|^\gamma}{\sigma}\right) \tag{8}$$

which only depends on temporal positions of the ith and j^{th} features, and where γ and σ are hyper-parameters to control the range of influence of distance relation. Likewise, X_{l+1}^L is the output of the $(l+1)^{th}$ localized layer.

4.3   Online Detection

As we mentioned, a violence detection system is not only applied for offline detection (Internet VCR), but also online detection (surveillance system). However, online detection by the above HL-Net is impeded by a major obstacle: HL-Net needs the whole video to obtain long-range dependencies. To jump out of the dilemma, we propose an HLC approximator, only taking previous video snippets as input, to generate precise predictions guided by HL-Net. Two stacked FC layers followed by ReLU and a 1D causal convolution layer constitute HLC approximator. The 1D causal convolution layer has kernel size 5 in time with stride 1, sliding convolutional filters over time. The 1D causal convolution layer also acts as the classifier, whose output is the violent activation denoted as C^S of shape T'. Even better, this operation introduces an additional branch named dynamic score branch to extend HL-Net, which depends on C^S.

Score Branch Implementation. The main role of this branch is to compute the response at a position as a weighted sum of the features at all positions, where weights depend on the closeness of the scores. Different from the relation matrices of holistic and localized branches, the relation matrix of score branch is updated in each iteration, and depends on predicted scores rather than the prior. Formally, the relation matrix of score branch is devised as follows,

$$A_{ij}^S = \rho\left(1 - \left|s(C_i^S) - s(C_j^S)\right|\right) \tag{9}$$

$$\rho(x) = \frac{1}{1 + exp(-\frac{x-0.5}{0.1})} \tag{10}$$

where function s is sigmoid, and function ρ is used to enhance (and weak) the pairwise relation where the closeness of the scores is greater (and less) than 0.5, and softmax is also used for the normalization.

Analogously, X_{l+1}^S is the output of the $(l+1)^{th}$ score layer, where $X_0^S = (X_0^H = X_0^L) = X^F$.

4.4   Training Based on MIL

We use an FC layer with 1 node to project the concatenation representations to the label space (1D space), and the violent activations we obtain after this projection can be represented as follows,

$$C^P = (X^H \| X^L \| X^S)W \tag{11}$$

where $\|$ denotes the concatenation operation, and $C^P \in \mathbb{R}^{T'}$ denotes the violent activations.

Following the principles of MIL [26,35], we use the average of K-max activation (C^P and C^S) over the temporal dimension rather than the whole activations to compute y^P and y^S, and K is defined as $\left\lfloor \frac{T'}{q} + 1 \right\rfloor$. The instances corresponding to the K-max activation in the positive bag is most likely to be true positive instances (violence). The instances corresponding to the K-max activation in the negative bag is hard instances. We expect these two types of instances to be as far as possible.

We define the classification loss, L_{BCE} and L_{BCE2}, as the binary cross-entropy between the predicted labels (y^P and y^S) and ground truth y. In addition, we also use the knowledge distillation loss to encourage the output of the HLC approximator to approximate the output of HL-Net.

$$L_{DISTILL} = \sum_{j=1}^{N} \left(-\sum_i s(C_i^P) log\left(s(C_i^S)\right) \right)_j \tag{12}$$

where N is the batch size. Finally, the total loss is the weighted sum of the above three loss, which is shown as follows,

$$L_{TOTAL} = L_{BCE} + L_{BCE2} + \lambda L_{DISTILL} \tag{13}$$

4.5   Inference

Aiming at different requests, our method can choose offline or online manners to efficiently detect violent events. Sigmoid functions follow the violent activations C^P and C^S and generate the violence confidence (score) that is bounded in the range of $[0, 1]$. Note that, in the online inference, only HLC approximator works, and HL-Net can be removed.

5   Experiments

5.1   Evaluation Metric

we utilize the frame-level precision-recall curve (PRC) and corresponding area under the curve (average precision, AP) [29] rather than receiver operating characteristic curve (ROC) and corresponding area under the curve (AUC) [42,43] since AUC usually shows an optimistic result when dealing with class-imbalanced data, and PRC and AP focus on positive samples (violence).

5.2   Implementation Details

Visual Features. We utilize two mainstream networks as our visual feature extractor F^V, namely, C3D [37] and I3D [3] networks. We extract *fc6* features from C3D that is pretrained on the Sports-1M dataset, and extract *global_pool* features from I3D pre-trained on Kinetics-400 dataset. I3D is a two-stream model, therefore, the visual feature has two versions, RGB and optical flow. We use the GPU implementation of TV-L1 [47] to compute the optical flow. We fix the frame rate to 24 FPS for all videos, and set the length of sliding window as 16 frames.

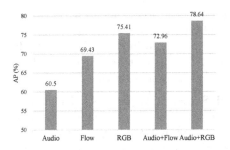
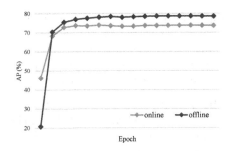

Fig. 5. AP comparison of different modalities.

Fig. 6. AP changing curves using models saved in different epoch.

Audio Features. we leverage the VGGish [6,17] network pretrained on a large YouTube dataset as the audio feature extractor F^A due to its remarkable performance on audio classification. The audio is divided into overlapped 960-ms segments, of which each segment has the only corresponding video snippet with aligned end time. The log-mel spectrogram patches of 96×64 bins, computed from segments, form the input to VGGish. Unless otherwise stated, we use the fusion of RGB feature of I3D and VGGish feature by default.

HL-Net Architecture. Generalized HL-Net is formed from holistic branch, localized branch, and an additional score branch. Each branch is a stack of two layers, where the number of output channels for each layer is 32. Furthermore, taking inspiration from [16,21,32], we add a residual connection for each layer, which enables GCN to reliably converge in the training stage.

Training Details. We implement the network based on PyTorch. For hyper-parameters, without otherwise stated, we set τ as 0.7, γ and σ of A^L as 1, q as 16, dropout rate as 0.7, and λ as 5. For network optimization, Adam is used as the optimizer. The initial learning rate is set as 10^{-3}, and is divided by 10 at the 10th epoch and 30th epoch. The network is trained for 50 epochs in total, and the batch size is 128.

5.3 Ablation Studies

The Effect of Modality. Most of work still focus on the unimodality for video event detection, and audio-visual input is one advantage of our work. Therefore, we conduct experiments on the XD-Violence dataset to verify the superiority of multimodality. We try five different inputs, i.e., audio modality, optical flow (Flow) modality, RGB modality, the fusion of audio and Flow modalities, the fusion of audio and RGB modalities, (For the unimodality, the two stacked FC layers of multimodal fusion module are still retained.) and report the results in Fig. 5. For the unimodality, a key observation is that visual modality is significantly superior to audio, which is not hard to understand: visual signals contain more rich information, which make algorithms see farther and more. More importantly, the fusion of multimodality shows clear advantage over the single modality, to be specific, the fusion of audio and Flow achieves clear improvements against the Flow-only input by 3.5% on AP, and the fusion of audio and RGB outperforms the RGB-only input by 3.2% on AP. This matches the expectation that audio provides complementary information for visual signals.

The Effect of Holistic, Localized and Score Branches. As depicted in Sect. 4, there are three parallel branches in our HL-Net. We manually delete one or two of the branches and show their performance in Table 2. We observe that: 1) three separate branches achieve similar performance; 2) removing any one of the three branches will harm the performance; 3) HL-Net achieves the best performance with all three branches work together. This demonstrates that all branches play an irreplaceable role in our HL-Net.

Online Detection vs Offline Detection. We show the performance comparison between online detection and offline detection in Table 3. It is observed that offline detection outperforms online detection by 5% on AP. The performance improvement benefits from the powerful ability of HL-Net, i.e., computing the response by three different types of feature aggregation. We also use the models saved in different training stages to inference, and show the performance changing curves. We can see from Fig. 6 that offline detection outperforms online detection except for in the initial training stage, we argue that this is because HLC approximator is a lightweight module that is easy to train and can find a good solution in the early stage, and HL-Net has relatively more parameters, needing more time to train.

5.4 Comparisons with State-of-the-Arts

We compare our method with several baselines on XD-Violence, and show the results in Table 3. It is obvious that our method can outperform current state-of-the-art methods. We also show the performance of the fusion of C3D and audio features in Table 3, and observe that C3D is inferior to I3D by a large margin in our violence detection task. Note that all baselines in Table 3 take the fusion of the RGB features of I3D and VGGish features as input. PRC on the XD-Violence and results on the UCF-Crime dataset are given in the supplementary material.

5.5 Qualitative Results

We present several qualitative examples in Fig. 7. (a)–(c) and (e)–(g) are violent videos, and our method successfully detects violent events. (d) and (h) are non-violent videos, and our method generates very low violence scores. As can be seen from the 1st row, multimodal input can localize violence more precisely than the RGB-only input with lower false positives and false negatives, especially in videos with audio. For instance, in (c), Audio+RGB detects the explosion according to not just fire, but explosive sound. From the 2nd row it is evident that online detection is slightly worse than offline detection, and with higher false alarms,

Table 2. AP comparison of different branches.

Holisitic branch	Localized branch	Score branch	AP(%)
✓			75.44
	✓		76.60
		✓	75.40
✓	✓		77.23
✓		✓	77.05
	✓	✓	77.70
✓	✓	✓	78.64

Table 3. AP comparison on the XD-Violence dataset.

Method	AP(%)
SVM baseline	50.78
OCSVM [30]	27.25
Hasan *et al.* [14]	30.77
Sultani *et al.* [35]	73.20
Ours (C3D)	67.19
Ours (Online)	73.67
Ours	78.64

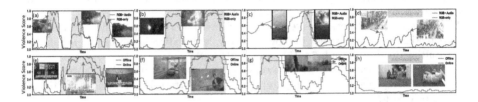

Fig. 7. Qualitative results of our method on test videos. Colored window shows the ground truth of violent regions. [Best viewed in color.] (Color figure online)

this is because of the lack of contextual information. Besides, an interesting finding is that our method considers there are violent events in the second half of (g), but this part is not marked as violence. After watching the video, we find out the reason: the accident process is finished, but there is still an overturned car. More qualitative results are provided in the supplementary material.

6   Conclusions

In this paper, we study the large-scale violence detection with audio-visual modalities under weak supervision. Due to lack of applicable datasets, we first release a large-scale violence dataset to fill the gap. Then we propose a method to explicitly modeling relationships among video snippets and learn powerful representations. Extensive experiments show, 1) our dataset is applicable; 2) multimodality significantly improves the performance; 3) explicitly exploiting relations is highly effective. In the further work, we will add some audio-dominated violence classes (e.g., scream), and our research will naturally extend to multiclass violence detection as XD-Violence is a multi-label dataset. In addition, more powerful online detection is yet to be further explored.

Acknowledgments. This work was supported in part by the Key Project of Science and Technology Innovation 2030 supported by the Ministry of Science and Technology of China under Grant 2018AAA0101302 and in part by the General Program of

National Natural Science Foundation of China (NSFC) under Grant 61773300. We sincerely thank Chao Wang, Chaolong Ying, Shihao Yuan, and Kaixin Yuan for their excellent annotation work.

References

1. Aytar, Y., Vondrick, C., Torralba, A.: SoundNet: learning sound representations from unlabeled video. In: Advances in Neural Information Processing Systems (NIPS), pp. 892–900 (2016)
2. Aytar, Y., Vondrick, C., Torralba, A.: See, hear, and read: deep aligned representations. arXiv preprint arXiv:1706.00932 (2017)
3. Carreira, J., Zisserman, A.: Quo vadis, action recognition? a new model and the kinetics dataset. In: Proceedings of the IEEE Conference on Computer Vision and Pattern Recognition (CVPR), pp. 6299–6308 (2017)
4. Demarty, C.H., Penet, C., Soleymani, M., Gravier, G.: VSD, a public dataset for the detection of violent scenes in movies: design, annotation, analysis and evaluation. Multimedia Tools Appl. **74**(17), 7379–7404 (2015)
5. Fu, J., et al.: Dual attention network for scene segmentation. In: Proceedings of the IEEE Conference on Computer Vision and Pattern Recognition (CVPR), pp. 3146–3154 (2019)
6. Gemmeke, J.F., et al.: Audio set: an ontology and human-labeled dataset for audio events. In: IEEE International Conference on Acoustics, Speech and Signal Processing (ICASSP), pp. 776–780. IEEE (2017)
7. Giannakopoulos, T., Makris, A., Kosmopoulos, D., Perantonis, S., Theodoridis, S.: Audio-visual fusion for detecting violent scenes in videos. In: Konstantopoulos, S., Perantonis, S., Karkaletsis, V., Spyropoulos, C.D., Vouros, G. (eds.) SETN 2010. LNCS (LNAI), vol. 6040, pp. 91–100. Springer, Heidelberg (2010). https://doi.org/10.1007/978-3-642-12842-4_13
8. Giannakopoulos, T., Pikrakis, A., Theodoridis, S.: A multi-class audio classification method with respect to violent content in movies using Bayesian networks. In: IEEE 9th Workshop on Multimedia Signal Processing, pp. 90–93. IEEE (2007)
9. Giannakopoulos, T., Pikrakis, A., Theodoridis, S.: A multimodal approach to violence detection in video sharing sites. In: 20th International Conference on Pattern Recognition (ICPR), pp. 3244–3247. IEEE (2010)
10. Ginosar, S., Bar, A., Kohavi, G., Chan, C., Owens, A., Malik, J.: Learning individual styles of conversational gesture. In: Proceedings of the IEEE Conference on Computer Vision and Pattern Recognition (CVPR), pp. 3497–3506 (2019)
11. Girdhar, R., Carreira, J., Doersch, C., Zisserman, A.: Video action transformer network. In: Proceedings of the IEEE Conference on Computer Vision and Pattern Recognition (CVPR), pp. 244–253 (2019)
12. Gu, C., et al.: Ava: a video dataset of spatio-temporally localized atomic visual actions. In: Proceedings of the IEEE Conference on Computer Vision and Pattern Recognition (CVPR), pp. 6047–6056 (2018)
13. Hanson, A., PNVR, K., Krishnagopal, S., Davis, L.: Bidirectional convolutional LSTM for the detection of violence in videos. In: Leal-Taixé, L., Roth, S. (eds.) ECCV 2018. LNCS, vol. 11130, pp. 280–295. Springer, Cham (2019). https://doi.org/10.1007/978-3-030-11012-3_24
14. Hasan, M., Choi, J., Neumann, J., Roy-Chowdhury, A.K., Davis, L.S.: Learning temporal regularity in video sequences. In: Proceedings of the IEEE Conference on Computer Vision and Pattern Recognition (CVPR), pp. 733–742 (2016)

15. Hassner, T., Itcher, Y., Kliper-Gross, O.: Violent flows: real-time detection of violent crowd behavior. In: IEEE Computer Society Conference on Computer Vision and Pattern Recognition Workshops, pp. 1–6. IEEE (2012)
16. He, K., Zhang, X., Ren, S., Sun, J.: Deep residual learning for image recognition. In: Proceedings of the IEEE Conference on Computer Vision and Pattern Recognition (CVPR), pp. 770–778 (2016)
17. Hershey, S., et al.: CNN architectures for large-scale audio classification. In: IEEE International Conference on Acoustics, Speech and Signal Processing (ICASSP), pp. 131–135. IEEE (2017)
18. Hu, H., Gu, J., Zhang, Z., Dai, J., Wei, Y.: Relation networks for object detection. In: Proceedings of the IEEE Conference on Computer Vision and Pattern Recognition (CVPR), pp. 3588–3597 (2018)
19. Kazakos, E., Nagrani, A., Zisserman, A., Damen, D.: Epic-fusion: audio-visual temporal binding for egocentric action recognition. In: Proceedings of the IEEE International Conference on Computer Vision (ICCV), pp. 5492–5501 (2019)
20. Kipf, T.N., Welling, M.: Semi-supervised classification with graph convolutional networks. arXiv preprint arXiv:1609.02907 (2016)
21. Li, G., Muller, M., Thabet, A., Ghanem, B.: DeepGCNs: can GCNs go as deep as CNNs? In: Proceedings of the IEEE International Conference on Computer Vision (ICCV), pp. 9267–9276 (2019)
22. Lin, J., Wang, W.: Weakly-supervised violence detection in movies with audio and video based co-training. In: Muneesawang, P., Wu, F., Kumazawa, I., Roeksabutr, A., Liao, M., Tang, X. (eds.) PCM 2009. LNCS, vol. 5879, pp. 930–935. Springer, Heidelberg (2009). https://doi.org/10.1007/978-3-642-10467-1_84
23. Mohammadi, S., Perina, A., Kiani, H., Murino, V.: Angry crowds: detecting violent events in videos. In: Leibe, B., Matas, J., Sebe, N., Welling, M. (eds.) ECCV 2016. LNCS, vol. 9911, pp. 3–18. Springer, Cham (2016). https://doi.org/10.1007/978-3-319-46478-7_1
24. Bermejo Nievas, E., Deniz Suarez, O., Bueno García, G., Sukthankar, R.: Violence detection in video using computer vision techniques. In: Real, P., Diaz-Pernil, D., Molina-Abril, H., Berciano, A., Kropatsch, W. (eds.) CAIP 2011. LNCS, vol. 6855, pp. 332–339. Springer, Heidelberg (2011). https://doi.org/10.1007/978-3-642-23678-5_39
25. Oh, T.H., et al.: Speech2face: learning the face behind a voice. In: Proceedings of the IEEE Conference on Computer Vision and Pattern Recognition (CVPR), pp. 7539–7548 (2019)
26. Paul, S., Roy, S., Roy-Chowdhury, A.K.: W-TALC: weakly-supervised temporal activity localization and classification. In: Ferrari, V., Hebert, M., Sminchisescu, C., Weiss, Y. (eds.) ECCV 2018. LNCS, vol. 11208, pp. 588–607. Springer, Cham (2018). https://doi.org/10.1007/978-3-030-01225-0_35
27. Peixoto, B., Lavi, B., Martin, J.P.P., Avila, S., Dias, Z., Rocha, A.: Toward subjective violence detection in videos. In: IEEE International Conference on Acoustics, Speech and Signal Processing (ICASSP), pp. 8276–8280. IEEE (2019)
28. Penet, C., Demarty, C.H., Gravier, G., Gros, P.: Multimodal information fusion and temporal integration for violence detection in movies. In: IEEE International Conference on Acoustics, Speech and Signal Processing (ICASSP), pp. 2393–2396. IEEE (2012)
29. Perez, M., Kot, A.C., Rocha, A.: Detection of real-world fights in surveillance videos. In: IEEE International Conference on Acoustics, Speech and Signal Processing (ICASSP), pp. 2662–2666. IEEE (2019)

30. Schölkopf, B., Williamson, R.C., Smola, A.J., Shawe-Taylor, J., Platt, J.C.: Support vector method for novelty detection. In: Advances in Neural Information Processing Systems (NIPS), pp. 582–588 (2000)

31. Senocak, A., Oh, T.H., Kim, J., Yang, M.H., So Kweon, I.: Learning to localize sound source in visual scenes. In: Proceedings of the IEEE Conference on Computer Vision and Pattern Recognition (CVPR), pp. 4358–4366 (2018)

32. Shi, L., Zhang, Y., Cheng, J., Lu, H.: Two-stream adaptive graph convolutional networks for skeleton-based action recognition. In: Proceedings of the IEEE Conference on Computer Vision and Pattern Recognition (CVPR), pp. 12026–12035 (2019)

33. Singh, A., Patil, D., Omkar, S.: Eye in the sky: real-time drone surveillance system (DSS) for violent individuals identification using scatternet hybrid deep learning network. In: Proceedings of the IEEE Conference on Computer Vision and Pattern Recognition Workshops (CVPRW), pp. 1629–1637 (2018)

34. Sudhakaran, S., Lanz, O.: Learning to detect violent videos using convolutional long short-term memory. In: IEEE International Conference on Advanced Video and Signal Based Surveillance (AVSS), pp. 1–6. IEEE (2017)

35. Sultani, W., Chen, C., Shah, M.: Real-world anomaly detection in surveillance videos. In: Proceedings of the IEEE Conference on Computer Vision and Pattern Recognition (CVPR), pp. 6479–6488 (2018)

36. Sun, C., Shrivastava, A., Vondrick, C., Murphy, K., Sukthankar, R., Schmid, C.: Actor-centric relation network. In: Ferrari, V., Hebert, M., Sminchisescu, C., Weiss, Y. (eds.) ECCV 2018. LNCS, vol. 11215, pp. 335–351. Springer, Cham (2018). https://doi.org/10.1007/978-3-030-01252-6_20

37. Tran, D., Bourdev, L., Fergus, R., Torresani, L., Paluri, M.: Learning spatiotemporal features with 3D convolutional networks. In: Proceedings of the IEEE International Conference on Computer Vision (ICCV), pp. 4489–4497 (2015)

38. Veličković, P., Cucurull, G., Casanova, A., Romero, A., Lio, P., Bengio, Y.: Graph attention networks. arXiv preprint arXiv:1710.10903 (2017)

39. Wang, X., Girshick, R., Gupta, A., He, K.: Non-local neural networks. In: Proceedings of the IEEE Conference on Computer Vision and Pattern Recognition (CVPR), pp. 7794–7803 (2018)

40. Wang, X., Gupta, A.: Videos as space-time region graphs. In: Ferrari, V., Hebert, M., Sminchisescu, C., Weiss, Y. (eds.) ECCV 2018. LNCS, vol. 11209, pp. 413–431. Springer, Cham (2018). https://doi.org/10.1007/978-3-030-01228-1_25

41. Wu, C.Y., Feichtenhofer, C., Fan, H., He, K., Krahenbuhl, P., Girshick, R.: Longterm feature banks for detailed video understanding. In: Proceedings of the IEEE Conference on Computer Vision and Pattern Recognition (CVPR), pp. 284–293 (2019)

42. Wu, P., Liu, J., Li, M., Sun, Y., Shen, F.: Fast sparse coding networks for anomaly detection in videos. Pattern Recogn. 107515 (2020)

43. Wu, P., Liu, J., Shen, F.: A deep one-class neural network for anomalous event detection in complex scenes. IEEE Trans. Neural Networks Learn. Syst. 31, 2609–2622 (2019)

44. Yan, S., Xiong, Y., Lin, D.: Spatial temporal graph convolutional networks for skeleton-based action recognition. In: Thirty-second AAAI Conference on Artificial Intelligence (AAAI) (2018)

45. Yao, T., Pan, Y., Li, Y., Mei, T.: Exploring visual relationship for image captioning. In: Ferrari, V., Hebert, M., Sminchisescu, C., Weiss, Y. (eds.) Computer Vision – ECCV 2018. LNCS, vol. 11218, pp. 711–727. Springer, Cham (2018). https://doi.org/10.1007/978-3-030-01264-9_42

46. Yue, K., Sun, M., Yuan, Y., Zhou, F., Ding, E., Xu, F.: Compact generalized non-local network. In: Advances in Neural Information Processing Systems (NIPS), pp. 6510–6519 (2018)
47. Zach, C., Pock, T., Bischof, H.: A duality based approach for realtime TV-L^1 optical flow. In: Hamprecht, F.A., Schnörr, C., Jähne, B. (eds.) DAGM 2007. LNCS, vol. 4713, pp. 214–223. Springer, Heidelberg (2007). https://doi.org/10.1007/978-3-540-74936-3_22
48. Zajdel, W., Krijnders, J.D., Andringa, T., Gavrila, D.M.: Cassandra: audio-video sensor fusion for aggression detection. In: IEEE International Conference on Advanced Video and Signal Based Surveillance (AVSS), pp. 200–205. IEEE (2007)
49. Zeng, R., et al.: Graph convolutional networks for temporal action localization. In: Proceedings of the IEEE International Conference on Computer Vision (ICCV), pp. 7094–7103 (2019)
50. Zhong, J.X., Li, N., Kong, W., Liu, S., Li, T.H., Li, G.: Graph convolutional label noise cleaner: train a plug-and-play action classifier for anomaly detection. In: Proceedings of the IEEE Conference on Computer Vision and Pattern Recognition (CVPR), pp. 1237–1246 (2019)
51. Zhou, B., Andonian, A., Oliva, A., Torralba, A.: Temporal relational reasoning in videos. In: Ferrari, V., Hebert, M., Sminchisescu, C., Weiss, Y. (eds.) ECCV 2018. LNCS, vol. 11205, pp. 831–846. Springer, Cham (2018). https://doi.org/10.1007/978-3-030-01246-5_49

SNE-RoadSeg: Incorporating Surface Normal Information into Semantic Segmentation for Accurate Freespace Detection

Rui Fan[1] , Hengli Wang[2] , Peide Cai[2] , and Ming Liu[2(✉)]

[1] UC San Diego, La Jolla, USA
rui.fan@ieee.org
[2] HKUST Robotics Institute, Hong Kong, China
{hwangdf,peide.cai,eelium}@ust.hk

Abstract. Freespace detection is an essential component of visual perception for self-driving cars. The recent efforts made in data-fusion convolutional neural networks (CNNs) have significantly improved semantic driving scene segmentation. Freespace can be hypothesized as a ground plane, on which the points have similar surface normals. Hence, in this paper, we first introduce a novel module, named surface normal estimator (SNE), which can infer surface normal information from dense depth/disparity images with high accuracy and efficiency. Furthermore, we propose a data-fusion CNN architecture, referred to as RoadSeg, which can extract and fuse features from both RGB images and the inferred surface normal information for accurate freespace detection. For research purposes, we publish a large-scale synthetic freespace detection dataset, named Ready-to-Drive (R2D) road dataset, collected under different illumination and weather conditions. The experimental results demonstrate that our proposed SNE module can benefit all the state-of-the-art CNNs for freespace detection, and our SNE-RoadSeg achieves the best overall performance among different datasets.

Keywords: Freespace detection · Self-driving cars · Data-fusion CNN · Semantic driving scene segmentation · Surface normal

Source Code, Dataset and Demo Video:
https://sites.google.com/view/sne-roadseg/home

1 Introduction

Autonomous cars are a regular feature in science fiction films and series, but thanks to the rise of artificial intelligence, the fantasy of picking up one such

---

R. Fan and H. Wang—These authors contributed equally to this work and are therefore joint first authors.

Electronic supplementary material The online version of this chapter (https://doi.org/10.1007/978-3-030-58577-8_21) contains supplementary material, which is available to authorized users.

A. Vedaldi et al. (Eds.): ECCV 2020, LNCS 12375, pp. 340–356, 2020.
https://doi.org/10.1007/978-3-030-58577-8_21

vehicle at your garage forecourt has turned into reality. Driving scene understanding is a crucial task for autonomous cars, and it has taken a big leap with recent advances in artificial intelligence [1]. Collision-free space (or simply freespace) detection is a fundamental component of driving scene understanding [27]. Freespace detection approaches generally classify each pixel in an RGB or depth/disparity image as drivable or undrivable. Such pixel-level classification results are then utilized by other modules in the autonomous system, such as trajectory prediction [4] and path planning [31], to ensure that the autonomous car can navigate safely in complex environments.

The existing freespace detection approaches can be categorized as either traditional or machine/deep learning-based. The traditional approaches generally formulate freespace with an explicit geometry model and find its best coefficients using optimization approaches [13]. [36] is a typical traditional freespace detection algorithm, where road segmentation is performed by fitting a B-spline model to the road disparity projections on a 2D disparity histogram (generally known as a v-disparity image) [12]. With recent advances in machine/deep learning, freespace detection is typically regarded as a semantic driving scene segmentation problem, where the convolutional neural networks (CNNs) are used to learn its best solution [34]. For instance, Lu et al. [25] employed an encoder-decoder architecture to segment RGB images in the bird's eye view for end-to-end freespace detection. Recently, many researchers have resorted to data-fusion CNN architectures to further improve the accuracy of semantic image segmentation. For example, Hazirbas et al. [19] incorporated depth information into conventional semantic segmentation via a data-fusion CNN architecture, which greatly enhanced the performance of driving scene segmentation.

In this paper, we first introduce a novel module named surface normal estimator (SNE), which can infer surface normal information from dense disparity/depth images with both high precision and efficiency. Additionally, we design a data-fusion CNN architecture named RoadSeg, which is capable of incorporating both RGB and surface normal information into semantic segmentation for accurate freespace detection. Since the existing freespace detection datasets with diverse illumination and weather conditions do not have either disparity/depth information or freespace ground truth, we created a large-scale synthetic freespace detection dataset, named Ready-to-Drive (R2D) road dataset (containing 11430 pairs of RGB and depth images), under different illumination and weather conditions. Our R2D road dataset is also publicly available for research purposes. To validate the feasibility and effectiveness of our introduced SNE module, we use three road datasets (KITTI [15], SYNTHIA [21] and our R2D) to train ten state-of-the-art CNNs (six single-modal CNNs and four data-fusion CNNs), with and without our proposed SNE module embedded. The experiments demonstrate that our proposed SNE module can benefit all these CNNs for freespace detection. Also, our method SNE-RoadSeg outperforms all other CNNs for freespace detection, where its overall performance is the second best on the KITTI road benchmark[1] [15].

---

[1] cvlibs.net/datasets/kitti/eval_road.php.

The remainder of this paper is structured as follows: Sect. 2 provides an overview of the state-of-the-art CNNs for semantic image segmentation. Section 3 introduces our proposed SNE-RoadSeg. Section 4 shows the experimental results and discusses both the effectiveness of our proposed SNE module and the performance of our SNE-RoadSeg. Finally, Sect. 5 concludes the paper.

2   Related Work

In 2015, Long *et al.* [24] introduced Fully Convolutional Network (FCN), a CNN for end-to-end semantic image segmentation. Since then, research on this topic has exploded. Based on FCN, Ronneberger *et al.* [26] proposed U-Net in the same year, which consists of a contracting path and an expansive path [26]. It adds skip connections between the contracting path and the expansive path to help better recover the full spatial resolution. Different from FCN, SegNet [3] utilizes an encoder-decoder architecture, which has become the mainstream structure for following approaches. An encoder-decoder architecture is typically composed of an encoder, a decoder and a final pixel-wise classification layer.

Furthermore, DeepLabv3+ [9], developed from DeepLabv1 [6], DeepLabv2 [7] and DeepLabv3 [8], was proposed in 2018. It employs depth-wise separable convolution in both atrous spatial pyramid pooling (ASPP) and the decoder, which makes its encoder-decoder architecture much faster and stronger [9]. Although the ASPP can generate feature maps by concatenating multiple atrous-convolved features, the resolution of the generated feature maps is not sufficiently dense for some applications such as autonomous driving [7]. To address this problem, DenseASPP [37] was designed to connect atrous convolutional layers (ACLs) densely. It is capable of generating multi-scale features that cover a larger and denser scale range, without significantly increasing the model size [37].

Different from the above-mentioned CNNs, DUpsampling [32] was proposed to recover the pixel-wise prediction by employing a data-dependent decoder. It allows the decoder to downsample the fused features before merging them, which not only reduces computational costs, but also decouples the resolutions of both the fused features and the final prediction [32]. GSCNN [30] utilizes a novel two-branch architecture consisting of a regular (classical) branch and a shape branch. The regular branch can be any backbone architecture, while the shape branch processes the shape information in parallel with the regular branch. Experimental results have demonstrated that this architecture can significantly boost the performance on thinner and smaller objects [30].

FuseNet [19] was designed to use RGB-D data for semantic image segmentation. The key ingredient of FuseNet is a fusion block, which employs element-wise summation to combine the feature maps obtained from two encoders. Although FuseNet [19] demonstrates impressive performance, the ability of CNNs to handle geometric information is limited, due to the fixed grid kernel structure [35]. To address this problem, depth-aware CNN [35] presents two intuitive and flexible operations: depth-aware convolution and depth-aware average pooling. These operations can efficiently incorporate geometric information into the CNN by leveraging the depth similarity between pixels [35].

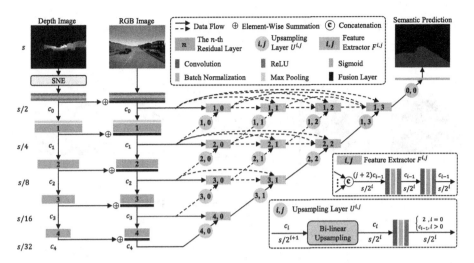

Fig. 1. The architecture of our SNE-RoadSeg. It consists of our SNE module, an RGB encoder, a surface normal encoder and a decoder with densely-connected skip connections. s represents the input resolution of the RGB and depth images. c_n represents the number of feature map channels at different levels.

MFNet [18] was proposed for semantic driving scene segmentation with the use of RGB-thermal vision data. In order to meet the real-time requirement of autonomous driving applications, MFNet focuses on minimizing the trade-off between accuracy and efficiency. Similarly, RTFNet [29] was developed to improve the semantic image segmentation performance using RGB-thermal vision data. Its main contribution is a novel decoder, which leverages short-cuts to produce sharp boundaries while keeping more detailed information [29].

3   SNE-RoadSeg

3.1   SNE

The proposed SNE is developed from our recent work three-filters-to-normal (3F2N) [14]. Its architecture is shown in Fig. 2. For a perspective camera model, a 3D point $\mathbf{P} = [X, Y, Z]^\top$ in the Euclidean coordinate system can be linked with a 2D image pixel $\mathbf{p} = [x, y]^\top$ using:

$$Z \begin{bmatrix} \mathbf{P} \\ 1 \end{bmatrix} = \mathbf{KP} = \begin{bmatrix} f_x & 0 & x_o \\ 0 & f_y & y_o \\ 0 & 0 & 1 \end{bmatrix} \mathbf{P}, \tag{1}$$

where $\mathbf{K}$ is the camera intrinsic matrix; $\mathbf{p_o} = [x_o, y_o]^\top$ is the image center; f_x and f_y are the camera focal lengths in pixels. The simplest way to estimate the surface normal $\mathbf{n} = [n_x, n_y, n_z]^\top$ of $\mathbf{P}$ is to fit a local plane:

$$n_x X + n_y Y + n_z Z + d = 0 \tag{2}$$

to $\mathbf{N_P^+} = [\mathbf{P}, \mathbf{N_P}]^\top$, where $\mathbf{N_P} = [\mathbf{Q}_1, \ldots, \mathbf{Q}_k]^\top$ is a set of k neighboring points of $\mathbf{P}$. Combining (1) and (2) results in [14]:

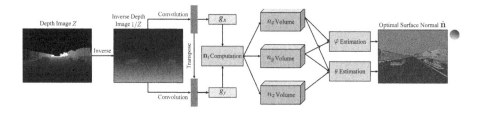

Fig. 2. The architecture of our proposed SNE module.

$$\frac{1}{Z} = -\frac{1}{d}\left(n_x \frac{x - x_\mathrm{o}}{f_x} + n_y \frac{y - y_\mathrm{o}}{f_y} + n_z\right). \tag{3}$$

Differentiating (3) with respect to x and y leads to:

$$g_x = \frac{\partial 1/Z}{\partial x} = -\frac{n_x}{df_x}, \quad g_y = \frac{\partial 1/Z}{\partial y} = -\frac{n_y}{df_y}, \tag{4}$$

which, as illustrated in Fig. 2, can be respectively approximated by convolving the inverse depth image $1/Z$ (or a disparity image, as disparity is in inverse proportion to depth) with a horizontal and a vertical image gradient filter [14]. Rearranging (4) results in the expressions of n_x and n_y as follows:

$$n_x = -df_x g_x, \quad n_y = -df_y g_y. \tag{5}$$

Given an arbitrary $\mathbf{Q}_i \in \mathbf{N_P}$, we can compute its corresponding n_{z_i} by plugging (5) into (2):

$$n_{z_i} = d\frac{f_x \Delta X_i g_x + f_y \Delta Y_i g_y}{\Delta Z_i}, \tag{6}$$

where $\mathbf{Q}_i - \mathbf{P} = [\Delta X_i, \Delta Y_i, \Delta Z_i]^\top$. Since (5) and (6) have a common factor of $-d$, the surface normal $\mathbf{n}_i$ obtained from $\mathbf{Q}_i$ and $\mathbf{P}$ has the following expression [34]:

$$\mathbf{n}_i = \left[f_x g_x, \quad f_y g_y, \quad -\frac{f_x \Delta X_i g_x + f_y \Delta Y_i g_y}{\Delta Z_i}\right]^\top. \tag{7}$$

A k-connected neighborhood system $\mathbf{N_P}$ of $\mathbf{P}$ can produce k normalized surface normals $\bar{\mathbf{n}}_1, \ldots, \bar{\mathbf{n}}_k$, where $\bar{\mathbf{n}}_i = \frac{\mathbf{n}_i}{\|\mathbf{n}_i\|_2} = [\bar{n}_{x_i}, \bar{n}_{y_i}, \bar{n}_{z_i}]^\top$. Since any normalized surface normals are projected on a sphere with center $(0,0,0)$ and radius 1, we believe that the optimal surface normal $\hat{\mathbf{n}}$ for $\mathbf{P}$ is also projected somewhere on the same sphere, where the projections of $\bar{\mathbf{n}}_1, \ldots, \bar{\mathbf{n}}_k$ distribute most intensively [13]. $\hat{\mathbf{n}}$ can be written in spherical coordinates as follows:

$$\hat{\mathbf{n}} = \left[\sin\theta\cos\varphi, \ \sin\theta\sin\varphi, \ \cos\theta\right]^\top, \tag{8}$$

where $\theta \in [0, \pi]$ denotes inclination and $\varphi \in [0, 2\pi)$ denotes azimuth. φ can be computed using:

$$\varphi = \arctan\left(\frac{f_y g_y}{f_x g_x}\right). \tag{9}$$

Similar to [13], we hypothesize that the angle between an arbitrary pair of normalized surface normals is less than $\pi/2$. $\hat{\mathbf{n}}$ can therefore be estimated by minimizing $E = -\sum_{i=1}^{k} \hat{\mathbf{n}} \cdot \bar{\mathbf{n}}_i$ [13]. $\frac{\partial E}{\partial \theta} = 0$ obtains:

$$\theta = \arctan\left(\frac{\sum_{i=1}^{k} \bar{n}_{x_i} \cos\varphi + \sum_{i=1}^{k} \bar{n}_{y_i} \sin\varphi}{\sum_{i=1}^{k} \bar{n}_{z_i}}\right). \tag{10}$$

Substituting θ and φ into (8) results in the optimal surface normal $\hat{\mathbf{n}}$, as shown in Fig. 2. The performance of our proposed SNE will be discussed in Sect. 4.

3.2   RoadSeg

U-Net [26] has demonstrated the effectiveness of using skip connections in recovering the full spatial resolution. However, its skip connections force aggregations only at the same-scale feature maps of the encoder and decoder, which, we believe, is an unnecessary constraint. Inspired by DenseNet [23], we propose RoadSeg, which exploits densely-connected skip connections to realize flexible feature fusion in the decoder.

As shown in Fig. 1, our proposed RoadSeg also adopts the popular encoder-decoder architecture. An RGB encoder and a surface normal encoder is employed to extract the feature maps from RGB images and from the inferred surface normal information, respectively. The extracted RGB and surface normal feature maps are hierarchically fused through element-wise summations. The fused feature maps are then fused again in the decoder through densely-connected skip connections to restore the resolution of the feature maps. At the end of RoadSeg, a sigmoid layer is used to generate the probability map for the semantic driving scene segmentation.

We use ResNet [20] as the backbone of our RGB and surface normal encoders, the structures of which are identical to each other. Specifically, the initial block consists of a convolutional layer, a batch normalization layer and a ReLU activation layer. Then, a max pooling layer and four residual layers are sequentially employed to gradually reduce the resolution as well as increase the number of feature map channels. ResNet has five architectures: ResNet-18, ResNet-34, ResNet-50, ResNet-101 and ResNet-152. Our RoadSeg follows the same naming rule of ResNet. c_n, the number of feature map channels (see Fig. 1) varies with respect to the adopted ResNet architecture. Specifically, c_0–c_4 are 64, 64, 128, 256 and 512, respectively, for ResNet-18 and ResNet-34, and are 64, 256, 512, 1024 and 2048, respectively, for ResNet-50, ResNet-101 and ResNet-152.

The decoder consists of two different types of modules: a) feature extractors $F^{i,j}$ and b) upsampling layers $U^{i,j}$, which are connected densely to realize flexible feature fusion. The feature extractor is employed to extract features from the fused feature maps, and it ensures that the feature map resolution is unchanged. The upsampling layer is employed to increase the resolution and decrease the feature map channels. Three convolutional layers in the feature extractor and the upsampling layer have the same kernel size of 3×3, the same stride of 1 and the same padding of 1.

4   Experiments

4.1   Datasets and Experimental Setup

In our experiments, we first evaluate the performance of our proposed SNE on the DIODE dataset [33], a public surface normal estimation dataset containing RGBD vision data of both indoor and outdoor scenarios. We utilize the average angular error (AAE), $e_{\text{AAE}} = \frac{1}{m} \sum_{k=1}^{m} \cos^{-1} \left(\frac{\langle \mathbf{n}_k, \hat{\mathbf{n}}_k \rangle}{\|\mathbf{n}_k\|_2 \|\hat{\mathbf{n}}_k\|_2} \right)$, to quantify our SNE's accuracy, where m is the number of 3D points used for evaluation; $\mathbf{n}_k$ and $\hat{\mathbf{n}}_k$ is the ground truth and estimated (optimal) surface normal, respectively. The experimental results are presented in Sect. 4.2.

Then, we carry out the experiments on the following three datasets to evaluate the performance of our proposed SNE-RoadSeg for freespace detection:

- The KITTI road dataset [15]: this dataset provides real-world RGB-D vision data. We split it into three subsets: a) training (173 images), b) validation (58 images), and c) testing (58 images).
- The SYNTHIA road dataset [21]: this dataset provides synthetic RGB-D vision data. We select 2224 images from it and group them into: a) training (1334 images), b) validation (445 images), and c) testing (445 images).
- Our R2D road dataset: along with our proposed SNE-RoadSeg, we also publish a large-scale synthetic freespace detection dataset, named R2D road dataset. This dataset is created using the CARLA[2] simulator [11]. Firstly, we mount a simulated stereo rig (baseline: 1.5 m) on the top of a vehicle to capture synchronized stereo images (resolution: 640 × 480 pixels) at 10 fps. The vehicle navigates in six different scenarios under different illumination and weather conditions (sunny, rainy, day and sunset). There are a total of 11430 pairs of stereo images with corresponding depth images and semantic segmentation ground truth. We split them into three subsets: a) training (6117 images), b) validation (2624 images), and c) testing (2689 images). Our dataset is publicly available at sites.google.com/view/sne-roadseg for research purposes.

We use these three datasets to train ten state-of-the-art CNNs, including six single-modal CNNs and four data-fusion CNNs. We conduct the experiments of single-modal CNNs with three setups: a) training with RGB images, b) training with depth images, and c) training with surface normal images (generated from depth images using our SNE), which are denoted as **RGB**, **Depth** and

---

[2] carla.org.

SNE-Depth, respectively. Similarly, the experiments of data-fusion CNNs are conducted using two setups: training using RGB-D vision data, with and without our SNE embedded, which are denoted as **RGBD** and **SNE-RGBD**, respectively. To compare the performances between our proposed RoadSeg and other state-of-the-art CNNs, we train our RoadSeg with the same setups as for the data-fusion CNNs on the three datasets. Moreover, we re-train our SNE-RoadSeg for the result submission to the KITTI road benchmark [15]. The experimental results are presented in Sect. 4.3. Additionally, the ablation study of our SNE-RoadSeg is provided in Sect. 4.4.

Five common metrics are used for the performance evaluation of freespace detection: accuracy, precision, recall, F-score and the intersection over union (IoU). Their corresponding definitions are as follows: Accuracy $= \frac{n_{\mathrm{tp}} + n_{\mathrm{tn}}}{n_{\mathrm{tp}} + n_{\mathrm{tn}} + n_{\mathrm{fp}} + n_{\mathrm{fn}}}$, Precision $= \frac{n_{\mathrm{tp}}}{n_{\mathrm{tp}} + n_{\mathrm{fp}}}$, Recall $= \frac{n_{\mathrm{tp}}}{n_{\mathrm{tp}} + n_{\mathrm{fn}}}$, F-score $= \frac{2n_{\mathrm{tp}}^2}{2n_{\mathrm{tp}}^2 + n_{\mathrm{tp}}(n_{\mathrm{fp}} + n_{\mathrm{fn}})}$ and IoU $= \frac{n_{\mathrm{tp}}}{n_{\mathrm{tp}} + n_{\mathrm{fp}} + n_{\mathrm{fn}}}$, where n_{tp}, n_{tn}, n_{fp} and n_{fn} represents the true positive, true negative, false positive, and false negative pixel numbers, respectively. In addition, the stochastic gradient descent with momentum (SGDM) optimizer is utilized to minimize the loss function, and the initial learning rate is set to 0.001. Furthermore, we adopt the early stopping mechanism on the validation subset to avoid over-fitting. The performance is then quantified using the testing subset.

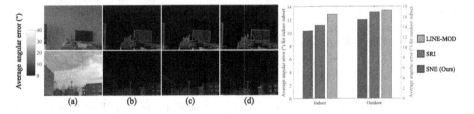

Fig. 3. Qualitative and quantitative results on the DIODE dataset: (a) RGB images; (b)–(d): the angular error maps obtained using our proposed SNE, SRI [2] and LINE-MOD [22], respectively.

4.2   Performance Evaluation of Our SNE

We simply set $g_x = \frac{1}{Z(x-1,y)} - \frac{1}{Z(x+1,y)}$ and $g_y = \frac{1}{Z(x,y-1)} - \frac{1}{Z(x,y+1)}$ to evaluate the accuracy of our proposed SNE. In addition, we also compare it with two well-known surface normal estimation approaches: SRI [2] and LINE-MOD [22]. The qualitative and quantitative comparisons are shown in Fig. 3. It can be observed that our proposed SNE outperforms SRI and LINE-MOD for both indoor and outdoor scenarios.

4.3   Performance Evaluation of Our SNE-RoadSeg

In this subsection, we evaluate the performance of our proposed SNE-RoadSeg-152 (abbreviated as SNE-RoadSeg) both qualitatively and quantitatively. Examples of the experimental results on the SYNTHIA road dataset [21] and our R2D road dataset are shown in Fig. 4. We can clearly observe that the CNNs with RGB images as inputs suffer greatly from poor illumination conditions. Moreover, the CNNs with our SNE embedded generally perform better than they do without our SNE embedded. The corresponding quantitative comparisons are given in Fig. 5 and Fig. 6. Readers can see that the IoU increases by approximately 2–12% for single-modal CNNs and by about 1–7% for data-fusion CNNs, while the F-score increases by around 1–7% for single-modal CNNs and by about 1–4% for data-fusion CNNs. We demonstrate that our proposed SNE can make the road areas become highly distinguishable, and thus, it will benefit all state-of-the-art CNNs for freespace detection.

Furthermore, from Fig. 5 and Fig. 6, we can observe that RoadSeg itself outperforms all other CNNs. We demonstrate that the densely-connected skip connections utilized in our proposed RoadSeg can help achieve flexible feature fusion and smooth the gradient flow to generate accurate freespace detection results. Also, RoadSeg with our SNE embedded performs better than all other CNNs with our SNE embedded. An increase of approximately 1.4–14.7% is witnessed on the IoU, while the F-score increases by about 0.7–8.8%.

In addition, we compare our proposed method with five state-of-the-art CNNs published on the KITTI road benchmark [15]. Examples of the experimental results are shown in Fig. 7. The quantitative comparisons are given in Table 1, which shows that our proposed SNE-RoadSeg achieves the highest MaxF (maximum F-score), AP (average precision) and PRE (precision), while LC-CRF [16] achieves the best REC (recall). Our freespace detection method is the second best on the KITTI road benchmark [15].

Figure 8 presents several unsatisfactory results of our SNE-RoadSeg on the KITTI road dataset [15]. Since the 3D points on freespace and sidewalks possess very similar surface normals, our proposed approach can sometimes mistakenly recognize part of sidewalks as freespace, especially when the textures of the road and sidewalks are similar. We believe this can be improved by leveraging surface normal gradient features, as there usually exists a clear boundary between freespace and sidewalks (due to their differences in height).

4.4   Ablation Study

In this subsection, we conduct ablation studies on our R2D road dataset to validate the effectiveness of the architecture for our RoadSeg. The performances of different architectures are provided in Table 2.

Firstly, we replace the backbone of RoadSeg with different ResNet architectures. The quantitative results are given in Table 2. The superior performance of our choice is as expected, because ResNet-152 has also presented the best image classification performance among the five ResNet architectures [20].

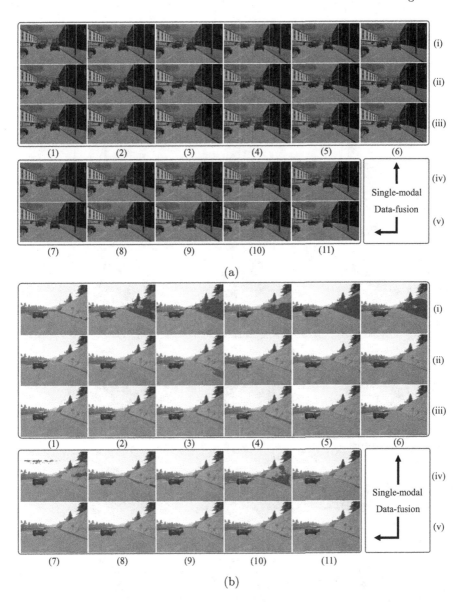

Fig. 4. Examples of the experimental results on (a) the SYNTHIA road dataset and (b) our R2D road dataset: (i) RGB, (ii) Depth, (iii) SNE-Depth (Ours), (iv) RGBD and (v) SNE-RGBD (Ours); (1) DeepLabv3+ [9], (2) U-Net [26], (3) SegNet [3], (4) GSCNN [30], (5) DUpsampling [32], (6) DenseASPP [37], (7) FuseNet [19], (8) RTFNet [29], (9) Depth-aware CNN [35], (10) MFNet [18] and (11) RoadSeg (Ours). The true positive, false negative and false positive pixels are shown in green, red and blue, respectively (Color figure online).

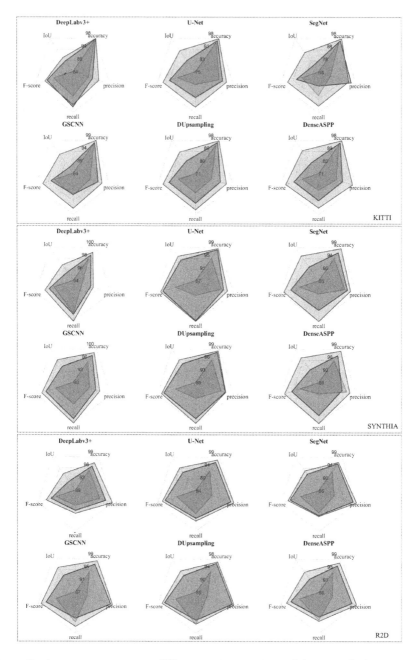

Fig. 5. Performance comparison (%) among DeepLabv3+ [9], U-Net [26], SegNet [3], GSCNN [30], DUpsampling [32] and DenseASPP [37] with and without our SNE embedded, where —— RGB, —— Depth, and —— SNE-Depth (Ours).

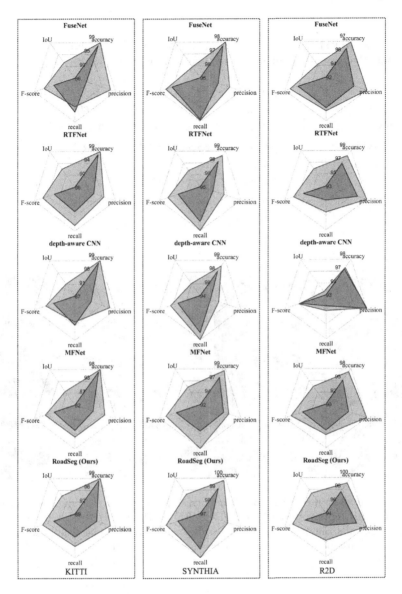

Fig. 6. Performance comparison (%) among FuseNet [19], RTFNet [29], depth-aware CNN [35], MFNet [18] and our RoadSeg with and without our SNE embedded, where —— RGBD and —— SNE-RGBD (Ours).

Table 1. The KITTI road benchmark results, where the best results are in bold type. Please note that we only compare our method with published works.

Method	MaxF (%)	AP (%)	PRE (%)	REC (%)	Rank
RBNet [10]	93.21	89.18	92.81	93.60	21
TVFNet [17]	95.34	90.26	95.73	94.94	16
LC-CRF [16]	95.68	88.34	93.62	**97.83**	13
LidCamNet [5]	96.03	93.93	96.23	95.83	7
RBANet [28]	96.30	89.72	95.14	97.50	6
SNE-RoadSeg (Ours)	**96.75**	**94.07**	**96.90**	96.61	**2**

Fig. 7. Examples on the KITTI road benchmark, where rows (a)–(f) show the freespace detection results obtained by RBNet [10], TVFNet [17], LC-CRF [16], LidCamNet [5], RBANet [28] and our proposed SNE-RoadSeg, respectively. The true positive, false negative and false positive pixels are shown in green, red and blue, respectively (Color figure online).

Fig. 8. Unsatisfactory results obtained using the KITTI road dataset. The true positive, false negative and false positive pixels are shown in green, red and blue, respectively (Color figure online).

Then, we remove one encoder from RoadSeg to evaluate its performance on single-modal vision data. We conduct five experiments: a) training with RGB images, denoted as **RGB**; b) training with depth images, denoted as **Depth**; c) training with depth images, denoted as **SNE-Depth**; d) training with four-channel RGB-D vision data, denoted as **RGBD-C**; and e) training with four-channel RGB-D vision data, denoted as **SNE-RGBD-C**. From Table 2, we can observe that our choice outperforms the single-modal architecture with respect to different modalities of training data, proving that the data fusion via a two-encoder architecture can benefit the freespace detection. It should be noted that although the single-modal architectures cannot provide competitive results, our proposed SNE still benefits them for better freespace detection performance.

Table 2. Performance comparison (%) among different architectures and setups on our R2D road dataset. The best results are shown in bold font.

Architecture	Setup	Accuracy	Precision	Recall	F-Score	IoU
RoadSeg-18	SNE-RGBD	93.6	93.5	91.3	92.4	85.9
RoadSeg-34		95.5	96.3	93.0	94.6	89.8
RoadSeg-50		96.8	97.5	95.2	96.3	92.9
RoadSeg-101		98.0	98.2	97.1	97.6	95.4
RoadSeg-152	RGB	94.0	91.9	93.8	92.8	86.6
	Depth	96.7	97.6	94.6	96.1	92.4
	SNE-Depth	97.6	98.9	95.5	97.2	94.5
	RGBD-C	95.1	92.8	95.6	94.2	89.0
	SNE-RGBD-C	97.0	97.5	95.3	96.4	93.0
RoadSeg-152-NSCs	SNE-RGBD	97.9	98.6	96.5	97.5	95.2
RoadSeg-152-SSCs		98.2	99.0	96.8	97.9	95.9
RoadSeg-152 (Ours)	SNE-RGBD	**98.6**	**99.1**	**97.6**	**98.3**	**96.7**

To further validate the effectiveness of our choice, we replace the densely-connected skip connections in the decoder with two different architectures: a) no skip connections (NSCs), which totally removes the skip connections; b) sparse skip connections (SSCs), which employs the skip connections only at the same-scale feature maps of the encoder and decoder (like U-Net). Table 2 verifies the superiority of the densely-connected skip connections, which helps to achieve flexible feature fusion and to smooth the gradient flow to generate accurate freespace detection results, as analyzed in Sect. 4.3.

5   Conclusion

The main contributions of this paper include: a) a module named SNE, capable of inferring surface normal information from depth/disparity images with both high precision and efficiency; b) a data-fusion CNN architecture named Road-Seg, capable of fusing both RGB and surface normal information for accurate freespace detection; and c) a publicly available synthetic dataset for semantic driving scene segmentation. To demonstrate the feasibility and effectiveness of the proposed SNE module, we embedded it into ten state-of-the-art CNNs and evaluated their performances for freespace detection. The experimental results illustrated that our introduced SNE can benefit all these CNNs for freespace detection. Furthermore, our proposed data-fusion CNN architecture RoadSeg is most compatible with our proposed SNE, and it outperforms all other CNNs when detecting drivable road regions.

Acknowledgements. This work was supported by the National Natural Science Foundation of China, under grant No. U1713211, and the Research Grant Council of Hong Kong SAR Government, China, under Project No. 11210017, awarded to Prof. Ming Liu.

References

1. Alvarez, J.M., Gevers, T., LeCun, Y., Lopez, A.M.: Road Scene Segmentation from a Single Image. In: Fitzgibbon, A., Lazebnik, S., Perona, P., Sato, Y., Schmid, C. (eds.) ECCV 2012. LNCS, vol. 7578, pp. 376–389. Springer, Heidelberg (2012). https://doi.org/10.1007/978-3-642-33786-4_28
2. Badino, H., Huber, D., Park, Y., Kanade, T.: Fast and accurate computation of surface normals from range images. In: 2011 IEEE International Conference on Robotics and Automation, pp. 3084–3091. IEEE (2011)
3. Badrinarayanan, V., Kendall, A., Cipolla, R.: Segnet: a deep convolutional encoder-decoder architecture for image segmentation. IEEE Trans. Pattern Anal. Mach. Intell. **39**(12), 2481–2495 (2017)
4. Cai, P., Wang, S., Sun, Y., Liu, M.: Probabilistic end-to-end vehicle navigation in complex dynamic environments with multimodal sensor fusion. IEEE Robot. Autom. Lett. **5**, 4218–4224 (2020)
5. Caltagirone, L., Bellone, M., Svensson, L., Wahde, M.: Lidar-camera fusion for road detection using fully convolutional neural networks. Robot. Autonomous Syst. **111**, 125–131 (2019)
6. Chen, L.C., Papandreou, G., Kokkinos, I., Murphy, K., Yuille, A.L.: Semantic image segmentation with deep convolutional nets and fully connected crfs. CoRR abs/1412.7062 (2014)
7. Chen, L.C., Papandreou, G., Kokkinos, I., Murphy, K., Yuille, A.L.: Deeplab: semantic image segmentation with deep convolutional nets, atrous convolution, and fully connected CRFS. IEEE Trans. Pattern Anal. Mach. Intell. **40**(4), 834–848 (2017)
8. Chen, L.C., Papandreou, G., Schroff, F., Adam, H.: Rethinking atrous convolution for semantic image segmentation. arXiv preprint arXiv:1706.05587 (2017)
9. Chen, L.C., Zhu, Y., Papandreou, G., Schroff, F., Adam, H.: Encoder-decoder with atrous separable convolution for semantic image segmentation. In: Proceedings of the European Conference on Computer Vision (ECCV), pp. 801–818 (2018)
10. Chen, Z., Chen, Z.: Rbnet: A deep neural network for unified road and road boundary detection. In: Liu, D., Xie, S., Li, Y., Zhao, D., El-Alfy, E.S. (eds.) International Conference on Neural Information Processing. pp. 677–687. Springer, Cham (2017). https://doi.org/10.1007/978-3-319-70087-8_70

11. Dosovitskiy, A., Ros, G., Codevilla, F., Lopez, A., Koltun, V.: CARLA: an open urban driving simulator. In: Levine, S., Vanhoucke, V., Goldberg, K. (eds.) Proceedings of the 1st Annual Conference on Robot Learning. Proceedings of Machine Learning Research, vol. 78, pp. 1–16. PMLR, 13–15 November 2017

12. Fan, R., Jiao, J., Pan, J., Huang, H., Shen, S., Liu, M.: Real-time dense stereo embedded in a UAV for road inspection. In: Proceedings of the IEEE/CVF Conference Computer Vision and Pattern Recognition Workshops (CVPRW), pp. 535–543 (2019)

13. Fan, R., Ozgunalp, U., Hosking, B., Liu, M., Pitas, I.: Pothole detection based on disparity transformation and road surface modeling. IEEE Trans. Image Process. **29**, 897–908 (2019)

14. Fan, R., Wang, H., Xue, B., Huang, H., Wang, Y., Liu, M., Pitas, I.: Three-filters-to-normal: an accurate and ultrafast surface normal estimator. arXiv preprint arXiv:2005.08165 (2020), under peer review

15. Fritsch, J., Kuehnl, T., Geiger, A.: A new performance measure and evaluation benchmark for road detection algorithms. In: International Conference on Intelligent Transportation Systems (ITSC) (2013)

16. Gu, S., Zhang, Y., Tang, J., Yang, J., Kong, H.: Road detection through CRF based lidar-camera fusion. In: 2019 International Conference on Robotics and Automation (ICRA), pp. 3832–3838. IEEE (2019)

17. Gu, S., Zhang, Y., Yang, J., Alvarez, J.M., Kong, H.: Two-view fusion based convolutional neural network for urban road detection. In: 2019 IEEE/RSJ International Conference on Intelligent Robots and Systems (IROS), pp. 6144–6149. IEEE (2019)

18. Ha, Q., Watanabe, K., Karasawa, T., Ushiku, Y., Harada, T.: Mfnet: Towards real-time semantic segmentation for autonomous vehicles with multi-spectral scenes. In: 2017 IEEE/RSJ International Conference on Intelligent Robots and Systems (IROS), pp. 5108–5115. IEEE (2017)

19. Hazirbas, C., Ma, L., Domokos, C., Cremers, D.: FuseNet: incorporating depth into semantic segmentation via fusion-based CNN architecture. In: Lai, S.-H., Lepetit, V., Nishino, K., Sato, Y. (eds.) ACCV 2016. LNCS, vol. 10111, pp. 213–228. Springer, Cham (2017). https://doi.org/10.1007/978-3-319-54181-5_14

20. He, K., Zhang, X., Ren, S., Sun, J.: Deep residual learning for image recognition. In: Proceedings of the IEEE Conference on Computer Vision and Pattern Recognition, pp. 770–778 (2016)

21. Hernandez-Juarez, D., et al.: Slanted stixels: representing san francisco's steepest streets. In: British Machine Vision Conference (BMVC) (2017)

22. Hinterstoisser, S., et al.: Gradient response maps for real-time detection of textureless objects. IEEE Trans. Pattern Anal. Mach. Intell. **34**(5), 876–888 (2011)

23. Huang, G., Liu, Z., Van Der Maaten, L., Weinberger, K.Q.: Densely connected convolutional networks. In: Proceedings of the IEEE Conference on Computer Vision and Pattern Recognition, pp. 4700–4708 (2017)

24. Long, J., Shelhamer, E., Darrell, T.: Fully convolutional networks for semantic segmentation. In: Proceedings of the IEEE Conference on Computer Vision and Pattern Recognition, pp. 3431–3440 (2015)

25. Lu, C., van de Molengraft, M.J.G., Dubbelman, G.: Monocular semantic occupancy grid mapping with convolutional variational encoder-decoder networks. IEEE Robot. Autom. Lett. **4**(2), 445–452 (2019)

26. Ronneberger, O., Fischer, P., Brox, T.: U-Net: convolutional networks for biomedical image segmentation. In: Navab, N., Hornegger, J., Wells, W.M., Frangi, A.F. (eds.) MICCAI 2015. LNCS, vol. 9351, pp. 234–241. Springer, Cham (2015). https://doi.org/10.1007/978-3-319-24574-4_28

27. Sless, L., El Shlomo, B., Cohen, G., Oron, S.: Road scene understanding by occupancy grid learning from sparse radar clusters using semantic segmentation. In: Proceedings of the IEEE International Conference on Computer Vision Workshops (2019)
28. Sun, J.Y., Kim, S.W., Lee, S.W., Kim, Y.W., Ko, S.J.: Reverse and boundary attention network for road segmentation. In: Proceedings of the IEEE International Conference on Computer Vision Workshops (2019)
29. Sun, Y., Zuo, W., Liu, M.: Rtfnet: RGB-thermal fusion network for semantic segmentation of urban scenes. IEEE Robot. Autom. Lett. **4**(3), 2576–2583 (2019)
30. Takikawa, T., Acuna, D., Jampani, V., Fidler, S.: Gated-SCNN: Gated shape CNNS for semantic segmentation. In: Proceedings of the IEEE International Conference on Computer Vision, pp. 5229–5238 (2019)
31. Thoma, J., Paudel, D.P., Chhatkuli, A., Probst, T., Gool, L.V.: Mapping, localization and path planning for image-based navigation using visual features and map. In: Proceedings of the IEEE Conference on Computer Vision and Pattern Recognition, pp. 7383–7391 (2019)
32. Tian, Z., He, T., Shen, C., Yan, Y.: Decoders matter for semantic segmentation: data-dependent decoding enables flexible feature aggregation. In: Proceedings of the IEEE Conference on Computer Vision and Pattern Recognition, pp. 3126–3135 (2019)
33. Vasiljevic, I., et al.: Diode: A dense indoor and outdoor depth dataset. arXiv preprint arXiv:1908.00463 (2019)
34. Wang, H., Fan, R., Sun, Y., Liu, M.: Applying surface normal information in drivable area and road anomaly detection for ground mobile robots. In: 2020 IEEE/RSJ International Conference on Intelligent Robots and Systems (IROS) (2020), to be published
35. Wang, W., Neumann, U.: Depth-aware CNN for RGB-D segmentation. In: Proceedings of the European Conference on Computer Vision (ECCV), pp. 135–150 (2018)
36. Wedel, A., Badino, H., Rabe, C., Loose, H., Franke, U., Cremers, D.: B-spline modeling of road surfaces with an application to free-space estimation. IEEE Trans. Intell. Transport. Syst. **10**(4), 572–583 (2009)
37. Yang, M., Yu, K., Zhang, C., Li, Z., Yang, K.: Denseaspp for semantic segmentation in street scenes. In: Proceedings of the IEEE Conference on Computer Vision and Pattern Recognition, pp. 3684–3692 (2018)

Modeling the Space of Point Landmark Constrained Diffeomorphisms

Chengfeng Wen⬥, Yang Guo⬥, and Xianfeng Gu$^{(\boxtimes)}$⬥

Stony Brook University, Stony Brook, NY 11794, USA
{chwen,yangguo,gu}@cs.stonybrook.edu

Abstract. Surface registration plays a fundamental role in shape analysis and geometric processing. Generally, there are three criteria in evaluating a surface mapping result: diffeomorphism, small distortion, and feature alignment. To fulfill these requirements, this work proposes a novel model of the space of point landmark constrained diffeomorphisms. Based on Teichmüller theory, this mapping space is generated by the Beltrami coefficients, which are infinitesimally Teichmüller equivalent to 0. These Beltrami coefficients are the solutions to a linear equation group. By using this theoretic model, optimal registrations can be achieved by iterative optimization with linear constraints in the diffeomorphism space, such as harmonic maps and Teichmüller maps, which minimize different types of distortion. The theoretical model is rigorous and has practical value. Our experimental results demonstrate the efficiency and efficacy of the proposed method.

Keywords: Teichmüller Map · Conformal geometry · Point landmark constrained diffeomorphism

1 Introduction

3D surface registration serves as a fundamental process in shape analysis and geometric processing tasks. In computer vision areas, such as human face registration and tracking [54,55], human body registration and tracking [3,16], and general surface registration [34], high-quality surface mappings are desirable. In medical imaging areas, such as brain morphometry [26,36,40,44] and virtual colonoscopy [28,48], the accuracy of shape classification and abnormality detection relies heavily on the quality of the surface registration results.

In this work, we propose a novel approach to model the space of point landmark constrained diffeomorphisms for 3D surface registration. The generators of this space are the Beltrami coefficients infinitesimally Teichmüller equivalent to 0. This theoretic result can be applied to optimize special energies in the point landmark constrained diffeomorphism space, such as harmonic energy and angle distortion, to obtain harmonic mappings and Teichmüller mappings. The computation of these mappings can be effectively accomplished by solving quadratic optimization problems. As shown in Fig. 6, given a male and a female facial

© Springer Nature Switzerland AG 2020
A. Vedaldi et al. (Eds.): ECCV 2020, LNCS 12375, pp. 357–373, 2020.
https://doi.org/10.1007/978-3-030-58577-8_22

surface with landmarks, there are infinite many diffeomorphisms between the two faces with landmark constraints. Conventional Tecihmüller map is only one of them. The current work allows us to perform optimization in this mapping space, for example, a diffeomorphism with landmark constraints with minimal elastic deformation energy (namely generalized harmonic energy).

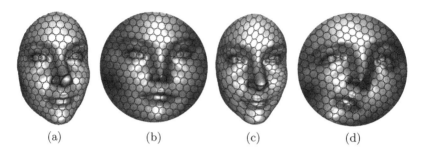

| (a) | (b) | (c) | (d) |

Fig. 1. Surface conformal mapping from (a) to (b) and quasi-conformal mapping from (c) to (d).

Criteria for Registrations. The following criteria are widely considered effective to evaluate the quality of a surface mapping. *i) Bijection* In most situations, a 1-1 correspondence is desired for surface registration purposes. *ii) Distortion* Surface registration will induce geometric distortions. In applications, it is highly preferred to minimize the distortion. *iii) Feature alignment* Surfaces, such as human faces and human bodies, have natural anatomical features like eye corners, nose tips, and joints. A high-quality surface registration should align these features accurately. *iv) Smoothness* In practice, surface registrations are required to be continuous and even smooth without folding or tearing. For surface registration purposes, an ideal mapping should be smooth, bijective, features aligned, and with least distortion. Thus, we propose an efficient algorithm to find point landmark constrained diffeomorphisms (smooth, bijective) with minimal distortions based on infinitesimal Teichmüller theory.

Space of Diffeomorphisms. Based on the quasi-conformal geometry theory [2,11,29], the mapping space of all diffeomorphisms between two surfaces is converted to a functional space defined on the source surface.

Consider a pair of Riemannian surfaces $(S, \mathbf{g})$ and $(T, \mathbf{h})$ with the same topology, as shown in the last two frames in Fig. 1, where S is the female face and T is the unit planar disk. A diffeomorphism $f : S \to T$ maps infinitesimal ellipses on the source to infinitesimal circles on the target. The shape of the ellipses (eccentricity and the orientation) is encoded into a complex function, the so-called Beltrami coefficient μ_f. The diffeomorphism f and its Beltrami coefficient μ_f mutually determine each other by the Beltrami equation in Eq. 4. The space of all diffeomorphisms between the two surfaces is essentially equivalent to the functional space of all Beltrami coefficients, whose norm is less than one almost everywhere (see e.g. [2]).

Point Landmark Constrained Diffeomorphism Space. Suppose some landmarks are given $\{p_i\}_{i=1}^n$ on the source surface S and $\{q_j\}_{j=1}^n$ on the target surface T, the landmark matching criteria requires the diffeomorphisms $f : S \to T$ maps each p_i to the corresponding q_i. Using the Beltrami coefficient representation of the mappings, the central question becomes how to choose μ, such that

$$f^\mu(p_i) = q_i, \ \forall i = 1, \ldots, n. \tag{1}$$

The diffeomorphisms satisfying the Eq. 1 form the point landmark constrained diffeomorphism space, denoted as $\mathcal{F}(S \setminus \Gamma, T \setminus \Lambda)$, where $\Gamma = \{p_i\}_{i=1}^n$ and $\Lambda = \{q_j\}_{j=1}^n$. Figure 2 shows one point landmark constrained diffeomorphism.

Suppose there are two Beltrami coefficients μ_1 and μ_2, such that f^{μ_k} satisfies the point landmark constraints in Eq. 1, then the composition $(f^{\mu_2})^{-1} \circ f^{\mu_1}$ is an automorphism of the source S, homotopic to identity and fixes all the landmarks p_i's. The point landmark constrained automorphisms form a group

$$G(S \setminus \Gamma) := \{f^\mu : \|\mu\|_\infty < 1, \ f^\mu \sim \mathrm{id}_S, \ f^\mu(p_i) = p_i, \forall i\}, \tag{2}$$

where Γ is the set of landmarks.

$G(S \setminus \Gamma)$ is an infinite dimensional Lie group, we can find a set of its "generators", the so-called infinitesimal Teichmüller trivial diffeomorphisms, $T^0(S \setminus \Gamma)$. Namely, for any $f^\mu \in G(S \setminus \Gamma)$, we can find a sequence of diffeomorphisms f^{μ_i}, $\mu_i \in T^0(S \setminus \Gamma)$, such that

$$\lim_{n \to \infty} f^{\mu_n/n} \circ f^{\mu_{n-1}/n} \circ \cdots f^{\mu_1/n} \to f^\mu,$$

where the space of infinitesimal Teichmüller trivial diffeomorphisms is given by (see e.g. [10])

$$T^0(S \setminus \Gamma) := \left\{ \mu : \|\mu\|_\infty < 1, \ \int_S \mu\varphi = 0, \ \forall \varphi \in \Omega(S \setminus \Gamma) \right\}. \tag{3}$$

Thus, the infinitesimal Teichmüller trivial diffeomorphisms $T^0(S \setminus \Gamma)$ generate the point landmark constrained automorphism group $G(S \setminus \Gamma)$. And $G(S \setminus \Gamma)$ gives all the point landmark constrained diffeomorphisms, namely solutions to Eq. 1.

Optimization in the Space of Diffeomorphisms. In practice, the optimal registration can be obtained by searching in the point landmark constrained diffeomorphism space (solutions to Eq. 1) for a solution that optimizes some specific energy. From the above discussion, the optimization can be carried out within $G(S \setminus \Gamma)$ or $T^0(S \setminus \Gamma)$ instead. Constraints described in Eq. 3 are linear, and given suitable energy, the optimization will become convex and can be solved by quadratic programming methods.

In this work, we compute point landmark constrained harmonic maps, which minimizes the elastic deformation energy in the point landmark constrained diffeomorphism space. Furthermore, we compute the point landmark constrained Teichmüller map that minimizes the angle distortion, namely the L^∞-norm of the Beltrami coefficient. Our experimental results demonstrate the efficacy and efficiency of the proposed method.

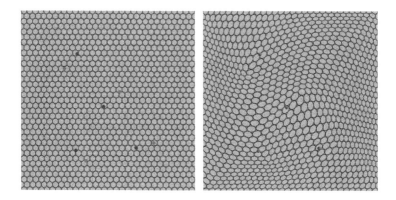

Fig. 2. Feature alignment illustration. The red dots are feature points and the blue dots represent the target position of the feature points. Namely, we are looking for a diffeomorphism that maps the red dots to the blue dots. The bijectivity is visualized by the circle patterns with no overlap or flip. The coincident blue and red dots represent the exact match of feature points (Color figure online).

Contributions. The proposed method has the following merits:

- Novel: We propose a novel approach to compute diffeomorphic surface mappings with point landmark constraints. Both point landmark constrained harmonic maps and Teichmüller maps can be computed under the same framework.
- Rigorous: We present a solid mathematical foundation to guarantee the existence of the Teichmüller map which is diffeomorphic, point landmarks aligned, and with least distortion.
- Practical: Based on infinitesimal Teichmüller theory, the feature aligned mapping is computed by iteratively solving quadratic optimization problems.

2   Related Works

3D surface registration methods have been extensively studied in recent years because of their fundamental importance. We list some of the most related work from more than abundant literature and readers are referred to [9,30,35,38] for surveys in surface registration and parameterization.

Point Cloud Registration. Point could registration methods can be applied on the vertices of 3D surfaces directly. Iterative Closest Points (ICP) [6,33] is one of the most well-known methods for point cloud registration. However, ICP is designed to deal with rigid motion situations. To alleviate this issue, multiple researches [4,5,14] presented non-rigid ICP methods that focus on non-rigid transformation. However, despite the popularity, these methods fail to guarantee the bijectivity and feature alignment constraints.

Conformal Parameterization. Conformal parameterization [13,39,49] is a powerful tool in delivering 1-to-1 correspondence between 3D surfaces and 2D domains while preserving local features. Several conformal parameterization registration methods are proposed in 3D facial surface registration and show good results [18,23,37,56]. Many computational approaches have been introduced such as least-square conformal mapping [20,21], holomorphic differentials based approaches [53] and Ricci flow techniques [17,18,49]. However, these conformal parameterization approaches cannot deal with feature constraints while preserving diffeomorphic from surfaces to surfaces. Gu *et al.* [13] proposed to compose an Möbius transformation to the conformal parameterization to minimize landmark mismatch energy. Similar ideas were presented in [42] as well. Lui *et al.* [23,24] and Choi *et al.* [7] improved this method by optimizing energy functionals consisting of harmonic energy and landmark mismatch energy, however, both approaches failed to either guarantee exact match of the landmarks or retain diffeomorphic given a large number of landmark constraints.

Quasiconformal Mapping. Quasiconformal mapping generalizes conformal mapping by allowing bounded angle distortion. Some early quasiconformal mapping algorithms were based on circle packing approaches [15] and were restricted between planar domains [8,29]. Zeng *et al.* [46,47] proposed to use the curvature flow approach with auxiliary metrics to compute quasiconformal maps for compact Riemann surfaces. Later, Zeng *et al.* [51] proposed a method to compute quasiconformal maps with curve landmark constraints by integrating quasi-holomorphic 1-forms, and in [52] graph-constrained diffeomorphisms are computed by computing harmonic maps with boundary constraints. To strictly enforce the landmark matching constraints, Lui *et al.* [27] optimized the registration using Beltrami holomorphic flow, where the surface diffeomorphism is obtained by adjusting Beltrami coefficients. Lui *et al.* [25] also applied a similar approach in surface registration compression. To compute general quasiconformal maps between arbitrary Riemann surfaces with similar ideas, Wong *et al.* [45] proposed a vector field approach named discrete Beltrami flow. Even though these approaches perform well in preserving bijectivity, they are less ideal in matching feature points accurately, especially when the shape deformation is large.

Teichmüller Mapping. Closely related to quasiconformal maps, Teichmüller map is the quasiconformal map where the L^∞ norm of the Beltrami coefficients is minimized. In other words, the Teichmüller map is the quasiconformal map as close to a conformal map as possible with certain constraints. Weber *et al.* [43] presented an algorithm to approximate the extremal quasiconformal map for genus zero surfaces with boundaries. Lui *et al.* [22] proposed an iterative algorithm by quasi-conformal iteration. A Beltrami holomorphic flow approach to compute Teichmüller extremal map for multiply-connected domains was introduced by Ng *et al.* [31]. In addition to algorithms, Teichmüller space and shape descriptor were applied in surface indexing and classification [19] and medical imaging [41,50]. For isogeometric analysis purposes, Nian and Chen [32] proposed an iterative algorithm to compute Teichmüller map based on alternating direction method of multipliers.

3   Theoretical Background and Definitions

In the following text, we refer to "point landmark(s)" as "landmark(s)" when there is no confusion.

Beltrami Equation. Suppose $f : \mathbb{D} \to \mathbb{D}$ is a complex function, and treated as a mapping from the unit disk $\mathbb{D}$ in the complex plane to itself. The *Beltrami coefficient.* μ is defined as

$$\frac{\partial f}{\partial \bar{z}} = \mu(z)\frac{\partial f}{\partial z}. \tag{4}$$

This equation is also called the *Beltrami equation*. The *dilatation* of f is defined as

$$K_f = \frac{1 + |\mu_f|}{1 - |\mu_f|} \tag{5}$$

Then the map f is said to be *quasiconformal* is K_f is bounded, and it's called K-quasiconformal is $K_f \leq K$. A map f is called *conformal (holomorphic)*, if μ_f is zero everywhere, or equivalently K_f is one everywhere. Intuitively, f maps infinitesimal ellipses to infinitesimal circles, the eccentricity of the ellipse is represented by the ratio between the major axis and the minor axis, which equals to K_f; the angle between the major axis of the ellipse and the horizontal direction is given by $1/2 \arg(\mu)$.

Measurable Riemann Mapping. Given a homeomorphism $f : \mathbb{D} \to \mathbb{D}$, its Beltrami coefficient μ_f can be computed by Eq. 4; inversely, given the Beltrami coefficient, there exists a corresponding map.

Theorem 1. (Measurable Riemann Mapping [1]). *Given a measurable complex function $\mu : \mathbb{D} \to \mathbb{C}$, such that $\|\mu\|_\infty < 1$, then there exists a homeomorphism $f : \mathbb{D} \to \mathbb{D}$ satisfying the Beltrami Eq. 4. Furthermore, two such kind of mappings differ by a Möbius transformation*

$$z \mapsto e^{i\theta}\frac{z - z_0}{1 - \bar{z}_0 z},$$

where $z_0 \in \mathbb{D}$.

The Beltrami differential is related to the Jacobian of the map f, $J(f)^2 = |f_z|^2(1 - |\mu_f|)^2$, hence if $\|\mu_f\|_\infty < 1$, then f is diffeomorphic. This shows the space of all automorphisms of the disk is equivalent to the space of Beltrami coefficients, quotient the Möbius transformation group,

$$\{\text{diffeomorphisms on } \mathbb{D}\} \cong \{\mu | \|\mu\|\infty < 1\}/\{Mobius\}$$

Riemann Surface. Suppose S is a topological surface, covered by a set of open sets $S \subset \bigcup U_\alpha$, each set U_α is mapped onto a complex domain $\varphi_\alpha : U_\alpha \to \mathbb{C}$, then $(U_\alpha, \varphi_\alpha)$ is a chart of S, $\{(U_\alpha, \varphi_\alpha)\}$ is an atlas of S. If $U_\alpha \cap U_\beta \neq \emptyset$, then the transition function is given by $\varphi_{\alpha\beta} := \varphi_\beta \circ \varphi_\alpha^{-1}$.

Definition 1 (Riemann Surface). *Suppose S is a topological surface with an atlas $\{(U_\alpha, \varphi_\alpha)\}$, if all transition functions are bi-holomorphic, then the atlas is called a conformal atlas, the surface is called a Riemann surface.*

Suppose $(S, \mathbf{g})$ is oriented, then for each point $p \in S$, one can find a small neighborhood U_p such that the isotherm parameterization φ_p exists inside U_p, then all (U_p, φ_p)'s form the conformal atlas, $(S, \mathbf{g})$ is a Riemann surface.

Holomorphic Quadratic Differential. Suppose S is a Riemann surface with a conformal atlas $\{(U_i, z_i)\}$, where z_i is the isothermal parameter inside U_i.

Definition 2 (Holomorphic Quadratic Differential). *A holomorphic quadratic differential on a Riemann surface S is an assignment of a function $\phi_i(z_i)$ on each local chart z_i such that if z_j is another local coordinate, we have*

$$\phi_i(z_i) = \phi_j(z_j) \left(\frac{dz_j}{dz_i}\right)^2.$$

We denote the space of all holomorphic differentials on S as $\Omega(S)$. Given a holomorphic quadratic differential $\varphi \in \Omega(S)$, a curve γ is called the horizontal trajectory of φ, if the integration of $\sqrt{\varphi}$ along γ is always a real number. Figure 3 illustrates the horizontal trajectories of holomorphic quadratic differentials on the cat surfaces.

$\Omega(S)$ is a complex linear space. For a genus $g > 1$ closed Riemann surface S, $\Omega(S)$ is $3g - 3$ dimensional. If S is a sphere with n punctures,

$$S = \mathbb{C} \cup \{\infty\} - \{a_1, a_2, \cdots, a_n\}$$

then every holomorphic quadratic differential has the form $\varphi(z)dz^2$, where

$$\phi(z) = \sum_{k=1}^{n} \frac{\rho_k}{z - a_k},$$

such that

$$\sum_{k=1}^{n} \rho_k = 0, \quad \sum_{k=1}^{k} \rho_k a_k = 0, \quad \sum_{k=1}^{n} \rho_k a_k^2 = 0.$$

Fig. 3. Holomorphic quadratic differentials.

For $S = \mathbb{D}\backslash\{z_1, z_2, ..., z_n\}$,

$$\phi_k(z) = \frac{\eta}{(z - z_k)}, 1 \leq k \leq n \tag{6}$$

form a basis of n dimensional complex vector space, where η is a constant such that $\|\phi\| = \int_S |\phi| = 1$

Beltrami Differential. Given a diffeomorphism between two Riemann surfaces $f : (S_1, \{z_i\}) \rightarrow (S_2, \{w_j\})$, the Beltrami differential can be defined as

$$\frac{\partial w_j}{\partial \bar{z}_i} d\bar{z}_i = \mu(z_i) \frac{\partial w_j}{\partial z_i} dz_i.$$

Then Beltrami differential $\mu(z_i) d\bar{z}_i / dz_i$ is invariant under the transition maps and thus is globally defined. The K-quasiconformal map can be generalized to the Riemann surface cases directly.

Teichmüller Equivalence. We are interested in such kind of homeomorphisms that fix the landmarks.

Definition 3 (Landmark Preserving Automorphism). *Suppose S is a Riemann surface, with landmarks $\Gamma = \{p_1, p_2, \cdots, p_n\}$, $f : S \rightarrow S$ is a diffeomorphism homotopic to the identity map, preserving the landmarks, $f(p_i) = p_i$, $i = 1, 2, \ldots, n$, then we say f is a landmark preserving automorphism.*

All the landmark preserving automorphisms form a group, denoted as $G(S \backslash \Gamma)$.

Definition 4 (Teichmüller Trivial). *Suppose μ is a Beltrami differential for a Riemann surface with landmarks, if f^μ is a landmark preserving automorphism, then μ is called Teichmüller equivalent to 0, or Teichmüller trivial, denoted as $\mu \sim 0$.*

The group of landmark preserving automorphism is isomorphic to the space of Teichmüller trivial Beltrami differentials,

$$G(S \backslash \Gamma) \cong \{\|\mu\|_\infty < 1, \mu \sim 0\}.$$

In practice, it is difficult to compute Teichmüller trivial Beltrami differential directly. Instead, we seek for infinitesimally teichmüller trivial differentials.

Definition 5 (Infinitesimal Teichmüller Equivalence). *Two Beltrami differentials μ and ν are called infinitesimally equivalent if $\forall \phi \in \Omega(S)$ with $\|\phi\| = 1$,*

$$\int_S \mu\phi = \int_S \nu\phi \tag{7}$$

The space of Beltrami differentials infinitesimally Teichmüller equivalent to ν is given by

$$T^\nu(S) := \left\{ \mu : \|\mu\|_\infty < 1, \ \int_S (\mu - \nu)\phi = 0, \forall \phi \in \Omega(S) \right\}. \tag{8}$$

Geometrically, if $\mu \in T^\nu(S)$ is infinitesimally Teichmüller equivalent to ν, then when $t \to 0$

$$f^{\nu+t(\mu-\nu)}(p_i) = f^\nu(p_i) + o(t), \ \forall p_i \in \Gamma.$$

Teichmüller Map. In general cases, the Teichmüller map is the one that minimizes the angle distortion.

Definition 6 (Extremal Map). *Let $f : S_1 \to S_2$ be a quasiconformal map between S_1 and S_2. f is said to be an extremal mapping if for any quasiconformal map $h : S_1 \to S_2$ isotopic to f relative to the boundary,*

$$K(f) \le K(h) \tag{9}$$

where $K(f) = (1 + \|\mu\|_\infty)/(1 - \|\mu\|_\infty)$ is the maximal dilation. It is uniquely extremal if the inequality in Eq. 9 is strict when $h \neq f$.

Definition 7 (Teichmüller Map). *Let $f : S_1 \to S_2$ be a quasiconformal map. f is said to be a Teichmüller map associated to the holomorphic quadratic differential ϕdz^2, if its associated Beltrami differential is of the form*

$$\mu(f) = k \frac{\bar{\phi}}{|\phi|} \tag{10}$$

for some constant $k < 1$ and quadratic differential with $\|\phi\| < \infty$.

Under general conditions, the Teichmüller map is the extremal quasiconformal map within its homotopy class.

4   Algorithm

Based on previous theories, optimization in the space of landmark constrained diffeomorphisms can be simplified to the optimization in the space of infinitesimally equivalent Beltrami coefficients.

In this section, we first present the general procedure of optimization in the space of infinitesimally equivalent Beltrami coefficients. Then we present an algorithm to compute Teichmüller map as a showcase of this general procedure.

Given $f_0 : \mathbb{D} \to \mathbb{D}$, with constraints $f_0(p_i) = q_i, p_i \in \Gamma$. The Beltrami coefficient of f_0 is μ_0, the problem we consider is

$$\min \qquad E(f, \mu)$$

$$\text{s.t. } \int_{\mathbb{D}} \mu \phi_j = \int_{\mathbb{D}} \mu_0 \phi_j, \forall \, \phi_j \in \Omega(\mathbb{D}) \tag{11}$$

where $\mu(z) = f_{\bar{z}}/f_z$.

In discrete setting, the domain $\mathbb{D}$ is represented as a triangle mesh $D = \bigcup \triangle_i$. The Beltrami coefficient is represented as a piecewise constant function on the triangle mesh $\mu = \sum \mu_i \triangle_i = (\mu_1, \mu_2, \cdots, \mu_T)^T$. The infinitesimal equivalence condition can be discretized as

$$\sum_i \mu_i a_{ji} = \sum_i \mu_{0(i)} a_{ji}$$

where $a_{ji} = \int_{\triangle_i} \phi_j$. These are linear constraints on μ.

Beside the constraints in 11, it's usually desirable to add more constraints in order to ensure μ as well as corresponding f have desired properties. For example, $|\mu| < 1$ is a common constraint to add to ensure the resulting map to be bijective. The energy E can have various forms depending on the diffeomorphism we want. Usually, we derive the energy form E based on properties of corresponding map f, which is the most challenging part of our algorithm.

The optimization problem can be solved iteratively. From initial μ_0, we solve the minimization problem 11 with either linear programming or quadratic programming to obtain $\nu = \arg\min_\mu E(f, \mu)$. Next, we have to ensure ν is indeed Beltrami coefficient of some f. We can solve the Beltrami equation to obtain f, or in some cases obtain f by closed form [12]. In either way, the f we obtained may slightly move the landmark. We can diffuse f locally to restore landmark constraints. Then we solve the minimization problem again to obtain a new ν. This procedure is performed iteratively until the optimal μ is attained as well as the optimal diffeomorphism f. Based on our experiments, this iterative procedure usually converges in a few iterations. For the Teichmüller problem we will introduce later, it's proved that the iterative procedure converges to a given precision in finite steps [12].

If landmarks and their targets are too far away, it's advisable to move landmarks to targets gradually. For a sequence of initial map $\{f_0^t, t = 1, 2, \cdots, T\}$, $f_0^t(p_i) = \frac{T-t}{T} p_i + \frac{t}{T} q_i$. We apply above optimization procedure for f_0^t to obtain f^t, then use f^t to initialize f_0^{t+1} by diffusing landmarks to next positions.

The algorithm is summarised as in Algorithm 1:

In Fig. 4, we illustrate the optimization within the space of landmark preserving diffeomorphisms. The initial map f_0 takes left to the middle with landmark constraints. For the map f from left to right, its Beltrami coefficient μ is infinitesimal equivalent to μ_0 of f_0. The landmark constraints are preserved.

Algorithm 1: Optimization in infinitesimal equivalence space

Result: Optimized μ and f
take some f_0 and μ^0;
while *stop criteria is not satisfied* **do**
    solve problem 11 to obtain μ;
    solve Beltrami equation 4 to obtain f;
    restore landmark constraints for f;
    compute $\mu = f_{\bar{z}}/f_z$;
    let $\mu^0 = \mu$;
end

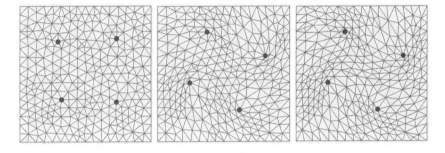

Fig. 4. Infinitesimal equivalence

4.1   Harmonic Map

Let $f : \mathbb{D} \to \mathbb{D}$ be a map that takes some feature points $\Gamma = \{p_1, p_2, \cdots, p_k\}$ to some target locations $Q = \{q_1, q_2, \cdots, q_k\}$, The harmonic energy of f can be defined as

$$E(f) = \int_{\mathbb{D}} (|f_z|^2 + |f_{\bar{z}}|^2) dz d\bar{z}$$

Without landmark constraints, by variational principle, the Euler–Lagrange equation for $E(f)$ is

$$\Delta f = 0 \tag{12}$$

This Laplace equation can be solved together with some boundary conditions. However with landmark constraints, if we simply enforce those landmark constraints, the solution to 12 generally is not diffeomorphic. The harmonic map with landmark constraints can be solved in the proposed space of landmark preserving diffeomorphisms.

Since $f_{\bar{z}} = \mu f_z$, we obtain

$$E(f, \mu) = \int_{\mathbb{D}} |f_z|^2 (1 + |\mu|^2) dz d\bar{z}$$

The map f is represented as a piecewise linear function, thus f_z on each triangle $\triangle_i$ is constant and so is μ. So the energy can be integrated as

$$E(f, \mu) = \sum_i |f_z|_i^2 (1 + |\mu_i|^2) A_i$$

where A_i is the triangle area of triangle $\triangle_i$. From initial map f, we can compute f_z and μ, then we optimize μ and f using Algorithm 1. Figure 5 shows an initial map with landmark constraints and harmonic map obtained from Algorithm 1.

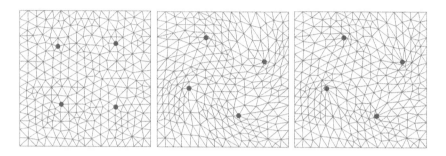

Fig. 5. Harmonic map. Left is a surface with landmarks. The initial map takes left to middle with landmark constraints. The harmonic map is from left to right.

4.2   Teichmüller Map

We apply previous general procedure to compute Teichmüller map via infinitesimal approach. Let $\mathbb{D}$ be extended complex plane or unit disk. Given an initial map f_0, which takes some feature points $\Gamma = \{p_1, p_2, \cdots, p_k\}$ to some target locations $Q = \{q_1, q_2, \cdots, q_k\}$, we find the Teichmüller map in the homotopy class of f_0 which preserves the feature points.

The quadratic differentials on $\mathbb{D}$ have closed form 6. Based on Teichmüller theory, we should minimize the L_∞ norm of μ.

$$\min E(f, \mu) = \|\mu\|_\infty$$

$$s.t. \int_{\mathbb{D}} \mu \phi_i = \int_{\mathbb{D}} \mu_0 \phi_i, \forall \; \phi_i \in \Omega(\mathbb{D}, \Gamma) \tag{13}$$

Quadratic differential $\phi_j(z)$ has a simple pole at p_j and analytic elsewhere, so it's integrable on every triangle and on $\mathbb{D}$. We denote the integral as

$$A = (a_{ji}) = \int_{\triangle_i} \phi_j \tag{14}$$

We further separate the real part and imaginary part and obtain

$$\begin{aligned} A_r x - A_i y &= b_r \\ A_i x + A_i y &= b_i \end{aligned} \tag{15}$$

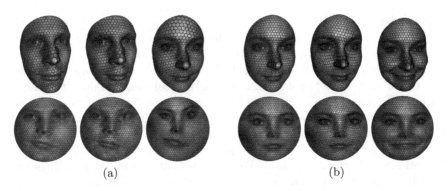

Fig. 6. The Teichmüller map between two human face surfaces (a) and the Teichmüller map between human face surfaces with neutral and smiling expressions (b).

where $\mu_i = x_i + iy_i, A = A_r + iA_i, b = b_r + ib_i = A\mu_0$. The L_∞ minimization problem can be solved by introducing an auxiliary variable z with constraints $|\mu_i|^2 = x_i^2 + y_i^2 \leq z$. So we get a equivalent minimization problem

$$
\begin{aligned}
\min \quad & z \\
s.t \quad & x_i^2 + y_i^2 < z \\
& A_r x - A_i y = b_r \\
& A_i x + A_i y = b_i
\end{aligned}
\tag{16}
$$

which is a linear programming problem with quadratic constraints. It can be efficiently solved using e.g., interior point method.

5   Experiment Results

Fig. 7. Heat map of L^∞-norm of μ. The first and third columns correspond to initial mappings and the second and fourth correspond to the optimized Teichmüller maps.

We show the optimized diffeomorphisms between 3D human faces. We manually select a few landmarks at the nose, eye corner and mouth corner to ensure meaningful matching. For 3D surfaces, we first conformally map them to the 2D unit disk. Then we compute the Teichmüller map mapping the unit disk to itself with prescribed landmark constraints. Figure 6(a) and Fig. 6(b) show the Teichmüller maps. In the two figures, the left and right columns are original surfaces textured with circle patterns. In the middle columns, we draw the ellipses which are deformed from each circle by the mappings. Note that in either case, all ellipses have the same eccentricities since Teichmüller map has the property of having the same $|\mu|$ almost everywhere (see Eq. 10).

In Fig. 7 we plot the L^∞-norm of μ corresponding to the initial maps and the optimized Teichmüller maps. We observe that the initial mappings have large distortions near landmark constraints, while in the Teichmüller maps the distortion is uniformly smoothed.

Our experiments are conducted on a desktop computer with Intel i7 4.3 GHz CPU. The human face surfaces in Fig. 6(a) and Fig. 6(b) have 40K triangles. The optimizations converge within 20 iterations and take 32 s. The convergence of optimization is shown in Fig. 8.

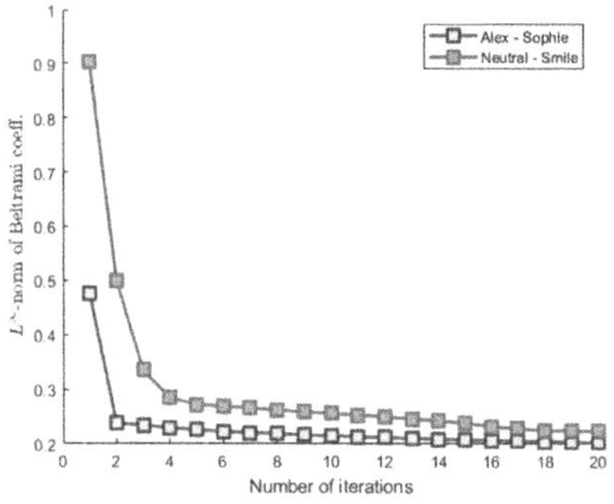

Fig. 8. Convergence of the optimization.

6   Conclusion

This work proposes a model of the space of diffeomorphisms with landmark constraints. The generators of the space consists of infinitesimal Teichmüller trival Beltrami coefficients. The harmonic map and Teichmüller maps can be obtained by solving convex optimizations with quadratic programming. The method is rigorous and practical. The experimental results demonstrate the efficacy and efficiency of the proposed method.

Acknowledgements. The project is partially supported by NSF CMMI-1762287, NSF DMS-1737812 and Ford URP.

References

1. Ahlfors, L., Bers, L.: Riemann's mapping theorem for variable metrics. Annals of Mathematics, pp. 385–404 (1960)
2. Ahlfors, L.V.: Lectures on quasiconformal mappings, vol. 38. American Mathematical Society (2006)
3. Aigerman, N., Poranne, R., Lipman, Y.: Seamless surface mappings. ACM Trans. Graphics (TOG) **34**(4), 72 (2015)
4. Allen, B., Curless, B., Curless, B., Popović, Z.: The space of human body shapes: reconstruction and parameterization from range scans. In: ACM Transactions on Graphics (TOG), vol. 22, pp. 587–594. ACM (2003)

5. Amberg, B., Romdhani, S., Vetter, T.: Optimal step nonrigid ICP algorithms for surface registration. In: 2007 IEEE Conference on Computer Vision and Pattern Recognition, pp. 1–8. IEEE (2007)
6. Besl, P., McKay, H.: A method for registration of 3-d shapes. In: IEEE Transactions on IEEE Trans Pattern Anal Mach Intelligent Pattern Analysis and Machine Intelligence, vol. 14, pp. 239–256, March 1992. https://doi.org/10.1109/34.121791
7. Choi, P.T., Lam, K.C., Lui, L.M.: Flash: fast landmark aligned spherical harmonic parameterization for genus-0 closed brain surfaces. SIAM J. Imag. Sci. **8**(1), 67–94 (2015)
8. Daripa, P.: A fast algorithm to solve the beltrami equation with applications to quasiconformal mappings. J. Computat. Phys. **106**(2), 355–365 (1993)
9. Floater, M.S., Hormann, K.: Surface parameterization: a tutorial and survey. In: Dodgson, N.A., Floater, M.S., Sabin, M.A. (eds.) Advances in multiresolution for geometric modelling, pp. 157–186. Springer, Heidelberg (2005). https://doi.org/10.1007/3-540-26808-1_
10. Gardiner, F.P.: Teichmüller theory and quadratic differentials. Bull. Amer. Math. Soc **19**, 494–498 (1988)
11. Gardiner, F.P., Lakic, N.: Quasiconformal teichmuller theory. No. 76, American Mathematical Soc. (2000)
12. Goswami, M., Gu, X., Pingali, V.P., Telang, G.: Computing teichmüller maps between polygons. Found. Comput. Math. **17**(2), 497–526 (2017)
13. Gu, X., Wang, Y., Chan, T.F., Thompson, P.M., Yau, S.T.: Genus zero surface conformal mapping and its application to brain surface mapping. IEEE Trans. Med. Imag. **23**(8), 949–958 (2004)
14. Haehnel, D., Thrun, S., Burgard, W.: An extension of the ICP algorithm for modeling nonrigid objects with mobile robots. IJCAI **3**, 915–920 (2003)
15. He, Z.X.: Solving beltrami equations by circle packing. Trans. Am. Math. Soc. **322**(2), 657–670 (1990)
16. Huang, P., Budd, C., Hilton, A.: Global temporal registration of multiple non-rigid surface sequences. In: CVPR 2011, pp. 3473–3480. IEEE (2011)
17. Jin, M., Kim, J., Gu, X.D.: Discrete surface Ricci Flow: theory and applications. In: Martin, R., Sabin, M., Winkler, J. (eds.) Mathematics of Surfaces 2007. LNCS, vol. 4647, pp. 209–232. Springer, Heidelberg (2007). https://doi.org/10.1007/978-3-540-73843-5_13
18. Jin, M., Kim, J., Luo, F., Gu, X.: Discrete surface Ricci flow. IEEE Trans. Visual. Comput. Graph. **14**(5), 1030–1043 (2008)
19. Jin, M., Zeng, W., Luo, F., Gu, X.: Computing tëichmuller shape space. IEEE Trans. Visual. Comput. Graph. **15**(3), 504–517 (2009)
20. Ju, L., Stern, J., Rehm, K., Schaper, K., Hurdal, M., Rottenberg, D.: Cortical surface flattening using least square conformal mapping with minimal metric distortion. In: IEEE International Symposium on Biomedical Imaging: Nano to Macro, 2004, pp. 77–80. IEEE (2004)
21. Lévy, B., Petitjean, S., Ray, N., Maillot, J.: Least squares conformal maps for automatic texture atlas generation. In: ACM Transactions on Graphics (TOG), vol. 21, pp. 362–371. ACM (2002)
22. Lui, L.M., Lam, K.C., Yau, S.T., Gu, X.: Teichmuller mapping (t-map) and its applications to landmark matching registration. SIAM J. Imag. Sci. **7**(1), 391–426 (2014)
23. Lui, L.M., Thiruvenkadam, S., Wang, Y., Thompson, P.M., Chan, T.F.: Optimized conformal surface registration with shape-based landmark matching. SIAM J. Imag. Sci. **3**(1), 52–78 (2010)

24. Lui, L.M., Wang, Y., Chan, T.F., Thompson, P.: Landmark constrained genus zero surface conformal mapping and its application to brain mapping research. Appl. Numer. Math. **57**(5–7), 847–858 (2007)

25. Lui, L.M., Wong, T.W., Thompson, P., Chan, T., Gu, X., Yau, S.T.: Compression of surface registrations using beltrami coefficients. In: 2010 IEEE Computer Society Conference on Computer Vision and Pattern Recognition, pp. 2839–2846. IEEE (2010)

26. Lui, L.M., Wong, T.W., Thompson, P., Chan, T., Gu, X., Yau, S.-T.: Shape-based diffeomorphic registration on hippocampal surfaces using beltrami holomorphic flow. In: Jiang, T., Navab, N., Pluim, J.P.W., Viergever, M.A. (eds.) MICCAI 2010. LNCS, vol. 6362, pp. 323–330. Springer, Heidelberg (2010). https://doi.org/10.1007/978-3-642-15745-5_40

27. Lui, L.M., et al.: Optimization of surface registrations using beltrami holomorphic flow. J. Sci. Comput. **50**(3), 557–585 (2012)

28. Ma, M., Marino, J., Nadeem, S., Gu, X.: Supine to prone colon registration and visualization based on optimal mass transport. Graph. Models **104**, 101031 (2019)

29. Mastin, C.W., Thompson, J.F.: Discrete quasiconformal mappings. Zeitschrift für angewandte Mathematik und Physik ZAMP **29**(1), 1–11 (1978)

30. Matabosch, C., Salvi, J., Pinsach, X., Pag, J.: A comparative survey on free-form surface registration. Image and Vision Computing, pp. 308–312 (2004)

31. Ng, T.C., Gu, X., Lui, L.M.: Computing extremal teichmüller map of multiply-connected domains via beltrami holomorphic flow. J. Sci. Comput. **60**(2), 249–275 (2014)

32. Nian, X., Chen, F.: Planar domain parameterization for isogeometric analysis based on teichmüller mapping. Comput. Meth. Appl. Mech. Eng. **311**, 41–55 (2016)

33. Rusinkiewicz, S., Levoy, M.: Efficient variants of the ICP algorithm. In: Proceedings Third International Conference on 3-D Digital Imaging and Modeling, pp. 145–152 (2001)

34. Salzmann, M., Moreno-Noguer, F., Lepetit, V., Fua, P.: Closed-form solution to non-rigid 3D surface registration. In: Forsyth, D., Torr, P., Zisserman, A. (eds.) ECCV 2008. LNCS, vol. 5305, pp. 581–594. Springer, Heidelberg (2008). https://doi.org/10.1007/978-3-540-88693-8_43

35. Sheffer, A., Praun, E., Rose, K., et al.: Mesh parameterization methods and their applications. Found. Trends® Comput. Graph. Vis. **2**(2), 105–171 (2007)

36. Shi, R., et al.: Hyperbolic harmonic mapping for constrained brain surface registration. In: Proceedings of the IEEE Conference on Computer Vision and Pattern Recognition, pp. 2531–2538 (2013)

37. Su, K., et al.: Area-preserving mesh parameterization for poly-annulus surfaces based on optimal mass transportation. Comput. Aided Geometric Des. **46**, 76–91 (2016)

38. Tam, G.K., et al.: Registration of 3d point clouds and meshes: a survey from rigid to nonrigid. IEEE Trans. Visual. Comput. Graph. **19**(7), 1199–1217 (2012)

39. Wang, S., Wang, Y., Jin, M., Gu, X.D., Samaras, D.: Conformal geometry and its applications on 3D shape matching, recognition, and stitching. IEEE Trans. Pattern Anal. Mach. Intell. **7**, 1209–1220 (2007)

40. Wang, Y., Chiang, M.C., Thompson, P.M.: Mutual information-based 3d surface matching with applications to face recognition and brain mapping. In: Tenth IEEE International Conference on Computer Vision (ICCV 2005), vol. 1, pp. 527–534. IEEE (2005)

41. Wang, Y., Dai, W., Gu, X., Chan, T., Toga, A., Thompson, P.: Studying brain morphology using teichmüller space theory. In: IEEE 12th International Conference on Computer Vision, ICCV. pp. 2365–2372 (2009)
42. Wang, Y., Lui, L.M., Chan, T.F., Thompson, P.M.: Optimization of brain conformal mapping with landmarks. In: Duncan, J.S., Gerig, G. (eds.) MICCAI 2005. LNCS, vol. 3750, pp. 675–683. Springer, Heidelberg (2005). https://doi.org/10.1007/11566489_83
43. Weber, O., Myles, A., Zorin, D.: Computing extremal quasiconformal maps. In: Computer Graphics Forum. vol. 31, pp. 1679–1689. Wiley Online Library (2012)
44. Wen, C., et al.: Surface foliation based brain morphometry analysis. In: Zhu, D., et al. (eds.) MBIA/MFCA -2019. LNCS, vol. 11846, pp. 186–195. Springer, Cham (2019). https://doi.org/10.1007/978-3-030-33226-6_20
45. Wong, T.W., Zhao, H.K.: Computation of quasi-conformal surface maps using discrete beltrami flow. SIAM J. Imag. Sci. **7**(4), 2675–2699 (2014)
46. Zeng, W., Gu, X.D.: Registration for 3d surfaces with large deformations using quasi-conformal curvature flow. In: CVPR 2011, pp. 2457–2464. IEEE (2011)
47. Zeng, W., Luo, F., Yau, S.-T., Gu, X.D.: Surface quasi-conformal mapping by solving beltrami equations. In: Hancock, E.R., Martin, R.R., Sabin, M.A. (eds.) Mathematics of Surfaces 2009. LNCS, vol. 5654, pp. 391–408. Springer, Heidelberg (2009). https://doi.org/10.1007/978-3-642-03596-8_23
48. Zeng, W., Marino, J., Gurijala, K.C., Gu, X., Kaufman, A.: Supine and prone colon registration using quasi-conformal mapping. IEEE Trans. Visual. Comput. Graph. **16**(6), 1348–1357 (2010)
49. Zeng, W., Samaras, D., Gu, D.: Ricci flow for 3d shape analysis. IEEE Trans. Pattern Anal. Mach. Intell. **32**(4), 662–677 (2010)
50. Zeng, W., Shi, R., Wang, Y., Yau, S.T., Gu, X., Initiative, A.D.N., et al.: Teichmüller shape descriptor and its application to alzheimer's disease study. Int. J. Comput. Vis. **105**(2), 155–170 (2013)
51. Giovinazzi, S., et al.: Towards a decision support tool for assessing, managing and mitigating seismic risk of electric power networks. In: Gervasi, O., et al. (eds.) ICCSA 2017. LNCS, vol. 10406, pp. 399–414. Springer, Cham (2017). https://doi.org/10.1007/978-3-319-62398-6_28
52. Zeng, W., Yang, Y.J., Razib, M.: Graph-constrained surface registration based on tutte embedding. In: Proceedings of the IEEE Conference on Computer Vision and Pattern Recognition Workshops, pp. 76–83 (2016)
53. Zeng, W., Yin, X., Zeng, Y., Lai, Y., Gu, X., Samaras, D.: 3D face matching and registration based on hyperbolic Ricci flow. In: IEEE Computer Society Conference on Computer Vision and Pattern Recognition Workshops, 2008. CVPRW 2008. pp. 1–8. IEEE (2008)
54. Zeng, W., Yin, X., Zhang, M., Luo, F., Gu, X.: Generalized koebe's method for conformal mapping multiply connected domains. In: 2009 SIAM/ACM Joint Conference on Geometric and Physical Modeling, pp. 89–100. ACM (2009)
55. Zeng, Y., Wang, C., Wang, Y., Gu, X., Samaras, D., Paragios, N.: Intrinsic dense 3d surface tracking. In: CVPR 2011, pp. 1225–1232. IEEE (2011)
56. Zheng, X., Wen, C., Lei, N., Ma, M., Gu, X.: Surface registration via foliation. In: Proceedings of the IEEE Conference on Computer Vision and Pa ERN Recognition, pp. 938–947 (2017)

PieNet: Personalized Image Enhancement Network

Han-Ul Kim[1], Young Jun Koh[2]($\boxtimes$), and Chang-Su Kim[1]

[1] Korea University, Seoul, Korea
hanulkim@mcl.korea.ac.kr, changsukim@korea.ac.kr
[2] Chungnam National University, Daejeon, Korea
yjkoh@cnu.ac.kr

Abstract. Image enhancement is an inherently subjective process since people have diverse preferences for image aesthetics. However, most enhancement techniques pay less attention to the personalization issue despite its importance. In this paper, we propose the first deep learning approach to personalized image enhancement, which can enhance new images for a new user, by asking him or her to select about 10–20 preferred images from a random set of images. First, we represent various users' preferences for enhancement as feature vectors in an embedding space, called preference vectors. We construct the embedding space based on metric learning. Then, we develop the personalized image enhancement network (PieNet) to enhance images adaptively using each user's preference vector. Experimental results demonstrate that the proposed algorithm is capable of achieving personalization successfully, as well as outperforming conventional general image enhancement algorithms significantly. The source codes and trained models are available at https://github.com/hukim1124/PieNet.

Keywords: Image enhancement · Personalization · Metric learning

1 Introduction

Nowadays, people take photographs casually but are often unsatisfied with them. Photos may be noisy because of limited camera sensors. Also, photos taken in uncontrolled environments may suffer from low dynamic ranges or distorted color tones [25,32]. Thus, image enhancement is required, which post-processes and edits photographs to satisfy user preferences. Professional softwares provide many tools to support manual image enhancement. However, the results of manual enhancement depend on users' skills and experience. Moreover, the manual process requires lots of efforts.

Many researches have been carried out to perform image enhancement automatically. But, image enhancement is a non-trivial problem partly due to the

Electronic supplementary material The online version of this chapter (https://doi.org/10.1007/978-3-030-58577-8_23) contains supplementary material, which is available to authorized users.

© Springer Nature Switzerland AG 2020
A. Vedaldi et al. (Eds.): ECCV 2020, LNCS 12375, pp. 374–390, 2020.
https://doi.org/10.1007/978-3-030-58577-8_23

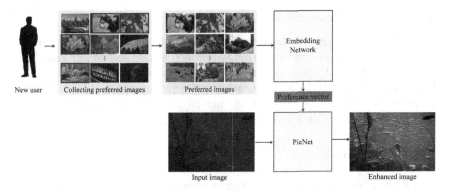

Fig. 1. Illustration of personalization for a new user. A user selects only about 10–20 preferred images from a random set of images. Then, the proposed algorithm analyzes the user's preference and enhances a new image according to the preferred style. Please see the supplemental video for this demonstration.

non-linear relationship between input and output images. Furthermore, it makes enhancement even more challenging that people have different preferences for images; image enhancement is a subjective process. In this regard, the deep-learning-based enhancement algorithms in [9, 19, 33, 35, 42, 45] have the common limitation that they cannot handle various user preferences. In [7, 24], personalized image enhancement systems have been developed. However, they are based on traditional enhancement techniques such as gamma-correction and S-curve. They may not yield output qualities that are as high as those of professionally enhanced images by experts.

We propose a deep learning algorithm for personalized image enhancement. First, we model diverse user preferences for image enhancement as feature vectors, called preference vectors, in an embedding space. More specifically, we perform metric learning to learn the embedding space, in which a preference vector conveys the preferred enhancement style of the corresponding user. Next, we propose a novel image enhancement network, referred to as PieNet, which employs the preference vector to achieve personalized enhancement. The proposed PieNet has an encoder-decoder architecture. The encoder part yields multi-scale features, representing local and global information for image enhancement. The decoder part includes personalized up-sample blocks, which employ the preference vector to produce personalized results. Experimental results demonstrate that the proposed algorithm outperforms the conventional deep learning algorithms [9, 14, 19, 35, 42] for general (*i.e.* non-personalized) image enhancement on the MIT-A5K dataset [5]. Moreover, it is shown that the proposed algorithm achieves personalization successfully. In particular, it is shown that the proposed algorithm achieves the personalization for a new user with the minimal effort of selecting only a few preferred images, as illustrated in Fig. 1.

This paper has three main contributions:

1. Development of PieNet to tackle the personalization issue in image enhancement, which is the *first* deep learning approach to the best of our knowledge.
2. Remarkable general image enhancement performance on MIT-Adobe 5K.
3. Excellent scalability of PieNet to achieve the personalization for a new user with the minimal effort of selecting only 10–20 preferred images. Please see the supplemental video for personalization demos.

2   Related Work

Image Enhancement: Early studies on image enhancement focused on improving image contrast. Histogram equalization [15] and its variants [2, 26, 29–31, 41, 44] modify the histogram of an image to improve its limited dynamic range. Also, retinex methods [6, 12, 13, 16, 20, 21, 43, 47] regard an image as the product of reflectance and illumination [28], and alter the illumination to enhance a poorly lit image. However, these methods may not reconstruct the complex mapping function between an image and its professionally enhanced version, edited by an expert.

An alternative approach to image enhancement is the data-driven one to learn the mapping between input and enhanced images from a large dataset. Bychkovsky *et al.* [5] introduced the MIT-A5K dataset, composed of 5,000 input and expert-retouched image pairs. They used the dataset to estimate mapping functions, based on regression schemes for predicting user adjustments. However, their method still may fail to reconstruct highly non-linear mapping functions between input and enhanced images.

Recently, motivated by the success of deep learning, several deep neural networks have been developed to deal with the non-linear image enhancement. Yan *et al.* [45] proposed a deep learning scheme, which uses image descriptors to predict a color mapping for each pixel. Lore *et al.* [33] developed a deep autoencoder to enhance low-light images. Gharbi *et al.* [14] proposed deep bilateral learning for real-time enhancement, which predicts local affine transforms in the bilateral space. Based on the retinex theory, Wang *et al.* [42] designed a deep network to predict an image-to-illumination mapping function instead of a direct image-to-image function. Also, to achieve unpaired learning for image enhancement, Park *et al.* [35] introduced the distort-and-recover approach that degraded high-quality images to generate pseudo paired data. Chen *et al.* [9] used two-way generative adversarial networks (GANs) for stable training. Deng *et al.* [11] developed an aesthetic-driven image enhancement algorithm. In [19, 46], an adversarial loss is integrated into reinforcement learning to learn to generate a sequence of enhancement operations. These deep learning algorithms provide promising performances, but are limited in that they do not consider different users' diverse preferences.

Personalization: Joshi *et al.* [23] proposed a personal photo enhancement algorithm, which uses a person's favorite photographs as examples to perform several tasks, including deblurring, super-resolution, and white-balancing. The proposed algorithm is, however, more related to the Kang *et al.*'s personalization

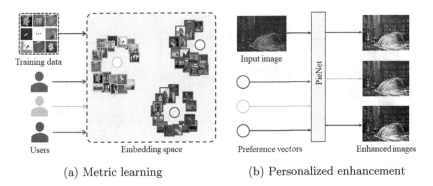

(a) Metric learning   (b) Personalized enhancement

Fig. 2. Overview of the proposed system, which has two stages. (a) Metric learning is performed to discover an embedding space, in which the characteristics of each user's preferred images are encoded into the preference vector. (b) PieNet employs the user's preference vector to yield personalized enhancement results.

system [24] for image enhancement. Their system asks a user to enhance 25 representative images by controlling a set of parameters. Given a new image, it finds the most 'similar' representative image. Then, it uses the corresponding set of parameters to enhance the new image. They adopted a metric learning scheme to define the 'similarity' between images such that it correlates well with the enhancement parameters. Caicedo *et al.* [7] extended the Kang *et al.*'s system to consider the enhancement results of other users with similar preferences based on collaborative filtering.

Compared to [7,24], the proposed algorithm has noticeable differences. First, while they demand a user to enhance training images by controlling parameters, the proposed algorithm requires a user to select only a few preferred images from a set of candidate images. Thus, the proposed algorithm needs much less user efforts. Second, we consider more complex mappings between input and enhanced images, by developing the *first* deep learning algorithm for personalized image enhancement.

Metric Learning: The objective of metric learning is to learn an embedding space, in which the distance between similar objects is shorter than the distance between dissimilar ones. In [4,10,17], a contrastive loss was employed to minimize the distances between objects in the same class, while constraining the distances between inter-class objects to be larger than a margin. Schroff *et al.* [37] proposed a triplet loss to encourage the distance between anchor and positive objects to be smaller than that between anchor and negative objects. To overcome slow convergence of the triplet loss, many extensions have been proposed [8,38–40].

Notice that, whereas the conventional personalized algorithms in [7,24] use metric learning to embed images with similar enhancement parameters tightly, the proposed algorithm performs it to directly embed preferred images of each user tightly and thus obtain the user's preference vector in the embedding space.

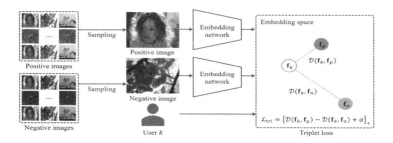

Fig. 3. Metric learning for the embedding space of preference vectors.

3 Proposed Algorithm

Figure 2 shows an overview of the proposed system for personalized image enhancement. First, we do metric learning to determine an embedding space, in which a user's preference for enhancing images is represented by a feature vector. We refer to this feature vector as the preference vector. Second, we develop PieNet that employs the preference vector to produce personalized enhanced results adaptively for the specific user.

3.1 Preference Vector

Let us consider an embedding space, in which a preference vector represents a user's preferred style for enhancement. We learn this embedding space to yield the preference vectors for multiple users based on metric learning. Each user provides two sets of preferred (positive) and non-preferred (negative) images. More specifically, we assume that there are N training images annotated by K users: $\{(I_i, \mathbf{y}_i)\}_{i=1}^{N}$, where I_i is the ith image and $\mathbf{y}_i = [y_{i1}, \ldots, y_{iK}]^T$ is its label vector. The kth element y_{ik} in $\mathbf{y}_i$ is 1 if user k likes I_i, and 0 otherwise.

Figure 3 illustrates how to determine the embedding space. We feed a pair of positive and negative images for each user into two identical embedding networks. Specifically, at each training iteration, we sample a triplet (k, I_p, I_n), where I_p and I_n are positive and negative images for user k, whose labels are $y_{p,k} = 1$ and $y_{n,k} = 0$. Each embedding network produces a 512-dimensional feature vector from an RGB color image of size 256×256. We employ ResNet-18 [18] as the twin embedding networks and perform L_2 normalization to each output feature vector.

Let $\mathbf{f}_p$ and $\mathbf{f}_n$ denote such feature vectors for positive and negative images, respectively, and $\mathbf{f}_k$ be the preference vector for user k. Then, we learn the embedding space, where the preference vector $\mathbf{f}_k$ is similar to the positive feature vector but dissimilar from the negative one. To this end, we compute the triplet loss [37], given by

$$\mathcal{L}_{\text{tri}}(\mathbf{f}_k, \mathbf{f}_p, \mathbf{f}_n) = \left[\mathcal{D}(\mathbf{f}_k, \mathbf{f}_p) - \mathcal{D}(\mathbf{f}_k, \mathbf{f}_n) + \alpha \right]_+ \tag{1}$$

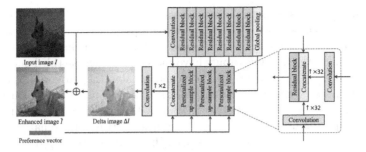

Fig. 4. Architecture of PieNet, composed of an encoder (top part) and a decoder (bottom part).

where $\mathcal{D}(\cdot)$ is the squared Euclidean distance, $[\cdot]_+$ is the rectifier, and α is a margin to be enforced between positive and negative images.

In the training phase, we initialize each preference vector with random values and the embedding networks with the pre-trained weights on ImageNet [36]. Then, we compute gradients to minimize the triplet loss,

$$\frac{\partial \mathcal{L}_{\mathrm{tri}}(\mathbf{f}_k, \mathbf{f}_p, \mathbf{f}_n)}{\partial \mathbf{f}_p} = 2(\mathbf{f}_p - \mathbf{f}_k) \cdot \mathbb{I}(\mathcal{L}_{\mathrm{tri}}(\mathbf{f}_k, \mathbf{f}_p, \mathbf{f}_n) > 0), \tag{2}$$

$$\frac{\partial \mathcal{L}_{\mathrm{tri}}(\mathbf{f}_k, \mathbf{f}_p, \mathbf{f}_n)}{\partial \mathbf{f}_n} = 2(\mathbf{f}_k - \mathbf{f}_n) \cdot \mathbb{I}(\mathcal{L}_{\mathrm{tri}}(\mathbf{f}_k, \mathbf{f}_p, \mathbf{f}_n) > 0), \tag{3}$$

$$\frac{\partial \mathcal{L}_{\mathrm{tri}}(\mathbf{f}_k, \mathbf{f}_p, \mathbf{f}_n)}{\partial \mathbf{f}_k} = 2(\mathbf{f}_n - \mathbf{f}_p) \cdot \mathbb{I}(\mathcal{L}_{\mathrm{tri}}(\mathbf{f}_k, \mathbf{f}_p, \mathbf{f}_n) > 0), \tag{4}$$

where $\mathbb{I}(\cdot)$ denotes the indicator function. Note that these gradients are back-propagated to update the preference vector $\mathbf{f}_k$ and the weight parameters in the embedding networks simultaneously. In this way, we determine the embedding space to yield the preference vectors for the K users, by employing the training images and their label vectors.

3.2 PieNet Architecture

Using the preference vectors, we perform personalized image enhancement. Figure 4 is the architecture of PieNet, which has an encoder and a decoder. The encoder takes an RGB color image as input. Its spatial resolution is 512×512. We also employ ResNet-18 to implement the encoder, which consists of one convolution layer, eight residual blocks, and one average pooling layer. The encoder extracts five multi-scale features from the convolution layer, the 2nd, 4th, 6th residual blocks, and the pooling layer, respectively. The intermediate features from the convolution layer and the residual blocks preserve detailed local information, while the global feature, extracted from the pooling layer, contains high-level information such as global brightness and scene category of the input image.

From the extracted features of the encoder, the decoder reconstructs a delta image ΔI, which is added to the input image I to enhance its quality. For the decoder, we develop the personalized up-sample block (PUB) to consider user preferences. In Fig. 4, each PUB takes three inputs: 1) the preference vector, 2) the output of the previous block, and 3) the intermediate feature of the encoder. It makes the first two inputs have the same size as the third one through convolution and up-sampling, and then concatenates the three data along the channel dimension. Then, the residual block in the PUB produces the output. By feeding the preference vector to every PUB, the decoder can satisfy the preferred style of the specific user. The output of the last PUB is concatenated with the output of the convolution layer in encoder. Then, the concatenated signal is up-sampled to be the same size as the input image and fed into the last convolution layer. The last convolution layer yields the delta image ΔI.

Finally, the enhanced image $\tilde{I}$ is obtained by adding the delta image ΔI to the input image I, given by

$$\tilde{I} = I + \Delta I. \tag{5}$$

Note that we predict the delta image instead of the enhanced image directly. This is because the down-sampling process in the network may lose image details. Even though the delta image loses some details, the enhanced image can restore those details from the input image.

3.3   PieNet Training

For user k, suppose that the preference vector $\mathbf{f}_k$ and an image pair (I, I_k^*) are available. Here, I_k^* is the ground-truth enhanced image that user k prefers to obtain from image I. We train PieNet using the preference vector and the image pair. Note that PieNet estimates a delta image ΔI_k and produces a personalized enhanced image $\tilde{I}_k$ via (5). We compare the estimated result with the ground-truth to train PieNet, by employing the loss function

$$\mathcal{L}(I, \Delta I_k, \tilde{I}_k, I_k^*) = \mathcal{L}_c(\tilde{I}_k, I_k^*) + \lambda_p \mathcal{L}_p(\tilde{I}_k, I_k^*) + \lambda_t \mathcal{L}_t(I, \Delta I_k) \tag{6}$$

where $\mathcal{L}_c$, $\mathcal{L}_p$, and $\mathcal{L}_t$ are color, perceptual, and total variation losses, respectively, and λ_p and λ_t are balancing parameters.

The color loss penalizes the mean absolute error between the predicted and ground-truth enhanced images, given by $\mathcal{L}_c(\tilde{I}_k, I_k^*) = \|\tilde{I}_k - I_k^*\|_1$. The perceptual loss [22] encourages the enhanced image and the ground-truth image to have similar features. Specifically, it is defined as $\mathcal{L}_p(\tilde{I}_k, I_k^*) = \|\tilde{\mathbf{f}}_k - \mathbf{f}_k^*\|_1$, where $\tilde{\mathbf{f}}_k$ and $\mathbf{f}_k^*$ denote the features for the estimated and ground-truth enhanced images, respectively, extracted from the embedding network in Sect. 3.1. Notice that the embedding network attempts to construct the embedding space, where the features of ground-truth enhanced images are compactly distributed near the preference vector. Hence, the perceptual loss constrains that the feature of the enhanced image should be near the preference vector.

Also, we use the total variation loss [1] to enforce the spatial smoothness of the delta image. To constrain neighboring pixels to exhibit similar delta values, the total variation loss is defined as

$$\mathcal{L}_t(I, \Delta I_k) = \|W_x \otimes \nabla_x(\Delta I_k)\|_1 + \|W_y \otimes \nabla_y(\Delta I_k)\|_1 \tag{7}$$

where $\otimes$ is the element-wise multiplication, and ∇_x and ∇_y denote the partial derivatives in the horizontal and vertical directions. Also, $W_x = \exp(-|\nabla_x I|)$ and $W_y = \exp(-|\nabla_y I|)$ are weight maps, which have large values in smooth regions in the input image. Thus, the total variation loss $\mathcal{L}_t$ imposes large penalties when neighboring pixels in smooth regions are assigned quite different delta values. On the contrary, near edges or complicated texture in the input image, delta values may be dissimilar from one another without causing large penalties.

3.4   Personalization for New Users

A critical issue in personalized enhancement is 'scalability,' which means the capability of accommodating the preference of a new user with minimal efforts. A straightforward approach is to repeat the entire training process, *i.e.* performing metric learning and training PieNet to consider a new user as well as the existing users. However, the fine-tuning of the embedding space and PieNet is a time-consuming process. Therefore, we decide the preference vector for the new user within the pre-trained embedding space and also use the pre-trained PieNet to produce personalized results. We consider two schemes to determine the preference vector for the new user.

The first scheme assumes that a new user provides two sets of positive and negative images. Then, given the pre-trained embedding space, the gradient in (4) is back-propagated to update the preference vector, while the other gradients in (2) and (3) are not back-propagated. Thus, we fix the embedding space while determining the preference vector. Then, using this preference vector, the pre-trained PieNet yields personalized enhanced images for the new user.

In the second scheme (which is computationally much simpler and is thus adopted in the default mode), a new user provides preferred images only. The pre-trained embedding network encodes these preferred images into feature vectors. Then, we determine the preference vector, by averaging the feature vectors. Note that the new user need not provide non-preferred images. Thus, the second scheme demands less user effort than the first scheme does. Moreover, it is shown in Sect. 4 that about 10 preferred images are sufficient for PieNet to yield desirable personalization results.

4   Experiments

4.1   Evaluation on MIT-Adobe 5K

Dataset and Metrics: We assess the proposed algorithm on the MIT-A5K dataset [5]. It consists of 5,000 input images, each of which was manually enhanced by five different photographers (A/B/C/D/E). Thus, there are five sets of 5,000 pairs of input and enhanced images, one set for each photographer. Among the 5,000 images, we randomly select 500 images to compose the test set as done in [9,42], and use the remaining 4,500 images as the training

Table 1. Comparison of the proposed algorithm with the conventional algorithms on MIT-Adobe 5K. For the 'single user' test, we use the photographer C's retouched images as the ground-truth. For the 'multiple users' test, we use the retouched images by the five photographers A/B/C/D/E as the ground-truth.

Method	Single user		Multiple users	
	PSNR	SSIM	mPSNR	mSSIM
WB [19]	18.36	0.810	17.83	0.799
D&R [35]	20.97	0.841	18.65	0.834
HDR [14]	23.44	0.882	21.64	0.872
DPE [9]	22.34	0.873	21.09	0.858
DUPE [42]	23.61	0.887	21.74	0.881
Proposed	**25.28**	**0.908**	**24.28**	**0.907**

set. For quantitative assessment, we employ PSNR and SSIM, which measure, respectively, color and structural similarity between predicted and ground-truth enhanced images.

Implementation Details: We jointly train the embedding network and the preference vectors for photographers A/B/C/D/E using the training set in MIT-Adobe 5K. For instance, when the preference vector for photographer A is trained, we regard his or her retouched images as the positive set, but the input images and the other photographers' retouched images as the negative set. This process is carried out for the other photographers similarly. We minimize the triplet loss in (1) using the Adam optimizer [27] with a learning rate of 1.0×10^{-4}. The training is iterated for 25,000 mini-batches, each of which includes 64 triplets. For data augmentation, we randomly rotate image pairs by multiples of 90 degrees. The margin α in (1) is set to 0.2.

To train PieNet, we use all image pairs in the training set, *i.e.* the five sets of image pairs for photographers A/B/C/D/E. In other words, we train PieNet for all five photographers using their preference vectors and image pairs. We also use the Adam optimizer to minimize the loss function in (6) with a learning rate of 1.0×10^{-4} for 100,000 mini-batches. The mini-batch size is 8. We randomly rotate images by multiples of $90°$. Also, we randomly perturb the preference vectors to make PieNet insensitive to small perturbations. Specifically, we add noise $\mathbf{n}$ to the preference vectors, where $\mathbf{n}$ is sampled from the hypersphere, $\|\mathbf{n}\|_2 = 0.1$. The parameters λ_p and λ_t in (6) are fixed to 0.4 and 0.01, respectively.

Experimental Results: Table 1 compares the proposed algorithm with the recent state-of-the-art algorithms in [9,14,19,35,42]. We obtain the results of the conventional algorithms using the source codes and parameters, provided by the respective authors. Note that these conventional algorithms are for general (*i.e.* non-personalized) image enhancement. Specifically, they attempt to mimic the retouching of photographer C only. In contrast, the proposed PieNet can provide enhanced images in five different styles using the preference vectors of photographers A/B/C/D/E.

In the 'single user' test, we compare the proposed algorithm with the conventional algorithms using the photographer C's enhanced images as the ground-truth. For this test, we only use the training images, retouched by C, to train the proposed algorithm for a fair comparison. WB [19] and D&R [35] provide poor performance than the other algorithms, since they use unpaired images for training. The proposed algorithm significantly outperforms the conventional algorithms [9,14,42], which conduct supervised learning directly using pairs of input and enhanced images. The proposed algorithm provides excellent performance due to two main factors. First, we adopt the effective network architecture. Second, combining different losses in (6) further improves the performance. Especially, we find that the perceptual loss $\mathcal{L}_\mathrm{p}$, based on the embedding network in Sect. 3.1, leads to a notable PSNR improvement.

The 'multiple users' test analyzes the personalization performance. The proposed algorithm can produce differently enhanced results according to the preference vectors. In this test, it yields personalized enhanced results for photographers A/B/C/D/E. We compare the personalized results with the corresponding ground-truth to compute PSNR and SSIM for each photographer, and then compute the average PSNR (mPSNR) and average SSIM (mSSIM) over the five photographers. In contrast, since the conventional algorithms provide only one enhanced result for an input image, we use the same enhanced result to compute PSNR and SSIM for each photographer.

By comparing the 'single user' and 'multiple users' tests, we see that the conventional algorithms experience significant degradation in the performance. This is because the conventional algorithms are designed to mimic the retouching style of photographer C only. Thus, they provide low PSNR and SSIM scores when compared with the other photographers' ground-truth. In contrast, the proposed algorithm experiences only minor degradation. It is worth pointing out that the proposed personalization performances in 'multiple users' even surpass the performances of all conventional algorithms in 'single user,' whose scores are computed for C only.

Notice that the proposed algorithm can produce personalized results for all five photographers without any additional training of PieNet, by changing only the preference vectors. On the contrary, for the conventional algorithms to provide reliable results for photographers A/B/D/E, they should retrain their networks four more times for the adaptation. The proposed algorithm hence achieves personalization more efficiently than the conventional algorithms.

Figure 5 compares the proposed algorithm with HDR, DPE, and DUPE qualitatively. In Fig. 5(b)–(d), these conventional algorithms provide reasonable results. They are designed to yield similar color tones to the photographer C's retouched image in Fig. 5(h). However, image enhancement is subjective, and photographer B prefers a different output in Fig. 5(f). As shown in Fig. 5(e) and (g), the proposed algorithm adaptively produces output images in B and C's styles, respectively. More experiments are available in the supplementary materials.

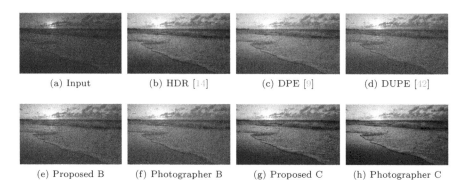

(a) Input	(b) HDR [14]	(c) DPE [9]	(d) DUPE [12]
(e) Proposed B	(f) Photographer B	(g) Proposed C	(h) Photographer C

Fig. 5. Qualitative comparison of enhanced images.

4.2 Personalization

Dataset and Metrics: We expand the MIT-A5K dataset to evaluate the personalization performances of the proposed algorithm for new users other than photographers A/B/C/D/E. However, it is an expensive task to collect 5,000 ground-truth images that are manually retouched by each new user. Therefore, for the expanded dataset, we enhance images using 28 conventional methods (including photographers A/B/C/D/E in [5] and predefined settings in Adobe Lightroom), instead of employing people to enhance images manually. These methods are regarded as users. Then, we divide the expanded dataset into the training and test sets:

- **Training (20 users):** 11 presets in Adobe Lightroom, 5 conventional methods [2,3,6,12,43], Photographers A/B/C/D [5]
- **Test (8 users):** 4 presets in Adobe Lightroom, 3 conventional methods [13,16,30], Photographer E [5]

Note that the enhancement methods in the training and test sets do not overlap. For each method in the training set, there are 4,500 pairs of input and enhanced images for training the embedding space and PieNet. On the other hand, for each method in the test set, there are 500 pairs of input and enhanced images, which are used to assess the personalization performances of the proposed algorithm.

Implementation Details: We train the embedding space and PieNet using the pairs of input and enhanced images in the training set. We use the same training settings in Sect. 4.1. In the test phase, we regard the methods in the test set as new users. Then, we compute the preference vector for each new user using the two schemes in Sect. 3.4: 1) triplet loss minimization and 2) feature vector average. For the first scheme, each method regards its enhanced images as positive or preferred images, while considering enhanced images of the other methods as negative images. For the second schemes, each method uses its enhanced images as preferred images.

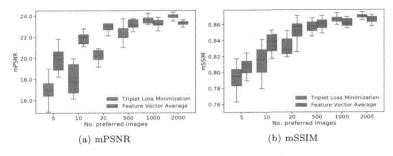

(a) mPSNR

(b) mSSIM

Fig. 6. Bar plots of the mPSNR and mSSIM scores for the 8 new users in the expanded MIT-A5K dataset. In each plot, we repeat ten experiments.

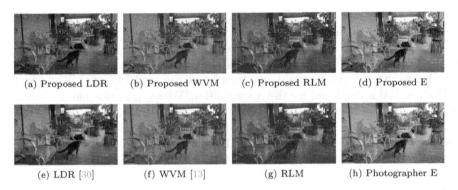

(a) Proposed LDR   (b) Proposed WVM   (c) Proposed RLM   (d) Proposed E

(e) LDR [30]   (f) WVM [13]   (g) RLM   (h) Photographer E

Fig. 7. Qualitative comparison of personalized enhancement results of the proposed algorithm with the ground-truth.

Experimental Results: Figure 6 shows the mPSNR and mSSIM scores for the 8 new users according to the number N_{pref} of preferred images, which are used to determine preference vectors. For the first scheme 'triplet loss minimization', we use three times as many negative images as preferred images. The preferred images and negative images are randomly selected 10 times for plotting the bar graphs. Given sufficiently many preferred images ($N_{pref} = 2,000$) to compute the preference vectors, the first scheme outperforms the second scheme. However, when there are only a limited number of preferred images, it experiences severe performance degradation. In contrast, the second scheme provides reliable personalization performances, even when only 20 preferred images are available. Therefore, in the following tests, we use the second scheme with $N_{pref} = 20$.

Figure 7 shows personalization results. The top row presents personalized enhancement results, while the bottom row shows the corresponding ground-truth generated by the four enhancement methods in the test set. Note that these four ground-truth images are considerably different from one another. For instance, photographer E in Fig. 7(h) boosts the brightness of the image, while the RedLiftMatte (RLM) preset in Lightroom in Fig. 7(g) renders the floor in different color tones. Nevertheless, the proposed algorithm successfully reflects

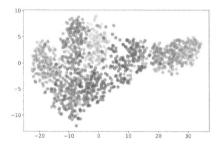

Fig. 8. Visualization of the embedding space. (Color figure online)

the tendency of each method in Fig. 7(e)–(h). The supplementary materials provide more personalization results.

Figure 8 visualizes the embedding space, in which the feature vectors of ground-truth images of the 8 test users are depicted as dots in different colors. For each user, 100 ground-truth images are randomly selected. The t-SNE technique [34] is employed for this visualization. We see that the feature vectors for the images are well clustered according to the users, which indicates that the embedding space is suitable for representing the preferred image styles of users. For the same reason, in this embedding space, we can easily construct the preference vector of a new user by averaging the feature vectors of only a few preferred images.

4.3   User Study

We conducted a user study with 10 participants to assess the personalization performance of the proposed algorithm for real people. It was designed as follows.

1. Each participant selected 20 preferred images from various pre-enhanced images in the test set of the expanded MIT-A5K dataset.
2. The proposed algorithm generated the preference vector for each participant using those preferred images.
3. Each participant was presented with six enhanced results of the same photograph, obtained by the proposed algorithm and the conventional algorithms [9,14,19,35,42], and was asked to vote for the algorithm yielding the most pleasing result. The photograph was also selected from the test set, but not used to generate the preference vector. This was repeated for 10 photographs.

For this user study, we pre-trained the proposed algorithm using the training set of the expanded dataset. All conventional algorithms were trained using the photographer E's retouched images, since it was the most often selected by the participants as their preferred method. Table 2 summarizes the voting results. The proposed algorithm gets the most votes by providing personalized results to each participant effectively.

Table 2. User study results. N_{vote} is the number of votes that each method gets.

	HDR [14]	DPE [9]	WB [19]	D&R [35]	DUPE [42]	Proposed
N_{vote}	101	64	17	2	71	245

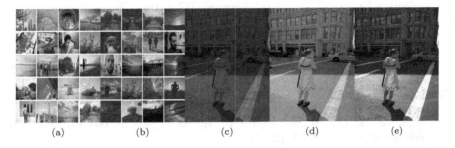

(a)   (b)   (c)   (d)   (e)

Fig. 9. Personalization results for two participants P1 and P2 in the user study: (a) P1's preferred images, (b) P2's preferred images, (c) input image, (d) enhanced image for P1, and (e) enhanced image for P2.

Figure 9 shows personalization results for two participants (P1 and P2) in the user study. P1 and P2, respectively, selected preferred images in (a) and (b). These preferred styles are represented by their preference vectors. PieNet enhances an input image in (c) using the preference vectors. We see that the enhancement results in (d) and (e) faithfully reflect their preferred styles.

5 Conclusions

We addressed the personalization issue in image enhancement. We trained an embedding space to obtain preference vectors based on metric learning, and developed PieNet to produce personalized results using the preference vectors. Experiments demonstrated that the proposed algorithm significantly outperforms the state-of-the-art algorithms on the MIT-A5K dataset. Also, it was demonstrated that a new user can obtain reliable results by providing only 10–20 preferred images to the proposed enhancement system.

Acknowledgements. This work was supported in part by the MSIT (Ministry of Science and ICT), Korea, under the ITRC (Information Technology Research Center) support program (IITP-2020-2016-0-00464) supervised by the IITP (Institute for Information & communications Technology Promotion), in part by the National Research Foundation of Korea (NRF) through the Korea Government (MSIP) under Grant NRF-2018R1A2B3003896, and in part by the research fund of Chungnam National University.

References

1. Aly, H.A., Dubois, E.: Image up-sampling using total-variation regularization with a new observation model. IEEE Trans. Image Process. **14**(10), 1647–1659 (2005)
2. Arici, T., Dikbas, S., Altunbasak, Y.: A histogram modification framework and its application for image contrast enhancement. IEEE Trans. Image Process. **18**(9), 1921–1935 (2009)
3. Aubry, M., Paris, S., Hasinoff, S.W., Kautz, J., Durand, F.: Fast local Laplacian filters: theory and applications. ACM Trans. Graph. **33**(5), 167 (2014)
4. Bromley, J., Guyon, I., LeCun, Y., Säckinger, E., Shah, R.: Signature verification using a "Siamese" time delay neural network. In: NIPS (1994)
5. Bychkovsky, V., Paris, S., Chan, E., Durand, F.: Learning photographic global tonal adjustment with a database of input/output image pairs. In: CVPR (2011)
6. Cai, B., Xu, X., Guo, K., Jia, K., Hu, B., Tao, D.: A joint intrinsic-extrinsic prior model for retinex. In: ICCV (2017)
7. Caicedo, J.C., Kapoor, A., Kang, S.B.: Collaborative personalization of image enhancement. In: CVPR (2011)
8. Chen, W., Chen, X., Zhang, J., Huang, K.: Beyond triplet loss: a deep quadruplet network for person re-identification. In: CVPR (2017)
9. Chen, Y.S., Wang, Y.C., Kao, M.H., Chuang, Y.Y.: Deep photo enhancer: unpaired learning for image enhancement from photographs with GANs. In: CVPR (2018)
10. Chopra, S., Hadsell, R., LeCun, Y.: Learning a similarity metric discriminatively, with application to face verification. In: CVPR (2005)
11. Deng, Y., Loy, C.C., Tang, X.: Aesthetic-driven image enhancement by adversarial learning. In: ACM MM (2018)
12. Fu, X., Liao, Y., Zeng, D., Huang, Y., Zhang, X.P., Ding, X.: A probabilistic method for image enhancement with simultaneous illumination and reflectance estimation. IEEE Trans. Image Process. **24**(12), 4965–4977 (2015)
13. Fu, X., Zeng, D., Huang, Y., Zhang, X.P., Ding, X.: A weighted variational model for simultaneous reflectance and illumination estimation. In: CVPR (2016)
14. Gharbi, M., Chen, J., Barron, J.T., Hasinoff, S.W., Durand, F.: Deep bilateral learning for real-time image enhancement. ACM Trans. Graph. **36**(4), 118 (2017)
15. Sundararajan, D.: Edge detection. In: Sundararajan, D. (ed.) Digital Image Processing, pp. 257–280. Springer, Singapore (2017). https://doi.org/10.1007/978-981-10-6113-4_9
16. Guo, X., Li, Y., Ling, H.: Lime: low-light image enhancement via illumination map estimation. IEEE Trans. Image Process. **26**(2), 982–993 (2016)
17. Hadsell, R., Chopra, S., LeCun, Y.: Dimensionality reduction by learning an invariant mapping. In: CVPR (2006)
18. He, K., Zhang, X., Ren, S., Sun, J.: Deep residual learning for image recognition. In: CVPR (2016)
19. Hu, Y., He, H., Xu, C., Wang, B., Lin, S.: Exposure: a white-box photo post-processing framework. ACM Trans. Graph. **37**(2), 26 (2018)
20. Jobson, D.J., Rahman, Z.U., Woodell, G.A.: A multiscale retinex for bridging the gap between color images and the human observation of scenes. IEEE Trans. Image Process. **6**(7), 965–976 (1997)
21. Jobson, D.J., Rahman, Z.U., Woodell, G.A.: Properties and performance of a center/surround retinex. IEEE Trans. Image Process. **6**(3), 451–462 (1997)

22. Johnson, J., Alahi, A., Fei-Fei, L.: Perceptual losses for real-time style transfer and super-resolution. In: Leibe, B., Matas, J., Sebe, N., Welling, M. (eds.) ECCV 2016. LNCS, vol. 9906, pp. 694–711. Springer, Cham (2016). https://doi.org/10.1007/978-3-319-46475-6_43
23. Joshi, N., Matusik, W., Adelson, E.H., Kriegman, D.J.: Personal photo enhancement using example images. ACM Trans. Graph. **29**(2), 12:1–12:15 (2010)
24. Kang, S.B., Kapoor, A., Lischinski, D.: Personalization of image enhancement. In: CVPR (2010)
25. Kim, J.H., Jang, W.D., Sim, J.Y., Kim, C.S.: Optimized contrast enhancement for real-time image and video dehazing. J. Vis. Commun. Image Represent. **24**, 410–425 (2013)
26. Kim, Y.T.: Contrast enhancement using brightness preserving bi-histogram equalization. IEEE Trans. Consum. Electro. **43**(1), 1–8 (1997)
27. Kingma, D.P., Ba, J.: Adam: a method for stochastic optimization. In: ICLR (2014)
28. Land, E.H.: The retinex theory of color vision. Sci. Am. **237**(6), 108–129 (1977)
29. Lee, C., Kim, J.H., Lee, C., Kim, C.S.: Optimized brightness compensation and contrast enhancement for transmissive liquid crystal displays. IEEE Trans. Circuits Syst. Video Technol. **24**, 576–590 (2014)
30. Lee, C., Lee, C., Kim, C.S.: Contrast enhancement based on layered difference representation of 2D histograms. IEEE Trans. Image Process. **22**(12), 5372–5384 (2013)
31. Lee, C., Lee, C., Lee, Y.Y., Kim, C.S.: Power-constrained contrast enhancement for emissive displays based on histogram equalization. IEEE Trans. Image Process. **21**(1), 80–93 (2011)
32. Lim, J., Heo, M., Lee, C., Kim, C.S.: Contrast enhancement of noisy low-light images based on structure-texture-noise decomposition. J. Vis. Commun. Image Represent. **45**, 107–121 (2017)
33. Lore, K.G., Akintayo, A., Sarkar, S.: LLNet: a deep autoencoder approach to natural low-light image enhancement. Pattern Recognit. **61**, 650–662 (2017)
34. van der Maaten, L., Hinton, G.: Visualizing data using t-SNE. J. Mach. Learn. Res. **9**, 2579–2605 (2008)
35. Park, J., Lee, J.Y., Yoo, D., Kweon, I.S.: Distort-and-recover: color enhancement using deep reinforcement learning. In: CVPR (2018)
36. Russakovsky, O., et al.: ImageNet large scale visual recognition challenge. Int. J. Comput. Vision **115**(3), 211–252 (2015)
37. Schroff, F., Kalenichenko, D., Philbin, J.: FaceNet: a unified embedding for face recognition and clustering. In: CVPR (2015)
38. Sohn, K.: Improved deep metric learning with multi-class N-pair loss objective. In: NIPS (2016)
39. Song, H.O., Jegelka, S., Rathod, V., Murphy, K.: Deep metric learning via facility location. In: CVPR (2017)
40. Song, H.O., Xiang, Y., Jegelka, S., Savarese, S.: Deep metric learning via lifted structured feature embedding. In: CVPR (2016)
41. Stark, J.A.: Adaptive image contrast enhancement using generalizations of histogram equalization. IEEE Trans. Image Process. **9**(5), 889–896 (2000)
42. Wang, R., Zhang, Q., Fu, C.W., Shen, X., Zheng, W.S., Jia, J.: Underexposed photo enhancement using deep illumination estimation. In: CVPR (2019)
43. Wang, S., Zheng, J., Hu, H.M., Li, B.: Naturalness preserved enhancement algorithm for non-uniform illumination images. IEEE Trans. Image Process. **22**(9), 3538–3548 (2013)

44. Wang, Y., Chen, Q., Zhang, B.: Image enhancement based on equal area dualistic sub-image histogram equalization method. IEEE Trans. Consum. Electro. **45**(1), 68–75 (1999)

45. Yan, Z., Zhang, H., Wang, B., Paris, S., Yu, Y.: Automatic photo adjustment using deep neural networks. ACM Trans. Graph. **35**(2), 11 (2016)

46. Yu, R., Liu, W., Zhang, Y., Qu, Z., Zhao, D., Zhang, B.: Deepexposure: learning to expose photos with asynchronously reinforced adversarial learning. In: NIPS (2018)

47. Yue, H., Yang, J., Sun, X., Wu, F., Hou, C.: Contrast enhancement based on intrinsic image decomposition. IEEE Trans. Image Process. **26**(8), 3981–3994 (2017)

Rotational Outlier Identification in Pose Graphs using Dual Decomposition

Arman Karimian$^{(\boxtimes)}$, Ziqi Yang, and Roberto Tron

Boston University, Boston, MA 02215, USA
{armandok,zy259,tron}@bu.edu

Abstract. In the last few years, there has been an increasing trend to consider Structure from Motion (SfM, in computer vision) and Simultaneous Localization and Mapping (SLAM, in robotics) problems from the point of view of *pose averaging* (also known as *global SfM*, in computer vision) or *Pose Graph Optimization* (PGO, in robotics), where the motion of the camera is reconstructed by considering only relative rigid body transformations instead of including also 3-D points (as done in a full Bundle Adjustment). At a high level, the advantage of this approach is that modern solvers can effectively avoid most of the problems of local minima, and that it is easier to reason about outlier poses (caused by feature mismatches and repetitive structures in the images). In this paper, we contribute to the state of the art of the latter, by proposing a method to detect incorrect orientation measurements prior to pose graph optimization by checking the geometric consistency of rotation measurements. The novel aspects of our method are the use of Expectation-Maximization to fine-tune the covariance of the noise in inlier measurements, and a new approximate graph inference procedure, of independent interest, that is specifically designed to take advantage of evidence on cycles with better performance than standard approaches (Belief Propagation). The paper includes simulation and experimental results that evaluate the performance of our outlier detection and cycle-based inference algorithms on synthetic and real-world data.

Keywords: Pose averaging · Outliers · Inference in graphical models

1 Introduction

Reconstructing a 3-D scene from a collection of ordered or unordered images or videos is one of the most prominent classical problems in computer vision and robotics. In computer vision, this task is known as *Structure from Motion* (SfM), and is traditionally performed using images alone. The typical solution pipeline for this problem [26,35,39] includes three steps: *1)* estimate relative poses between pairs of images using matched features [7,17,31] and robust fitting techniques [21,24]; *2)* combine the pairwise estimates, either in sequential stages [3,4,22,43,44], or by combining poses alone (without considering a 3-D

ⓒ Springer Nature Switzerland AG 2020
A. Vedaldi et al. (Eds.): ECCV 2020, LNCS 12375, pp. 391–407, 2020.
https://doi.org/10.1007/978-3-030-58577-8_24

structure) in a *pose-averaging* [1,5,14,25,33,50,51] or *pose-graph* [11] approach; *3)* use Bundle Adjustment (BA) [18,26,49], which minimizes the reprojection error by considering jointly the motion and the structure.

In robotics, a very similar task is known as *Simultaneous Localization and Mapping* (SLAM, [10]), and it usually includes the use of additional information such as wheel odometry, inertial measurements, or laser scans. Visual SLAM is a variant of the SLAM problem where only visual information obtained from a camera is used for the task [48]. Similarly to the case of SfM, the state of the art approach for SLAM is based on a pose graph formulation where nodes represent robot poses at different times, and edges represent relative pose measurements between pairs of nodes. One typical difference between typical SfM and SLAM applications is that, in the latter, the images are mostly ordered; hence, edges in the graph can be divided into two categories: *ego motion* edges which correspond to temporally close measurements; e.g., visual odometry measurements (for which temporal correlations can easily predict the presence of outliers), and *loop closure* edges which correspond to temporally distant measurements, e.g., when the same physical location is revisited at different times.

In both pose averaging for SfM, and PGO for SLAM, the absolute poses (nodes) in the graph are estimated from all the measurements (edges) via a Maximum Likelihood (ML) formulation [16,39], which typically involves solving a nonlinear least squares error minimization problem, and is highly sensitive to initialized values and the unavoidable presence of outlier measurements. For the problem of initialization, the most effective solutions use relaxations based on eigenvector computations or semi-definite programming [5,11,33,51]. More recent techniques can certify the global optimality of their ML estimates [9,20,27,41].

For the problem of outliers, traditional approaches rely on local optimization from an initial guess, and either discount outliers using robust cost functions [2,30,38]), or attempt to directly identify them [12,23,29,46,47]. In the latter group, there exist techniques based on inference on graphical models [53]. Empirically, these methods work well when the outliers are only few, and embedded in a dense graph of otherwise valid measurements; their performance decreases in more challenging regimes, such as in the alignment of multiple maps in SLAM, where many of the associations (loop closures) can be erroneous, and, for instance, finding a good initial guess for the alignment is more challenging. Existing solutions for this problem are limited to either a single map (as the optimization based approach in [28]) or two maps (as the set maximization approach of [32]).

Paper Contributions. We propose a probabilistic approach for outlier detection between any number of maps. Our algorithm checks for the geometric consistency of the rotation measurements in loops within the graph of poses, and decides if each edge is an inlier or outlier without relying on a trajectory estimate. We use a Gaussian additive noise model for rotation measurements and use the rotational error over cycles as evidence to infer the inlier/outlier probabilities. We use the Expectation-Maximization algorithm to fine-tune the parameters of the distribution of measurement errors and present simulation results. For the inference step required by our algorithm, we first apply Belief Propagation (BP), and highlight its shortcomings in this setting, next we present a novel inference algorithm based on a novel cycle-based dual decomposition and the Alternating Direction Method of Multipliers (ADMM) which has local convergence guarantees. We evaluate the performance of our proposed solution using simulations and in the alignment of four real-world maps produced by a standard SLAM algorithm.

Paper Outline. In the remainder of the paper, we first review a probabilistic graphical model for error propagation on the space of rotations, and errors on cycles of poses (Sect. 2). We then review Belief Propagation for performing inference on the graphical model, describe our ADMM-based alternative (Sect. 3); and show how this inference can be embedded in an Expectation-Maximization procedure to estimate the variances of the inliers and the outliers (Sect. 4). Finally, we present our simulations and experiments (Sect. 5).

2 Probabilistic Model

In this section, we describe the approximate additive Gaussian noise model on rotations used for modeling the errors on single edges and along graph cycles, as well as the graphical model used to relate the inlier versus outlier probabilities for each edge with the evidence provided by the geometric consistency of cycles.

2.1 Gaussian Noise Model and Uncertainty Propagation

We denote the graph of poses as $\mathcal{G} = (\mathcal{V}, \mathcal{E}, \mathcal{T})$ with vertices $\mathcal{V} = \{1, \ldots, n\}$ representing absolute poses that need to be estimated, and edges $\mathcal{E} \subseteq \mathcal{V} \times \mathcal{V}$ representing the existence of measured relative transformations $\tilde{\mathbf{T}}_{ij} \in \mathcal{T}$ between them, i.e., $\tilde{\mathbf{T}}_{ij} \approx \mathbf{T}_j \mathbf{T}_i^{-1}$. Each pose $\mathbf{T}_i$ is represented as a member of a matrix Lie group, i.e., a group whose elements and group operation are representable by square matrices, and that is also a smooth differentiable manifold.

In this paper, we limit our attention to SO(3), leaving the applications of our methods to other Lie groups (e.g., SE(d) or Sim(d)) as future work. As we will show, this choice already provides significant benefits in the detection of outliers.

We model errors over rotations through a Gaussian distribution in local exponential coordinates, i.e., the distribution is defined in the tangent space at the mean, and mapped to the Lie group via the exponential map. Formally:

$$\epsilon \sim \mathcal{N}(\mathbf{0}, \mathbf{\Sigma})$$
$$\tilde{\mathbf{R}} = \exp(\hat{\epsilon})\,\mathbf{R} \tag{1}$$

where $\epsilon \in \mathbb{R}^3$ is a zero-mean Gaussian random variable with covariance matrix $\mathbf{\Sigma} \in \mathbb{R}^{3 \times 3}$, and $\hat{\epsilon} \in \mathfrak{so}(3)$ is a skew-symmetric matrix given by the *hat operator*

$$\hat{\epsilon} = \begin{bmatrix} 0 & -\epsilon_3 & \epsilon_2 \\ \epsilon_3 & 0 & -\epsilon_1 \\ -\epsilon_2 & \epsilon_1 & 0 \end{bmatrix}. \tag{2}$$

We assume that, for inlier measurements, the magnitude of the vector ϵ, which represents the magnitude of the noise, is small, thus justifying the following.

Lemma 1 ([6], 7.3). *The first order approximation of the uncertainty in the composition of two rotations* $\tilde{\mathbf{R}}_1 \sim \mathcal{N}_{SO(3)}(\mathbf{R}_1, \boldsymbol{\Sigma}_1)$ *and* $\tilde{\mathbf{R}}_2 \sim \mathcal{N}_{SO(3)}(\mathbf{R}_2, \boldsymbol{\Sigma}_2)$ *is:*

$$\tilde{\mathbf{R}}_2\tilde{\mathbf{R}}_1 \sim \mathcal{N}_{SO(3)}(\mathbf{R}_2\mathbf{R}_1, \boldsymbol{\Sigma}_2 + \mathbf{R}_2\boldsymbol{\Sigma}_1\mathbf{R}_2^\top) \tag{3}$$

This approximation, which comes from the truncation of the BCH formula, is justified by our assumption that the inlier errors are relatively small. In addition, we make the following assumption about the noise covariance $\boldsymbol{\Sigma}$:

Assumption 1. *The rotation distributions are isotropic, i.e.* $\boldsymbol{\Sigma}_i = \sigma_i^2\mathbf{I}_3$, *where* $\mathbf{I}_3$ *is the identity matrix.*

Combining Assumption 1 with (3), the distribution of the composition of a subset $\mathcal{S} \subset \mathcal{V}$ of noisy rotations is given by:

$$\prod_{i \in \mathcal{S}} \tilde{\mathbf{R}}_i \sim \mathcal{N}_{SO(3)}\left(\prod_{i \in \mathcal{S}} \mathbf{R}_i, \left(\sum_{i \in \mathcal{S}} \sigma_i^2\right)\mathbf{I}_3\right). \tag{4}$$

If all the variances σ_i are equal, the resultant covariance matrix is given by $m\sigma^2\mathbf{I}_3$, where $m = |\mathcal{S}|$. Since the expected length of a zero-mean spherical Gaussian random variable $\boldsymbol{\varepsilon} \sim \mathcal{N}(\mathbf{0}, \varsigma^2\mathbf{I}_d)$ is tightly bounded as $\frac{d}{\sqrt{d+1}}\varsigma \leq \mathbb{E}(\|\boldsymbol{\varepsilon}\|) \leq \sqrt{d}\varsigma$ [13, Definition 3.1], for small enough m and σ the expected value of noise is proportional to $\sqrt{m}$; this was experimentally validated in [19, Figure 3].

2.2   Inlier and Outlier Gaussian Mixture Model

We model the distribution for each measurement R_e along an edge $e \in \mathcal{E}$ with a Gaussian mixture model with two modes, one for inliers and the other for outliers. We use the Bernoulli indicator variable $x_e \in \{0,1\}$ to denote e as an inlier ($x_e = 0$) or an outlier ($x_e = 1$), with respective (user-defined) prior probabilities $p(x_e = 0) = \pi_e$ and $p(x_e = 1) = 1 - \pi_e \doteq \bar{\pi}_e$. Building upon Assumption 1, we assume that every inlier edge has uncertainty $\sigma^2\mathbf{I}_3$ and every outlier edge has uncertainty $\bar{\sigma}^2\mathbf{I}_3$, where $\bar{\sigma} \gg \sigma$; note that a sufficiently large value of $\bar{\sigma}$ in practice leads to an approximation of the uniform distribution.

2.3   Graphical Model for Evidence over Cycles

A simple cycle is a closed chain of edges where each edge appears only once. Every simple cycle c in our pose graph corresponds to an ordered set of rotation measurements along the edges of the cycle, and the composition $\tilde{\mathbf{R}}_c$ of these rotations $\tilde{\mathbf{R}}_c = \prod_{e \in c} \tilde{\mathbf{R}}_e$ should, ideally, be close to the identity (i.e., by transforming a reference frame along a cycle return it to its initial pose). Defining z_c to be the geodesic distance of $\tilde{\mathbf{R}}_c$ from the identity, i.e.,

$$z_c = \frac{1}{\sqrt{2}}\| \log(\tilde{\mathbf{R}}_c)\|_F = \arccos\left(\frac{\operatorname{tr}(\tilde{\mathbf{R}}_c) - 1}{2}\right), \tag{5}$$

where $\|\cdot\|_F$ is the Frobenius norm, we can use (4) to obtain a probabilistic mode (distribution) of $\tilde{\mathbf{R}}_c$. Note that the variance of $\tilde{\mathbf{R}}_c$ mainly depends on the length and the number of outliers of the cycle.

Similarly to previous work that aims to use geometric relations in cycles in Structure from Motion [19,53], we model the relation between errors on edges and cycles by using a Bayesian network in which every edge $e \in \mathcal{E}$, and every cycle $c \in \mathcal{C}$ of the original pose graph is modeled by a node in the Bayesian network, and each edge e is connected to the cycles c to which it belongs (Fig. 1b); the cycles serve as evidence for inferring the hidden inlier/outlier state random variables x_e for each edge $e \in \mathcal{E}$. The joint probability distribution given by this model for hidden states $\boldsymbol{x} \in \{0,1\}^{|\mathcal{E}|}$ and cycle-consistency errors $\boldsymbol{z} \in \mathbb{R}^{|\mathcal{C}|}$ is

$$p(\boldsymbol{x}, \boldsymbol{z}) = \prod_{e \in \mathcal{E}} p(x_e) \prod_{c \in \mathcal{C}} p(z_c \mid \boldsymbol{x}_c), \qquad (6)$$

where $\mathcal{C}$ is a set of cycles in $\mathcal{G}$, and $p(x_e)$ is the prior probability of edge e, and $\boldsymbol{x}_c$ is the vector containing x_e values for every $e \in c$. Equation (6) can be graphically represented using a factor graph (Fig. 1c).

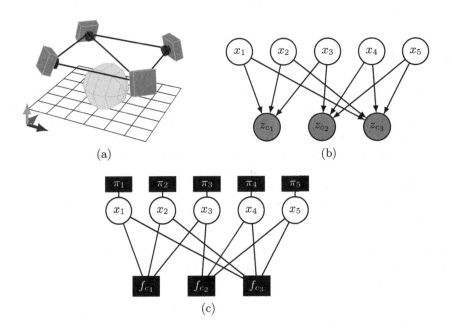

(a) (b)

(c)

Fig. 1. (a) Example of a small problem with four poses, five measurements, and three cycles; (b) A Bayesian network representation where the upper nodes correspond to the edges and the bottom nodes correspond to cycles (shaded in gray because they are observed variables); (c) The factor graph representation, with $\{\pi_e\}$ representing the edge prior probabilities, and $\{f_c\}$ the cycle evidence.

Letting $s_c = \mathbf{1}^\top \boldsymbol{x}_c$ be the number of outliers in c for the configuration $\boldsymbol{x}$, the distribution $p(z_c \mid \boldsymbol{x}_c)$ is obtained from (4), where the covariance matrix is given by $\varsigma_c^2(\boldsymbol{x}_c)\mathbf{I}_3 = (s_c\bar{\sigma}^2 + (|c| - s_c)\sigma^2)\mathbf{I}_3$ and $|c|$ is the length of the cycle.

Note that in robotics applications (see also our experiments), we can limit the inference of x_e to loop closure edges $\mathcal{E}_{lc}$; in modern systems, ego motion edges are unlikely to contain outlier measurements, hence we set the priors $\pi_e = 1$ for any ego motion edge $e \in \mathcal{E}$. Moreover, using all possible cycles is neither necessary nor practical for this task. The total number of cycles in a graph, in general, grows combinatorially with the size of the graph, leading to a proportional increase in the computational cost. To deal with this issue, we restrict ourselves to cycles from a *Minimum Cycle Basis* $\mathcal{C}_{min} \in 2^{\bar{\mathcal{C}}}$ of the pose graph, obtained using the de Pina's method [34]. This reduces the number of cycles to $\mathcal{O}(|\mathcal{E}_{lc}|)$, covers all the edges in bi-connected components of the pose-graph, and every other cycle can be obtained as a combination of cycles in the basis. Moreover, the MCB, which is minimal in the sense of the number of times each edge appears in cycles in $\mathcal{C}_{min}$, has the benefits of *1)* reducing the number of connections in the Bayesian graphical model (Fig. 1b); and *2)* short cycles reduce the uncertainty in the observations z_c along cycles with only inliers (see the discussion in Sect. 2.1). In future work, we will explore the option of finding a basis that is minimal in the sense of the sum of the errors z_c.

3   Inference for Graphical Models

In this section, we assume that the set of parameters $\Theta = \{\sigma, \bar{\sigma}, \Pi\}$ where $\Pi = \{\pi_e\}_{e\in\mathcal{E}_{lc}}$ is given, and that we aim to find the marginal probabilities from (6) for each edge, i.e., $\gamma_e \triangleq p(x_e|\boldsymbol{z})$, $e \in \mathcal{E}$ (in Sect. 4, we will extend the procedure to concurrently estimate σ, $\bar{\sigma}$ from the same data). An exact solution to this probabilistic inference problem can easily become intractable, as the complexity increases exponentially with the number of edges. Resorting to approximation methods, we consider two options *1)* Loopy Belief Propagation (BP), which represents the standard traditional choice for approximate inference in graphs, although it is not guaranteed to converge for general graphs; and *2)* our novel inference algorithm based on dual decomposition along cycles with the Alternating Direction Method of Multipliers (ADMM), which instead has local convergence guarantees. Section 5 shows that, in our setting, the latter is superior in terms of outlier detection.

3.1   Belief Propagation

Belief Propagation is one of the most well known inference algorithms for finding marginal and conditional probabilities; it is a Variational Inference approach based on the minimization of the Bethe free energy [52]. For graphical models with loops, vanilla BP may fail to converge, and, even if it converges, the solution is generally not exact.

We review here the factor graph version of BP via the example of Fig. 1. In BP, messages are sent between neighboring variables and factors according to the following equations [52]:

$$n_{e \to f_c}(x_e) = \prod_{f \in N(e) \backslash f_c} m_{f \to e}(x_e), \qquad m_{f_c \to e}(x_e) = \sum_{\boldsymbol{x}_c \backslash x_e} f_c(\boldsymbol{x}_c) \prod_{i \in N(f_c) \backslash e} n_{i \to f_c}(x_i), \qquad (7)$$

where $n_{e \to f_c}$ is the message from variable e to factor f_c, and $m_{f_c \to e}$ is the message from factor f_c to variable e; $N(e)$ and $N(f_c)$ denote the factors that are connected to the random variable x_e and the factor f_c, respectively (the former includes the prior π_e, which is constant). These messages are passed in an asynchronous order until convergence of beliefs (approximate marginals), which are computed as:

$$b_e(x_e) \propto \prod_{f \in N(e)} m_{f \to e}(x_e), \qquad b_c(\boldsymbol{x}_c) \propto f_c(\boldsymbol{x}_c) \prod_{e \in N(f_c)} n_{e \to f_c}(x_e), \qquad (8)$$

where b_e is the belief for the indicator variable of an edge and an approximation of $\gamma_e = p(x_e|\boldsymbol{z})$, and b_c is the belief of all random variables connected to factor f_c, and an approximation of $\gamma_c \triangleq p(\boldsymbol{x}_c|\boldsymbol{z})$ (note that γ_c is used in the Expectation-Maximization procedure in Sect. 5). To force convergence of the BP iterations, we introduce a damping factor as suggested in [40, Chapter 22] (we use 0.5 in our experiments).

3.2 Alternating Direction Method of Multipliers

The Alternating Direction Method of Multipliers (ADMM) provides a robust and decomposable algorithm for optimization problems by breaking them into smaller and easier to handle sub-problems [8]. For convex problems, ADMM guarantees global linear convergence rate [37]. It can also be used in non-convex problems, although in that case it will convergence to a local minimum.

In order to estimate γ_e and γ_c, instead of marginalizing $p(\boldsymbol{x}, \boldsymbol{z})$ over each edge directly, we propose to marginalize over the each cycle, i.e.,

$$p_c(\boldsymbol{x}_c, z_c) = p(z_c|\boldsymbol{x}_c) \prod_{e \in c} p(x_e), \qquad (9)$$

and then force the marginals of each edge e from different overlapping cycles to agree on a common value. Intuitively, our approximation strategy aims to preserve the statistical correlation (joint distribution) between edges in the same cycle, while ignoring the correlations across cycles.

More in detail, we can implement this strategy by solving a *consensus* problem with ADMM [8, Chapter 7]. We denote as $\hat{\boldsymbol{v}}_c \in \mathbb{R}^{2^{|c|}}$ the vector containing all probabilities $p_c(\boldsymbol{x}_c|z_c)$ obtained from (9) evaluated over all possible values of $\boldsymbol{x}_c \in \{0,1\}^{|c|}$. For each cycle c, we try to estimate a vector $\boldsymbol{v}_c$ such that $1)$ $\boldsymbol{v}_c$ is close to $\hat{\boldsymbol{v}}_c$, and $2)$ when two distributions $\boldsymbol{v}_c$, $\boldsymbol{v}_{c'}$ for two overlapping cycles c, c' are marginalized with respect to a common edge $e \in (c \cap c')$, the two results agree. We will parameterize the marginal distribution γ_e by keeping track of the inlier probability alone, denoted as $w_e = p(x_e = 0|z)$. We can then formulate the following minimization problem:

$$\min_{w,\{v_c\}} \sum_{c \in \mathcal{C}} h_c(v_c),$$
$$\text{subject to } \mathbf{p}_{e,c}^\top v_c = w_e, \forall c \in \mathcal{C}, e \in c, \tag{10}$$
$$0 \le w \le 1,$$

where $w \in \mathbb{R}^{|\mathcal{E}_{lc}|}$ is the vector of all $\{w_e\}$, and the indicator vectors $\mathbf{p}_{e,c} \in \{0,1\}^{2^{|c|}}$ are a vectorial representation for obtaining the marginal inlier probability w_e given the cycle distribution v_c. In (10), each h_c (i.e., each cycle), is considered a subproblem with its own local constraints that can be solved in a distributed fashion. As stated earlier, we want v_c to be close to $\hat{v}_c$ with respect to some metric. If we choose the 2-Wasserstein metric, h_c will be formulated as follows:

$$h_c(v_c) = \begin{cases} \|v_c - \hat{v}_c\|^2 & \text{if } \mathbf{1}^\top v_c = 1, 0 \le v_c \le 1, \\ +\infty & \text{otherwise.} \end{cases} \tag{11}$$

(In future work, we plan to evaluate other measures of similarity between c and $\hat{c}$, such as the Kullback–Leibler divergence.) Subproblems c, c' that share an edge are forced to agree through the constraints $\mathbf{p}_{e,c}^\top v_c = \mathbf{p}_{e,c'}^\top v_{c'} = w_e$. This problem formulation is very similar to a consensus optimization problem, with the difference being that a linear combination of the variables v_c should reach consensus instead of the full vector, plus the global constraint $0 \le w \le 1$. To apply ADMM, we write (10) using the indicator function $g(w)$ which returns $+\infty$ if the global constraint $0 \le w \le 1$ is violated,

$$\min_{w_e} \sum_{c \in \mathcal{C}} h_c(v_c) + g(w),$$
$$\text{subject to } \mathbf{P}_c v_c = w_c, \forall c \in \mathcal{C}, \tag{12}$$

where the vector $w_c \in \mathbb{R}^{|c|}$ contains the elements w_e of w for every $e \in c$ and $\mathbf{P} \in \mathbb{R}^{|c| \times 2^{|c|}}$ is obtained by horizontally stacking the vectors $\mathbf{p}_{e,c}^\top$ column-wise. The augmented Lagrangian for (12) is:

$$L_\rho = \sum_{c \in \mathcal{C}} \left(h_c(v_c) + y_c^\top (\mathbf{P}_c v_c - w_c) + \frac{\rho}{2} \|\mathbf{P}_c v_c - w_c\|^2 \right) + g(w), \tag{13}$$

with dual variables $y_c \in \mathbb{R}^{|c|}$, and penalty parameter ρ. The ADMM iterations for this problem are given by [8]:

$$v_c^{k+1} = \underset{v_c}{\text{argmin}} \left(h_c(v_c) + y_c^{k\top} \mathbf{P}_c v_c + \frac{\rho}{2} \|\mathbf{P}_c v_c - w_c^k\|^2 \right)$$
$$w^{k+1} = \underset{w}{\text{argmin}} \left(g(w) + \sum_{c \in \mathcal{C}} \left(-y_c^{k\top} w_c + \frac{\rho}{2} \|\mathbf{P}_c v_c - w_c\|^2 \right) \right) \tag{14}$$
$$y_c^{k+1} = y_c^k + \rho(\mathbf{P}_c v_c - w_c^k).$$

Note that the variables v_c and y_c can be updated in parallel for each cycle. The solution for v_c is obtained by solving a quadratic programming problem (which can be done efficiently), while the solution for the global consensus variable w is given by:

$$w_e^{k+1} = \max(0, \min(1, \omega_e^{k+1})), \tag{15}$$

$$\omega_e^{k+1} = \frac{\sum_{c;e\in c}\left(\mathbf{P}_{e,c}^\mathsf{T} v_c^{k+1} + \frac{1}{\rho}[y_c^k]_e\right)}{\sum_{c;e\in c} 1}. \tag{16}$$

The denominator in (16) is the number of times edge e appears in different cycles, and therefore ω_e^{k+1} is the average of marginalized values for edge e plus the component of y_c^k that corresponds to e over cycles that contain that edge. In (15), the values of ω^{k+1} are projected to be between zero and one.

This problem will reach (local) optimality when the primal residual r^k and dual residuals t^k converge to zero, where:

$$
\begin{aligned}
r^k &= \sum_{c\in\mathcal{C}} \left\| \mathbf{P}_c v_c^k - w_c^k \right\|^2 \\
t^k &= \rho^2 \sum_{e\in\mathcal{E}_{lc}} \sum_{c;e\in c} (w_e^k - w_e^{k-1})^2
\end{aligned}
\tag{17}
$$

The penalty parameter ρ plays a very important role in the convergence speed of this method. Intuitively, small ρ allows intermediate solutions to have a much lower cost while somewhat ignoring the primal feasibility, and makes the solution less impacted by initial value and easier to escape from the local minima, whereas a large ρ will place a large penalty on violating the consensus constraints, but tends to produce small primal residuals. As suggested in [8, Chapter 3], we start with a small ρ, and gradually change the value of ρ based on primal and dual residual, using the following dynamic update rule:

$$
\rho^{k+1} = \begin{cases} \tau^{incr}\rho^k & \text{if } r^k \le \mu t^k \\ \rho^k/\tau^{decr} & \text{if } t^k \le \mu r^k \\ \rho^k & \text{otherwise,} \end{cases}
\tag{18}
$$

where $\mu > 1$, $\tau^{decr} > 1$, and $\tau^{decr} > 1$ are constant parameters.

A disadvantage of our method is that the local variables $\{v_c\}$ have dimensions that grow exponentially with the length of the cycles; however, in our experiments we noted that cycles of length up to $|c| = 15$ remain tractable, and longer cycles could be discarded, since they are likely to provide only very weak evidence.

4   Expectation-Maximization

In the previous sections, we assumed that the parameters $\Theta = \{\sigma, \bar{\sigma}, \Pi\}$ (the inlier and outlier standard deviations, and the edge prior probabilities, respectively) were given. However, this assumption is not true and these parameters need to be estimated. By including parameters in the distribution, we rewrite (6) as:

$$p(\boldsymbol{x}, \boldsymbol{z} | \Theta) = \prod_{e \in \mathcal{E}} p(x_e | \pi_e) \prod_{c \in \mathcal{C}} p(z_c | \varsigma_c(\boldsymbol{x}_c)) \tag{19}$$

where the first term is a given by Bernoulli distribution. With some abuse of notation, we assume π_e is $p(x_e = 0)$ and $\bar{\pi}_e = 1 - \pi_e$ which yields $p(x_e | \pi_e) = \pi_e^{1-x_e} \bar{\pi}_e^{x_e}$. The second term is a wrapped Gaussian mixture distribution:

$$p(z_c | \varsigma_c(\boldsymbol{x}_c)) = \frac{1}{\psi_c} \frac{\varsigma_c^{-3}}{\phi(\varsigma_c)} \exp\left(\frac{-z_c^2}{2\varsigma_c^2}\right) \tag{20}$$

with $\varsigma_c(\boldsymbol{x}) = \sqrt{(\mathbf{1}^\mathsf{T} \boldsymbol{x}_c)\bar{\sigma}^2 + (|c| - \mathbf{1}^\mathsf{T} \boldsymbol{x}_c)\sigma^2}$, $\phi(\varsigma_c)$ is a normalizing constant for the wrapped Gaussian, ψ_c normalizes over all the possible values for $\boldsymbol{x}$,

$$\psi_c = \sum_{s=0}^{|c|} \binom{|c|}{s} \frac{\varsigma_c^{-3}(s)}{\phi(\varsigma_c(s))} \exp\left(\frac{-z_c^2}{2\varsigma_c^2(s)}\right), \tag{21}$$

and the term ς_c^{-3} comes from the denominator of the Gaussian probability density function, $\sqrt{\det(\varsigma_c^2 \mathbf{I}_3)}$. The value of $\varsigma_c(\boldsymbol{x}_c)$ only depends on the number of outliers $s_c = \mathbf{1}^\mathsf{T} \boldsymbol{x}_c$, hence we can denote it is $\varsigma_c(s_c)$. The log-likelihood function is:

$$\mathcal{L}(\Theta; \boldsymbol{x}, \boldsymbol{z}) = \log(p(\boldsymbol{x}, \boldsymbol{z} | \Theta))$$

$$= \sum_{e \in \mathcal{E}} (1 - x_e) \log(\pi_e) + x_e \log(\bar{\pi}_e) + \sum_{c \in \mathcal{C}} -3 \log \varsigma_c - \frac{z_c^2}{2\varsigma_c^2} - \log\left(\psi_c \phi(\varsigma_c)\right) \tag{22}$$

In the Expectation step, we find the expectation of the log likelihood of Θ^i with respect to the current distribution of $\boldsymbol{x}$ given $\boldsymbol{z}$ and previous estimate of parameters Θ^{i-1}:

$$Q(\Theta^{(i)} | \Theta^{(i-1)}) = \mathbb{E}_{\boldsymbol{x} | \boldsymbol{z}, \Theta^{(i-1)}} [\mathcal{L}] = \sum_{\boldsymbol{x} \in \mathbb{Z}_2^{|\mathcal{E}|}} \mathcal{L}(\Theta^{(i)}; \boldsymbol{x}, \boldsymbol{z}) p(\boldsymbol{x} | \boldsymbol{z}, \Theta^{(i-1)}) \tag{23}$$

We use $\gamma_e^i = p(x_e | \boldsymbol{z}, \Theta^i)$ for the marginal of edge e, and $\gamma_c^i = p(\boldsymbol{x}_c | \boldsymbol{z}, \Theta^i)$ for the marginal of cycle c, given the parameters Θ^i (these are approximated via either BP or ADMM). Now, by expanding (23) we get:

$$Q(\Theta^{(i)} | \Theta^{(i-1)}) = \sum_{e \in \mathcal{E}} \sum_{x_e \in \mathbb{Z}_2} p(x_e | \boldsymbol{z}, \Theta^{(i-1)}) \log p(x_e | \pi_e^{(i)}) \tag{24a}$$

$$+ \sum_{c \in \mathcal{C}} \sum_{\boldsymbol{x}_c \in \mathbb{Z}_2^{|c|}} p(\boldsymbol{x}_c | \boldsymbol{z}, \Theta^{(i-1)}) \log p(z_c | \boldsymbol{x}_c, \sigma^{(i)}, \bar{\sigma}^{(i)}) \tag{24b}$$

In the Maximization step, we find $\Theta^{(i)} = \arg\max_\Theta Q(\Theta | \Theta^{(i-1)})$. For $\Pi^{(i)}$, we have $\pi_e^{(i)} = \gamma_e^{(i-1)}$, but for $\sigma^{(i)}$ and $\bar{\sigma}^{(i)}$ it is not as straightforward. Each term in the summation in (24b) is a quasiconcave function, but their sum need not be quasiconcave. Therefore, we use a grid-search to find σ and $\bar{\sigma}$ at each iteration.

5   Simulations and Experiments on Map Merging

In this section, we provide performance results of our outlier detection algorithm over synthetic and real data. For the synthetic data, we repeatedly generate a pose graph with two maps of 15 nodes each and random poses. At every iteration, m edges are added between the two maps, where m varies from 10 to 200 with increments of 5. For every given m, from 1 to $m - 1$ edges are selected to be outliers (with unitary increments). Inlier and outlier edges are given a random noise rotation with a random direction, and the magnitude of noise uniformly selected within $2.4° \leq \|\epsilon\| \leq 3.6°$ for inliers and $72° \leq \|\bar{\epsilon}\| \leq 108°$ for outliers (although we obtained similar results with different outlier distributions). The total number of generated graphs is $8,112$ and both BP and ADMM inference algorithms were used on the same graphs.

In Fig. 2a, we plot each simulation as a point on the precision-recall plane. Figure 2b, the ratio of detected outliers is plotted versus the ratio of the outlier edges to total loop closure edges. It is clear that our ADMM inference performs better than BP, as it has overall higher precision and recall. In addition, as the ratio of outliers to loop closure edges increase, the performance of BP continuously deteriorates, while ADMM presents a V-shaped curve; we hypothesize that this is due to the fact that situations with a majority of inliners or outliers represent easier cases (there is little discrepancy between the results of the local inferences over the different cycles), while mixed situations are more difficult to reconcile.

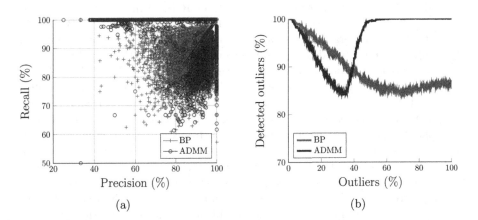

(a)   (b)

Fig. 2. (a) Precision ($\frac{TP}{TP+FP}$)-Recall ($\frac{TP}{TP+FN}$) for each simulated case using a threshold for γ_e of 0.5; points toward the top and right boundaries are better. (b) The ratio of detected outliers to the total number of outliers (Recall) versus the ratio of outlier loop closure edges to total loop closure edges; higher is better.

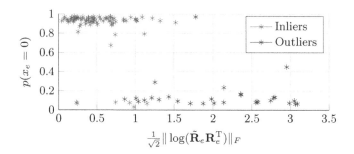

Fig. 3. Error in relative pairwise orientation measurement $\tilde{\mathbf{R}}_e$ w.r.t ground-truth $\mathbf{R}_e$ (x−axis) versus the probability of detection as inlier (y−axis) by our EM-ADMM algorithm. (Color figure online)

In Fig. 3 we present an SfM experiment on the Castle-P30 dataset [45]. We obtain 83 pairwise relative rotation measurements by estimating the essential matrix (shown in red), and further add 30 random pairwise relative measurements by sampling a random vector in $SO(3)$ with a magnitude uniformly sampled between 0 and π (shown in blue). The given results indicate that most of the edges with small noise in their measurements are classified as inliers and those with high noise are classified as outliers.

In Fig. 4 we present the result of implementing our classifier on actual data obtained from an office environment and compare its performance with the method in [53]. Four independent sequences of RGB-D images were obtained using an Intel RealSense D435 camera, and were processed with ORB-SLAM2 [36]. The result of merging maps is shown with and without removing outliers, after initial alignment and pose graph optimization using GTSAM [15].

Data associations between the maps are obtained using the ORB-SLAM2's place recognition module in addition to an object detector (MobileNet-SSD [42]). In Table 1a the number of image pairs (RGB and Depth) for each map and the total number of images is given. A subset of these images is picked as Keyframes and constitute nodes in the joint pose graph. In Table 1b, the number of loop closure edges between the maps is given. There are no loop closure edges within each map. In Table 1c, the length of the cycles and frequency of cycles of those lengths are given from $\mathcal{C}_{min}$ which is used by our algorithm and also by the algorithm from [53]. We removed cycles from $\mathcal{C}_{min}$ with length greater than 15 due to increased complexity in the subproblem 11 and reduced the quality of evidence gathered from long cycles. As stated before, the length of the cycle is the number of loop-closure edges (total of 247) that appear in a bigger cycle which includes ego motion edges of robots' pose graphs. The outlier detection algorithm from [53] detects 13 outliers, whereas our algorithm finds 21 outliers.

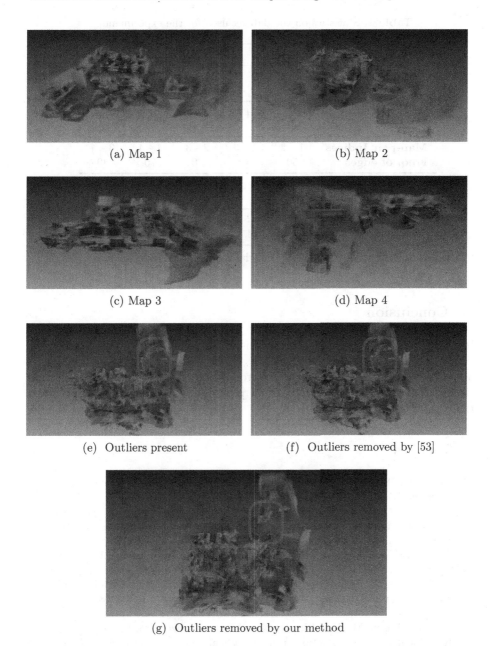

(a) Map 1   (b) Map 2

(c) Map 3   (d) Map 4

(e) Outliers present   (f) Outliers removed by [53]

(g) Outliers removed by our method

Fig. 4. In (a)–(d), point clouds from four different sequences of images are depicted. The final point cloud made from joining all the four pointclouds without outlier detection is shown in (e) and with outlier detection is shown in (f) (from [53]) and (g) (our method). In (e) and (f), phantoms can be observed (e.g., see marked area) and the overall shape of the environment is not correct due to misalignment caused by outliers, whereas with out method in (g) these issues are considerably less severe.

Table 1. Statistics on the dataset used for the experiments.

	Map 1	Map 2	Map 3	Map 4	Total
Nodes	637	471	447	220	1,775
Images	3,233	2,289	2,641	926	9,089

(a) Images and nodes in each map and their total numbers from our dataset.

Map-pair Indices	1 - 2	1 - 3	2 - 3	1 - 4	3 - 4
Freq. of edges	50	8	70	70	49

(b) Frequency of loop-closure edges between map pairs.

Cycle length	3	4	5	6	7	12	15
Cycle freq.	80	102	45	8	2	1	1

(c) Length of the cycles in our dataset versus their frequency.

6 Conclusion

In this paper, we presented a probabilistic outlier detection algorithm which detects outliers based on the geometric consistency of rotation measurements over the cycles of a pose graph. We introduced a novel discrete inference algorithm with convergence guarantees that performed better than Belief Propagation.

References

1. Aftab, K., Hartley, R., Trumpf, J.: Generalized Weiszfeld algorithms for Lq optimization. IEEE Trans. Pattern Anal. Mach. Intell. **37**(4), 728–745 (2015)
2. Agarwal, P., Tipaldi, G.D., Spinello, L., Stachniss, C., Burgard, W.: Robust map optimization using dynamic covariance scaling. In: 2013 IEEE International Conference on Robotics and Automation, pp. 62–69. Citeseer (2013)
3. Agarwal, S., et al.: Building Rome in a day. Commun. ACM **54**(10), 105–112 (2011)
4. Agarwal, S., Snavely, N., Seitz, S.M., Szeliski, R.: Bundle adjustment in the large. In: Daniilidis, K., Maragos, P., Paragios, N. (eds.) ECCV 2010. LNCS, vol. 6312, pp. 29–42. Springer, Heidelberg (2010). https://doi.org/10.1007/978-3-642-15552-9_3
5. Arie-Nachimson, M., Kovalsky, S., Kemelmacher-Shlizerman, I., Singer, A., Basri, R.: Global motion estimation from point matches. In: International Conference on 3D Imaging, Modeling, Processing, Visualization and Transmission, pp. 81–88 (2012)
6. Barfoot, T.D.: State Estimation for Robotics, 1st edn. Cambridge University Press, Cambridge (2017)
7. Bay, H., Ess, A., Tuytelaars, T., Gool, L.V.: Speeded-up robust features (SURF). Comput. Vis. Image Underst. **110**(3), 346–359 (2008)

8. Boyd, S., Parikh, N., Chu, E., Peleato, B., Eckstein, J., et al.: Distributed opti-mization and statistical learning via the alternating direction method of multipliers. Found. Trends® Mach. Learn. **3**(1), 1–122 (2011)

9. Briales, J., Gonzalez-Jimenez, J.: Cartan-sync: fast and global SE(d)-synchronization. IEEE Robot. Autom. Lett. **2**(4), 2127–2134 (2017)

10. Cadena, C., et al.: Past, present, and future of simultaneous localization and map-ping: toward the robust-perception age. IEEE Trans. Robot. **32**(6), 1309–1332 (2016)

11. Carlone, L., Tron, R., Daniilidis, K., Dellaert, F.: Initialization techniques for 3D SLAM: a survey on rotation estimation and its use in pose graph optimization. In: IEEE International Conference on Robotics and Automation (2015)

12. Carlone, L., Censi, A., Dellaert, F.: Selecting good measurements via l1 relaxation: a convex approach for robust estimation over graphs. In: 2014 IEEE/RSJ Interna-tional Conference on Intelligent Robots and Systems (IROS 2014), pp. 2667–2674. IEEE (2014)

13. Chandrasekaran, V., Recht, B., Parrilo, P.A., Willsky, A.S.: The convex geometry of linear inverse problems. Found. Comput. Math. **12**(6), 805–849 (2012)

14. Chatterjee, A., Govindu, V.M.: Efficient and robust large-scale rotation averaging. In: IEEE International Conference on Computer Vision, pp. 521–528 (2013)

15. Dellaert, F.: Factor graphs and GTSAM: a hands-on introduction. Technical report, Georgia Institute of Technology (2012)

16. Dellaert, F., Kaess, M.: Square root SAM: simultaneous localization and mapping via square root information smoothing. Int. J. Robot. Res. **25**(12), 1181–1203 (2006)

17. Dong, J., Soatto, S.: Domain-size pooling in local descriptors: DSP-SIFT. In: IEEE Conference on Computer Vision and Pattern Recognition, pp. 5097–5106 (2015)

18. Engels, E.C., Stewénius, H., Nistér, D.: Bundle adjustment rules. In: Photogram-metric Computer Vision, vol. 2, pp. 124–131 (2006)

19. Enqvist, O., Kahl, F., Olsson, C.: Non-sequential structure from motion. In: 2011 IEEE International Conference on Computer Vision Workshops (ICCV Work-shops), pp. 264–271. IEEE (2011)

20. Eriksson, A., Olsson, C., Kahl, F., Chin, T.J.: Rotation averaging and strong dual-ity. In: Proceedings of the IEEE Conference on Computer Vision and Pattern Recognition, pp. 127–135 (2018)

21. Fischler, M.A., Bolles, R.C.: Random sample consensus: a paradigm for model fitting with applications to image analysis and automated cartography. Commun. ACM **24**(6), 381–395 (1981)

22. Frahm, J.-M., et al.: Building Rome on a cloudless day. In: Daniilidis, K., Mara-gos, P., Paragios, N. (eds.) ECCV 2010. LNCS, vol. 6314, pp. 368–381. Springer, Heidelberg (2010). https://doi.org/10.1007/978-3-642-15561-1_27

23. Graham, M.C., How, J.P., Gustafson, D.E.: Robust incremental slam with consistency-checking. In: 2015 IEEE/RSJ International Conference on Intelligent Robots and Systems (IROS), pp. 117–124. IEEE (2015)

24. Hartley, R., Li, H.: An efficient hidden variable approach to minimal-case camera motion estimation. IEEE Trans. Pattern Anal. Mach. Intell. **34**(12), 2303–2314 (2012)

25. Hartley, R., Trumpf, J., Dai, Y., Li, H.: Rotation averaging. Int. J. Comput. Vision **103**(3), 267–305 (2013)

26. Hartley, R.I., Zisserman, A.: Multiple View Geometry in Computer Vision, 2nd edn. Cambridge University Press, Cambridge (2004)

27. Kasten, Y., Geifman, A., Galun, M., Basri, R.: Algebraic characterization of essential matrices and their averaging in multiview settings. In: Proceedings of the IEEE International Conference on Computer Vision, pp. 5895–5903 (2019)
28. Lajoie, P.Y., Hu, S., Beltrame, G., Carlone, L.: Modeling perceptual aliasing in slam via discrete-continuous graphical models. IEEE Robot. Autom. Lett. **4**(2), 1232–1239 (2019)
29. Latif, Y., Cadena, C., Neira, J.: Robust loop closing over time for pose graph slam. Int. J. Robot. Res. **32**(14), 1611–1626 (2013)
30. Lee, G.H., Fraundorfer, F., Pollefeys, M.: Robust pose-graph loop-closures with expectation-maximization. In: 2013 IEEE/RSJ International Conference on Intelligent Robots and Systems, pp. 556–563. IEEE (2013)
31. Lowe, D.G.: Distinctive image features from scale-invariant keypoints. Int. J. Comput. Vision **60**(2), 91–110 (2004)
32. Mangelson, J.G., Dominic, D., Eustice, R.M., Vasudevan, R.: Pairwise consistent measurement set maximization for robust multi-robot map merging. In: 2018 IEEE International Conference on Robotics and Automation (ICRA), pp. 2916–2923. IEEE (2018)
33. Martinec, D., Pajdla, T.: Robust rotation and translation estimation in multiview reconstruction. In: IEEE Conference on Computer Vision and Pattern Recognition, pp. 1–8 (2007)
34. Mehlhorn, K., Michail, D.: Implementing minimum cycle basis algorithms. J. Exp. Algorithmics (JEA) **11**, 2–5 (2007)
35. Moulon, P., Monasse, P., Marlet, R.: Global fusion of relative motions for robust, accurate and scalable structure from motion. In: Proceedings of the IEEE International Conference on Computer Vision, pp. 3248–3255 (2013)
36. Mur-Artal, R., Tardós, J.D.: ORB-SLAM2: an open-source slam system for monocular, stereo, and RGB-D cameras. IEEE Trans. Robot. **33**(5), 1255–1262 (2017)
37. Nishihara, R., Lessard, L., Recht, B., Packard, A., Jordan, M.I.: A general analysis of the convergence of ADMM. arXiv preprint arXiv:1502.02009 (2015)
38. Olson, E., Agarwal, P.: Inference on networks of mixtures for robust robot mapping. Int. J. Robot. Res. **32**(7), 826–840 (2013)
39. Olson, E., Leonard, J., Teller, S.: Fast iterative alignment of pose graphs with poor initial estimates. In: Proceedings 2006 IEEE International Conference on Robotics and Automation, ICRA 2006, pp. 2262–2269. IEEE (2006)
40. Robert, C.: Machine learning, a probabilistic perspective (2014)
41. Rosen, D.M., Carlone, L., Bandeira, A.S., Leonard, J.J.: A certifiably correct algorithm for synchronization over the special euclidean group. In: Goldberg, K., Abbeel, P., Bekris, K., Miller, L. (eds.) Algorithmic Foundations of Robotics XII: Proceedings of the Twelfth Workshop on the Algorithmic Foundations of Robotics, pp. 64–79. Springer, Cham (2020). https://doi.org/10.1007/978-3-030-43089-4_5
42. Sandler, M., Howard, A., Zhu, M., Zhmoginov, A., Chen, L.C.: Mobilenetv 2: inverted residuals and linear bottlenecks. In: Proceedings of the IEEE Conference on Computer Vision and Pattern Recognition, pp. 4510–4520 (2018)
43. Snavely, N., Seitz, S.M., Szeliski, R.: Photo tourism: exploring photo collections in 3D. ACM Trans. Graph. **25**, 835–846 (2006)
44. Snavely, N., Seitz, S.M., Szeliski, R.: Skeletal graphs for efficient structure from motion. In: IEEE Conference on Computer Vision and Pattern Recognition, vol. 1, p. 2 (2008)

45. Strecha, C., Von Hansen, W., Van Gool, L., Fua, P., Thoennessen, U.: On bench-marking camera calibration and multi-view stereo for high resolution imagery. In: 2008 IEEE Conference on Computer Vision and Pattern Recognition, pp. 1–8. IEEE (2008)
46. Sünderhauf, N., Protzel, P.: Switchable constraints for robust pose graph slam. In: 2012 IEEE/RSJ International Conference on Intelligent Robots and Systems, pp. 1879–1884. IEEE (2012)
47. Sünderhauf, N., Protzel, P.: Towards a robust back-end for pose graph slam. In: 2012 IEEE International Conference on Robotics and Automation, pp. 1254–1261. IEEE (2012)
48. Taketomi, T., Uchiyama, H., Ikeda, S.: Visual slam algorithms: a survey from 2010 to 2016. IPSJ Trans. Comput. Vis. Appl. **9**(1), 16 (2017)
49. Triggs, B., McLauchlan, P.F., Hartley, R.I., Fitzgibbon, A.W.: Bundle adjustment — a modern synthesis. In: Triggs, B., Zisserman, A., Szeliski, R. (eds.) IWVA 1999. LNCS, vol. 1883, pp. 298–372. Springer, Heidelberg (2000). https://doi.org/10.1007/3-540-44480-7_21
50. Tron, R., Vidal, R.: Distributed 3-D localization of camera sensor networks from 2-D image measurements. IEEE Trans. Autom. Control **59**(12), 3325–3340 (2014)
51. Wang, L., Singer, A.: Exact and stable recovery of rotations for robust synchro-nization. Inf. Inference **2**(2), 145–193 (2013)
52. Yedidia, J.S., Freeman, W.T., Weiss, Y.: Constructing free-energy approximations and generalized belief propagation algorithms. IEEE Trans. Inf. Theory **51**(7), 2282–2312 (2005)
53. Zach, C., Klopschitz, M., Pollefeys, M.: Disambiguating visual relations using loop constraints. In: 2010 IEEE Conference on Computer Vision and Pattern Recognition (CVPR), pp. 1426–1433. IEEE (2010)

Speech-Driven Facial Animation Using Cascaded GANs for Learning of Motion and Texture

Dipanjan Das, Sandika Biswas$^{(\boxtimes)}$, Sanjana Sinha, and Brojeshwar Bhowmick

Embedded Systems and Robotics, TCS Research, Kolkata, India
{dipanjan.da,biswas.sandika,sanjana.sinha,b.bhowmick}@tcs.com

Abstract. Speech-driven facial animation methods should produce accurate and realistic lip motions with natural expressions and realistic texture portraying target-specific facial characteristics. Moreover, the methods should also be adaptable to any unknown faces and speech quickly during inference. Current state-of-the-art methods fail to generate realistic animation from any speech on unknown faces due to their poor generalization over different facial characteristics, languages, and accents. Some of these failures can be attributed to the end-to-end learning of the complex relationship between the multiple modalities of speech and the video. In this paper, we propose a novel strategy where we partition the problem and learn the motion and texture separately. Firstly, we train a GAN network to learn the lip motion in a canonical landmark using DeepSpeech features and induce eye-blinks before transferring the motion to the person-specific face. Next, we use another GAN based texture generator network to generate high fidelity face corresponding to the motion on person-specific landmark. We use meta-learning to make the texture generator GAN more flexible to adapt to the unknown subject's traits of the face during inference. Our method gives significantly improved facial animation than the state-of-the-art methods and generalizes well across the different datasets, different languages, and accents, and also works reliably well in presence of noises in the speech.

Keywords: Realistic facial animation · Meta-learning · Cascaded GAN

1 Introduction

Speech-driven facial animation can be used for many applications such as video games, virtual assistants, animation movies, etc. and has thus garnered broad

---

D. Das and S. Biswas—Equal contribution.

Electronic supplementary material The online version of this chapter (https://doi.org/10.1007/978-3-030-58577-8_25) contains supplementary material, which is available to authorized users.

A. Vedaldi et al. (Eds.): ECCV 2020, LNCS 12375, pp. 408–424, 2020.
https://doi.org/10.1007/978-3-030-58577-8_25

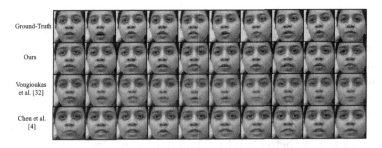

Fig. 1. Recent state-of-the-art methods [4,32] (Evaluated using their publicly available pre-trained models trained on LRW and TCD-TIMIT datasets respectively.) for speech-driven facial animation fail to accurately capture the mouth shapes and detailed facial texture on an unknown test subject whose facial characteristics differ from the training data. In these methods, the generated face can appear to be very different from the given target identity [32], or there can be a significant blur in the mouth region [4], leading to unrealistic face animation. On the other hand, our generated facial texture and mouth shapes can accurately resemble the ground-truth animation sequence.

interest. The problem of generating such facial animation is multifaceted, requiring accurate lip-sync, a natural expression like eye blinks, head orientations, capturing subject-specific traits like identity, lip deformations, etc. Also, the generation of such animation should not be overly dependent on the training set, and the method, therefore, should be adaptable to unknown faces and speeches quickly. Existing end-to-end learning methods [32,36] show poor adaptability given an unknown speech or face resulting in implausible animation. In order to overcome the problems of generating images directly from speech, Chen *et al.* [4] learn an intermediate high-level representation of motion from audio followed by texturing. Although this method preserves the identity but fails to produce accurate and realistic lip synchronization, as shown in Fig. 1 (the last row). On the other hand, [32] produces plausible lip motion but renders incorrect identity, as shown in Fig. 1 (third row). Therefore, the key challenges existing in the talking face problem are i) accurate lip synchronization along with identity preservation, ii) presence of natural expression like eye blinks, iii) fast adaptation to unknown subjects, and speeches for all practical purposes. Figure 1 shows that none of the most recent state-of-the-art methods produce animations which solves all the above challenges.

In this paper, we propose a novel strategy to solve the above-mentioned challenges. In essence, our method partitions the problem into four stages. First, we design a GAN network for learning motion on canonical (person-independent) landmark from DeepSpeech features obtained from audio. GAN is powerful in learning the subtle deformations in lips due to speech, and learning motion in a canonical face makes the method invariant to the person-specific face geometry. Along with this, DeepSpeech features alleviates the problems due to different accents and noises. With all these together, our method is able to learn motion from speech robustly and also adaptable to the unknown speech. Next, we impose eye blinks predicted from a separate network and transfer this learned canonical facial landmark motion to person-specific landmark motion using Procrustes alignment [29]. Subsequently, we train another GAN network for texture generation conditioning with the person-specific landmark. For better adaptation to the unknown subject and unknown head orientation, we meta-learn this GAN network using Model-Agnostic-Meta-Learning (MAML) algorithm [12]. At test time, we fine-tune the meta-learned model with few samples (20 images) to adapt quickly (approx. 100 secs fine-tuning) to the unseen subject. Our method produces significantly better results (Fig. 1, second row) with more accurate lip synchronization, better identity preservation, and easy adaptation to the unseen subjects over the state-of-the-art techniques. Figure 2 shows a conceptual diagram of our approach. The contributions of our work can be summarized as follows:

1. We design a GAN network for learning canonical facial landmark motion from a speech by using DeepSpeech features. The use of GAN helps to learn subtle deformations in lips accurately. DeepSpeech and motion learning in canonical face alleviates the problems in learning due to the variety of person-specific faces and speeches. Therefore the method is more robust to noises, accent, and different face geometry.
2. We use model-agnostic-meta-learning to train another GAN for texture generation conditioned on the person-specific texture. GAN produces high fidelity face images from given landmarks and because the network is meta-learned it provides quick as well as a better adaptation to the unseen subject using a few examples at the fine-tuning stage.

2 Related Work

Speech-Driven Face Animation: In recent years many researchers have focused on the synthesis of 2D talking face video from audio input [3,4,7,28, 30,32,36]. The methods which are most relevant to us are [4,7,28,31,32,36,37] which animate an entire face from speech. Earlier methods that learn subject-specific 2D facial animation [11,13,30] require a large amount of training data of the target subject. The first subject-independent learning method [7] achieves good lip synchronization, but images generated require additional de-blurring. Hence GAN-based methods [4,5,28,31,32,36] were proposed for generating sharp facial texture in speech-driven 2D facial animation. Although these methods animate the entire face, they mainly target lip synchronization with audio [4,5,28,36], by learning disentangled audio representations [22] for robustness to noise and emotional content in audio, and disentangled audio-visual representations [36] to segregate identity information from speech [4,36]. However, these methods have not addressed the other aspects for the realism of synthesized face video, such as natural expressions, identity preservation of target, etc.

Beyond Lip Synchronization - Realistic Facial Animation: The absence of spontaneous movements such as eye blinks in synthesized face videos can

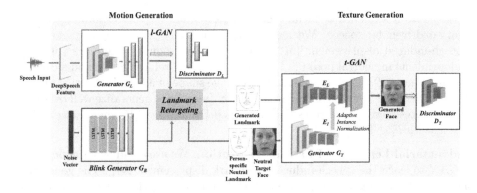

Fig. 2. Block diagram of our proposed method for speech-driven facial animation.

be easily perceived as fake [21]. Recent works [31,32] have tried to address the problem of video realism by using adversarial learning of spontaneous facial gestures such as blinks. However, the generated videos with natural expressions may still imperfectly resemble the target identity, which can also be perceived as being fake. To retain facial identity information from the given identity image of target, image attention has been learnt with the help of facial landmarks in a hierarchical approach [4]. In this approach [4] the audio is used to generate motion on 2D facial landmarks, and the image texture is generated by conditioning on the landmarks. Although the generated texture in static facial regions can retain the texture from the identity image, the generated texture in regions of motion, especially the eyes and mouth, can differ from the target identity. Hence identity-specific texture generation is needed for realistic rendering of a target's talking face.

3   Proposed Methodology

Given an arbitrary speech and a set of images of a target face, our objective is to synthesize speech synchronized realistic animation of the target face. Inspired by [4], we capture facial motion in a lower dimension space represented by 68 facial landmark points and synthesize texture conditioned on the motion of predicted landmarks. To this end, we use a GAN based cascaded learning approach consisting of the following: (1) Learning speech-driven motion on $2D$ facial landmarks independent of identity, (2) Learning eye blink motion on landmarks, (3) Landmark retargeting to generate target-specific facial shape along with motion, (4) Generating facial texture from the motion of landmarks. Figure 2 shows our overall approach.

3.1   Speech-Driven Motion Generation on Facial Landmarks

Let, A be an audio signal represented by a series of overlapping audio windows $\{W_t | t \in [0, T]\}$ with corresponding feature representations $\{F_t\}$. Our goal is to generate a sequence of facial landmarks $\{\ell_t \in \mathbb{R}^{68 \times 2}\}$ corresponding to the motion driven by speech. We learn a mapping $\mathcal{M}_L : F_t \rightarrow \delta\ell_t^m$ to generate speech-induced displacement $\{\delta\ell_t^m \in \mathbb{R}^{68 \times 2}\}$ on a canonical landmark (person-independent) in neutral pose ℓ_p^m. Learning the speech-related motion on a canonical landmark ℓ_p^m, which represents the average shape of a face, is effective due to the invariance of any specific facial structure. In order to generalize well over different voices, accent etc. we use a pre-trained DeepSpeech [15] model to extract the feature $F_t \in \mathbb{R}^{6 \times 29}$.

Adversarial Learning of Landmark Motion. We use an adversarial network *l-GAN* to learn the speech-induced landmark displacement $\mathcal{M}_L$. The generator network G_L generates displacements $\{\delta\ell_t^m\}$ of a canonical landmark from a neutral pose ℓ_p^m. Our discriminator D_L takes the resultant canonical landmarks $\{\ell_t^m = \ell_p^m + \delta\ell_t^m\}$ and the ground-truth canonical landmarks as inputs to learn the real against fake. The loss functions used for training *l-GAN* are as follows: *Distance loss:* This is L_2 loss between generated canonical landmarks $\{\ell_t^m\}$ and ground-truth landmarks $\{\ell_t^{m*}\}$ for each frame t.

$$\mathcal{L}_{dist} = ||\ell_t^m - \ell_t^{m*}||_2^2 \tag{1}$$

Regularization loss: We use L_2 loss between consecutive frames for ensuring temporal smoothness in predicted landmarks.

$$\mathcal{L}_{reg} = ||\ell_t^m - \ell_{t-1}^m||_2^2 \tag{2}$$

Direction Loss: We also impose a consistency in the motion vectors $(\overrightarrow{\delta\ell_t^m})$ by:

$$\mathcal{L}_{dir} = ||\overrightarrow{\delta\ell_t^m} - \overrightarrow{\delta\ell_t^{m*}}||_2^2 \quad where, \quad \overrightarrow{\delta\ell_t^m} = \begin{cases} 1, & if \ \ell_{t+1}^m > \ell_t^m \\ 0, & otherwise \end{cases} \tag{3}$$

GAN Loss: We use an adversarial loss for capturing detailed mouth deformations.

$$\mathcal{L}_{gan} = \mathbb{E}_{\ell_t^{m*}}[log(D_L(\ell_t^{m*}))] + \mathbb{E}_{F_t}[log(1 - D_L(G_L(\ell_p^m, F_t)))] \tag{4}$$

The final objective function which is to be minimized is as follows:

$$\mathcal{L}_{motion} = \lambda_{dist}\mathcal{L}_{dist} + \lambda_{reg}\mathcal{L}_{reg} + \lambda_{dir}\mathcal{L}_{dir} + \lambda_{gan}\mathcal{L}_{gan} \tag{5}$$

where, λ_{dist}, λ_{reg}, λ_{dir}, λ_{gan} are experimentally set to 1, 0.5, 0.5 and 1, as presented in the ablation study (Sect. 4.3).

3.2   Spontaneous Eye Blink Generation on Facial Landmarks

Eye blinks are essential for realism of synthesized face animation, but not dependent on speech. Therefore, we propose an unsupervised method for generation of realistic eye blinks through learning a mapping $\mathcal{M}_B : Z_t \rightarrow \delta\ell_t^e$ from a random

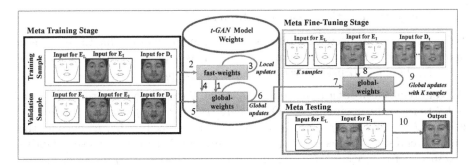

Fig. 3. State transitions of fast-weights (FW) and global-weights (GW) of *t-GAN* during meta-training. The sequence of training schedule: (1) copying FW to GW to keep global state unchanged during the training, (2)–(3) update the FW in iterations, (4)–(5) compute validation loss using FW, (6) update GW using total validation loss, (7) copy GW for the fine-tuning, (8)–(9) updating the GW using K sample images, (10) using the updated GW to produce target subject's face.

noise $Z_t \sim \mathcal{N}(\mu, \sigma^2)|t \in (0, T)$ to eye landmark displacements $\{\delta\ell_t^e \in \mathbb{R}^{22 \times 2}\}$. Our blink generator network G_B learns the blink pattern and duration through the mapping $\mathcal{M}_B$ and generates a sequence of eye landmark displacements $\{\delta\ell_t^e\}$ on the canonical face by minimizing the MMD (Maximum Mean Discrepancy) [14] loss defined as follows:

$$L_{MMD} = \mathbb{E}_{X,X' \sim p}\mathcal{K}(X, X') + \mathbb{E}_{Y,Y' \sim q}\mathcal{K}(Y, Y') - 2\mathbb{E}_{X \sim p, Y \sim q}\mathcal{K}(X, Y) \quad (6)$$

where, $\mathcal{K}(x, y)$ is defined as $exp(-\frac{|x-y|^2}{2\sigma})$, p and q represents samples from distributions X and Y of GT $\{\delta\ell_t^{e*}\}$ and generated eye landmark motion $\{\delta\ell_t^e\}$ respectively. We also use a Min-max regularization to ensure that the range of the generated landmarks matches with the average range of average displacements present in the training data. We augment the eye blink with the speech-driven canonical landmark motion (Sect. 3.1) and retarget (Sect. 3.3) the combined landmarks $\ell_t^M = \{\ell_t^m \bigcup \ell_t^e\}$, where $\{\ell_t^e = \ell_p^e + \delta\ell_t^e\}$, to generate the person-specific landmarks $\{\ell_t\}$ for subsequent use for texture generation.

3.3 Landmark Retargeting

We retarget the canonical landmarks $\{\ell_t^M\}$ generated by G_L and G_B, to person-specific landmarks $\{\ell_t\}$ (used for texture generation) as follows:

$$\ell_t = \ell_p + \delta\ell_t \text{ where, } \delta\ell_t = \delta\ell_t' * S(\ell_t)/S(\mathcal{T}(\ell_t^M)) \text{ ; } \delta\ell_t' = \mathcal{T}(\ell_t^M) - \mathcal{T}(\ell_p^m) \quad (7)$$

where, ℓ_p is the person-specific landmark in neutral pose (extracted from the target image), $S(\ell) \in \mathbb{R}^2$ is the scale (height $\times$ width) of ℓ and $\mathcal{T} : \ell \rightarrow \ell'$ represents a Procrustes (rigid) alignment of ℓ with ℓ_p.

3.4   Image Generation from Landmarks

We use the person-specific landmarks $\{\ell_t\}$ containing motion due to the speech and the eye blink to synthesize animated face images $\{I_t\}$ by learning a mapping $\mathcal{M}_T : (\ell_t, \{\mathcal{I}^n\}) \rightarrow I_t$ using given target images $\{\mathcal{I}^n | n \in [0, N]\}$.

Adversarial Generation of Image Texture. We use an adversarial network t-GAN to learn the mapping $\mathcal{M}_T$. Our generator network G_T consists of a texture encoder E_I and landmark encoder-decoder E_L influenced by E_I. E_I encodes the texture representation as $e = E_I(\mathcal{I}^n)$ for the input N images. We use Adaptive Instance Normalization [17] to modulate the bottleneck of E_L using e. Finally we use a discriminator network D_T to discriminate the real images from the fake. The losses for training the t-GAN are as follows:

Reconstruction Loss: L_2 distance between synthesized $\{I_t\}$ and GT images $\{I_t^*\}$,

$$\mathcal{L}_{pix} = ||I_t - I_t^*||_2^2 \tag{8}$$

Adversarial Loss: For sharpness of the texture an adversarial loss is minimized.

$$\mathcal{L}_{adv} = \mathbb{E}_{I_t^*}[log(D_T(\mathcal{I}^n, I_t^*))] + \mathbb{E}_{\ell_t}[log(1 - D_T(\mathcal{I}^n, G_T(\ell_t, \mathcal{I}^n)))] \tag{9}$$

Perceptual Loss: We use a perceptual loss [18] which is the difference in feature representations vgg_1 and vgg_2 of generated and ground truth images obtained using pre-trained $VGG19$ and $VGGFace$ [26] respectively.

$$\mathcal{L}_{feat} = \alpha_1||vgg_1(I_t) - vgg_1(I_t^*)||_2^2 + \alpha_2||vgg_2(I_t) - vgg_2(I_t^*)||_2^2 \tag{10}$$

The total loss minimized for training G_T network is defined as,

$$\mathcal{L}_{texture} = \lambda_{pix}\mathcal{L}_{pix} + \lambda_{adv}\mathcal{L}_{adv} + \lambda_{feat}\mathcal{L}_{feat} \tag{11}$$

Meta-Learning. We use model-agnostic meta-learning (MAML) [12] to train our t-GAN for quick adaptation to the unknown face at inference time using few images. MAML trains on a set of tasks T called episodes. For each task, the number of samples for training and validation is d_{trn} and d_{qry}, respectively. For our problem, we define subject specific task as $T^s = (I_i^s, l_j^s) \cdots (I_{i_{d_{trn}+d_{qry}}}^s, l_{j_{d_{trn}+d_{qry}}}^s)$ of task set $\{T^s\}$, where s is the subject index, I_i^s is the ith face image for subject s, l_j^s is the jth landmark for the same subject s. During meta-training, MAML store the current weights of the t-GAN into global-weights and train the t-GAN with d_{trn} samples for m iteration using a constant step size. During each iteration, it measures the loss L^i with the validations samples d_{qry}. Then the total loss $L = L^1 + L^2 \cdots + L^m$ is used to update global-weights as shown in Fig 3. The resultant direction of the global-weights encodes a global information of the t-GAN network for all the tasks, which is used as an initialization for fine-tuning during inference.

During fine-tuning, we initialize the t-GAN from the global-weights and update the weights by minimizing the loss as described in Eq. 11. We use a few ($K = 20$) example images of the target face for the fine-tuning.

4   Experimental Results

In this section, we present the experimental results of our proposed method on different datasets along with the network ablation study. We also show that the accuracy of our cascaded GAN based approach is quite higher than an alternate regression-based motion and texture generation. Our meta-learning based texture generation strategy makes our method to be more adaptable to any unknown faces. The combined result is a significantly better facial animation from speech than the state-of-the-art methods in terms of both quantitative and qualitative results. In what follows, we present detailed experiments for each of the building blocks of our pipeline.

4.1   Datasets

We use TCD-TIMIT [16], GRID [9], and Voxceleb [24] datasets for our experiments. We train our model only on TCD-TIMIT and test the model on GRID as well as our own recorded data for showing the efficacy of our method on cross datasets with completely unknown faces. Our training split contains 3378 videos from 49 subjects with around 6913 sentences uttered in a limited variety of accents. Test split (same as [32]) of TCD-TIMIT and GRID datasets contains 1631 and 9957 videos respectively.

4.2   Motion Generation on Landmarks

Network Architecture of *l-GAN*: The architecture of the generator network G_L of *l-GAN* is built upon the encoder-decoder architecture used in [10] for generating mesh vertices. LeakyReLU [33] activation is used after each layer of the encoder network. The input DeepSpeech features are encoded to a 33 dimensional vector (PCA coefficients), which is decoded to obtain the canonical landmark displacements from the neutral pose. The discriminator network D_L consists of 2 linear layers, which re-encodes the predicted or ground-truth landmarks into PCA coefficients to discriminate between real and fake. We initialize weights of the last layer of the decoder in G_L and the first layer of D_L with 33 PCA components computed over the landmark displacements in training data.

Network Architecture of Blink Generator G_B: We use RNN to predict a sequence of displacements $\mathbb{R}^{n \times 75 \times 44}$, i.e x, y coordinates of eye landmarks $\{\ell_t^e \in \mathbb{R}^{22 \times 2}\}$ over 75 timestamps from given noise vector $z \sim \mathcal{N}(\mu, \sigma^2)$ with $z \in \mathbb{R}^{n \times 75 \times 10}$. Similar to the G_L of our *l-GAN* network, the last linear layer weights are initialized with PCA components (with 99% variants) computed over ground-truth eye landmark displacements.

Training Details: We extract audio features from the second last layer (before softmax) of the DeepSpeech [15] network. We consider sliding windows of Δt features for providing a temporal context to each video frame. To compute accurate ground-truth facial landmark required for our training, we experiment with different existing state-of-the-art methods [1, 19, 34] and find that the combination of OpenFace [1] and face segmentation [34] to be most effective for our purpose. Our speech-driven motion generation network is trained on the TCD-TIMIT dataset. The canonical landmarks used for training *l-GAN* are generated by an inverse process of the landmark retargeting method, as described in Sect. 3.3. We train our *l-GAN* network with a batch size of 6. Losses saturate after 40 epochs, which takes around 3 h on a single GPU of Quadro P5000 system. We use Adam [20] optimization with a learning rate of $2e - 4$ for training both of our *l-GAN* and blink generator network.

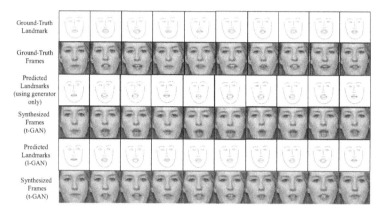

Fig. 4. Performance comparison of *l-GAN* using only generator (third row) and the complete GAN (fifth row). The regression-based approach cannot capture the finer details like "a" and "o" of lip motion without the help of the discriminator.

Quantitative Results: We present our quantitative results in Table 1 and 2. For comparative analysis we use publicly availabe pre-trained models of state-of-the-art methods [4, 32, 36]. Our model is trained on TCD-TIMIT [16], while models of [4] and [36] are pre-trained on LRW [8] dataset. [32] is trained on both TCD-TIMIT and GRID separately.

For evaluating and comparing the accuracy of lip synchronization produced by our method, we use a) LMD, Landmark Distance (as used in [3,4]) and b) Audio-Visual synchronization metrics (AV Offset and AV confidence produced by Syncnet [6]). For all methods, LMD is computed using lip landmarks extracted from the final generated frames. A lower value of LMD and AV offset with higher AV confidence indicates better lip synchronization. Our method shows better accuracy compared to state-of-the-art methods. Our models trained on TCD-TIMIT also shows good generalization capability in cross-dataset evaluation on GRID dataset (Table 2). Although [4] also generates facial landmarks from audio features (MFCC), unlike their regression-based approach, our use of

DeepSpeech features, landmark retargeting, and adversarial learning results in improved accuracy of landmark generation.

Moreover, our facial landmarks contain natural eye blink motion for added realism. We detect eye blinks using a sharp drop in EAR (Eye Aspect Ratio) signal [32] calculated using landmarks of eye corners and eyelids. Blink duration is calculated as the number of consecutive frames between the start and end of the sharp drop in the EAR. The average blink duration and blink frequencies generated from our method is similar to that of natural human blinks. Our method produces a blink rate of 0.3 blinks/s and 0.38 blinks/s (Table 1 and 2) for TCD-TIMIT and GRID datasets respectively which is similar to

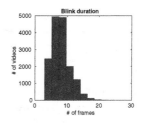

Fig. 5. Statistics of average blink duration.

the average human blink rate of 0.28–0.4 blinks/s. Also, we achieve an average blink duration of 0.33 s and 0.4 s, which is similar to as reported in ground-truth (Table 1 and 2). In Fig. 5 we present distribution of blink durations (in no. of frames) in synthesized videos of GRID and TCD-TIMIT datasets. So, our method can produce realistic eye blinks similar to [32], but with better identity-preserved texture, due to our decoupled learning of eye blinks on landmarks.

Ablation Study: An ablation study of window size Δt (Fig. 6) has indicated a value of $\Delta t = 6$ frames (duration of around 198 ms) results in the lowest LMD. In Fig. 6 we also present an ablation study for different losses used for training our motion prediction network. It is seen that the proposed loss L_{motion} achieves the best accuracy. Use of L_2 regularization loss helps to achieve temporal smoothness and consistency on predicted landmarks over consecutive frames. We use direction loss (Eq. 3) to capture the relative movements of landmarks over consecutive frames. Using direction loss helps to achieve faster convergence of our landmark prediction network. Use of DeepSpeech features helps us to achieve robustness in lip synchronization even for audios with noise, different accents, and different languages (Please refer to supplementary video). We experiment to evaluate the robustness of our *l-GAN* with different levels of noise by adding synthetic noise in the audio input. Figure 6 shows upto -30 dB, the lip motion does not get affected by the noise and starts degrading afterward. In Fig. 4 we present a qualitative result of the landmark generation network on the TCD-TIMIT dataset. It shows the effectiveness of using discriminator in *l-GAN*.

4.3 Texture Generation from Landmark Motion:

Network Architecture of *t-GAN*: We adapt a similar approach of an image-to-image translation method proposed by [18] for implementation of our texture generator G_T. Our landmark encoder-decoder network E_L takes generated person-specific landmarks represented as images of size $\mathbb{R}^{3 \times 256 \times 256}$ and E_I takes channel-wise concatenated face images with corresponding landmark images of the target subject. We use six downsampling layers for both E_I and the encoder

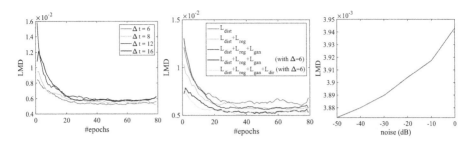

Fig. 6. Left: Landmark Distance (LMD) with varying context window (Δt) of deep-speech features. **Middle:** LMD with different losses used for training speech-driven motion generation network. **Right:** Error in lip synchronization (LMD) with different noise levels.

of E_L and six upsampling layers for the decoder of the E_L. To generate high fidelity images, we use residual block for our downsampling and upsampling layers similar to [2]. We use instance normalization for the residual blocks and adaptive instance normalization on the bottle-neck layer of the E_L using the activation produced by the last layer of E_I. Moreover, to generate sharper images, we use a similar self-attention method as [35] at the 32×32 layer activation of downsampling and upsampling layers. Our discriminator network D_T consists of 6 residual blocks similar to E_I, followed by a max pooling and a fully connected layer. To stabilize our GAN training, we use spectral normalization [23] for both generator and discriminator network.

Training and Testing Details: We meta-train our t-GAN network using ground-truth landmarks following the teacher forcing strategy. We use fixed step size [12] $1e-3$ and Adam as the meta-optimizer [12] with learning rate $1e-4$. The values of α_1, α_2, λ_{pix}, λ_{adv} λ_{feat} and are experimentally set to $1e-1$, $2e-3$, 0.5, 1.0 and 0.3 respectively. At test time, we use 5 images of the target identity, and the person-specific landmark generated by the l-GAN to produce the output images. Before testing, we perform a fine-tuning of the meta-trained network using 20 images of the target person and the corresponding ground-truth landmarks. We use a clustered GPU of NVIDIA Tesla V100 for meta-training and Quadro P5000 for fine-tuning our network.

Table 1. Comparative results on TCD-TIMIT [16].

Methods	Trained on	PSNR	SSIM	CPBD	LMD	ACD (10^{-4})	FaceNet	AVOff.	AVConf.	Blink/s	Blink Dur. (s)
[32]	TCD-TIMIT	24.2	0.73	0.30	2.1	1.76	0.578	1	5.5	0.19	0.33
[4]	LRW	20.31	0.59	0.16	1.71	1.1	0.409	1	3.91	NA	NA
[36]	LRW	23.82	0.63	0.14	1.9	1.24	0.472	1	1.94	NA	NA
Ours	TCD-TIMIT	**30.7**	**0.74**	**0.61**	**1.4**	**0.98**	**0.377**	1	**5.91**	**0.3**	**0.33**

Table 2. Comparative results on GRID [9] (our cross-dataset evaluation).

Methods	Trained on	PSNR	SSIM	CPBD	LMD	ACD (10^{-4})	FaceNet	WER (%)	AVOff.	AVConf.	Blink/s	Blink Dur. (s)
[32]	GRID	27.1	0.81	0.26	1.56	1.47	0.802	23.1	1	7.4	0.45	0.36
[4]	LRW	23.98	0.76	0.06	1.29	1.57	0.563	31.1	1	5.85	NA	NA
[36]	LRW	22.79	0.76	0.04	1.49	1.78	0.628	36.72	2	4.29	NA	NA
Ours	TCD-TIMIT	**29.9**	**0.83**	**0.29**	**1.22**	**1.12**	**0.466**	**19.33**	1	**7.72**	0.38	0.4

Quantitative Results: Here, we present the comparative performance of our GAN-based texture generation network with the most recent state-of-the-art methods [4,36] and [32]. Similar to *l-GAN*, the *t-GAN* is trained on TCD-TIMIT and evaluated on the test split of GRID, TCD-TIMIT and the unknown subjects. We compute the performance metrics PSNR, SSIM (Structural Similarity), CPBD (Cumulative Probability Blur Detection) [25], ACD (Average Content Distance) [32] and similarity between FaceNet [27] features for reference identity image (1st frame of ground truth video) and the predicted frames. Our method outperforms (Table 1 and 2) the state-of-the-art methods for all the datasets indicating better image quality. Due to inaccessibility of LRW [8] dataset we have evaluated our texture generation method on Voxceleb [24] dataset which gives average PSNR, SSIM and CPBD of 25.2, 0.63, 0.11 respectively. Our method does not produce head motion and synthesizes texture with frontal face. Hence, for Voxceleb, our method gives poor performance than that of TCD-TIMIT and GRID.

Qualitative Results: Figure 10 shows qualitative comparison against [4,36] and [32]. It can be seen that [32] and [36] fail to preserve the identity of the test subject over frames in the synthesized video. Although [4] can preserve the identity, there is a significant blur, especially around the mouth region. Also, it lacks any natural movements over face except lip or jaw motion yielding an unrealistic face animation. On the other hand, our method can synthesize high fidelity images (256×256) with preserved identity and natural eye motions. Figure 7 shows the qualitative comparison of our GAN based texture generation against a regression-based (without discriminator) network output where it is evident that our GAN based network gives more accurate lip deformation with similar motion as ground-truth.

Ablation Study: We show a detailed ablation study on the TCD-TIMIT dataset to find out the effect of different losses (Table 3). Among channel-wise concatenation and adaptive instance normalization, which are the two different approaches in neural style transfer, adaptive instance normalization works better for our problem. Figure 7 and quantitative result (Table 3) show that GAN based method produces more accurate lip deformation than the regression-based method, which always produces an overly smooth outcome. Figure 9 shows the ablation study for the number of images required for fine-tuning. Using single image for fine-tune yields average PSNR, SSIM and CPBD values of 27.95,

Fig. 7. Qualitative comparison between our *t-GAN* based method (Row 2) against the regression based generator G_T (Row 3) method. Use of GAN results in more accurate mouth shape.

Table 3. Ablation Study of our model. CC = channel wise concatenation.

Methods	PSNR	SSIM	CPBD	LMD
Model+CC+L_{pix}	27.2	0.62	0.51	1.65
Model+ADIN+L_{pix}	28.3	0.66	0.56	1.57
Model+ADIN+L_{pix} $+L_{feat}$	28.9	0.70	0.58	1.5
Model+ADIN+L_{pix} $+L_{feat}+L_{adv}$	**30.7**	**0.74**	**0.61**	**1.4**

Table 4. Epoch-wise quantitative analysis in fine-tuning.

DataSet	Epoch	PSNR	SSIM	CPBD	LMD
GRID	1	21.5	0.58	0.04	6.70
	5	27.3	0.76	0.08	1.47
	10	29.8	0.83	0.29	1.22
TCD-TIMIT	1	20.6	0.59	0.38	7.80
	5	28.1	0.70	0.58	1.64
	10	30.7	0.74	0.61	1.4

0.82, and 0.27 respectively for GRID dataset. Our method can produce accurate motion and texture after 10 epochs (Table 4) of fine-tuning with $K = 20$ sample images.

Meta-Learning vs. Transfer-Learning: We compare the performance of MAML [12] and transfer-learning for our problem. To this end, we train a model with the

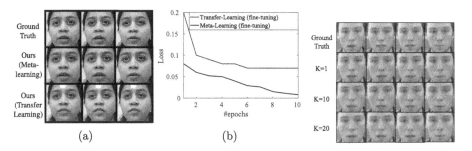

(a)   (b)

Fig. 8. Comparison between the fine-tuning stage of meta-learning and transfer-learning. Meta-learning (black) provides better initialization than the transfer-learning (blue). (Color figure online)

Fig. 9. Ablation study for no. of images during fine-tuning on GRID dataset.

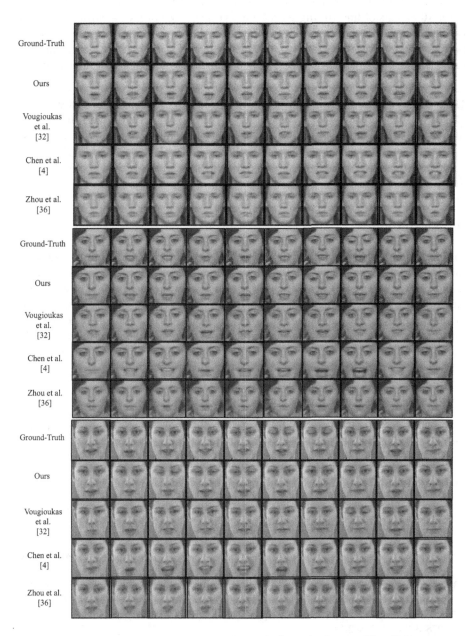

Fig. 10. Qualitative comparison with the latest SoA methods on TCD-TIMIT dataset (Upper 10 rows) and GRID dataset (Lower 5 rows). Our results indicate improved identity preservation of the subject, good lip synchronization, detailed texture (such as teeth), lesser blur, and presence of randomly introduced eye blinks.

same model architecture until it converges to similar loss values as meta-learning. After 10 epochs of fine-tuning with 20 images, the loss of meta-learning is much lower (Fig. 8b) than the transfer-learning (fine-tuning) and produces significantly better visual results (Fig. 8a) than transfer-learning. Moreover, fine-tuning of meta-learned network takes nearly 10 epochs with 20 images, which is much smaller than the transfer-learning based fine-tuning.

User Study: We assess the realism of our animation through a user study, where 25 users are asked to rate between 1(fake)–10(real) for 30 (10 videos from each of the methods) synthesized videos randomly selected from TCD-TIMIT and GRID. Our method achieves better realism scores with an average of 72.76% compared to state-of-the-art methods [4] and [32] with average scores 58.48% and 61.29% respectively.

5   Conclusion

In this paper, we present a novel strategy for speech driven facial animation. Our method produces realistic facial animation for unknown subjects with different languages and accents in speech showing generalization capability. We attribute this advantage due to our separate learning of motion and texture generator GANs with meta-learning capability. As a combined result, our method outperforms state-of-the-art methods significantly. In the future, we would like to study the effect of meta-learning for learning landmark motion from speech to mimic personalized speaking styles.

Acknowledgement. We would like to thank Dr. Angshul Majumdar, Professor at IIIT Delhi, India for helping us to get access of TCD-TIMIT dataset required for this research. We would also like to thank Govind Gopal from our infrastructure team for his immense support for creating a clustered GPU setup for training.

References

1. Baltrusaitis, T., Zadeh, A., Lim, Y.C., Morency, L.P.: Openface 2.0: facial behavior analysis toolkit. In: 2018 13th IEEE International Conference on Automatic Face & Gesture Recognition (FG 2018), pp. 59–66. IEEE (2018)
2. Brock, A., Donahue, J., Simonyan, K.: Large scale GAN training for high fidelity natural image synthesis. arXiv preprint arXiv:1809.11096 (2018)
3. Chen, L., Li, Z., Maddox, R.K., Duan, Z., Xu, C.: Lip movements generation at a glance. In: Ferrari, V., Hebert, M., Sminchisescu, C., Weiss, Y. (eds.) ECCV 2018. LNCS, vol. 11211, pp. 538–553. Springer, Cham (2018). https://doi.org/10.1007/978-3-030-01234-2_32
4. Chen, L., Maddox, R.K., Duan, Z., Xu, C.: Hierarchical cross-modal talking face generation with dynamic pixel-wise loss. In: Proceedings of the IEEE Conference on Computer Vision and Pattern Recognition pp. 7832–7841 (2019)
5. Chen, L., Srivastava, S., Duan, Z., Xu, C.: Deep cross-modal audio-visual generation. In: Proceedings of the on Thematic Workshops of ACM Multimedia. pp. 349–357. ACM (2017)

6. Chung, J.S., Zisserman, A.: Out of time: automated lip sync in the wild. In: Workshop on Multi-view Lip-reading, ACCV (2016)
7. Chung, J.S., Jamaludin, A., Zisserman, A.: You said that? arXiv preprint arXiv:1705.02966 (2017)
8. Chung, J.S., Zisserman, A.: Lip reading in the wild. In: Lai, S.-H., Lepetit, V., Nishino, K., Sato, Y. (eds.) ACCV 2016. LNCS, vol. 10112, pp. 87–103. Springer, Cham (2017). https://doi.org/10.1007/978-3-319-54184-6_6
9. Cooke, M., Barker, J., Cunningham, S., Shao, X.: An audio-visual corpus for speech perception and automatic speech recognition. J. Acoust. Soc. Am. **120**(5), 2421–2424 (2006)
10. Cudeiro, D., Bolkart, T., Laidlaw, C., Ranjan, A., Black, M.J.: Capture, learning, and synthesis of 3d speaking styles. In: Proceedings of the IEEE Conference on Computer Vision and Pattern Recognition, pp. 10101–10111 (2019)
11. Fan, B., Wang, L., Soong, F.K., Xie, L.: Photo-real talking head with deep bidirectional lstm. In: 2015 IEEE International Conference on Acoustics, Speech and Signal Processing (ICASSP), pp. 4884–4888. IEEE (2015)
12. Finn, C., Abbeel, P., Levine, S.: Model-agnostic meta-learning for fast adaptation of deep networks. In: Proceedings of the 34th International Conference on Machine Learning-Volume 70, pp. 1126–1135. JMLR. org (2017)
13. Garrido, P., et al.: VDUB: modifying face video of actors for plausible visual alignment to a dubbed audio track. In: Computer Graphics Forum, vol. 34, pp. 193–204. Wiley Online Library (2015)
14. Gretton, A., Borgwardt, K., Rasch, M., Schölkopf, B., Smola, A.J.: A kernel method for the two-sample-problem. In: Advances in Neural Information Processing Systems, pp. 513–520 (2007)
15. Hannun, A., et al.: Deep speech: scaling up end-to-end speech recognition. arXiv preprint arXiv:1412.5567 (2014)
16. Harte, N., Gillen, E.: TCD-TIMIT: an audio-visual corpus of continuous speech. IEEE Trans. Multimedia **17**(5), 603–615 (2015)
17. Huang, X., Belongie, S.: Arbitrary style transfer in real-time with adaptive instance normalization. In: Proceedings of the IEEE International Conference on Computer Vision, pp. 1501–1510 (2017)
18. Johnson, J., Alahi, A., Fei-Fei, L.: Perceptual losses for real-time style transfer and super-resolution. In: Leibe, B., Matas, J., Sebe, N., Welling, M. (eds.) ECCV 2016. LNCS, vol. 9906, pp. 694–711. Springer, Cham (2016). https://doi.org/10.1007/978-3-319-46475-6_43
19. Kazemi, V., Sullivan, J.: One millisecond face alignment with an ensemble of regression trees. In: Proceedings of the IEEE Conference on Computer Vision and Pattern Recognition, pp. 1867–1874 (2014)
20. Kingma, D.P., Ba, J.: Adam: a method for stochastic optimization. arXiv preprint arXiv:1412.6980 (2014)
21. Li, Y., Chang, M.C., Lyu, S.: In ICTU oculi: exposing AI generated fake face videos by detecting eye blinking. arXiv preprint arXiv:1806.02877 (2018)
22. Mittal, G., Wang, B.: Animating face using disentangled audio representations. In: The IEEE Winter Conference on Applications of Computer Vision, pp. 3290–3298 (2020)
23. Miyato, T., Kataoka, T., Koyama, M., Yoshida, Y.: Spectral normalization for generative adversarial networks. arXiv preprint arXiv:1802.05957 (2018)
24. Nagrani, A., Chung, J.S., Zisserman, A.: VoxCeleb: a large-scale speaker identification dataset. arXiv preprint arXiv:1706.08612 (2017)

25. Narvekar, N.D., Karam, L.J.: A no-reference perceptual image sharpness metric based on a cumulative probability of blur detection. In: 2009 International Workshop on Quality of Multimedia Experience, pp. 87–91. IEEE (2009)

26. Parkhi, O.M., Vedaldi, A., Zisserman, A., et al.: Deep face recognition. In: BMVC, vol. 1, 6 (2015)

27. Schroff, F., Kalenichenko, D., Philbin, J.: FaceNet: a unified embedding for face recognition and clustering. In: Proceedings of the IEEE Conference on Computer Vision and Pattern Recognition, pp. 815–823 (2015)

28. Song, Y., Zhu, J., Wang, X., Qi, H.: Talking face generation by conditional recurrent adversarial network. arXiv preprint arXiv:1804.04786 (2018)

29. Srivastava, A., Joshi, S.H., Mio, W., Liu, X.: Statistical shape analysis: clustering, learning, and testing. IEEE Trans. Pattern Anal. Mach. Intell. **27**(4), 590–602 (2005)

30. Suwajanakorn, S., Seitz, S.M., Kemelmacher-Shlizerman, I.: Synthesizing Obama: learning lip sync from audio. ACM Trans. Graphics (TOG) **36**(4), 95 (2017)

31. Vougioukas, K., Center, S.A., Petridis, S., Pantic, M.: End-to-end speech-driven realistic facial animation with temporal GANs. In: Proceedings of the IEEE Conference on Computer Vision and Pattern Recognition Workshops, pp. 37–40 (2019)

32. Vougioukas, K., Petridis, S., Pantic, M.: Realistic speech-driven facial animation with GANs. Int. J. Comput. Vision pp. 1–16 (2019)

33. Xu, B., Wang, N., Chen, T., Li, M.: Empirical evaluation of rectified activations in convolutional network. arXiv preprint arXiv:1505.00853 (2015)

34. Yu, C., Wang, J., Peng, C., Gao, C., Yu, G., Sang, N.: BiSeNet: bilateral segmentation network for real-time semantic segmentation. In: Ferrari, V., Hebert, M., Sminchisescu, C., Weiss, Y. (eds.) ECCV 2018. LNCS, vol. 11217, pp. 334–349. Springer, Cham (2018). https://doi.org/10.1007/978-3-030-01261-8_20

35. Zhang, H., Goodfellow, I., Metaxas, D., Odena, A.: Self-attention generative adversarial networks. arXiv preprint arXiv:1805.08318 (2018)

36. Zhou, H., Liu, Y., Liu, Z., Luo, P., Wang, X.: Talking face generation by adversarially disentangled audio-visual representation. In: Proceedings of the AAAI Conference on Artificial Intelligence, vol. 33, pp. 9299–9306 (2019)

37. Zhu, H., Zheng, A., Huang, H., He, R.: High-resolution talking face generation via mutual information approximation. arXiv preprint arXiv:1812.06589 (2018)

Solving Phase Retrieval with a Learned Reference

Rakib Hyder[iD], Zikui Cai[iD], and M. Salman Asif[(✉)][iD]

University of California, Riverside, CA 92521, USA
{rhyde001,zcai032,sasif}@ucr.edu

Abstract. Fourier phase retrieval is a classical problem that deals with the recovery of an image from the amplitude measurements of its Fourier coefficients. Conventional methods solve this problem via iterative (alternating) minimization by leveraging some prior knowledge about the structure of the unknown image. The inherent ambiguities about shift and flip in the Fourier measurements make this problem especially difficult; and most of the existing methods use several random restarts with different permutations. In this paper, we assume that a known (learned) reference is added to the signal before capturing the Fourier amplitude measurements. Our method is inspired by the principle of adding a reference signal in holography. To recover the signal, we implement an iterative phase retrieval method as an unrolled network. Then we use back propagation to learn the reference that provides us the best reconstruction for a fixed number of phase retrieval iterations. We performed a number of simulations on a variety of datasets under different conditions and found that our proposed method for phase retrieval via unrolled network and learned reference provides near-perfect recovery at fixed (small) computational cost. We compared our method with standard Fourier phase retrieval methods and observed significant performance enhancement using the learned reference.

1 Introduction

The problem of *phase retrieval* refers to the challenge of recovering a real- or complex-valued signal from its amplitude measurements. This problem arises in diffraction imaging, X-ray crystallography, and ptychography [14,15,21,35,43]. Fourier phase retrieval is a special class of phase retrieval problems aimed at the recovery of a signal from the amplitude of its Fourier coefficients. Let us assume that Fourier amplitude measurements are given as

$$y = |Fx| + \eta, \tag{1}$$

where F denotes the Fourier transform operator, x denotes the unknown signal or image, and η denotes the measurement noise. Our goal is to recover x given y.

R. Hyder and Z. Cai—Equal contribution.

© Springer Nature Switzerland AG 2020
A. Vedaldi et al. (Eds.): ECCV 2020, LNCS 12375, pp. 425–441, 2020.
https://doi.org/10.1007/978-3-030-58577-8_26

Fourier phase retrieval is essential in many applications, especially in optical coherent imaging. Classical methods for phase retrieval utilize the prior knowledge about the support and positivity of the signals [14,15]. Subsequent work has considered the case where the unknown signal is *structured* and belongs to a low-dimensional manifold that is known *a priori*. Examples of such low-dimensional structures include sparsity [27,46], low-rank [12,26], or neural generative models [25,28]. Other techniques like Amplitude flow [47] and Wirtinger flow use alternating minimization [7]. Many of these newer algorithms involve solving a *non-convex* problem using iterative, gradient-based methods; therefore, they need to be carefully initialized. The initialization technique of choice is spectral initialization, first proposed in the context of phase retrieval in [36], and extended to the sparse signal case in [27,46].

Fourier phase retrieval problem does not satisfy the assumptions needed for successful spectral initialization and remains highly sensitive to the initialization choice. Furthermore, Fourier amplitude measurements have the so-called trivial ambiguities about possible shifts and flips of the images. Therefore, many Fourier phase retrieval methods test a number of random initializations with all possible flips and shifts and select the estimate with the best recovery error [34].

In this paper, we assume that a known (learned) reference is added to the signal before capturing the Fourier amplitude measurements. The main motivation for this comes from the empirical observation that knowing a part of the image can often help resolve the trivial ambiguities [3,18,22]. We extend this concept and assume that a known reference signal is added to the target signal and aim to recover the target signal from the Fourier amplitude of the combined signal. Adding a reference may not feasible in all cases, but our method will be applicable whenever we can add a reference or split the target signal into known and unknown parts. We can describe the Fourier amplitude (phaseless) measurements with a known reference signal u as

$$y = |F(x + u)| + \eta. \tag{2}$$

Similar reference-based measurements and phase retrieval problems also arise in holographic optical coherence imaging [37].

Our goal is to recover the signal x from the amplitude measurements in (2). To do that, we implement a gradient descent method for phase retrieval. We present the algorithm as an unrolled network for a general system in Fig. 1. Every layer of the network implements one step of the gradient descent update. To minimize the computational complexity of the recovery algorithm, we seek to minimize the number of iterations (hence the layers in the network). In addition, we seek to learn the reference u to maximize the accuracy of the recovered signal for a given number of iterations. The learned u and reconstruction results for different datasets are summarized in Fig. 2.

1.1 Our Contributions

We present an iterative method to efficiently recover a signal from the Fourier amplitude measurements using a fixed number of iterations. To achieve this

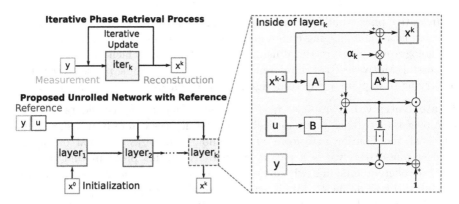

Fig. 1. Our proposed approach for learning reference signal by solving phase retrieval using an unrolled network. Unrolled network has K layers. Each layer$_k$ gets amplitude measurements y, reference u, and estimate x^{k-1} as inputs, and updates the estimate to x^k. The operations inside layer$_k$ are shown in the dashed box on the right, where A and B are both linear measurement operators, and A^* is the adjoint operator of A.

goal, we first learn a reference signal that can be added to the phaseless Fourier measurements to enable the exact solution of the phase retrieval problem. We demonstrate that the reference learned on a very small training set perform remarkably well on the test dataset.

Our main contributions can be summarized as follows.

- The proposed method uses a fixed number of gradient descent iterations (i.e., fixed computational cost) to solve the Fourier phase retrieval problem.
- We formulate the gradient descent method as an unrolled network that allows us to learn a robust reference signal for a class of images. We demonstrate that reference learned on a very small dataset performs remarkably well on diverse and large test datasets. To the best of our knowledge, this is the first work on learning a reference for phase retrieval problems.
- We tested our method extensively on different challenging datasets and demonstrated the superiority of our method.
- We demonstrate the robustness of our approach by testing it with the noisy measurements using the reference that was trained on noise-free measurements.

2   Related Work

Holography. Digital holography is an interferometric imaging technique that does not require the use of any imaging lens. Utilizing the theory of diffraction of light, a hologram can be used to reconstruct three-dimensional (3D) images [39]. With this advantage, holography can be used to perform simultaneous imaging of multidimensional information, such as 3D structure, dynamics, quantitative phase, multiple wavelengths, and polarization state of light [44]. In the computational imaging community, many attempts have been made in solving holographic phase retrieval using references, among which [3] has been very successful. Motivated by the reference design for holographic phase retrieval, we are trying to explore a way to design references for general phase retrieval.

Phase Retrieval. The phase retrieval problem has drawn considerable attention over the years, as many optical detection devices can only measure amplitudes of the Fourier transform of the underlying object (signal or image). Fourier phase retrieval is a particular instance of this problem that arises in optical coherent imaging, where we seek to recover an image from its Fourier modulus [14,15,33,35,41,43]. Existing algorithms for solving phase retrieval can be broadly classified into convex and non-convex approaches [23]. Convex approaches usually solve a constrained optimization problem after lifting the problem. The PhaseLift algorithm [8] and its variations [17], [6] belong to this class. On the other hand, non-convex approaches usually depend on Amplitude flow [45,46] and Wirtinger flow [5,7,11,52]. If we know some structure of the signal a priori, it helps in the reconstruction. Sparsity is a very popular signal prior. Some of the approaches for sparse phase retrieval include [2,5,24,32,36,38,46]. Furthermore, [23,27,36] used minimization (AltMin)-based approach and [10] used total variation regularization to solve phase retrieval. Recently, various researchers have explored the idea of replacing the sparsity priors with generative priors for solving inverse problems. Some of the generative prior-based approaches can be found in [20,23,28,42].

Data-Driven Approaches for Phase Retrieval. The use of deep learning-based methods to solve computational imaging problems such as phase retrieval is becoming popular. Deep learning methods leverage the power of huge amounts of data and tend to provide superior performance compared to traditional methods while also run significantly faster with the acceleration of GPU devices. A few examples demonstrating the benefit of the data-driven approaches include [34] for robust phase retrieval, [30] for Fourier ptychographic microscopy, and [40] for holographic image reconstruction.

Unrolled Network for Inverse Problem. Unrolled networks, which are constructed by unrolled iterations of a generic non-linear reconstruction algorithm, have also been gaining popularity for solving inverse problems in recent years [4,13,16,19,29,31,48,50]. Iterative methods usually terminate the iteration when the condition satisfies theoretical convergence properties, thus rendering the number of iterations uncertain. An unrolled network has a fixed number of iterations (and cost) by construction and they produce good results in a small number of steps while enabling efficient usage of training data.

Reference Design. Fourier phase retrieval faces different trivial ambiguities because of the structure of Fourier transformation. As a phase shift in the Fourier domain results in a circular shift in the spatial domain, we will get the same Fourier amplitude measurements for any circular shift of the original signal. In recent papers [3,18,22,51], authors tried to use side information with sparsity prior to mitigate these ambiguities. However, in those studies, the reference and target signal are separated by some margin. If the separation between target and reference is large enough, then the nonlinear PR problem simplifies to a linear inverse problem [1,3].

In this paper, we consider the reference signal to be additive and overlapping with the target signal. To the best of our knowledge, there has not been any study on such unrestricted reference design. While driven by data, our approach for reference design uses training samples in a very efficient way. The number of training images required by our network is parsimonious without limiting its generalizability. The reference learned by our network provides robust recovery test images with different sizes. Apart from the great flexibility, our unrolled network uses a well-defined routine in each layer and demonstrates excellent interpretability as opposed to black-box deep neural networks.

3   Proposed Approach

We use the general formulation for the phase retrieval from amplitude measurements. The formulation can be extended for phase retrieval with squared amplitude measurement as well. In our setup, we model amplitude measurements of a target signal x and a reference signal u as $y = |Ax + Bu|$, where A and B are linear measurement operators. Our goal is to learn a reference signal that provides us the best recovery of the target signal. We formulate this overall task as the following optimization problem:

$$\underset{\hat{x}(u)}{\text{minimize}} \; \|x - \hat{x}(u)\|_2^2 \quad \text{s.t.} \;\; y = |A\hat{x}(u) + Bu|, \tag{3}$$

where $\hat{x}(u)$ denotes the solution of the phase retrieval problem for a given reference u. Our approach to learn u and solve (3) can be divided into two nested steps: (1) Outer step updates u to minimize the recovery error for phase retrieval and (2) inner step uses the learned u to recover target images by solving phase retrieval.

To solve the (inner step) of phase retrieval problem, we use an unrolled network. Figure 1 depicts the structure of our phase retrieval algorithm. In the unrolled phase retrieval network, we have K blocks to represent K iterations of the phase retrieval algorithm. We minimize the following loss to solve the phase retrieval problem:

$$L_x(x, u) = \|y - |Ax + Bu|\|_2^2. \tag{4}$$

Every block of the unrolled phase retrieval network is equivalent to one gradient descent step for (4). For some value of reference estimate, u, we can represent the target signal estimate after $k + 1^{th}$ block of the unrolled network as

$$x^{k+1} = x^k - \alpha_k \nabla_x L_x(x^k, u), \tag{5}$$

where $\nabla_x L_x(x^k, u)$ is the gradient of L_x with respect to x at the given values of x^k, u. As the loss function in (4) is not differentiable, we can redefine it as

$$L_x(x, u) = \|y \odot p - (Ax + Bu)\|_2^2, \tag{6}$$

where $p = \angle(Ax^k + Bu) = (Ax^k + Bu)/|Ax^k + Bu|$. The expression of gradient can be written as

$$\nabla_x L_x(x^k, u) = 2A^*[p \odot (p^* \odot (Ax^k + Bu) - y)], \tag{7}$$

where A^* denotes the adjoint of A. After K blocks, we get the estimate of the target signal that we denote as $\hat{x}(u) = x^K$.

In the learning phase, we are given a set of training signals, $\{x_1, x_2, ..., x_N\}$, which share the same distribution as our target signals. We initialize x^0 and u^0 with some initial (feasible) values. First we minimize the following loss with respect to u:

$$L_u(u) = \sum_{i=1}^{N} \|x_i - \hat{x}_i\|_2^2 = \sum_{i=1}^{N} \|x_i - x_i^K\|_2^2. \tag{8}$$

We can rewrite (8) using the gradient recursion in (5) as

$$L_u(u) = \sum_{i=1}^{N} \|x_i - x_i^0 + \sum_{k=0}^{K-1} \alpha_k \nabla_x L_x(x_i^k, u)\|_2^2. \tag{9}$$

We can then use gradient descent to to minimize $L_u(u)$. We can represent the $j + 1^{th}$ iteration of gradient descent step as

$$u^{j+1} = u^j - \beta \nabla_u L_u(u^j). \tag{10}$$

The expression for $\nabla_u L_u(u)$ can be written as

$$\nabla_u L_u(u) = 2 \sum_{i=1}^{N} \left[\sum_{k=0}^{K-1} \alpha_k J_u(x_i^k, u) \right] \left[x_i - x_i^0 + \sum_{k=0}^{K-1} \alpha_k \nabla_x L_x(x_i^k, u) \right], \tag{11}$$

where $J_u(x_i^k, u) = \nabla_u \nabla_x L_x(x_i^k, u)$ is a Jacobian matrix with rows and columns of the same size as u and x, respectively. The measurement vector $y = |Ax + Bu|$ is a function of u during training. Since we model $\hat{x}(u)$ as an unrolled network, we can think of the gradient step as a backpropagation step. To compute $\nabla_u L_u(u)$, we backpropagate through the entire unrolled network. At the end of Jth outer iteration, we will get our learned reference $\hat{u} = u^J$.

Once we have learned a reference, $\hat{u}$, we can use it to capture (phaseless) amplitude measurements as $y = |Ax^* + B\hat{u}|$ for target signal x^*. To solve the phase retrieval problem, we perform one forward pass through the unrolled network. Pseudocodes for training and testing are provided in Algorithms 1,2.

In our Fourier phase retrieval experiments $A = B = F$, where F is the Fourier transform operation. To implement similar method for squared amplitude measurements, we can simply replace $p = \angle(Ax^k + Bu^j)$ with $p = Ax^k + Bu^j$. In all our experiments, we initialized x^0 as a zero vector whenever $\hat{u} \neq 0$. We can also add additional constraints on the reference while minimizing the loss function in (9). In our experiments, we used target signals with intensity values in the range $[0, 1]$; therefore, we restricted the range of entries in u to $[0, 1]$ as well. We discuss other constraints in the experiment section.

Algorithm 1. Learning Reference Signal

Input: Training signals $\{x_1, x_2, ..., x_N\}$, measurement operators, A and B.
Initialize $\{x_1^0, x_2^0, ..., x_N^0\}, u^0$
for $j = 0, 1, ..., J - 1$ **do**
  for $i = 1, 2, ..., N$ **do**
    $y_i = |Ax_i^* + Bu^j|$
    for $k = 0, 1, ..., K - 1$ **do**
      $L_x(x_i^k, u^j) = \||y_i - |Ax_i^k + Bu^j|\||_2^2$
      $x_i^{k+1} \leftarrow x_i^k - \alpha_k \nabla_x L_x(x_i^k, u^j)$
    end for
  end for
  $L_u(u^j) = \sum_{i=1}^N \||x_i^* - x_i^0 + \sum_{k=1}^K \alpha_k \nabla_x L_x(x_i^{k-1}, u^j)\||_2^2$
  $u^{j+1} \leftarrow u^j - \beta \nabla_u L_u(u^j)$
end for
Output: Optimal reference, $\hat{u} = u^J$

Algorithm 2. Solving Phase Retrieval via Unrolled Network

Input: Measurements y, learned reference $\hat{u}$, measurement operators, A and B.
Initialize x^0
for $k = 0, 1, ..., K - 1$ **do**
  $L_x(x^k, \hat{u}) = \||y - |Ax^k + B\hat{u}|\||_2^2$
  $x^{k+1} \leftarrow x^k - \alpha_k \nabla_x L_x(x^k, \hat{u})$
end for
Output: Estimation of target signal $\hat{x} = x^K$

---

4 Experiments

Datasets. We have used MNIST digits, EMNIST letters, Fashion MNIST, CIFAR10, SVHN, CelebA datasets, and different well-known standard images for our experiments. We convert all images to grayscale and resize 28×28 images to 32×32. Although there are tens of thousands training images in MNIST, EMNIST letters, Fashion MNIST, CIFAR10, and SVHN dataset, we have used only a few (e.g., 32) of them in training. We have shown that the references learned on the small number of training images perform remarkably well on the entire test dataset. MNIST, Fashion MNIST, and CIFAR10 test datasets contain 10000 test images each; EMNIST letters dataset contains 24800 test images; SVHN test dataset contains 26032 test images. We used 1032 images from CelebA and center-cropped and resized all of them to 200×200. We selected 32 images for training and the rest for testing.

We present the results for these different datasets using references learned from 32 images from the same dataset in Fig. 2. We present results for six standard images of size 512×512 from [34] using a resized reference learned from CelebA dataset in Fig. 3.

Measurements. We simulated amplitude measurements of the 2D Fourier transform. We performed 4 times oversampling in the spatial domain for both

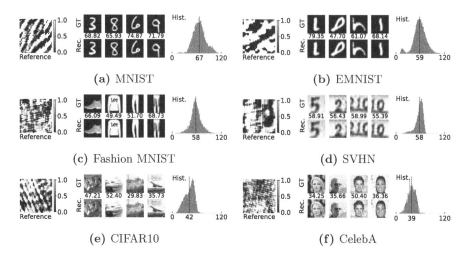

(a) MNIST

(b) EMNIST

(c) Fashion MNIST

(d) SVHN

(e) CIFAR10

(f) CelebA

Fig. 2. Reconstruction results using learned references. Each block **(a)**–**(f)** shows results for a different dataset: (**left**) learned reference with a colorbar; (**middle**) sample original images and reconstruction with PSNR on top; (**right**) histogram of PSNR over the entire test dataset (vertical dashed line represents the mean PSNR).

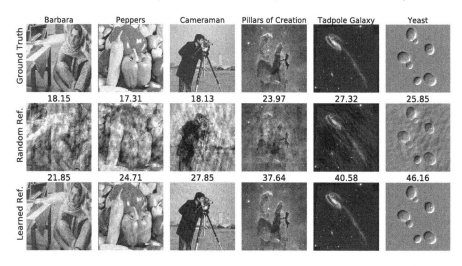

Fig. 3. Phase retrieval results using learned and random references. **First Row:** Original 512×512 test images. **Second Row:** Reconstruction using random references with uniform distribution between $[0, 1]$ **best result out of 100 trials. Third Row:** Reconstruction using the reference learned on CelebA dataset and resized from 200×200 to 512×512. (PSNR shown on top of images.)

reference and target signal. Unless otherwise mentioned, we consider our measurements to be noise-free. We also report results for noisy measurements.

4.1 Configurations of Reference (u)

The reference signal u, which we are trying to learn, has a number of hyper-parameters that inherently affect the performance of the phase retrieval process. We considered several constraints on u, including the support, size, range, position, and sparsity.

We tested reference signals with both complex and real values and found that u has comparable results in the two domains. Since it is easy to physically create amplitude or phase-only reference signals, we constrain u to be in the real domain; thus, $u \in \mathbb{R}^{m \times n}$ and m, n represent height and width, respectively. The height and width of u determine the overlapping area between the target signal and the reference. We found that u with larger size tends to have better performance, especially when the value of u is constrained to a small range. The intensity values of u play a major role in its performance. If we constrain the value of u to be within a certain range: $u[i,j] \in [u_{min}, u_{max}]$, for all i,j, we observed that bigger range of u yields better performance. This is because when u is unconstrained then we can construct a u with a large norm. Consider the noiseless setting with quadratic measurements $|F(x+u)|^2 = |Fx|^2 + |Fu|^2 + 2\text{Re}(Fx \odot Fu)$, the last term is the real value of the element-wise product of target and reference Fourier transforms. We can remove $|Fu|^2$ because it is known. If u is large compared to x, then we can also ignore the quadratic term $|Fx|^2$ and recover x in a single iteration if all entries of Fu are nonzero. To avoid this situation and make the problem stable in the presence of noise, we restricted the values in the reference u to be in $[0,1]$ range.

4.2 Setup of Training Samples and Sample Size

We observed that we can learn the reference signal from a small number of training images. In Table 1, we report test results for different reference signals learned on first N images from MNIST training dataset for $N = 32, 128, 512$. We kept the signal and reference strength (i.e., the range of the signal) equal for this experiment. We observe that increasing the training size improves test performance. However, we can get reasonable reconstruction performance on large test datasets (10k+ images) with reference learned using only 32 images.

Table 1. PSNR for different training sizes

TRAIN/TEST	MNIST	EMNIST	F. MNIST	SVHN	CIFAR10
TRAINING SIZE=32	66.54	58.72	57.81	57.51	41.60
TRAINING SIZE=128	76.25	64.16	55.86	59.50	44.34
TRAINING SIZE=512	79.14	62.34	52.01	59.78	48.90

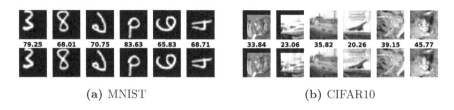

(a) MNIST   (b) CIFAR10

Fig. 4. Test results on shifted/flipped/rotated images using the reference learned on upright and centered (canonical) training images. (PSNR shown on top of images.)

4.3 Generalization of Reference on Different Classes

We are interested in evaluating the generalization of our learned reference. (i.e., how the reference performs when trained on one dataset and tested on another). In the comparison study, we took the reference u trained on each dataset and then tested them on the remaining 4 datasets. The value range of the reference is between $[0, 1]$, the number of steps in the unrolled network is $K = 50$. We observed that when the datasets share great similarity (e.g., MNIST and EMNIST are both sparse digits or letters), the reference signal tends to work well on both datasets. Even when the datasets differ greatly in their distributions, the reference trained on one dataset provides good results on other datasets (with only a few dB of PSNR decrease in performance).

We also tested our method on shifted and rotated versions of test images. Results in Fig. 4 demonstrate that even though the reference was trained on upright and centered images, we can perfectly recover shifted and rotated images.

Our key insight about this generalization phenomenon is that the main challenge in Fourier phase retrieval methods is initialization and ambiguities that arise because of symmetries. We are able to solve these issues using a learned reference because of the following reasons: (1) A reference gives us a good initialization for the phase retrieval iterations. (2) The presence of a reference breaks the symmetries that arise in Fourier amplitude measurements. Moreover, we are not learning to solve the phase retrieval problem in an end-to-end manner or learn a signal-dependent denoiser to solve the inverse problem [34,40]. We are learning reference signals to primarily help a predefined phase retrieval algorithm to recover the true signal from the phaseless measurements. Thus, the references learned on one class of images provide good results on other images, see Table 2. This study shows that the reference learned using our network has the ability to generalize to new datasets, thus making our method suitable for real-life applications where new test cases keep emerging.

4.4 Noise Response

To test the robustness of our method in the presence of noise, we added Gaussian and Poisson noise at different levels to the measurements. Poisson noise or shot noise is the most common in the practical systems. We model the Poisson noise

Table 2. PSNR with references trained and tested on different datasets

TRAIN/TEST	MNIST	EMNIST	F. MNIST	SVHN	CIFAR10
MNIST	66.54	55.12	40.87	41.87	31.72
EMNIST	**72.84**	**58.72**	52.18	55.42	48.16
F. MNIST	40.87	55.67	**57.81**	50.70	42.85
SVHN	41.87	46.76	49.60	**57.51**	**51.54**
CIFAR10	31.72	38.93	36.40	40.36	41.60

following the same approach as in [34]. We simulate the measurements as

$$y(i) = |z(i)| + \eta(i) \quad \text{for all } i = 1, 2, \ldots, m, \tag{12}$$

where $\eta(i) \sim \mathcal{N}(0, \sigma^2)$ for Gaussian noise and $\eta(i) \sim \mathcal{N}(0, \lambda|z(i)|^2)$ for Poisson noise with $z = Ax + Bu$. We varied σ, λ to generate noise at different signal-to-noise ratios. Poisson noise affects the larger measurements with higher strength than the smaller measurements. As the sensors can measure only positive measurements, we kept the measurements positive by applying ReLU function after noise addition. We can observe the effect of noise in Fig. 5. Even though we did not add noise during training, we get reasonable reconstruction and performance degrades gracefully with increased noise.

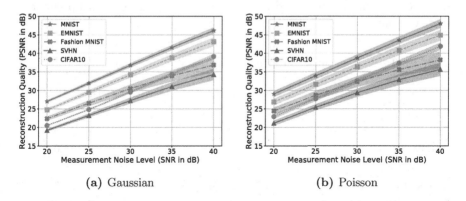

(a) Gaussian   (b) Poisson

Fig. 5. Reconstruction quality of the test images vs noise level of the measurements for different datasets. We learned the reference using noise-free measurements.

4.5   Random Reference Versus Learned Reference

To demonstrate the advantage of the learned reference signal, we compared the performance of learned reference and random reference on some standard images. The results are shown in Fig. 3. The learned reference is trained using 32 images from CelebA dataset which we resized to 200×200. The test images used in Fig. 3 are 512×512, so we resized the learned reference from 200×200 to 512×512. For random reference, we selected the entries of the reference uniformly at random from $[0, 1]$. We selected the best result out of 100 trials for every test image with random reference. We can observe from the results that our learned reference significantly outperforms the random reference even though the test image distribution is distinct from the training data. The number of steps of the unrolled network is $K = 50$.

4.6 Comparison with Existing Phase Retrieval Methods

We have shown comparison with other approaches in Table 3. We selected Kaczmarz [49] and Amplitude flow [11] for comparison using PhasePack package [9]. We also show Hybrid Input Output (HIO), which is similar to our phase retrieval routine without any reference. We observe that our approach with learned reference can outperform all other approaches on all the datasets. All the traditional phase retrieval methods suffer from the trivial circular shift, rotation, and flip ambiguities, thus produce significantly worse reconstruction than our method does. Our method uses a reference signal to simplify the initialization and removes the shift/reflect ambiguities. To mathematically explain this fact, a shifted or flipped version of x would not give us the same Fourier measurements as $|F(x + u)|$ if u is chosen appropriately as we do with the learning procedure. As we showed in Fig. 5, our method can perfectly recover the shifted and flipped versions of the images using the reference that was trained with upright and centered images.

Table 3. Comparison with existing phase retrieval methods

METHODS	MNIST	EMNIST	F. MNIST	SVHN	CIFAR10
HIO	9.04	8.42	9.65	19.87	14.70
AMPLITUDE FLOW	9.99	9.79	11.90	20.25	15.04
KACZMARZ	11.81	11.47	13.44	19.48	15.01
FLAT REFERENCE	18.21	17.24	16.56	20.89	15.81
RANDOM REFERENCE	36.87	28.41	27.27	36.45	25.57
LEARNED REFERENCE (OURS)	**66.54**	**58.72**	**57.81**	**57.51**	**41.60**

4.7 Effects of Number of Layers (K)

We tested our unrolled network with different numbers of layers (i.e., K) at training and test time. The results are summarized in Fig. 6. We first used the same values of K for training and testing. We observed that as K increases, the reconstruction quality (measured in PSNR) improves. Then we fixed $K = 1$ or

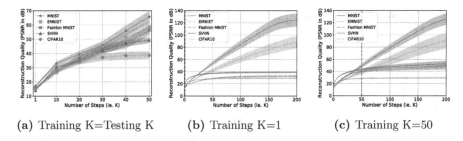

(a) Training K=Testing K   (b) Training K=1   (c) Training K=50

Fig. 6. Reconstruction PSNR vs the number of blocks (K) in the unrolled network at training and testing. (a) K is same for training and testing (shaded region shows ± 0.25 times **std** of PSNR). (b) $K = 1$ and (c) $K = 50$, but tested using different K.

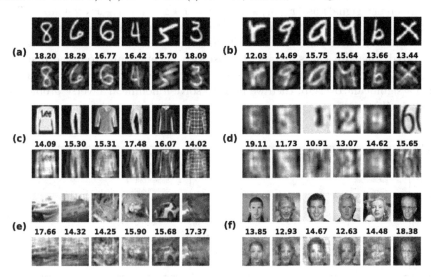

Fig. 7. Single step reconstruction with reference in range $[0,1]$. Each of the **6 sets** (a)–(f) has the ground truth in the first row. Second row is the reconstruction (PSNR shown on top of images.)

$K = 50$ at training, but used different values of K at testing. We observed that if we increase K at the test time, PSNR improves up to a certain level and then it plateaus. The PSNR achieved with reference trained with $K = 50$ is better than what the referenced trained with $K = 1$ provided. These results provide us a trade-off between the reconstruction speed and quality. As we increase K, the reconstruction quality improves but the reconstruction requires more steps (computations and time).

Finally, we learned a reference using $K = 1$ and tested it on different images with $K = 1$. To our surprise, our method was able to produce reasonable quality reconstruction with this extreme setting. We present some single-step reconstructions of each data set in Fig. 7.

4.8 Localizing the Reference

We also evaluated the effect of localizing the reference to a small region. For example, the reference is constrained to be within a small block in the corner or the center of the target signal. We restricted u to be an 8×8 block and placed it in different positions. We found that corner positions provide better results as shown in Fig. 8. As we bring the reference support closer to the center, the quality of reconstruction deteriorates. This observation is related to the method in [1,3,18], where if the known reference signal is separated from the target signal, then the phase retrieval problem can be solved as a linear inverse problem.

Note that signal recovery from Fourier phase retrieval is equivalent to signal recovery from its autocorrelation. We can write the autocorrelation of target plus reference signals as $(x + u) \star (x + u) = x \star x + u \star u + x \star u + u \star x$. The first term is a quadratic function of x, the second term is known, and the last two terms are linear functions of x. If the supports for x and u are sufficiently separated, then we can separate the last two linear terms from the first two quadratic terms and recover x by solving a linear problem. However, if x and u have a significant overlap, then we need to solve a nonlinear inverse problem as we do in this paper.

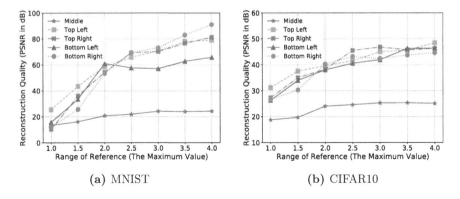

(a) MNIST   (b) CIFAR10

Fig. 8. Performance of our method if the reference is an 8×8 block placed at different positions. Fixing the minimum value at 0, we increased the maximum value of the reference we learn. We observe that the small reference placed in the corners performs better than the ones placed in the center.

5 Conclusion

We presented a framework for learning a reference signal to solve the Fourier phase retrieval problem. The reference signal is learned using a small number of training images using an unrolled network as a solver for the phase retrieval problem. Once learned, the reference signal serves as a prior which significantly improves the efficiency of the signal reconstruction in the phase retrieval process. The learned reference generalizes to a broad class of datasets with different distribution compared to the training samples. We demonstrated the robustness and efficiency of our method through extensive experiments.

Acknowledgment. The first two authors contributed equally in this work. This research was supported in parts by an ONR grant N00014-19-1-2264, DARPA REVEAL Program, and a Google Faculty Award.

References

1. Arab, F., Asif, M.S.: Fourier phase retrieval with arbitrary reference signal. In: ICASSP 2020–2020 IEEE International Conference on Acoustics, Speech and Signal Processing (ICASSP), pp. 1479–1483. IEEE (2020)
2. Bahmani, S., Romberg, J.: Efficient compressive phase retrieval with constrained sensing vectors. In: Proceedings of the Advances in Neural Information Processing Systems (NeurIPS), pp. 523–531 (2015)
3. Barmherzig, D., Sun, J., Li, P., Lane, T., Candès, E.: Holographic phase retrieval and reference design. Inverse Problems (2019)
4. Bostan, E., Kamilov, U.S., Waller, L.: Learning-based image reconstruction via parallel proximal algorithm. IEEE Sig. Process. Lett. **25**(7), 989–993 (2018)
5. Cai, T., Li, X., Ma, Z., et al.: Optimal rates of convergence for noisy sparse phase retrieval via thresholded wirtinger flow. Ann. Stat. **44**(5), 2221–2251 (2016)
6. Candes, E., Li, X., Soltanolkotabi, M.: Phase retrieval from coded diffraction patterns. Appl. Comput. Harmon. Anal. **39**(2), 277–299 (2015)
7. Candes, E., Li, X., Soltanolkotabi, M.: Phase retrieval via wirtinger flow: theory and algorithms. IEEE Trans. Inform. Theory **61**(4), 1985–2007 (2015)
8. Candes, E., Strohmer, T., Voroninski, V.: Phaselift: exact and stable signal recovery from magnitude measurements via convex programming. Comm. Pure Appl. Math. **66**(8), 1241–1274 (2013)
9. Chandra, R., Zhong, Z., Hontz, J., McCulloch, V., Studer, C., Goldstein, T.: Phasepack: a phase retrieval library. In: Asilomar Conference on Signals, Systems, and Computers (2017)
10. Chang, H., Lou, Y., Ng, M., Zeng, T.: Phase retrieval from incomplete magnitude information via total variation regularization. SIAM. J. Sci. Comput. **38**(6), A3672–A3695 (2016)
11. Chen, Y., Candes, E.: Solving random quadratic systems of equations is nearly as easy as solving linear systems. In: Proceedings of the Advances in Neural Information Processing Systems (NeurIPS), pp. 739–747 (2015)
12. Chen, Z., Jagatap, G., Nayer, S., Hegde, C., Vaswani, N.: Low rank fourier ptychography. In: 2018 IEEE International Conference on Acoustics, Speech and Signal Processing (ICASSP), pp. 6538–6542, April 2018
13. Diamond, S., Sitzmann, V., Heide, F., Wetzstein, G.: Unrolled optimization with deep priors (2017). arXiv preprint arXiv:1705.08041
14. Fienup, J.R.: Phase retrieval algorithms: a comparison. Appl. Opt. **21**(15), 2758–2769 (1982)
15. Gerchberg, R.W.: A practical algorithm for the determination of phase from image and diffraction plane pictures. Optik **35**, 237–246 (1972)

16. Gregor, K., LeCun, Y.: Learning fast approximations of sparse coding. In: Proceedings of the 27th International Conference on International Conference on Machine Learning, pp. 399–406 (2010)
17. Gross, D., Krahmer, F., Kueng, R.: Improved recovery guarantees for phase retrieval from coded diffraction patterns. Appl. Comput. Harmon. Anal. **42**(1), 37–64 (2017)
18. Guizar-Sicairos, M., Fienup, J.: Holography with extended reference by autocorrelation linear differential operation. Opt. Express **15**(26), 17592–17612 (2007)
19. Hammernik, K., Klatzer, T., Kobler, E., Recht, M.P., Sodickson, D.K., Pock, T., Knoll, F.: Learning a variational network for reconstruction of accelerated MRI data. Magn. Reson. Med. **79**(6), 3055–3071 (2018)
20. Hand, P., Leong, O., Voroninski, V.: Phase retrieval under a generative prior. In: Proceedings of the Advances in Neural Information Processing Systems (NeurIPS), pp. 9154–9164 (2018)
21. Harrison, R.: Phase problem in crystallography. JOSA a **10**(5), 1046–1055 (1993)
22. Hyder, R., Hegde, C., Asif, M.: Fourier phase retrieval with side information using generative prior. In: Proceedings of the Asilomar Conf. Signals, Systems, and Computers. IEEE (2019)
23. Hyder, R., S., V., Hegde, C., Asif, M.: Alternating phase projected gradient descent with generative priors for solving compressive phase retrieval. In: Proceedings of the IEEE International Conference Acoustics, Speech, and Signal Processing (ICASSP), pp. 7705–7709. IEEE (2019)
24. Jaganathan, K., Oymak, S., Hassibi, B.: Recovery of sparse 1-D signals from the magnitudes of their fourier transform. In: Proceedings of the International Symposium on Information Theory Proceedings (ISIT), pp. 1473–1477. IEEE (2012)
25. Jagatap, G., Chen, Z., Hegde, C., Vaswani, N.: Sub-diffraction imaging using fourier ptychography and structured sparsity. In: 2018 IEEE International Conference on Acoustics, Speech and Signal Processing (ICASSP), pp. 6493–6497, April 2018
26. Jagatap, G., Chen, Z., Nayer, S., Hegde, C., Vaswani, N.: Sample efficient fourier ptychography for structured data. IEEE Trans. Comput. Imaging **6**, 344–357 (2020)
27. Jagatap, G., Hegde, C.: Fast, sample-efficient algorithms for structured phase retrieval. In: Advances in Neural Information Processing Systems, pp. 4917–4927 (2017)
28. Jagatap, G., Hegde, C.: Algorithmic guarantees for inverse imaging with untrained network priors. In: Advances in Neural Information Processing Systems, pp. 14832–14842 (2019)
29. Kamilov, U.S., Mansour, H.: Learning optimal nonlinearities for iterative thresholding algorithms. IEEE Sig. Process. Lett. **23**(5), 747–751 (2016)
30. Kellman, M., Bostan, E., Chen, M., Waller, L.: Data-driven design for fourier ptychographic microscopy. In: International Conference for Computational Photography, pp. 1–8 (2019)
31. Kellman, M.R., Bostan, E., Repina, N.A., Waller, L.: Physics-based learned design: optimized coded-illumination for quantitative phase imaging. IEEE Trans. Comput. Imaging **5**(3), 344–353 (2019)
32. Li, X., Voroninski, V.: Sparse signal recovery from quadratic measurements via convex programming. SIAM J. Math. Anal. **45**(5), 3019–3033 (2013)
33. Maiden, A., Rodenburg, J.: An improved ptychographical phase retrieval algorithm for diffractive imaging. Ultramicroscopy **109**(10), 1256–1262 (2009)

34. Metzler, C.A., Schniter, P., Veeraraghavan, A., Baraniuk, R.G.: prDeep: robust phase retrieval with a flexible deep network. In: Proceedings of the International Conference on Machine Learning (2018)
35. Millane, R.: Phase retrieval in crystallography and optics. JOSA A **7**(3), 394–411 (1990)
36. Netrapalli, P., Jain, P., Sanghavi, S.: Phase retrieval using alternating minimization. In: Proceedings of the Advance in Neural Information Processing Systems (NeurIPS), pp. 2796–2804 (2013)
37. Nolte, D.D.: Optical Interferometry for Biology and Medicine, vol. 1. Springer Science & Business Media, New York (2011). https://doi.org/10.1007/978-1-4614-0890-1
38. Ohlsson, H., Yang, A., Dong, R., Sastry, S.: CPRL-an extension of compressive sensing to the phase retrieval problem. In: Proceedings of the Advance in Neural Information Processing System (NeurIPS), pp. 1367–1375 (2012)
39. Park, I., Middleton, R., Coggrave, C.R., Ruiz, P.D., Coupland, J.M.: Characterization of the reference wave in a compact digital holographic camera. Appl. Opt. **57**(1), A235–A241 (2018)
40. Rivenson, Y., Zhang, Y., Günaydın, H., Teng, D., Ozcan, A.: Phase recovery and holographic image reconstruction using deep learning in neural networks. Light Sci. Appl. **7**(2), 17141–17141 (2018)
41. Rodenburg, J.M.: Ptychography and related diffractive imaging methods. Adv. Imaging Electron Phys. **150**, 87–184 (2008)
42. Shamshad, F., Ahmed, A.: Robust compressive phase retrieval via deep generative priors (2018). arXiv preprint arXiv:1808.05854
43. Shechtman, Y., Eldar, Y., Cohen, O., Chapman, H., Miao, J., Segev, M.: Phase retrieval with application to optical imaging: a contemporary overview. IEEE Sig. Process. Mag. **32**(3), 87–109 (2015)
44. Tahara, T., Quan, X., Otani, R., Takaki, Y., Matoba, O.: Digital holography and its multidimensional imaging applications: a review. Microscopy **67**(2), 55–67 (2018)
45. Wang, G., Giannakis, G.: Solving random systems of quadratic equations via truncated generalized gradient flow. In: Processing Advance in Neural Information Processing System (NeurIPS), pp. 568–576 (2016)
46. Wang, G., Zhang, L., Giannakis, G.B., Akcakaya, M., Chen, J.: Sparse phase retrieval via truncated amplitude flow. IEEE Trans. Sig. Process. **66**, 479–491 (2018)
47. Wang, G., Giannakis, G., Saad, Y., Chen, J.: Solving most systems of random quadratic equations. In: Advances in Neural Information Processing Systems, pp. 1867–1877 (2017)
48. Wang, S., Fidler, S., Urtasun, R.: Proximal deep structured models. In: Advances in Neural Information Processing Systems, pp. 865–873 (2016)
49. Wei, K.: Solving systems of phaseless equations via Kaczmarz methods: a proof of concept study. Inverse Prob. **31**(12), 125008 (2015)
50. Yang, Y., Sun, J., Li, H., Xu, Z.: Deep ADMM-net for compressive sensing MRI. In: Advances in Neural Information Processing Systems, pp. 10–18 (2016)
51. Yuan, Z., Wang, H.: Phase retrieval with background information. Inverse Prob. **35**(5), 054003 (2019)
52. Zhang, H., Liang, Y.: Reshaped wirtinger flow for solving quadratic system of equations. In: Proceedings of the Advance in Neural Information Processing System (NeurIPS), pp. 2622–2630 (2016)

Dual Grid Net: Hand Mesh Vertex Regression from Single Depth Maps

Chengde Wan[1(✉)], Thomas Probst[2], Luc Van Gool[2], and Angela Yao[3]

[1] Facebook Reality Labs, Pittsburgh, USA
vgrasp@fb.com
[2] Computer Vision Laboratory, ETH Zürich, Zürich, Switzerland
[3] National University of Singapore, Singapore, Singapore

Abstract. We aim to recover the dense 3D surface of the hand from depth maps and propose a network that can predict mesh vertices, transformation matrices for every joint and joint coordinates in a single forward pass. Use fully convolutional architectures, we first map depth image features to the mesh grid and then regress the mesh coordinates into real world 3D coordinates. The final mesh is found by sampling from the mesh grid refit in closed-form based on an articulated template mesh. When trained with supervision from sparse key-points, our accuracy is comparable with state-of-the-art on the NYU dataset for key point localization, all while recovering mesh vertices and dense correspondences. Under multi-view settings for training, our framework can also learn through self-supervision by minimizing a set of data-fitting terms and kinematic priors. Our approach is competitive with strongly supervised methods and showcases the potential for self-supervision in dense mesh estimation.

1 Introduction

We consider the problem of estimating 3D shape and pose of articulated objects from single depth images. Specifically, we want to estimate the position of surface mesh vertices of the human hand model. Unlike skeleton joints, dense mesh vertices encode both pose and shape of the hand and enable a much wider range of virtual and mixed reality applications. For example, one can directly put the virtual hand in a VR game, or overlay a user's hand surface with another texture map in mixed reality. Furthermore, manipulation of virtual objects can naturally be modelled through interaction of dense surface representations.

Estimating mesh vertices is significantly more challenging than estimating skeleton joints. First, the scale of the problem increases by several magnitudes. To reasonably represent a human hand, one needs thousands of mesh vertices, as opposed to tens of joint positions and angles. Secondly, getting accurate 3D ground truth for the thousands of vertices from real-world data is extremely difficult even though having large amounts of labelled training data is crucial for data-driven learning based methods.

© Springer Nature Switzerland AG 2020
A. Vedaldi et al. (Eds.): ECCV 2020, LNCS 12375, pp. 442–459, 2020.
https://doi.org/10.1007/978-3-030-58577-8_27

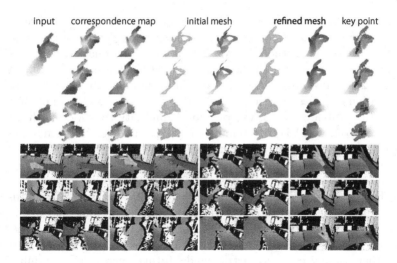

input   correspondence map   initial mesh   **refined mesh**   key point

Fig. 1. Upper rows: qualitative results on NYU [55]. In each group, upper rows are results supervised with key-point annotations and lower rows with self-supervision. We visualize the correspondence map with each mesh coordinate, the rendered shading and depth map of the initial estimated mesh model and refined ones, as well as key-points. Bottom rows: qualitative results from real-world data with multiple users and view points showing the estimated mesh and corresponding keypoints.

The most recent works that estimate mesh vertices leverage deep methods such as VoxelNet [57], graph convolutions [13,37], or parametric models [5,23,68]. These approaches have made significant advances for hand pose estimation but are not without drawbacks. They tend to be restricted to fixed mesh topologies, have a very large number of network parameters, are difficult to train, and or are limited in spatial resolution. The use of parametric models such as SMPL [22] and MANO [38] has made 3D mesh estimation highly accessible. The models are highly compact; for example, MANO has 19 [16] dimensions for each hand. But by directly estimating shape parameters and joint angles of the mesh, these parametric approaches may not capture finer spatial details. They are also sensitive to perturbations, since small offsets from a single dimension of an estimate easily propagates to many mesh vertices.

We were motivated to develop a method that disentangles hand pose from shape estimation and is able to explicitly enforce estimated pose aligned with pre-calibrated hand shapes when available. Since both captured inputs and meshes are inherently surfaces, it is natural to consider them as a 2D embedding in a 3D Euclidean space. To this end, we propose solving mesh vertex regression with a fully 2D convolutional architecture that learns the extrinsic geometric properties of 3D inputs as well as intrinsics of the mesh model. Our approach is easy to train, highly efficient, and flexible enough to handle different mesh topologies and templates. Moreover, we can also capture very fine spatial detailing through per-pixel correspondences to a mesh model, thereby allowing finer spatial resolution and for better alignment between the mesh model and depth observations.

At the core of our method are two 2D fully convolutional networks (FCNs), applied to the image and mesh estimates consecutively (see Fig. 2). Linking the FCNs is a 2D embedding that propagates gradients directly from the irregular representation of a mesh to the regular and ordered representation of an image. To refine the estimated mesh, we solve for a similarity transform with singular value decomposition (SVD) to a template hand mesh model. We then re-pose the template mesh based on the transform to yield a denoised mesh surface together with key points. Since SVD has closed form solutions and is a differentiable operator, one can also place supervision on the estimated key points.

We first pre-trained our network on a synthetic dataset. Afterwards, the network can be fine tuned to real-world data by either feeding sparse key-point annotations or by directly minimizing the reconstruction error between the mesh estimation and observations. For the latter case, we propose a self-supervision scheme that minimizes a geometric model-fitting energy as a training loss. The model's accuracy steadily improves with increasing amounts of data seen, even without any human-provided labels. Finally, since correspondences between observed hand pixels and the mesh are estimated in a differentiable way, we can optimize the correspondences jointly with the disparity between the correspondence pairs during model-fitting. This differs from and complements standard ICP optimization methods. Such a self-supervision scheme greatly improves the accuracy trained by synthetic data only. To further resolve the self-occlusion, a multi-view consistency term can be optionally added when a multi-view camera setup is available. In the multi-view camera setup, the proposed self-supervision method can achieve competitive accuracy to supervised state-of-the-art.

Our contributions can be summarized as follows,

- We propose a novel fully convolutional network architecture for regressing thousands of mesh vertices in an end-to-end manner.
- A self-learning scheme is proposed for training the network; without any human labels, our network achieves competitive results when compared to fully supervised state-of-the-art. Such a learning approach offers a new and accurate way of annotating real-world data and thereby solves one of the key difficulties in making progress for hand pose estimation.
- Our method bridges a gap between data-driven discriminative methods and optimization-based model-fitting and benefits from both: accuracy that improves with the amount of data shown, while not needing human annotations.

2   Related Works

Hand Pose Estimation. Deep learning has significantly advanced state-of-the-art for hand pose estimation. The general trend has been to develop deeper and more complex network architectures [7,8,11,14,24,27,61,63]. Such progress has hinged on having large amounts of annotated data [43,55,67]. Obtaining accurate annotations, even for simple 3D joint coordinates, is extremely difficult and time consuming. Annotations generated by manually initializing trackers [28,55] require carefully designed interfaces for 3D annotation and there is often large discrepancies between human annotators [48]. Motion-capture rigs [43] and auxiliary sensors [67] are fully automatic but have limited deployment environments. To mitigate the lack of annotations, semi-supervised approaches [6,33,60] and approaches coupling real and synthetic data [32,36,42] have also been proposed.

An alternative line of work [18,25,35,40,46,49,51,53,54] estimates pose by minimizing a model-fitting error. Model-fitting needs little to no human labels, but the accuracy is heavily dependent on the careful design of the energy function. A recent trend bridges data-driven and model-fitting approaches [10,13,56, 59] by using a differentiable renderer and incorporating the model-fitting error as a part of the training loss. Our work continues in this trend, but differs from previous methods in two key respects. First, we re-parameterize the mesh with a 2D embedding, which allows us to use a 2D fully convolutional network architecture. Secondly, we apply self-supervision on both the image grid and the mesh grid, leading to efficient gradient flows during back-propagation.

Human Mesh Model Recovery From Single Image. Data-driven methods have greatly advanced the 3D reconstruction of shape and pose of the full body [3,19,30,31,39,50,52,56,57,62,65], face [17,21,37,66] and hands [5, 13,16,17,23,54,68]. Earlier works focused on landmark detection[3], segmentation [54], and finding correspondences [17,25,52,62,66], and performed a model-based optimization to fit the mesh in a subsequent step. Recently, trends have shifted to end-to-end learning of the mesh with neural networks. Several works [5,16,19,23,30,31,56,65,68] favour parametric models like SMPL [22] and MANO [38].

Various encoder-decoder frameworks have also been used, applying graph convolution to mesh vertices [13,37], VoxelNet to 3D occupancy grids [57], and fully connected and transposed convolutions to silhouettes [50] and texture and mesh vertices [21]. Unlike these works, our approach is based on correspondence estimation. Yet we also differ from other correspondence-based methods [1,17, 52,62,66] in that we directly estimate mesh vertices with a single forward pass.

3D Network Architectures. It is highly intuitive to parameterize 3D inputs and outputs as an occupancy grid or distance field and use a 3D architecture [12, 24,57]. Networks such as VoxelNet however are parameter heavy and severely limited in spatial resolution. PointNet [34] is a light-weight alternative and while it can interpret 3D inputs a set of un-ordered points, it also largely ignores spatial contexts which may be important downstream.

Since captured 3D inputs are inherently object surfaces, it is natural to consider them as 2D embeddings in 3D space. Several works [9,20,37] have modeled mesh surfaces as a graph and have applied graph network architectures to capture intrinsic and extrinsic geometric properties of the mesh. Our method also works on the hand surface, but it is a simpler and more flexible network architecture which is easier to train. Our method most resembles [2,47] by mapping high dimension data to a 2D grid. However, instead of just working on points from the depth map, we use dual grids, enabling the mapping of heterogeneous data from Euclidean space to mesh surfaces and vice versa.

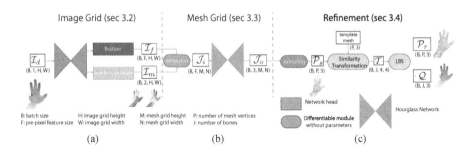

Fig. 2. System Framework. Starting from a depth map of the segmented hand as input, we estimate a dense correspondence map to the mesh model for every point on the image grid (Sect. 3.2). This correspondence maps features from the image grid to the mesh grid and allows us to recover the 3D coordinates of all the mesh vertices (Sect. 3.3) on the mesh grid. Finally, coordinates are refined by skinning a template mesh model with respect to the recovered vertices (Sect. 3.4).

3   Dual Grid Net

Our Dual Grid Net (DGN) is an efficient fully convolutional network architecture for mesh vertex estimation. At its core are consecutive 2D convolutions on two grids – an image grid and a mesh grid – where features from one grid can be mapped to another differentiably. We assume we are provided a canonical hand mesh model which is generic and applicable to all users' hands. In a given depth map, every pixel on the hand's surface has a correspondence to the mesh surface. Finding these correspondences is equivalent to mapping pixel coordinates from the image grid to the mesh grid (Sect. 3.1). Armed with a dense correspondence (Sect. 3.2) we map features from the image grid to the mesh grid and recover the 3D coordinates of all the mesh vertices (Sect. 3.3). We further refine these coordinates by skinning a template mesh model with respect to the recovered mesh vertices (Sect. 3.4). The entire process is illustrated in Fig. 2.

3.1   Mesh Model

We use a triangle mesh model (see Fig. 3(a)) with 1721 vertices. Every point on the mesh surface has a pair of *"intrinsic"* coordinates which depend only on its position on the mesh and is therefore invariant to hand pose, shape, or view point. In addition, we consider *"extrinsic"* properties of points on the mesh surface such as texture, colour, or its 3D coordinates in the camera. Both the

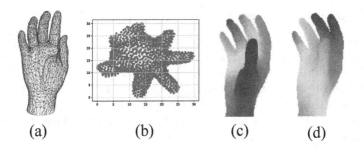

(a)   (b)   (c)   (d)

Fig. 3. (a) Triangular mesh model used in this work; (b) 2D MDS embedding of the mesh vertices; (c, d) mesh coordinates on mesh surface corresponding to 2D MDS embedding.

intrinsics and extrinsics of each mesh vertex can be approximated via linear interpolation of neighbouring points on the mesh surface.

A common way to parameterize mesh coordinates is via UV maps [1]. We follow a similar approach and use multidimensional scaling (MDS) [4] to parameterize the mesh. For any two points on a mesh surface, MDS aims to keep their Cartesian distance *w.r.t.* the mesh coordinates to be as close as possible to the geodesic distance on the mesh surface. We set the dimension of mesh coordinates (a.k.a. the intrinsic coordinates) to 2, to allow for 2D convolutions on the mesh grid. The learned MDS embedding used in this work is shown in Fig. 3(b), and the corresponding mesh coordinates projected onto the 3D mesh surface in Fig. 3(c) and (d) respectively.

3.2 Mesh Coordinate Estimation

Similar to [1], we first estimate the 2D mesh coordinates for all pixels from the hand region. We adopt an hourglass network [26] (see Fig. 2) as the backbone architecture and apply it in two heads. The first head estimates the 2D mesh coordinates $\mathcal{I}_m$ for all depth pixels while the second head estimates a generic feature map $\mathcal{I}_f$ which is later mapped to the mesh grid. Unlike [17], which performs classification followed by residual regression, we adopt a direct regression approach, which we find achieves sufficient accuracy.

Previous works [5,13,23,68] encoded image inputs as a fixed-size latent vector. Our approach, by using dense mesh coordinates, has two major advantages. Firstly, it allows us to use an FCN architecture. This important difference means we can maintain spatial resolution but also has advantages of efficiency and translational invariance. It is also much easier for learning, since one can directly apply pixel-wise supervision on both image grid and mesh grid. Secondly, the estimated mesh coordinates establishes a dense correspondence map between the captured hand surface and the mesh surface. The correspondence map, as we will show in Sect. 4.1, allows us to directly embed a lifting energy [18], which is beneficial to minimizing the model-fitting error in a self-supervised setting.

3.3   Mapping from Image Grid to Mesh Grid

In this section, we describe the recovery of all mesh vertices, including occluded ones, from the estimated per-pixel mesh coordinates and features on the image grid. Based on the estimated mesh coordinates, feature maps computed from the depth image can be mapped from the image grid to the mesh grid. Similar to [2], we call this process *extension* (see Fig. 4).

More specifically, for any pixel p belonging to the hand surface, we can regress its coordinate on the mesh grid $m = (m_x, m_y) \in \mathcal{R}^2$ as well as its corresponding feature $f \in \mathcal{R}^d$ as obtained by the feature head as described in Sect. 3.2. f is propagated to the mesh grid via soft assignment to the neighbours of m:

$$f \xrightleftharpoons[\text{sampling}]{\text{propagation}} \sum_{n \in \Omega(m)} w_n \cdot f_n. \tag{1}$$

f is propagated to the grid point n with a weighting determined by the softmax of its distance to m as follows, where $\sigma = 0.5$:

$$w_n = \frac{e^{-\sigma(n-m)^2}}{\sum_l e^{-\sigma(l-m)^2}}. \tag{2}$$

We adopt a second hourglass network on the mesh grid to recover all mesh vertices. Given that every mesh vertex is associated with a fixed mesh coordinate, the output features of hourglass network is aggregated according to their mesh coordinates of vertices. In turn, this process is named as *sampling* (see Fig. 4).

Note that propagated features will only partially occupy the mesh grid due to occlusions. However, the sampling process requires features from all over the mesh grid. This resembles an image in-painting process and we leverage the encoder-decoder structure of the hourglass to utilize both global and local context when filling in these values.

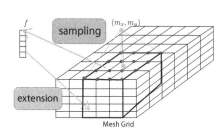

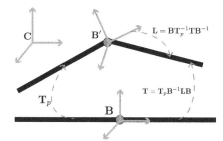

Fig. 4. Illustrations of the extension and sampling process, where $f \in \mathcal{R}^f$ is the mapped feature and $(m_x, m_y) \in \mathcal{R}^2$ is its corresponding coordinate on the mesh grid. The black box indicates the kernel size of extension and sampling.

Fig. 5. The relationship between local transformation $\mathbf{L}$ *w.r.t.* the local bone frame $\mathbf{B}$ and global transformation $\mathbf{T}$ *w.r.t.* the camera frame $\mathbf{C}$.

3.4   Refining Mesh Vertices

Post-sampling, the initial mesh estimate is not very accurate (see Fig. 1). But given that our interest is to work with a specific model, *i.e.*, that of the (canonical) hand, it is excessive to add further network structures for more accurate estimates. Instead, we propose to refine the vertices with a kinematic module. We align the initial mesh estimate with a template mesh model and solve for a rigid transformation via a closed form solution.

More specifically, given the correspondence between estimated vertices $\mathcal{P}_s$ and vertices from the template model $\mathcal{Q}$ for each hand part (palm or finger bone), we estimate a similarity transformation matrix $\mathbf{T}$ by minimizing the Euclidean distance between correspondence points $p_i \in \mathcal{P}_s$ and $q_i \in \mathcal{Q}$ as

$$\mathbf{T}^* = \mathrm{argmin}_{\mathbf{T}} \sum_i \|p_i - \mathbf{T}q_i\|. \tag{3}$$

The refined mesh results from posing the template mesh with the similarity transformation matrices through linear blend skinning (LBS). Note that Eq. 3 is a least squares minimization and that $\mathbf{T}^*$ can be found in closed form [44] *e.g.*, with singular value decomposition (SVD).

By using a closed form solution, the mesh can be refined with a single forward pass through the network. Coordinates of key points can also be obtained from the transformation matrices in a similar way as mesh vertices. And because SVD is differentiable, supervision can also be placed on top of the key-point coordinates. As will be shown in Sect. 5, when given only the supervision of these sparse key-points, our method can accurately recover dense meshes.

3.5   Supervised Training Loss

We apply MSE to the correspondence estimation $\mathcal{I}_m$ and refined mesh vertices $\mathcal{P}_r$, to optimize network parameters θ, where $\widehat{\mathcal{I}_m^{(i)}}$ and $\widehat{\mathcal{P}_r^{(i)}}$ are the ground-truth correspondence map and mesh vertex coordinates for the ith sample respectively:

$$L(\theta) = \sum_i \|\mathcal{I}_m^{(i)} - \widehat{\mathcal{I}_m^{(i)}}\|^2 + \alpha\|\mathcal{P}_r^{(i)} - \widehat{\mathcal{P}_r^{(i)}}\|^2. \tag{4}$$

3.6   Implementation Details

The hand region is first localized with the segmentation network of [54]. The image input to the hourglass network on the image grid is 64×64; the size of the mesh grid is set as 16×16. To further reduce computation, we adopt pixel shuffling techniques [41] to decrease the spatial resolution by a factor of 2 on both the image grid and mesh grid. While the number of input and output feature channels are increased by a factor of 4, the number of feature channels in hidden layers are unchanged. The kernel size of extension and sampling are both 8×8.

4   Self-supervision on Unlabelled Real Data

Training of the network proposed in Sect. 3 with direct supervision would require labels in the form of dense correspondences and vertex locations. This is impossible to annotate for real-world data. Yet training with only synthetic data is also not an option. As shown later in the experiments and also observed in the literature [32,36,59], the large domain gap between real and synthesized depth maps gives rise to compromised accuracy. Since our network essentially performs a (differentiable) rendering, the natural question that arises is whether we can incorporate a model-fitting loss into training for self-supervised learning.

The self-supervision term is similar to conventional model-fitting energy functions and is formulated as follows,

$$L(\theta) = \sum_i l_{\text{data}}^{(i)}(\theta) + \lambda_1 l_{\text{prior}}^{(i)}(\theta) + \lambda_2 l_{\text{mv}}^{(i)}(\theta) \qquad (5)$$

where θ is the network parameter and $l^{(i)}$ is the loss for the ith sample. For notation simplicity, we omit the superscript in the rest of this section. The term l_{data} measures how much the rendered depth map resembles the input depth map. Priors l_{prior} constrain the estimate to be kinematically feasible. Finally, a multi-view consistency term l_{mv} which can be used in calibrated multi-camera setups to handle self-occlusion. The λ's are associated weighting hyperparameters.

4.1   Data Terms

For l_{data}, we use only an ICP and a lifting energy term:

$$l_{\text{data}}(\theta) = l_{\text{ICP}}(\theta) + \omega l_{\text{lifting}}(\theta). \qquad (6)$$

The **ICP term** measures the disparity between points to their projections onto the mesh surface:

$$l_{\text{ICP}}(\theta) = \sum_{i \in \mathcal{I}} \min_{j \in m(\{\mathbf{T}\}|\theta)} d(i, j), \qquad (7)$$

where $m(\{\mathbf{T}\}|\theta)$ is the skinned mesh surface, where $\{\mathbf{T}\}$ is a set of per-joint transformation matrices, which are estimated as per Sect. 3.4. $l_{\text{ICP}}(\theta)$ approximates the point to surface distance by finding the nearest vertex from the mesh model based on the distance function d. For $d(\cdot, \cdot)$, we use a smooth L_1 loss. Similar to [49], we restrict the points to find only correspondences on the frontal surface of the mesh.

We also leverage the correspondence map and minimize the distance between points and their estimated correspondences on the mesh surface via a **lifting term**:

$$l_{\text{lifting}}(\theta) = \sum_{i \in \mathcal{I}} d(i, f(i|\theta)), \qquad (8)$$

where $f(i|\theta)$ estimates the 3D coordinates of the correspondence of i on the mesh surface, given the estimated mesh coordinate of i through the sampling process (see Fig. 4). The lifting term simultaneously optimizes over the correspondence map $\mathcal{I}_m$ on the image grid and the coordinate map $\mathcal{J}_o$ on the mesh grid (see Fig. 2); this introduces more efficient gradient flows to different network stages.

4.2 Kinematic Priors

The kinematic priors are defined as

$$l_{\text{prior}}(\theta) = l_{\text{collision}}(\theta) + \kappa_1 l_{\text{arap}} + \kappa_2 l_{\text{offset}}(\theta). \tag{9}$$

The **collision term** $l_{\text{collision}}(\theta)$ penalizes collisions between any pair of joints:

$$l_{\text{collision}}(\theta) = \sum_{i,j} \max(t - \|p_i - p_j\|, 0), \tag{10}$$

where p_i and p_j are the 3D coordinate of the corresponding joints. We set the threshold $t = 5\,\text{mm}$ for all pair of joints.

The **as rigid as possible term** $L_{\text{arap}}(\theta)$ [45] constrains local deformations of estimated mesh surfaces to be rigid:

$$l_{\text{arap}} = \|\mathcal{P}_r - \mathcal{P}_s\|^2, \tag{11}$$

where $\mathcal{P}_s$ are the originally estimated mesh vertices. $\mathcal{P}_r$ are the refined vertices through linear blend skinning and are guaranteed to be rigid for each part.

Section 3.4 described how to estimate the similarity transformation $\mathbf{T}$ with respect to the camera frame for each hand part. $\mathbf{T}$ transforms the bone from a rest pose[1] to the observed pose with respect to the camera frame. From the perspective of forward kinematics, $\mathbf{T}$ can be defined as

$$\mathbf{T} = \mathbf{T}_p \cdot \mathbf{B}^{-1} \cdot \mathbf{L} \cdot \mathbf{B}, \tag{12}$$

where $\mathbf{T}_p$ is the parent transformation matrix, $\mathbf{B}$ is the bone frame in the neutral pose (see Fig. 5) . $\mathbf{L}$ is the local transformation matrix with respect to the bone frame $\mathbf{B}$. Since $\mathbf{B}$ is given in the original mesh model and $\mathbf{T}_p$ is known from previous estimates, $\mathbf{L}$ can be recovered with a closed form solution.

We rewrite $\mathbf{L}$ as $[\mathbf{SR}|t]$, where $\mathbf{S} \in R^{3 \times 3}$ is a diagonal matrix scaling the matrix, $\mathbf{R} \in R^{3 \times 3}$ is the rotation matrix, $t \in R^3$ is the translation. Note that except for the wrist, there is no translation on the remaining joints. We thus penalize translations in the finger's local transformation with an **offset term**

$$l_{\text{offset}} = \sum_{i \in \mathcal{F}} \|t_i\|^2, \tag{13}$$

where $\mathcal{F}$ represents all the finger joints.

As the joint angles can be calculated from local transformation $\mathbf{L}$ with a closed-form solution, further constraints such as push constraints can easily be added. We find this to be unnecessary since synthetic data with supervision is also fed to the network to regularize the estimates (see Sect. 4.4).

---

[1] Defined by placing origin at the joint and aligning the z-axis with its parent bone.

4.3   Multiple View Consistency

To handle severe self-occlusions and holes in noisy depth inputs, we add consistency constraints l_{mv} applied to real data captured on a multi-camera rig:

$$l_{mv}(\theta) = l_{vertex}(\theta) + \eta_1 l_{ICP}(\theta) + \eta_2 l_{lifting}(\theta). \tag{14}$$

By calibrating the extrinsics of the camera, the **vertex term** l_{vertex} minimizes the distance between mesh vertices to their robust average (median in this paper) in the canonical frame. l_{ICP} and $l_{lifting}$ work similarly to the aforementioned single-view cases, except that the estimated mesh model is first mapped to another camera frame and then matched against the corresponding depth map.

4.4   Active Data Augmentation by Estimation

Since the proposed method could recover the hand mesh, we propose a strategy to actively feed synthesized data given the estimated mesh on real data to the network. The supervision from the synthesized data provides more realistic poses and helps the network to better recover from wrong estimates. From our experiments, we find this strategy to be useful to stabilize the self-supervision training and further decrease the model fitting error on unlabelled training data.

5   Experimentation

5.1   Dataset and Evaluation Protocols

We evaluate on the NYU Hand Pose Dataset [55]. It is currently the only publicly available multi-view depth dataset and features sequences captured by 3 calibrated and synchronized PrimeSense cameras. It consists of 72757×3 frames for training and 8252×3 for testing. NYU is highly challenging as the depth maps are noisy and the sequences cover a wide range of hand poses. Additionally, we synthesize a dataset of 20K depth maps of various hand poses with random holes and noise to evaluate the trained network's ability to generalize to new synthesized samples. We follow [54] to detect hands (~ 1 ms per frame). In total, our method is highly efficient and achieves 59.2 FPS on an Nvidia 1080Ti GPU.

While our framework is flexible to any hand model, *e.g.*, the MANO model [38], we follow [55] and use the LibHand model from [58] in the following experiments. This provides for an unbiased quantitative analysis since the definition of the palm center differs in different skeleton models. Note that the original hand shape from LibHand is different from either subject in the NYU dataset. Following the protocol of [55] and previous works, we quantitatively evaluate a subset of 14 joints with two standard metrics: mean joint position error (in mm) averaged over all joints and frames, and the percentage of success frames, *i.e.*, frames where all predictions are within a certain threshold [52].

5.2   Training with only Synthesized Data

We first evaluate how a network trained on synthesized data can generalize to newly synthesized data and real data (see second to sixth row in Table 1). The synthesized data is rendered from a mesh model with various poses and shapes and then corrupted with random depth noise and holes. Data is synthesized in an on-line manner and around 7.2 million samples are fed into the network for training. Our proposed kinematic module successfully reduces the average error over all mesh vertices from 14.75 mm to 7.65 mm. The network can also generalize to newly synthesized samples and achieves a high accuracy with only 7.1 mm mean joint position error. However, the error increases almost three-fold to 23.21 mm when testing on real-world depth maps. This shows that even though the network encounters data augmented with random noise, it readily over-fits to the rasterization artifacts and hand shapes of synthesized depth maps.

5.3   Ablation Studies

Variations in Training Data. We investigate how different training data and different supervision impacts the accuracy. First, we train only with the 8252×3 testing samples to check how well self-supervision can fit the mesh model to depth maps. We then trained with all training data, but in a single view setting to check how a multi-view set up impacts performance. Finally, we also look into supervision with sparse key-points to check if the proposed network accurately recover the mesh vertices and the key-points on unseen samples in testing set.

According to Table 1, self-supervision based fine tuning on real data significantly reduces the mean joint error from 23.21 mm of synthetic data trained network to 16.96 mm. Similar improvements can also be found in Fig. 6a with 15%–20% more successful frames on the error thresholds between 20 mm to 40 mm after fine tuning. However, single view only is not adequate to address the challenges from noisy depth map and severe self occlusion. To this end, we find leveraging multiple view consistency as additional constraints (see Sect. 4.3) further improve the self-supervision results (see Table 1 and Fig. 6a).

Our estimates are highly accurate, with only 8.5 mm mean joint position error (see Table 1). Furthermore, 67.8% of frames have a maximum error below 20 mm and 85.3% below 30 mm respectively (see Fig. 6a). Interestingly, training directly on the test samples gives rise to a higher mean joint error than training on a larger training set excluding the test samples (14.50 mm vs 13.09 mm, see Table 1). We attribute this to the poor initialization of the network when trained on synthesized data. The learning likely gets trapped in local minima since first-order based optimization is used during back-propagation. However, if the amount of training data increases, mean joint position error decreases. This justifies the benefits of data-driven approaches over conventional model-based trackers which optimizes each frame independently.

As shown in Fig. 1, our method can accurately reconstruct the 3D mesh model given only sparse key-point supervision. When it comes to mean joint position error, the estimation is highly accurate with only 8.5 mean joint position error (see Table 1). Furthermore, 67.8% of frames have a maximum error below 20 mm and 85.3% below 30 mm respectively (see Fig. 6a).

Studying the **impact of self-supervision loss terms.** We study the individual contributions of the different self-supervision loss terms by training without the L_{lifting}, $L_{\text{collision}}$, L_{arap}, L_{offset} and active augmentation techniques. The contributions of each of the terms are validated as we observe similar decreases in accuracy when they are omitted (see Table 1 and Fig. 6b). Notable is the fact that without the lifting energy term, the average error increases by 1.41 mm from 13.09 mm to 14.50 mm. The percentage of successful frames drops by 7% from 64% to 57% on the error threshold of 30 mm.

5.4   Comparison to State-of-the-art

We compare to recent state-of-the-art in Table 2. When trained with keypoint annotations, our method outperforms all other methods except [24] and [36] with respect to mean joint position error. In addition, according to Fig. 6c, our method performs similarly to [14,32] when the error threshold is larger than 10 mm and outperforms all other methods except [36]. We note however that [24] report an ensemble prediction result. This is impractical for real time use; in comparison, our method is highly efficient and runs at 59.2 FPS on an NVidia 1080Ti GPU. Furthermore, we out-perform [24] when compared its single model result. The work of [36] leverages domain adaptation techniques to better utilize synthesized data. This is complementary to our proposed method and beyond our current scope. It is also worth noting that key-point estimation is a byproduct of our proposed method. Our method is not designed to learn key-points; rather, the primary aim of our work is to recover mesh vertices.

We also compare our self-supervision method with [10], which to best of our knowledge is the only other unsupervised method. As is shown in Fig. 6c, our network outperforms [10] by a large margin for the percentage of successful frames at error thresholds higher than 25 mm. We achieve a higher accuracy for two reasons. First, our mesh parameterization allows the method to be robust to small estimation offsets while [10] uses joint angles, which tend to propagate errors from parent joints to children joints. Second, there are no gradients in their *depth term* (Eq. 6 in [10]) associated with unexplained points from the depth map which we handle with our proposed data term.

We further compare our self-supervision method with fully supervised deep learning methods. Surprisingly, when trained without any human label, our self-supervision based method achieves competitive results and even out-performs several fully supervised methods [12,15,23,29,60,64,69]. This highly encouraging results suggests that our method can be applied to provide labels for RGB datasets with weak supervision from depth maps.

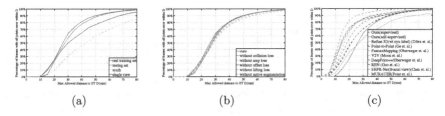

(a)   (b)   (c)

Fig. 6. (a) Impact of data used for self-supervision; (b) Impact of different loss terms and active data augmentation on self-supervised learning; (c) Comparison to fully supervised (dashed line) and self-supervised (solid line) state-of-arts.

Table 1. Ablation study and self comparison. We report mean joint error averaged over all joints and frames.

Method	Mean joint error	Method	Mean joint error
Ours (fully supervised)	8.5 mm	ours (self-supervised)	13.09 mm
Variations on training data		Impact of loss terms	
Trained on synt:		Without active augmentation	14.52 mm
Key-points (tested on real)	23.21 mm	Without $L_{lifting}$	14.50 mm
Key-points (tested on sync)	7.10 mm	Without $L_{collision}$	13.85 mm
Mesh vertices (tested on sync)	14.75 mm	Without L_{arap}	14.06 mm
refined mesh vertices (tested on sync)	7.65 mm	Without L_{offset}	14.12 mm
Self-supervised learning			
Trained on test set	14.50 mm		
Trained with single view	16.96 mm		

Table 2. Comparison with fully supervised state-of-the-art. We report mean joint error averaged over all joints and frames. All methods are tested on the NYU [55] test set. We show the comparison for reference, but would like to stress that results are not directly comparable as our method is primarily designed for mesh vertex recovery and not keypoint accuracy.

Method	Mean joint error	Method	Mean joint error
Ours (supervised)	8.5 mm	Ours (self-supervised)	13.1 mm
A2J [63]	8.6 mm	FeatureMapping [36]	7.4 mm
V2V(ensemble) [24]	8.4 mm	V2V(single model) [24]	9.2 mm
Point-to-Point [14]	9.0 mm	SHPR(three views) [8]	9.4 mm
MURAUER [32]	9.5 mm	DenseReg [61]	10.2 mm
Pose-REN [7]	11.8 mm	DeepPrior++ [27]	12.2 mm
REN-4 × 6 × 6 [15]	13.4 mm	3DCNN [12]	14.1 mm
DeepHPS(fine-tuned) [23]	14.2 mm	Lie-X [64]	14.5 mm
CrossingNet [60]	15.5 mm	Feedback [29]	15.9 mm

6   Conclusion and Discussion

We have presented a new network architecture to regress mesh vertices from single depth map with efficient 2D fully convolutional network. At its core is re-parameterization of the mesh model. We demonstrate on-par performance to state-of-arts method in the supervised setting and competitive self-supervision results with multi-camera setup. As future work, we will check how explicit hand shape calibration as proposed in [18] can be incorporated into current framework, as well as extension to RGB inputs.

Acknowledgement. The authors gratefully acknowledge supports from ETH Computer Vision Lab's institutional funding, the Chinese Scholarship Council and the NUS Startup Grant R-252-000-A40-133.

References

1. Alp Guler, R., Trigeorgis, G., Antonakos, E., Snape, P., Zafeiriou, S., Kokkinos, I.: Densereg: fully convolutional dense shape regression in-the-wild. In: CVPR (2017)
2. Atzmon, M., Maron, H., Lipman, Y.: Point convolutional neural networks by extension operators. ACM Transactions on Graphics (TOG) (2018)
3. Bogo, F., Kanazawa, A., Lassner, C., Gehler, P., Romero, J., Black, M.J.: Keep it SMPL: automatic estimation of 3D human pose and shape from a single image. In: Leibe, B., Matas, J., Sebe, N., Welling, M. (eds.) ECCV 2016. LNCS, vol. 9909, pp. 561–578. Springer, Cham (2016). https://doi.org/10.1007/978-3-319-46454-1_34
4. Borg, I., Groenen, P.J.F.: Modern Multidimensional Scaling. SSS. Springer, New York (2005). https://doi.org/10.1007/0-387-28981-X
5. Boukhayma, A., de Bem, R., Torr, P.H.: 3D hand shape and pose from images in the wild. In: CVPR (2019)
6. Cai, Y., Ge, L., Cai, J., Yuan, J.: Weakly-supervised 3D hand pose estimation from monocular RGB images. In: Ferrari, V., Hebert, M., Sminchisescu, C., Weiss, Y. (eds.) ECCV 2018. LNCS, vol. 11210, pp. 678–694. Springer, Cham (2018). https://doi.org/10.1007/978-3-030-01231-1_41
7. Chen, X., Wang, G., Guo, H., Zhang, C.: Pose guided structured region ensemble network for cascaded hand pose estimation (2017). arXiv preprint arXiv:1708.03416
8. Chen, X., Wang, G., Zhang, C., Kim, T.K., Ji, X.: SHPR-net: deep semantic hand pose regression from point clouds. IEEE Access **6**, 43425–43439 (2018)
9. Defferrard, M., Bresson, X., Vandergheynst, P.: Convolutional neural networks on graphs with fast localized spectral filtering. In: Advances in Neural Information Processing Systems (2016). https://arxiv.org/abs/1606.09375
10. Dibra, E., Wolf, T., Oztireli, C., Gross, M.: How to refine 3d hand pose estimation from unlabelled depth data? In: 3D Vision (3DV) (2017)
11. Ge, L., Cai, Y., Weng, J., Yuan, J.: Hand pointnet: 3D hand pose estimation using point sets. In: CVPR (2018)
12. Ge, L., Liang, H., Yuan, J., Thalmann, D.: 3D convolutional neural networks for efficient and robust hand pose estimation from single depth images. In: CVPR. vol. 1, p. 5 (2017)
13. Ge, L., Ren, Z., Li, Y., Xue, Z., Wang, Y., Cai, J., Yuan, J.: 3D hand shape and pose estimation from a single RGB image. In: CVPR (2019)

14. Ge, L., Ren, Z., Yuan, J.: Point-to-point regression pointnet for 3D hand pose estimation. In: ECCV (2018)
15. Guo, H., Wang, G., Chen, X., Zhang, C., Qiao, F., Yang, H.: Region ensemble network: improving convolutional network for hand pose estimation. In: Image Processing (ICIP) (2017)
16. Hasson, Y., et al.: Learning joint reconstruction of hands and manipulated objects. In: CVPR, June 2019
17. Joo, H., Simon, T., Sheikh, Y.: Total capture: a 3D deformation model for tracking faces, hands, and bodies. In: CVPR, pp. 8320–8329 (2018)
18. Joseph Tan, D., et al.: Fits like a glove: rapid and reliable hand shape personalization. In: CVPR (2016)
19. Kanazawa, A., Black, M.J., Jacobs, D.W., Malik, J.: End-to-end recovery of human shape and pose. In: Computer Vision and Pattern Regognition (CVPR) (2018)
20. Kostrikov, I., Jiang, Z., Panozzo, D., Zorin, D., Joan, B.: Surface networks. In: 2018 IEEE Conference on Computer Vision and Pattern Recognition CVPR 2018 (2018)
21. Lombardi, S., Saragih, J., Simon, T., Sheikh, Y.: Deep appearance models for face rendering. ACM Trans. Graph. (TOG) **37**, 1–13 (2018)
22. Loper, M., Mahmood, N., Romero, J., Pons-Moll, G., Black, M.J.: SMPL: a skinned multi-person linear model. ACM Trans. Graph. **34**(6), 248:1–248:16 (2015). (Proc. SIGGRAPH Asia)
23. Malik, J., et al.: DeepHPS: end-to-end estimation of 3d hand pose and shape by learning from synthetic depth. In: 2018 International Conference on 3D Vision (3DV) (2018)
24. Moon, G., Chang, J.Y., Lee, K.M.: V2V-posenet: voxel-to-voxel prediction network for accurate 3d hand and human pose estimation from a single depth map. In: CVPR (2018)
25. Mueller, F., et al.: Real-time pose and shape reconstruction of two interacting hands with a single depth camera. ACM Trans. Graph. (TOG) **38**(4), 49 (2019)
26. Newell, A., Yang, K., Deng, J.: Stacked hourglass networks for human pose estimation. In: Leibe, B., Matas, J., Sebe, N., Welling, M. (eds.) ECCV 2016. LNCS, vol. 9912, pp. 483–499. Springer, Cham (2016). https://doi.org/10.1007/978-3-319-46484-8_29
27. Oberweger, M., Lepetit, V.: Deepprior++: improving fast and accurate 3D hand pose estimation. In: ICCV workshop (2017)
28. Oberweger, M., Riegler, G., Wohlhart, P., Lepetit, V.: Efficiently creating 3D training data for fine hand pose estimation. In: CVPR, pp. 4957–4965 (2016)
29. Oberweger, M., Wohlhart, P., Lepetit, V.: Training a feedback loop for hand pose estimation. In: ICCV (2015)
30. Omran, M., Lassner, C., Pons-Moll, G., Gehler, P., Schiele, B.: Neural body fitting: unifying deep learning and model based human pose and shape estimation. In: 2018 International Conference on 3D Vision (3DV), pp. 484–494. IEEE (2018)
31. Pavlakos, G., Zhu, L., Zhou, X., Daniilidis, K.: Learning to estimate 3D human pose and shape from a single color image. In: CVPR (2018)
32. Poier, G., Opitz, M., Schinagl, D., Bischof, H.: Murauer: mapping unlabeled real data for label austerity. In: 2019 IEEE Winter Conference on Applications of Computer Vision (WACV), pp. 1393–1402. IEEE (2019)
33. Poier, G., Schinagl, D., Bischof, H.: Learning pose specific representations by predicting different views. In: CVPR (2018)
34. Qi, C.R., Su, H., Mo, K., Guibas, L.J.: Pointnet: Deep learning on point sets for 3d classification and segmentation (2016). arXiv preprint arXiv:1612.00593

35. Qian, C., Sun, X., Wei, Y., Tang, X., Sun, J.: Realtime and robust hand tracking from depth. In: CVPR (2014)
36. Rad, M., Oberweger, M., Lepetit, V.: Feature mapping for learning fast and accurate 3D pose inference from synthetic images. In: CVPR (2018)
37. Ranjan, A., Bolkart, T., Sanyal, S., Black, M.J.: Generating 3D faces using convolutional mesh autoencoders. In: Proceedings of the European Conference on Computer Vision (ECCV), pp. 704–720 (2018)
38. Romero, J., Tzionas, D., Black, M.J.: Embodied hands: modeling and capturing hands and bodies together. ACM Trans. Graph. **36**(6), 245 (2017). (Proc. SIGGRAPH Asia)
39. Saito, S., Huang, Z., Natsume, R., Morishima, S., Kanazawa, A., Li, H.: PIFu: pixel-aligned implicit function for high-resolution clothed human digitization (2019). arXiv preprint arXiv:1905.05172
40. Sharp, T., et al.: Accurate, robust, and flexible real-time hand tracking. In: Proceedings of the 33rd Annual ACM Conference on Human Factors in Computing Systems (2015)
41. Shi, W., et al.: Real-time single image and video super-resolution using an efficient sub-pixel convolutional neural network. In: CVPR (2016)
42. Shrivastava, A., Pfister, T., Tuzel, O., Susskind, J., Wang, W., Webb, R.: Learning from simulated and unsupervised images through adversarial training. In: CVPR (2017)
43. Simon, T., Joo, H., Matthews, I.A., Sheikh, Y.: Hand keypoint detection in single images using multiview bootstrapping. In: CVPR (2017)
44. Sorkine, O.: Least-squares rigid motion using SVD. Technical notes (2009)
45. Sorkine, O., Alexa, M.: As-rigid-as-possible surface modeling. In: Proceedings of the Fifth Eurographics Symposium on Geometry Processing (2007)
46. Sridhar, S., Mueller, F., Zollhöfer, M., Casas, D., Oulasvirta, A., Theobalt, C.: Real-time joint tracking of a hand manipulating an object from RGB-D input. In: Leibe, B., Matas, J., Sebe, N., Welling, M. (eds.) ECCV 2016. LNCS, vol. 9906, pp. 294–310. Springer, Cham (2016). https://doi.org/10.1007/978-3-319-46475-6_19
47. Su, H., et al.: SPLATNet: sparse lattice networks for point cloud processing. In: CVPR, pp. 2530–2539 (2018)
48. Supancic, J.S., Rogez, G., Yang, Y., Shotton, J., Ramanan, D.: Depth-based hand pose estimation: data, methods, and challenges. In: ICCV (2015)
49. Tagliasacchi, A., Schroeder, M., Tkach, A., Bouaziz, S., Botsch, M., Pauly, M.: Robust articulated-ICP for real-time hand tracking. Comput. Graph. Forum **34**(5), 101–114 (2015). (Symposium on Geometry Processing)
50. Tan, J., Budvytis, I., Cipolla, R.: Indirect deep structured learning for 3D human body shape and pose prediction. In: Proceedings of the BMVC, London, UK, pp. 4–7 (2017)
51. Tang, D., Taylor, J., Kohli, P., Keskin, C., Kim, T.K., Shotton, J.: Opening the black box: hierarchical sampling optimization for estimating human hand pose. In: ICCV (2015)
52. Taylor, J., Shotton, J., Sharp, T., Fitzgibbon, A.: The vitruvian manifold: inferring dense correspondences for one-shot human pose estimation. In: CVPR (2012)
53. Taylor, J., et al.: User-specific hand modeling from monocular depth sequences. In: CVPR (2014)
54. Taylor, J., et al.: Articulated distance fields for ultra-fast tracking of hands interacting. ACM Trans. Graph. (TOG) **36**, 1–12 (2017)

55. Tompson, J., Stein, M., Lecun, Y., Perlin, K.: Real-time continuous pose recovery of human hands using convolutional networks. ACM Trans. Graph. (ToG) **33**, 1–10 (2014)
56. Tung, H.Y., Tung, H.W., Yumer, E., Fragkiadaki, K.: Self-supervised learning of motion capture. In: Advances in Neural Information Processing Systems (NIPS) (2017)
57. Varol, G., et al.: Bodynet: volumetric inference of 3D human body shapes. In: Proceedings of the European Conference on Computer Vision (ECCV), pp. 20–36 (2018)
58. Šarić, M.: Libhand: A library for hand articulation (2011). http://www.libhand.org/. version 0.9
59. Wan, C., Probst, T., Gool, L.V., Yao, A.: Self-supervised 3D hand pose estimation through training by fitting. In: The IEEE Conference on Computer Vision and Pattern Recognition (CVPR), June 2019
60. Wan, C., Probst, T., Van Gool, L., Yao, A.: Crossing nets: combining GANs and VAEs with a shared latent space for hand pose estimation. In: CVPR (2017)
61. Wan, C., Probst, T., Van Gool, L., Yao, A.: Dense 3D regression for hand pose estimation. In: CVPR (2018)
62. Wei, L., Huang, Q., Ceylan, D., Vouga, E., Li, H.: Dense human body correspondences using convolutional networks. In: The IEEE Conference on Computer Vision and Pattern Recognition (CVPR), June 2016
63. Xiong, F., et al.: A2J: anchor-to-joint regression network for 3D articulated pose estimation from a single depth image. In: ICCV (2019)
64. Xu, C., Govindarajan, L.N., Zhang, Y., Cheng, L.: Lie-X: depth image based articulated object pose estimation, tracking, and action recognition on lie groups. Int. J. Comput. Vis. **123**, 454–478 (2017). https://doi.org/10.1007/s11263-017-0998-6
65. Xu, Y., Zhu, S.C., Tung, T.: Denserac: joint 3D pose and shape estimation by dense render-and-compare. In: ICCV (2019)
66. Yu, R., Saito, S., Li, H., Ceylan, D., Li, H.: Learning dense facial correspondences in unconstrained images. In: CVPR (2018)
67. Yuan, S., Ye, Q., Stenger, B., Jain, S., Kim, T.K.: Bighand2.2M benchmark: hand pose dataset and state of the art analysis. In: CVPR (2017)
68. Zhang, X., Li, Q., Zhang, W., Zheng, W.: End-to-end hand mesh recovery from a monocular RGB image. In: ICCV (2019)
69. Zhou, X., Wan, Q., Zhang, W., Xue, X., Wei, Y.: Model-based deep hand pose estimation (2016). arXiv preprint arXiv:1606.06854

Author Index